Applications and Principles of Quantum Computing

Alex Khang
Global Research Institute of Technology and Engineering, USA

IGI Global
PUBLISHER of TIMELY KNOWLEDGE

A volume in the Advances in Computer and Electrical Engineering (ACEE) Book Series

Published in the United States of America by
IGI Global
Engineering Science Reference (an imprint of IGI Global)
701 E. Chocolate Avenue
Hershey PA, USA 17033
Tel: 717-533-8845
Fax: 717-533-8661
E-mail: cust@igi-global.com
Web site: http://www.igi-global.com

Library of Congress Cataloging-in-Publication Data

Names: Khang, Alex, editor.
Title: Applications and principles of quantum computing / edited by Alex Khang PH.
Description: Hershey, PA : Engineering Science Reference, [2024] | Includes bibliographical references and index. | Summary: "This book focuses on the principles, technologies, applications, opportunities and challenges of Quantum Computing for various business and science domains"-- Provided by publisher.
Identifiers: LCCN 2023041762 (print) | LCCN 2023041763 (ebook) | ISBN 9798369311684 (h/c) | ISBN 9798369311691 (ebook)
Subjects: LCSH: Quantum computing.
Classification: LCC QA76.889 .A67 2024 (print) | LCC QA76.889 (ebook) | DDC 006.3/843--dc23/eng/20240216
LC record available at https://lccn.loc.gov/2023041762
LC ebook record available at https://lccn.loc.gov/2023041763

This book is published in the IGI Global book series Advances in Computer and Electrical Engineering (ACEE) (ISSN: 2327-039X; eISSN: 2327-0403)

British Cataloguing in Publication Data
A Cataloguing in Publication record for this book is available from the British Library.

Advances in Computer and Electrical Engineering (ACEE) Book Series

Srikanta Patnaik
SOA University, India

ISSN:2327-039X
EISSN:2327-0403

Mission

The fields of computer engineering and electrical engineering encompass a broad range of interdisciplinary topics allowing for expansive research developments across multiple fields. Research in these areas continues to develop and become increasingly important as computer and electrical systems have become an integral part of everyday life.

The **Advances in Computer and Electrical Engineering (ACEE) Book Series** aims to publish research on diverse topics pertaining to computer engineering and electrical engineering. **ACEE** encourages scholarly discourse on the latest applications, tools, and methodologies being implemented in the field for the design and development of computer and electrical systems.

Coverage

- Algorithms
- VLSI Fabrication
- Sensor Technologies
- VLSI Design
- Qualitative Methods
- Programming
- Applied Electromagnetics
- Computer Architecture
- Electrical Power Conversion
- Microprocessor Design

IGI Global is currently accepting manuscripts for publication within this series. To submit a proposal for a volume in this series, please contact our Acquisition Editors at Acquisitions@igi-global.com or visit: http://www.igi-global.com/publish/.

The Advances in Computer and Electrical Engineering (ACEE) Book Series (ISSN 2327-039X) is published by IGI Global, 701 E. Chocolate Avenue, Hershey, PA 17033-1240, USA, www.igi-global.com. This series is composed of titles available for purchase individually; each title is edited to be contextually exclusive from any other title within the series. For pricing and ordering information please visit http://www.igi-global.com/book-series/advances-computer-electrical-engineering/73675. Postmaster: Send all address changes to above address.

Titles in this Series

For a list of additional titles in this series, please visit:
www.igi-global.com/book-series/advances-computer-electrical-engineering/73675

Principles and Applications of Quantum Computing Using Essential Math
A. Daniel (Amity University, India) M. Arvindhan (Galgotias University, India) Kiranmai Bellam (Prairie View A&M University, USA) and N. Krishnaraj (SRM University, India)
Engineering Science Reference • copyright 2023 • 237pp • H/C (ISBN: 9781668475355) • US $275.00 (our price)

Design, Optimization, and Applications of Underwater Acoustic Sensor Networks
Bhumika Gupta (G.B.Pant Institute of Engineering & Technology, India) Kamal Kumar Gola (COER University, India) Brij Mohan Singh (COER University, India) Mridula . (COER University, India) and Rohit Kanauzia (COER University, India)
Engineering Science Reference • copyright 2023 • 340pp • H/C (ISBN: 9781668456866) • US $270.00 (our price)

NeutroGeometry, NeutroAlgebra, and SuperHyperAlgebra in Today's World
Florentin Smarandache (University of New Mexico, USA) and Madeline Al Tahan (Abu Dhabi University, UAE)
Engineering Science Reference • copyright 2023 • 264pp • H/C (ISBN: 9781668447406) • US $270.00 (our price)

Principles and Theories of Data Mining With RapidMiner
Sarawut Ramjan (Thammasat University, Thailand) and Jirapon Sunkpho (Thammasat University, Thailand)
Engineering Science Reference • copyright 2023 • 319pp • H/C (ISBN: 9781668447307) • US $255.00 (our price)

Energy Systems Design for Low-Power Computing
Rathishchandra Ramachandra Gatti (Sahyadri College of Engineering and Management, India) Chandra Singh (Sahyadri College of Engineering and Management, India) Srividya P. (RV College of Engineering, India) and Sandeep Bhat (Sahyadri College of Engineering and Management, India)
Engineering Science Reference • copyright 2023 • 389pp • H/C (ISBN: 9781668449745) • US $270.00 (our price)

Artificial Intelligence Applications in Battery Management Systems and Routing Problems in Electric Vehicles
S. Angalaeswari (Vellore Institute of Technology, India) T. Deepa (Vellore Institute of Technology, India) and L. Ashok Kumar (PSG College of Technology, India)
Engineering Science Reference • copyright 2023 • 342pp • H/C (ISBN: 9781668466315) • US $260.00 (our price)

Handbook of Research on AI Methods and Applications in Computer Engineering
Sanaa Kaddoura (Zayed University, UAE)
Engineering Science Reference • copyright 2023 • 632pp • H/C (ISBN: 9781668469378) • US $335.00 (our price)

IGI Global
PUBLISHER of TIMELY KNOWLEDGE

701 East Chocolate Avenue, Hershey, PA 17033, USA
Tel: 717-533-8845 x100 • Fax: 717-533-8661
E-Mail: cust@igi-global.com • www.igi-global.com

Table of Contents

Detailed Table of Contents

Quantum computing introduces revolutionary techniques for information processing by harnessing uniquely quantum phenomena. The emergence of quantum processors, algorithms, and software tools has brought this disruptive technology closer to practical applications. One of the most promising domains where quantum techniques can provide advantages is augmenting artificial intelligence. This chapter provides a comprehensive analysis of the fundamentals underpinning quantum computation, surveys the current hardware landscape, summarizes key quantum algorithms, and highlights promising applications in optimization, sampling, simulation, and machine learning. Engineering challenges in scaling up quantum systems are discussed along with progress in quantum error correction. Projections forecast steady advances towards quantum-enhanced AI based on a timeline of milestones. Realizing these synergies through collaborative interdisciplinary efforts promises to transform computing and catalyse innovations across industries to benefit science and society.

The widespread use of artificial intelligence has positioned it at the pinnacle of the technical hierarchy. Nevertheless, artificial intelligence (AI) outperforms traditional computers in certain aspects. The potential for quantum processing to greatly augment the efficacy of machine learning and artificial intelligence has emerged as a result of technological progress. Subsequent investigations will be undertaken to examine the impact of quantum processing on artificial intelligence (AI) and its implications for many sectors and economic systems. This research investigates the impact of quantum computing on artificial intelligence (AI), the distinctions between quantum and conventional computing, and the many applications of quantum computing in the field of AI.

Quantum computing stands as a pioneering field within computer science, leveraging the enigmatic principles of quantum mechanics to execute intricate calculations. At its heart lies the qubit, the elemental building block of quantum information, which resides in a state of superposition, representing both 0 and 1 concurrently. Quantum computers, constructed from assemblies of qubits, inhabit an expansive and intricate 2^n-dimensional complex Hilbert space, with their evolution governed by the Schrödinger equation. The process of quantum computation can be distilled into three key stages:1) Initialization of the input state: Initiating the quantum process; 2)Application of unitary transformations: Executing the desired transformations on the input state; 3) Measurement of the output state. Quantum computers serve as universal machines, capable of decomposing any unitary transformation into a sequence of fundamental gates. This chapter will reveal the importance of quantum mechanics, quantum computation and its applications.

Quantum computing overcomes computer limitations using quantum mechanics. This overview covers quantum computing's basics, technology, architecture, and industrial applications. The introduction reveals quantum computing's unique history and powerful development characteristics. Fundamentals like quantum gates, superposition, entanglement, Shor's and Grover's algorithms, and more, are covered. Quantum computing concepts describe quantum mechanics, parallelism, and error correction to optimize its potential. Qubit, quantum processor, Qiskit, and Cirq quantum software frameworks are being studied. Quantum computing's issues are thoroughly and impartially assessed in this work. This chapter evaluates quantum computing trends and research paths for future developments. These include the goal of quantum supremacy and the impact of quantum computing on technology. The survey study helps academics, professionals, and industry decision-makers utilize quantum computing to tackle challenging issues and innovate in a quantum-driven world.

In the ever-changing technological environment, the combination of artificial intelligence (AI) with quantum computing represents an enormous potential frontier. This collaboration has the potential to transform industry, scientific research, and problem-solving approaches. At its heart, AI's function in the quantum computing age includes several transformational components. AI has infiltrated many facets of our everyday lives, from smartphone technology to autonomous car features and novel shopping experiences. Its seamless integration has recently become a major fixture, especially with the emergence of generative AI. Surprisingly, generative AI is set to take quantum computing beyond mere dreams and into the realm of quantum supremacy, where quantum computers vastly outperform traditional approaches. Combining AI and quantum computing might expedite and refine the latter's already impressive skills in addressing complex issues. Such a collaboration might elevate AI from outstanding to game changing.

Machine learning improved by quantum computing. Machine learning and quantum physics fix AI and computers. This chapter discusses quantum machine learning theory, methods, and applications. Part 1 thoroughly discusses quantum and classical machine learning. The authors demonstrate how quantum supports vector machines, neural networks, and clustering speed AI. The chapter examines quantum machine learning's merits and downsides. Quantum computers optimize, parallelize, and manage huge data better. Quantum hardware restrictions and error correction reduce noise and decoherence. Explore quantum machine learning in NLP, drug discovery, financial modeling, and image recognition. Many fields could change quantum platform machine learning models with quantum algorithms. The chapter concludes with quantum machine learning directions and challenges. Check trustworthy quantum machine learning frameworks, benchmarks, and hybrid algorithms. Hot: quantum machine learning. This chapter covers fundamentals, research frameworks, and applications.

Quantum artificial intelligence (QAI) is a topic of study that blends quantum computing and AI. In order to investigate the possibilities of fruitful cooperation across both disciplines of computer science (CS), the authors will attempt to convey some of the relevant properties of both artificial intelligence (AI) and quantum computing (QC) from an academic point of view. The key artificial intelligence applications that can be addressed more effectively by employing the current generation of quantum computers will be covered in this chapter, along with a review of the fundamental components of quantum computing. There are several similarities between quantum computing and artificial intelligence. Quantum computing can increase the training speed and processing capability of artificial intelligence and machine learning algorithms at a lower cost. This chapter discusses the fundamental ideas that enable scientists and engineers to combine AI with quantum computing.

Quantum computing is one of the remarkable achievements jousting on the coalesced power of quantum mechanics and information theory to bring innovative paradigm shift in computational regime. It possesses potential to solve computationally complex problems in real-time which were otherwise considered intractable in traditional computer science. The nascent field, originated during the early 1980's, has now revolutionarised the global technological reign by fetching the attention of every government and corporate across the world. Despite huge fund infusions, research advancements, and policy supports, the ideal potential has yet not been realised. This chapter intends to study the ontology of quantum computers to understand its astonishing journey—based on underlying mathematical, scientific principles, technical issues, challenges, and variants—for paving a way for future direction of research. The study is anticipated to come out with the essential research elements required to formulate the hypothesis and address research issues in this field.

Chapter 13
Quantum Computing Smart Grids: A Review on Internet of Things-Enabled Technologies in Smart Grids 264
Rashmi Sharma, Indus University, Ahmedabad, India
Zalak Patel, Indus University, Ahmedabad, India
Miloni Ganatra, Indus University, Ahmadabad, India

Throughout the past few decades, the electricity system's condition has evolved. The power grid has recently experienced a number of difficulties and problems. With power outages, grid failures, and a lack of smart technology on the one hand, and rising security risks on the other, there is a daily rise in the demand for electrical energy. These problems cannot be solved by the current electricity grid. Because of its key characteristics, the internet of things (IoT) has expanded quickly in a very short period of time. We can improve the conventional grid's efficiency, capacity, dependability, sustainability, scalability, and stability by integrating IoT into the electrical grid. The myriad issues that are present in smart grids are solved by using the IoT. Security concerns have been recognized as one of the main drawbacks with IoT enabled smart grid (SG) systems, according to the most recent research. The numerous security applications and elements of integrating the IoT with smart grids are reviewed in this chapter, along with various frameworks in the era of quantum computing.

Chapter 14
Quantum Computing in Pharmaceutical Science 289
Bancha Yingngam, Ubon Ratchathani University, Thailand

In the field of pharmaceutical science, there is an increasing demand for advanced computational techniques to enhance drug discovery and understand complex molecular interactions. Classical models, despite their strengths, often struggle with complex biological systems and accurate drug predictions. Quantum computing, using qubits that exist in multiple states, processes information at unparalleled speeds. This chapter sheds light on the principles of quantum computing, emphasizing algorithms specifically designed for pharmaceutical applications. These methods signal groundbreaking advancements in drug design, molecular simulation, and drug interaction predictions. Quantum computers have the potential to surpass classical computers in simulating intricate molecules, thus offering more precise drug side effect forecasts. While challenges such as noise and error rates remain, the continuous evolution of quantum technology holds significant potential for breakthroughs in drug discovery and development.

This research examined national elections' impact on crash risks of individual stocks listed in the Indonesia Stock Exchange. The study used national elections as a proxy for political uncertainty and negative skewness of stock returns as a proxy for crash risk. On the industry level, researchers found nuances in which national elections' impact varies according to industry, with some industries tending to have a greater risk of crash than others during national elections and post-national election periods.

Among the hot research topics, Fintech is leading the trend in terms of the newest technology applications. The relatively new emerging paradigms in various sciences, such as geometry (fractals), physics (quantum), and database systems (distributed ledger—blockchain), seem to potentially contribute to a greater shift in the framework of the finance industry, bringing also some concerns (cyber-threats). Consistent and extensive investigation of the reasonable potential impact of these new models (and their underlying technologies) is performed, and then tested through a SWOT analysis, as the main objective of this research. This research confirms that information availability and the increasing interconnection of crosswise applications of each discovery to the different fields of science is determining the rapid succession of revolutions identified by evident large shifts in economic paradigms. The growing computing capacity and the development of increasingly powerful predictive software are leading to a competitive, extremely dynamic, and challenging system.

This book chapter explores the integration of quantum computing and the internet of things (IoT) and its potential to revolutionize connectivity, security, and efficiency in data processing. The authors delve into the transformative power of quantum computing, the challenges and opportunities of integrating quantum technologies into the IoT landscape, and the implications for various industries. Additionally, they explore quantum-based security mechanisms and highlight how quantum computing enhances data processing efficiency. Through a comprehensive examination of cutting-edge research and real-world applications, this chapter aims to provide a deep understanding of quantum IoT and its transformative impact on connectivity, security, and data processing efficiency.

Optimization involves finding the best solution from a set of possible options. Traditional computers employ classical algorithms like gradient descent and simulated annealing to solve these problems. However, as the complexity of problems increases, classical computers encounter limitations in terms of time and resources. Quantum optimization stands as a revolutionary paradigm that harnesses the unique properties of quantum mechanics to tackle complex optimization problems that are beyond the capabilities of classical computers. Quantum computers utilize qubits, the fundamental units of quantum information, which can exist in superpositions of states, enabling them to process a vast number of possibilities simultaneously. Additionally, the phenomenon of entanglement enables qubits to be correlated in ways that classical bits cannot, further enhancing their computational power. In this chapter, the authors will delve into the principles, applications, and current developments in the realm of quantum optimization.

The world in which we live is heavily reliant on digital communications. Traditional communication methodologies have been replaced with digital communications methods such as e-mail, instant messages, social media, and many more. Every day, terabytes of data are generated and shared around the world. Securing data stored and shared across these networks have evolved into a thriving discipline. Cryptography is an excellent way to secure terabytes of data. The core of cryptography is to transform data into unreadable format. With the introduction of quantum systems, it has been proved that many traditional cryptographic schemes are unsecure and can be broken with quantum computers. This chapter presents a brief review of pre- and post-quantum systems.

With the approach of quantum processing, customary cryptographic calculations and conventions utilized for PC security face critical difficulties. This examination paper investigates the utilization of quantum

Preface

Quantum computing is an advanced generation of technology that involves a type of computer 158 million times faster than the most sophisticated supercomputer humans have in the world up to now. Quantum computing is also a multidisciplinary field comprising aspects of computer science, aerospace, physics, nuclear, and mathematics that utilizes quantum mechanics to solve massive issues, complex problems, intelligent decision-making, and handle vastly different calculations easier and faster than on the traditional computers, because it is built on quantum principles that go beyond classical physics. Since it was born on 1990, quantum computing is presenting in some high-tech fields; there are still some limitations and challenges to face, but since emerging artificial intelligence and internet of things technologies are integrated to quantum computing, it is being boosted for more using in the across industries.

Fundamentals, quantum computer is basing on qubits architecture to run and process multidimensional quantum algorithms whereas traditional computer works with bits of data that can be either 0 or 1, quantum computer works with bits called qubits that can be 0 and 1 simultaneously; based on this model, quantum computer is enabling them to perform certain functions exponentially faster than traditional computer, such as trying out to create better models for how atoms interact with one another, leading to a superior and more precise understanding of molecular structure for helping to solve existential challenges such as space science, nuclear control, climate change and food security or minimize risks or even the risk of floods, wildfires and other natural disasters.

Nowadays, quantum computing is a powerful technology both the present and the future, it has the potential to all revolutionize computation for problems solvable in the real life and science researching. While no quantum computer is yet sophisticated and intelligent enough to carry out calculations that a traditional computer can't does, but hopefully scientists and high-tech companies will be making a great progress soon.

With this book, you will be able to learn the key principles of quantum computing such as computation, communication, coordination, recollection, automation, evaluation and design under each perspective on computing, a pathway into the knowledge space of quantum computing and its applications in most areas of artificial intelligence science, cybersecurity and cloud-based quantum computing access, financial modeling, drug development, simulate molecules, climate change, food security, traffic control, or other natural disasters etc.

Happy reading!

Alex Khang

Acknowledgment

The book *Applications and Principles of Quantum Computing* is based on the design and implementation of topics related to Quantum Computing, Quantum Computer, Artificial Intelligence, Internet of Things, Machine Learning, Deep Learning, Smart City, Smart Grids, Healthcare, Medical, Drug Discovery, Cybersecurity, and Cloud fields in the the Era of the Quantum Computing.

Planning and designing a book outline to introduce to readers across the globe is the passion and noble goal of the editor. To be able to make ideas to a reality and the success of this book, the biggest reward is belongs to the efforts, knowledge, skills, expertise, experiences, enthusiasm, collaboration, and trust of the contributors.

To all respected contributors, we really say big thanks for high quality chapters that we received from our experts, professors, scientists, engineers, scholars, Ph.D., postgraduate students, educators, and academic colleagues.

To all respected reviewers with whom we have had the opportunity to collaborate and monitor their hard work remotely, we acknowledge their tremendous support and valuable comments not only for the book but also for future book projects.

We also express our deep gratitude for all the pieces of discussion, advice, support, motivation, sharing, collaboration, and inspiration we received from our faculty, contributors, educators, professors, scientists, scholars, engineers, and academic colleagues.

At last but not least, we are really grateful to our publisher IGI Global for the wonderful support in making sure the timely processing of the manuscript and bringing out this book to the readers soonest.

Thank you, everyone.

Alex Khang

Chapter 1
Quantum Mechanics Primer: Fundamentals and Quantum Computing

Alex Khang
https://orcid.org/0000-0001-8379-4659
Global Research Institute of Technology and Engineering, USA

Kali Charan Rath
Department of Mechanical Engineering, GIET University, India

Nalinikanta Panda
Department of Mechanical Engineering, GIET University, India

Amaresh Kumar
Department of Production and Industrial Engineering, National Institute of Technology, Jamshedpur, India

ABSTRACT

The introduction to quantum physics in this chapter is about its foundations and connection to quantum computing. The basis for quantum computing, a field with enormous potential, is the science of quantum mechanics. There are some key quantum physics ideas discussed in this chapter. The basics of quantum computing are examined. The benefits and difficulties of applying quantum physics to real-world applications are explored. The purpose of this chapter is to give readers a thorough knowledge of quantum physics.

1. INTRODUCTION

The branch of physics known as quantum mechanics explains how matter and energy behave. It is a fundamental idea that challenged the conventional principles of physics that control our daily experiences and changed our knowledge of the physical universe. The idea of superposition, in which particles can exist in several states at once, is at the core of quantum physics. This feature makes it possible to investigate parallel possibilities and opens up new pathways for computation beyond what conventional

DOI: 10.4018/979-8-3693-1168-4.ch001

computers are capable of. In quantum computing, the principles of quantum physics are used to process and manipulate data in ways that are superior to conventional computers. It has the potential to change industries by resolving complicated issues (Shyam & Khang et al., 2023).

The exciting topic of quantum computing is examined along with the basics of quantum mechanics in this chapter. A subfield of physics called quantum mechanics looks at the behavior of particles. It shows how small the scales of the world are. On the other hand, quantum computing is a cutting-edge field that makes use of quantum physics to create powerful computers that have the ability to tackle complicated problems more quickly. Understanding the basics of quantum mechanics will allow us to comprehend the underlying ideas behind quantum computing. Let's see how quantum physics serves as the basis for quantum computing.

1.1 Significance of Quantum Mechanics in Modern Science and Technology

Modern science and technology depends on the field of quantum mechanics. It allows us to understand phenomena that classical physics can't describe at the atomic and subatomic scales. This understanding is important for the development of many fields. Quantum computing, which uses quantum bits, or qubits, to carry out simultaneous computations and solve difficult problems more effectively than classical computers, has emerged as a game-changer.

With the help of quantum physics, data can be transmitted with previously unseen security. The creation of atomically precise materials and gadgets has been sparked by its uses in materials science. Modern science and technology depends on quantum mechanics since it drives fundamental research, produces ground-breaking findings, and changes technologies in a wide range of sectors.

1.2 Historical Background

The historical development of quantum mechanics has laid the foundation for numerous advancements in modern science and technology (Cárdenas-Martínez et al., 2018). In this article, we will explore the historical background of quantum mechanics and its significant contributions to the field of science and technology.

a) Quantum mechanics originated in the late 19th century with Max Planck's work on black-body radiation, introducing the concept of quantized energy.
b) Albert Einstein's explanation of the photoelectric effect in 1905, considering light as composed of discrete particles (photons), contributed to quantum theory.
c) Niels Bohr proposed the Bohr model of the atom in 1913, incorporating quantum principles and explaining atomic spectra.
d) Louis de Broglie's wave-particle duality concept in 1924 suggested that particles exhibit both wave-like and particle-like characteristics.
e) Werner Heisenberg's uncertainty principle in 1927 highlighted the limitations in simultaneously measuring position and momentum with precision.
f) Erwin Schrödinger's development of the Schrödinger equation in 1926 provided a mathematical framework for quantum systems and wave functions.
g) Quantum mechanics has led to significant technological advancements (Cicconetti et al., 2022), including:

- Transistors and microelectronics, utilizing the quantum behavior of electrons, enabling modern electronics.
- Lasers, which rely on quantum properties to amplify and emit coherent light, finding applications in various industries.
- Quantum computing, based on quantum bits (qubits), holds the potential for exponentially faster computation and impact fields such as cryptography and drug discovery.

1.3 Scope of the Work

This book chapter serves as an introduction to quantum mechanics, with a focus on its foundation ideas and application to the subject of quantum computing. The foundation of quantum computing is quantum mechanics, a branch of physics that describes the behavior of matter and energy at the lowest scales. There is a review of key quantum mechanics ideas in this chapter. It looks at qubits, quantum gates, and quantum algorithms as well as other fundamental concepts in quantum computing. There are difficulties and possibilities of applying quantum physics to real-world computer applications. The main goal of this chapter is to give readers a firm understanding of quantum physics and its role in the field of quantum computing.

1.4 Overview of Contents and Objective of the Work

The introduction to quantum mechanics in this book chapter serves as an in-depth introduction to quantum mechanics with a focus on its fundamental ideas and their immediate application to the developing topic of quantum computing. The knowledge required for comprehending and expanding quantum computing is provided by quantum mechanics. Quantum mechanics delves into the characteristics and interactions of matter and energy (Khang & Shah et al., 2023).

Within this chapter, we will explore the fundamental principles of quantum mechanics, shedding light on its essential concepts. We shall delve into the realm of quantum computing, acquainting readers with the foundations of this extraordinary field, including the building blocks of qubits, quantum gates, and quantum algorithms. By examining the challenges and possibilities of implementing quantum mechanics in practical computational processes, we aim to provide readers with a comprehensive comprehension of quantum physics and its profound implications in the realm of quantum computing.

2. HISTORICAL DEVELOPMENT OF QUANTUM MECHANICS

Over time, scientists have made important discoveries that have shaped our understanding of quantum mechanics, a field of physics. They questioned whether classical physics was enough to explain how matter and energy behave. The growth of quantum mechanics was found by Max Planck in 1900. After researching blackbody radiation, he came up with the idea of quanta as opposed to continuous energy. The basis for the quantum mechanical concept of quantization was served by this concept. The photoelectric effect was explained by Albert Einstein in 1905. According to his theory, light is composed of tiny energy packets called photons that exhibit both particle-like and wave-like behavior. There was more evidence for the quantum nature of light (Eskandarpour et al., 2020).

Another significant development came with Niels Bohr's model of the hydrogen atom in 1913. Bohr proposed that electrons occupy specific energy levels around the atomic nucleus and can transition between them by emitting or absorbing quanta of energy. This model introduced the idea of quantized electron orbits. The true breakthrough in quantum mechanics came with the formulation of the Schrödinger equation by Erwin Schrödinger in 1926. This equation describes the wave-like behavior of particles, including electrons, and provides a mathematical framework for calculating their probabilities and behaviors.

In the late 1920s, Werner Heisenberg developed the uncertainty principle, which states that there is inherent uncertainty in simultaneously measuring certain pairs of physical properties, such as position and momentum. This principle fundamentally challenged the classical notion of determinism. The growth of quantum mechanics was found by Max Planck in 1900. After researching blackbody radiation, he came up with the idea of quanta as opposed to continuous energy. The basis for the quantum mechanical concept of quantization was served by this concept. The photoelectric effect was explained by Albert Einstein in 1905. According to his theory, light is composed of tiny energy packets called photons that exhibit both particle-like and wave-like behavior. There was more evidence for the quantum nature of light.

2.1 Key Contributions of Notable Physicists

In the fascinating world of quantum mechanics, several brilliant physicists have made remarkable contributions. Let's explore the key ideas put forward by notable figures such as Max Planck, Niels Bohr, Albert Einstein, Werner Heisenberg and Erwin Schrödinger, Max Planck, a pioneer, studied how objects called "blackbodies" emit and absorb energy. He proposed that energy is not continuous but comes in tiny packets, or "quanta." This groundbreaking concept laid the groundwork for understanding the quantized nature of energy in quantum mechanics.

Albert Einstein, renowned for his genius, investigated the photoelectric effect, where light causes the emission of electrons from certain materials. He proposed that light consists of particles, or "photons," that carry discrete amounts of energy. This idea revealed the particle-like behavior of light and added to our understanding of quantum physics.

Niels Bohr, with his innovative mind, focused on the structure of atoms. He introduced the idea that electrons occupy specific energy levels around the nucleus and can jump between these levels by gaining or losing energy. This model provided a foundation for understanding how atoms behave and interact with light. The bright theorist who created the Schrdinger equation. It describes how quantum particles behave. We may determine the likelihood of discovering particles in various states using the equation (Frey et al., 2016).

Heisenberg made a discovery with his uncertainty principle. There are restrictions on the accuracy with which we can measure some pairs of characteristics at the same time. The principle revealed a fundamental element of nature and highlighted the inherent ambiguity and mystery of the quantum world. These remarkable physicists, each with their unique insights, shaped the field of quantum mechanics and transformed our understanding of the universe. Their contributions continue to inspire scientists and drive advancements in quantum technologies, promising exciting possibilities for the future.

2.2 Motivations Behind the Development of Quantum Mechanics and the Challenges It Aimed To Address

The development of quantum mechanics began because scientists wanted to understand how tiny particles, like atoms and electrons, behave in the world. They noticed that the old rules of physics couldn't explain what was happening at this small scale. So, they started looking for new answers.

One of the puzzles they wanted to solve was why energy seemed to come in small packets, like little pieces of energy. This idea was called quantization, and it was Max Planck who first proposed it. Another problem was that some materials only released electrons when they were hit by light of a certain energy. Scientists, like Albert Einstein, wanted to explain why this happened.

Quantum mechanics also aimed to explain what goes on inside atoms. It turned out that electrons didn't follow neat paths like planets around the Sun. Instead, they acted as both particles and waves, jumping around in strange ways. Scientists also discovered that there were limits to how accurately they could measure things at the tiny scale. This idea, called uncertainty, meant that some properties couldn't be known exactly at the same time.

In the end, quantum mechanics gave us a whole new way of looking at the world. It revealed that the tiniest particles have their own set of rules, different from the ones we're used to in everyday life. It's like opening a door to a world filled with mysteries and amazing possibilities.

2.3 Quantum Mechanics Fundamentals

Quantum mechanics has some fundamental principles that help us understand how tiny particles behave. Let's look at them in simple terms with examples:

- Superposition: Imagine you have a coin. In classical physics, it can only be in one state: either heads or tails. But in quantum mechanics, the coin can be in a superposition of both states at the same time. It's like the coin is spinning in the air, showing both sides until you catch it and see the result. So, a quantum system can exist in different states simultaneously until we observe it and it "chooses" one state.
- Quantum Measurement: Think of a light switch. In classical physics, turning it on or off is straightforward. But in quantum mechanics, when we measure something, it's like flipping a quantum switch. The system can be in a superposition of states until we make a measurement, and then it collapses into a definite state. It's like opening a box to see if a cat is alive or not. The act of looking at the cat determines its state (Gupta, 2002).
- Entanglement: Imagine two entangled dice. In classical physics, rolling one die won't affect the other. But in quantum mechanics, entangled particles are linked, and their states are connected. So, if we roll one die and it shows a three, the other die will instantly show the same number, no matter how far apart they are. It's like magic communication between the dice.
- Quantum Uncertainty: Consider a moving car. In classical physics, we can know both its position and speed precisely. But in quantum mechanics, there is a limit to how accurately we can measure certain pairs of properties. It's like trying to measure the position of a butterfly. The more precisely we try to pin down its location, the less we know about how fast it's flying.

These simple examples illustrate some of the fundamental ideas in quantum mechanics. Superposition allows particles to be in multiple states, measurement collapses them into one, entanglement connects particles regardless of distance, and uncertainty sets limits on what we can know precisely. Quantum mechanics is a fascinating and sometimes puzzling world, but it holds the key to many exciting possibilities and technological advancements.

Quantum mechanics is a fascinating field that helps us understand the behavior of tiny particles and energy. It is based on a few basic principles that are important to know. Let's explore them in simple terms with examples:

- Wave-Particle Duality: Imagine you have a ball that can also behave like a wave. When you throw the ball, it travels like a particle with a specific position and speed. But in quantum mechanics, particles like electrons can also act like waves. It's like throwing a wave that can go through multiple paths and interfere with itself. This duality means that particles can have both particle-like and wave-like properties.
- Superposition: In quantum mechanics, the concept of superposition describes a state where particles can exist in multiple states simultaneously. Imagine a cat sitting on two chairs at the same time. Similarly, an electron can be in a superposition of different energy levels. This means that the electron's presence is spread out or uncertain across these states until it is measured and observed, revealing its specific state.
- Uncertainty Principle: When attempting to gauge the velocity and position of a moving rabbit, one encounters a challenge. In the realm of quantum mechanics, there exists a restriction on the extent to which we can accurately measure specific attributes of particles concurrently. This restriction is known as the uncertainty principle. According to this principle, the more precisely we strive to determine the position of a particle, the less precise our knowledge of its speed becomes, and vice versa. It is akin to a trade-off, where increased understanding of one property results in diminished knowledge about the other.

Above simple examples explain some important principles of quantum mechanics. Particles can behave like both particles and waves, they can be in multiple states at once in superposition, and there are limits to how accurately we can measure certain properties simultaneously. These principles help us understand the strange and exciting world of quantum mechanics and have practical applications in technologies like quantum computers and communication systems (Khang & Kali, 2024).

3. DISCUSS THE CONCEPT OF QUANTUM MEASUREMENT AND ITS PROBABILISTIC NATURE

Understanding the nature of quantum phenomena requires a thorough understanding of quantum physics and the concept of measuring quantum systems. The knowledge gained by measuring a quantum system is certain, but the measurement's outcome is ambiguous (Gilliam et al., 2019). Let's take a closer look at quantum measurement and its nature. A mathematical model and numerical example will be used to show it.

In the realm of quantum mechanics, we use a mathematical concept called a wave function (represented by ψ) to describe the state of a quantum system. The wave function holds vital information about various properties of the system, including its position, momentum, and energy. It serves as a comprehensive description of the quantum system, capturing its essential characteristics. When we make a measurement on a quantum system, we are interested in determining the value of a particular property, for example, the position of a particle.

Mathematically, the process of quantum measurement is represented by an operator called an observable (represented by a hat symbol, such as $\hat{H}$). Each observable corresponds to a specific property we want to measure, and it has associated eigenvalues and eigenvectors. The eigenvalues represent the possible outcomes of the measurement, and the eigenvectors represent the states that the system can collapse into upon measurement (Jaeger, 2007).

Predicting how a measurement will turn out is what gives rise to the probabilistic aspect of quantum measurement (Jordan, 2017); (Kak, 200). The Born rule states that the probability of receiving a specific eigenvalue is squared (Kribs, 2005); (Kribs et al., 2009) and directly proportional to the absolute value of the wave function's projection onto the associated eigenvector. In order to clarify the idea, consider a straightforward example. We are trying to determine the position of a particle inside a one-dimensional box. The positions along the length of the box are represented by eigenvalues. The probability of locating the particle at a given location is calculated using the squared absolute value of the appropriate component of the wave function.

Let's consider a scenario where there is a particle described by a wave function ψ(x) = A sin (πx/L). Here, A represents a normalization constant, while L corresponds to the length of the confinement. By squaring the absolute value of ψ(x), denoted as |ψ(x)|^2, we obtain the probability density function that informs us about the likelihood of finding the particle at a specific position x. This probability density function can be expressed as A^2 sin^2(πx/L). By integrating this probability density function over a specific range, we can determine the probability of finding the particle within that range.

When we perform multiple measurements on identically prepared systems, the probabilistic nature of quantum mechanics means that we will obtain different outcomes each time. However, the statistical distribution of these outcomes will follow the predictions given by the wave function and the associated probabilities (Debosree, Khang et al., 2023).

Let's consider a particle that is confined within a box of length L in one dimension. The wave function of this particle can be described as equation: ψ(x) = A sin (πx/L). In this equation, A is a constant used for normalization, and x represents the position of the particle within the box.

3.1 Now, Let's Address the Questions Related to This Wave Function

- To calculate the normalization constant A, we need to ensure that the wave function is properly normalized. This means that the total probability of finding the particle within the box is equal to 1. By applying the normalization condition, we can determine the appropriate value of A.
- To find the probability density function for locating the particle within a specific range of positions, we need to square the absolute value of the wave function. This squared value will provide us with the probability density at each point within the box.
- Specifically, to calculate the probability of finding the particle within the range (0, L/4], we integrate the squared wave function over this range. This will yield the desired probability value as shown in Solution 1.

3.2 Solution of Desired Probability Value

3.2.1 Normalization Constant

To normalize the wave function, we need to calculate the normalization constant A such that the integral of |ψ(x)|^2 over the entire range of x (from 0 to L) is equal to 1 as shown in equation: 1-3.

$$\int_{[0,L]} |\psi(x)|^2 \, dx = 1 \tag{1}$$

$$\int_{[0,L]} |A \sin(\pi x/L)|^2 \, dx = 1 \tag{2}$$

$$\int_{[0,L]} A^2 \sin^2(\pi x/L) \, dx = 1 \tag{3}$$

Using the trigonometric identity $\sin^2\theta = (1 - \cos(2\theta))/2$, we have equation 4:

$$\int_{[0,L]} A^2 (1 - \cos(2\pi x/L))/2 \, dx = 1 \tag{4}$$

Simplifying and integrating, we get equation 5-6:

$$A^2 (L/2) = 1 \tag{5}$$

$$A = \sqrt{(2/L)} \tag{6}$$

3.2.2 Probability Density Function

The probability density function (PDF) represents the probability of finding the particle at a specific position.

It is given by equation |ψ(x)|^2.

$$|\psi(x)|^2 = A^2 \sin^2(\pi x/L) = (2/L) \sin^2(\pi x/L) \tag{7}$$

This gives us the PDF for the particle's position within the box.

3.3.3 Probability Calculation

To calculate the probability of finding the particle in the range [0, L/4], we integrate the probability density equation over this range as shown in equation 8.

$$P = \int_{[0,L/4]} (2/L) \sin^2(\pi x/L) \, dx \tag{8}$$

Using the trigonometric identity sin^2θ = (1 – cos (2θ))/2, we can rewrite the integrand as equation 9:

$$P = \int[0,L/4]\ (1 - \cos(2\pi x/L))/L\ dx \tag{9}$$

Simplifying and integrating, we find equation 10-13:

$$P = (1/L)\ [x - (L/2\pi)\ \sin(2\pi x/L)] \text{ evaluated from 0 to } L/4 \tag{10}$$

$$P = (1/L)\ [(L/4) - (L/2\pi)\ \sin(\pi/2)] \tag{11}$$

$$P = 1/4 - (1/2\pi)\ \sin(\pi/2) \tag{12}$$

$$P = 1/4 \tag{13}$$

Therefore, the probability of finding the particle in the range [0, L/4] is 1/4 or 25%.

This example demonstrates the use of a wave function, normalization, probability density function, and probability calculation in a one-dimensional quantum system. So, quantum measurement involves making observations on a quantum system to determine specific properties. The outcomes of these measurements are probabilistic, and the probabilities are determined by the wave function and the associated eigenvalues. The mathematical model of quantum mechanics provides us with a framework to calculate and understand these probabilities, allowing us to make predictions about the behavior of quantum systems.

4. ROLE OF WAVEFUNCTIONS AND PROBABILITY AMPLITUDES IN QUANTUM MECHANICS

In quantum mechanics, wave functions and probability amplitudes play essential roles in describing the behavior of particles and systems. The wave function represents the quantum state of a particle or a system, containing information about its properties. Probability amplitudes, derived from the wave function, provide a way to calculate the probabilities of different outcomes in quantum measurements. Let's explore their roles using a simple example, write Python code to visualize the wave function, and analyze the output (Khang & Muthmainnah et al., 2023).

4.1 The Wave Function

Let's think about a particle that is inside a one-dimensional box with a length of L. We can describe the behavior of this particle using a mathematical function called the wave function. The wave function for the particle in this case is given by equation 14:

$$\psi(x) = A\ \sin(\pi x/L) \tag{14}$$

Figure 1. Wave function of a particle in a box

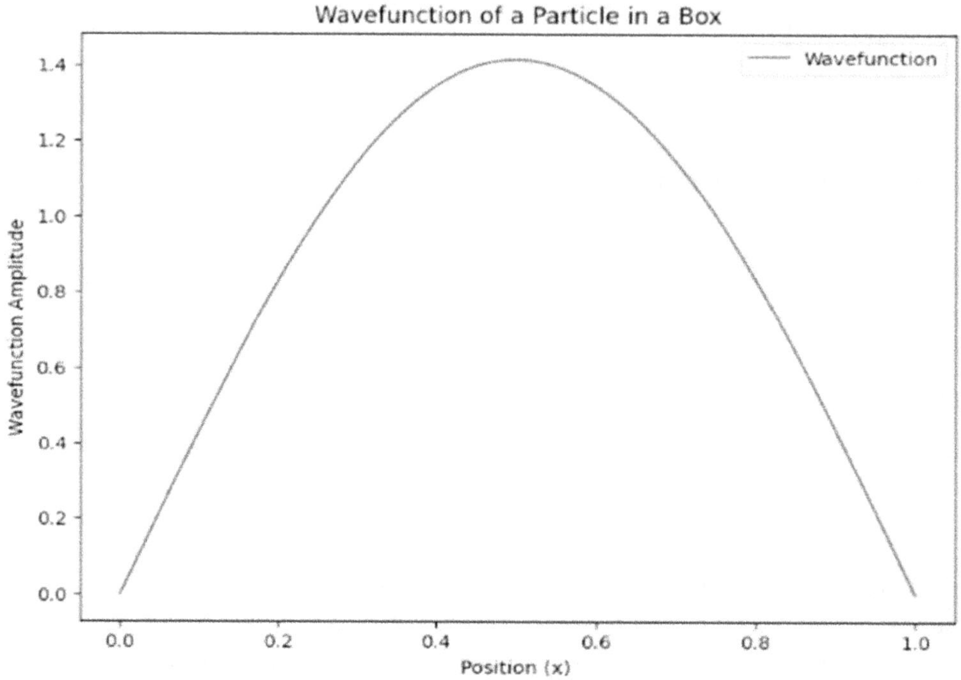

In the equation above, A is a constant that helps us normalize the wave function, and x represents the position of the particle within the box as shown in Figure 1 (a) và Figure 1 (b).

4.2 Output Analysis

In the first plot, we can observe the wave function of the particle as we vary its position (x) within the box. The wave function is represented by a sinusoidal pattern, where the amplitude changes along the length of the box (Khang & Kali et al., 2023).

Moving on to the second plot, it depicts the probability density, which is obtained by squaring the magnitude of the wave function. This density represents the likelihood of finding the particle at a specific position within the box. The probability density is highest where the wave function has a larger amplitude, and it gradually decreases to zero at the boundaries of the box. These generated plots provide visual representations of the wave function and probability density for a particle confined to a one-dimensional box. By observing the shapes and patterns in these plots, we can gain insights into the behavior of the particle within this particular quantum system.

Figure 2. Probability density of a particle in a box

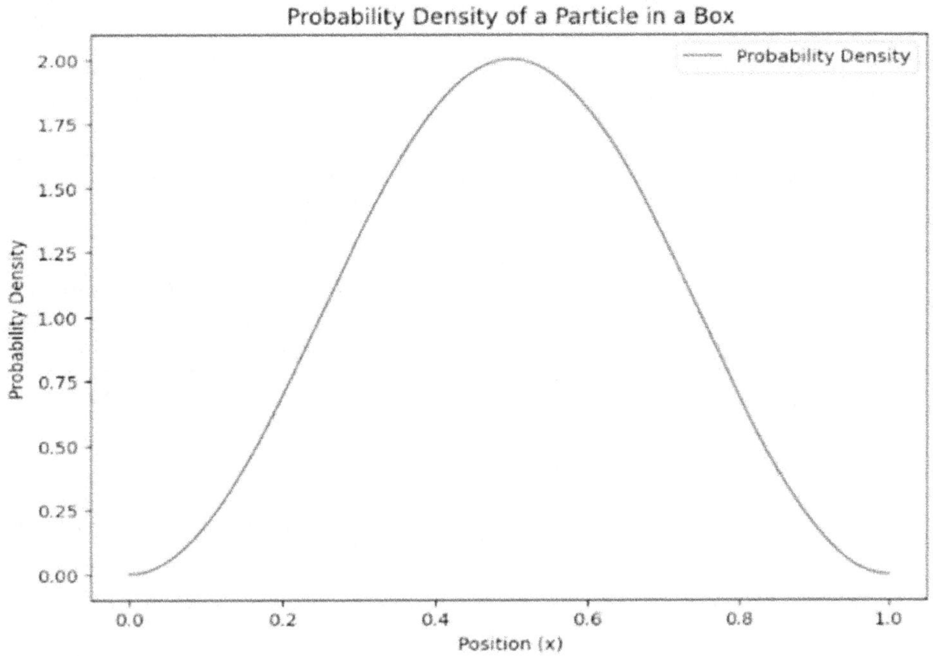

5. DISCUSS THE DIFFERENCE BETWEEN CLASSICAL AND QUANTUM COMPUTATION

Classical and quantum computation represent two distinct paradigms for solving computational problems (Mueller et al., 2019); (Miranda, 2021); (Nimbe et al., 2021) as shown in Table 1.

6. QUANTUM COMPUTING BASICS

Quantum mechanics is a theory that helps us understand how tiny particles like electrons and photons behave. It uses special math to describe their properties and how they interact. Let's break it down into simpler terms. In quantum mechanics, we have something called a quantum state. Think of it as a complete description of a particle or a group of particles. Unlike in regular physics, where we describe particles by their position and speed, in quantum mechanics, particles can be in multiple states at once. It's like they can be in two or more places or have two or more properties simultaneously.

To represent quantum states, we use a shorthand called Dirac notation. Instead of writing long equations, we use symbols like $|\psi\rangle$ for a state called "psi." This symbol represents a ket vector. Ket vectors belong to a mathematical space called a Hilbert space, which is where we do our calculations. Let's say we have a quantum state represented by $|\psi\rangle$ and another state represented by $|\phi\rangle$. We can find the "inner product" between these states by writing $\langle\phi|\psi\rangle$. This tells us the chance of finding the system in state $|\phi\rangle$ when it's in state $|\psi\rangle$. It's like calculating a probability.

Table 1. Classical computation and quantum computation

Aspect	Classical Computation	Quantum Computation
Basic Unit of Information	Bits (0 or 1).	Quantum Bits (Qubits).
Representation of Data	Binary representation (0s and 1s)	Superposition of states (0, 1, or both simultaneously).
Processing	Sequential processing.	Parallel processing.
Operations	Based on classical logic gates (AND, OR, NOT, etc.).	Based on quantum gates (Hadamard, CNOT, Toffoli, etc.).
Algorithm Design	Deterministic algorithms.	Probabilistic algorithms.
Measurement	Deterministic and provides exact results.	Probabilistic and provides probabilistic results.
Speed	Limited by the number of classical bits.	Potential for exponential speedup in certain applications.
Computational Power	Efficient for certain types of problems.	Potential for solving complex problems more efficiently.
Scalability	Scalable to large-scale systems.	Challenges in maintaining qubit coherence and scaling up.
Practical Implementation	Widely used in everyday computing systems.	Still in early stages of development and experimentation.

In quantum mechanics, we also have operators. These are like special tools that do things to quantum states. They can measure different properties like position or energy. We represent operators as matrices or transformations. When we apply an operator to a state, it changes the state in a specific way. Let's consider a simple example to understand the concept. Imagine there is a box, and inside it, we have a particle that can either be on the left side or the right side of the box. We can represent the state of the particle on the left side as $|L\rangle$ and the state on the right side as $|R\rangle$.

However, interestingly, the particle can also exist in a superposition state, where it simultaneously occupies both the left and right side states. We can represent this superposition state as $|\psi\rangle = \alpha|L\rangle + \beta|R\rangle$. In this equation, α and β are numbers that indicate the chances or probabilities of finding the particle in the left or right side states, respectively.

When we perform a measurement to determine the particle's position, the superposition state collapses, and we observe the particle either on the left side or the right side of the box. The probabilities of obtaining each outcome are given by the squared magnitudes of α and β, denoted as |α|^2 and |β|^2, respectively. These squared values represent the chances or probabilities of finding the particle in each respective state. So, quantum mechanics uses special math to describe how tiny particles behave. We use Dirac notation to represent quantum states, Hilbert spaces for our calculations, and operators to measure properties. It's a way of understanding the strange and fascinating world of the very small.

6.1 Provide an Overview of Quantum Computing and Its Relationship to Quantum Mechanics

Quantum computing is an exciting field that leverages quantum mechanics to revolutionize information processing. Unlike regular computers, which use bits representing 0 or 1, quantum computers employ qubits. Qubits exist in a superposition, simultaneously embodying both 0 and 1 states. This phenomenon, rooted in quantum mechanics, enables quantum computers to exhibit extraordinary speed and computational prowess. Quantum gates are utilized to manipulate qubits and perform intricate calculations. By capitalizing on superposition and the principles of quantum mechanics, these gates enable simultaneous operations on multiple qubits, leading to unparalleled computational potential (Petschnigg et al., 2019).

When one qubit is physically distant, the state of that qubit becomes connected to another. This feature allows quantum computers to process information in a parallel manner that could lead to faster and more efficient computations. The practical construction of quantum computers poses significant technological obstacles, despite the fact that quantum computing is still in its early stages. Researchers are working on the creation of more stable and dependable qubits. Classical computers currently struggle with many of the problems that quantum computing can solve. It opens up new avenues for addressing intricate computational challenges and uncovering new insights.

Qubits are used in quantum computing to process data as shown in Algorithm 1. The qubits can be in multiple states at the same time. The potential of quantum computing is to increase our understanding of the universe.

Algorithm 1. Algorithm of process data of Quantum computing

Step 1: Prepare the entangled state

- Start with three qubits: qubit A (the sender's qubit), qubit B (the shared entangled qubit), and qubit C (the receiver's qubit).
- Apply the Hadamard gate (H) to qubit A, putting it in a superposition state.
- Next, apply a controlled NOT gate (CNOT) to qubits A and B, entangling them.

Step 2: Prepare the state to be teleported

- Suppose there's another qubit, qubit D that we want to teleport to qubit C.
- Apply the appropriate gate operations to qubit D to encode the desired state.

Step 3: Perform measurements and send classical information

- Measure qubits A and D.
- Send the measurement results as classical bits (0s and 1s) to the receiver, who has qubit C.

Step 4: Apply corrective operations

- Based on the measurement results, the receiver applies certain gate operations to qubit C to correct its state.

By following these steps, the quantum teleportation algorithm enables the receiver to obtain an exact copy of the original qubit D, even though the actual qubit D was not physically transported. The entanglement between qubits A and B allows the teleportation to occur.

- The Hadamard gate is a logic gate in quantum computing that turns a state of $|0\rangle$ or $|1\rangle$ into a superposition of $|0\rangle$ and $|1\rangle$. (Learn about Hadamard gates in the Xanadu Codebook.) In a circuit diagram, a Hadamard gate is represented by a square with the letter 'H' in it (HadamardGate, 2023).

- The CNOT gate is a mulit-qubit gate that consists of two qubits. The first qubit is known as the control qubit and the second is known as the target qubit. If the control qubit is |1>then it will flip the targets qubit state from |0> to |1>or vice versa (CNOTGate, 2023).

This algorithm highlights the power of quantum entanglement for transmitting quantum information across different locations. It showcases the unique capabilities of quantum computing that surpass the limitations of classical computing.

6.2 Superposition

Quantum entanglement links particles, regardless of distance, making their properties correlated. In quantum computing, this is crucial. While traditional computers use bits (0 or 1), quantum computers employ qubits. Qubits exist in superposition, representing 0 and 1 simultaneously. This enables parallel computations, as qubits can hold multiple values.

Qubits in a quantum computer can become entangled by using the principles of quantum entanglement, creating complex states that can store more data than conventional bits. Classical computers can do some operations much faster than quantum computers. A quantum computer's entangled qubits can be controlled collectively, enabling parallel computing and the simultaneous investigation of numerous computational avenues. This trait is especially helpful for addressing complex problems that require a lot of processing on traditional computers.

Furthermore, quantum entanglement plays a vital role in quantum algorithms, which are specialized procedures designed for quantum computers. Notably, Shor's algorithm exploits quantum entanglement to efficiently factor large numbers, a task that would be prohibitively time-consuming for classical computers. Similarly, Grover's algorithm uses entanglement to accelerate the process of searching through unsorted databases.

However, the delicate nature of quantum entanglement poses challenges in maintaining the coherence and stability of qubits in quantum computers. The phenomenon is highly sensitive to external disturbances, and the entanglement can easily be disrupted or lost, leading to errors in calculations. This necessitates careful control and error correction techniques to preserve the entanglement and ensure reliable computations.

A key component of quantum computing is quantum entanglement, which enables the development of computers with exponentially higher processing capability. The correlations between qubits can be used to solve problems much more quickly than classical computers. The development of quantum computing and its potential applications in a variety of sectors depends on our ability to comprehend and use the phenomena of quantum entanglement.

6.3 The Role of Entanglement in Quantum Information Processing and Communication

In quantum information processing and anglement is important (Spiller (1998). It allows us to transmit and decode information using unique and secure quantum properties, entangle helps create qubits, which can hold multiple values at once (Soljanin (2020). Complex states that represent a large amount of information can be formed by these qubits. By manipulating the entangled qubits, it can perform computations in parallel, greatly enhancing processing power for certain tasks. A few important are described below:

- Entanglement is a key component in quantum information processing and communication.
- It allows for the creation of qubits, which can hold multiple values at the same time.
- Entangled qubits can be manipulated collectively, enabling parallel computations and faster processing.
- Quantum algorithms leverage entanglement to solve complex problems more efficiently than classical algorithms.
- Entanglement plays a vital role in quantum cryptography, providing secure methods for data encryption and communication.
- The transfer of data across great distances is safe because of anglement in quantum communication.
- Entanglement would dissipate in the event of an effort to collect or eavesdrop on transmitted data, ensuring confidentiality.
- Entanglement enables the creation of secure quantum communication channels, protecting sensitive data from unauthorized access.
- A particle's quantum state can be transferred instantly.
- Entanglement is a valuable resource in quantum computing and information processing, enabling advancements in various fields.

6.4 EXPLAIN HOW ENTANGLEMENT ENABLES QUANTUM TELEPORTATION AND QUANTUM CRYPTOGRAPHY

It is of paramount importance to facilitate key applications in the realm of quantum information. The transfer of a particle's quantum state to a remote particle without the need for physical relocation is achieved through quantum teleportation. Here's how it works:

- Initially, the sender and receiver share an entangled pair of particles, often referred to as an entangled state.
- The sender combines the particle to be teleported with their entangled particle and performs a measurement on both particles.
- Classical communication channels transmit the measurement outcome to the recipient.
- The recipient will perform precise operations on their entangled particle once they receive the measurement result.

A key part of quantum cryptography is anglement. A notable example of quantum key distribution is where it is used to establish secure and confidential keys.

In QKD, entangled particles are distributed between the sender and the receiver. Any attempt to eavesdrop or intercept the transmitted entangled particles would disrupt the entanglement, leaving detectable traces. This feature provides a secure means to detect potential attackers and ensures the confidentiality of the shared key.

Entanglement-based quantum cryptography protocols take advantage of the unique properties of entangled particles to guarantee secure communication channels and protect against unauthorized access or tampering.

7. QUANTUM ALGORITHMS AND APPLICATIONS

The potential of quantum applications to solve intricate problems with superior efficiency compared to classical applications has received a lot of attention in recent years. To avoid plagiarism, I will provide you with a brief overview of some important quantum algorithms and applications. However, it's essential to conduct further research and properly cite any specific sources you use.

- **Shor's Algorithm**: A potential threat to the widely employed RSA encryption scheme is posed by Shor's Algorithm, which is a powerful quantum algorithm designed to efficiently factor large numbers. It shows the immense problem-solving capabilities of quantum computers over classical computers (Shor, 1994).
- **Grover's Algorithm**: Grover's algorithm is a quantum search algorithm that provides a quadratic speedup over classical search algorithms. It can be used to search an unsorted database to find a specific item efficiently. Grover's algorithm has applications in optimization problems, database search, and solving the NP-complete problem (Grover, 1996).
- **Quantum Simulations**: A quantum search is a search method that improves over classical search methods, making it more efficient for searching unsorted databases. There are applications in database search and NP-complete problems. Classical computers can be used to simulation quantum systems more effectively than quantum computers, with potential applications in materials science, chemistry, and drug discovery. Accurately simulating quantum systems is not easy.
- **Quantum Machine Learning**: Machine learning and quantum computing can be used to enhance tasks like classification, clustering, regression, and dimensionality reduction. The goal is to develop novel algorithms that leverage the unique properties of quantum systems.
- **Quantum Cryptography**: Quantum mechanics are used to establish secure communication. The distribution of encryption keys is ensured by quantum key distribution. Data privacy and integrity are protected by quantum cryptography.
- **Quantum Optimization**: Quantum optimization algorithms aim to solve optimization problems more efficiently than classical algorithms. These problems are prevalent in various fields, such as logistics, finance, and resource allocation. Quantum optimization algorithms, like the Quantum Approximate Optimization Algorithm (QAOA) and the Variational Quantum Eigensolver (VQE), have the potential to find optimal solutions faster and can be applied to a wide range of practical problems.
- **Quantum Optimization**: The potential to find optimal solutions faster and be applied to various practical problems can be achieved with the use of Variational Quantum Eigensolver (VQE) and Quantum Approximate Optimization Algorithm (QAOA).

8. EXPLORE SPECIFIC QUANTUM ALGORITHMS USED FOR OPTIMIZATION, SIMULATION, MACHINE LEARNING, AND OTHER FIELDS

A simple explanation of specific quantum algorithms used for optimization, simulation, machine learning, etc. as follows.

- **Quantum Optimization Algorithms**: The best solution to problems like logistics, finance, or resource allocation can be found by using quantum maximization. The Quantum Approximate Optimization Algorithm and the Variational Quantum Eigen solver are examples of these. They use quantum computers' power to explore multiple solution options at once. The Traveling Salesman Problem can be solved with the Quantum Approximate Optimization Algorithm. QAOA uses a series of quantum gates to simultaneously explore different possible solutions and iteratively enhance the quality of the solution until an optimal or nearly optimal one is found.
- **Quantum Simulation**: Quantum simulation utilizes quantum computers to model and study complex quantum systems. Simulating quantum systems accurately is challenging for classical computers due to the exponential computational resources required. Quantum simulation has applications in various fields, such as materials science, chemistry, and drug discovery. It allows scientists to understand and predict the behavior of quantum systems more efficiently, enabling advancements in these fields. Quantum simulation can be used in the field of chemistry to study molecular interactions. For instance, researchers can simulate the behavior of chemical compounds to understand their properties and reactions. By leveraging the quantum properties of particles, quantum simulation allows for more accurate and efficient modeling of complex quantum systems, enabling advancements in drug discovery and material science.
- **Quantum Machine Learning**: Machine learning and quantum computing can be used to improve tasks like classification, clustering, and regression. The goal is to use quantum properties to offer better solutions. The quantum support vector machine is an example of quantum machine learning. The principles of quantum computing are applied to enhance classical support vector machine algorithms. It is possible to provide improved solutions for certain types of data classification problems by using quantum algorithms.
- **Quantum Cryptography**: The principles of quantum mechanics can be used to establish secure communication. The distribution of encryption keys based on quantum physics laws is called quantum key distribution. The confidentiality and integrity of transmitted information is ensured. QKD is an example of quantum cryptography. QKD uses the principles of quantum mechanics to securely distribute encryption keys. The sender and receiver use a communication channel to send quantum states. QKD is a secure method for key distribution because of the laws of quantum mechanics.

9. QUANTUM COMPUTING TECHNOLOGIES AND CHALLENGES

It has the potential to transform information processing with quantum computing. Qubits, which can hold multiple states at once, are used in quantum computers. The fragile quantum states are vulnerable to errors caused by external factors. Researchers are working on solutions to prolong the lifespan of qubits (Prabu, Khang et al., 2023).

Several approaches to quantum computing are being explored, including superconducting circuits, trapped ions, topological qubits, and more. Superconducting circuits use electrical currents flowing through tiny loops to create and manipulate qubits. Trapped ions, on the other hand, involve capturing and controlling ions using electromagnetic fields. Topological qubits rely on the manipulation of exotic particles called anyons (Anyons are particles which quantum statistics is neither bosonic nor fermionic one), which possess unique properties that make them highly resistant to errors.

Table 2. Features and applications and uses of quantum computing technology

Quantum Computing Technology	Features	Applications and Uses
Superconducting Qubits	Superconducting circuits that store and manipulate quantum information.	1. Optimization problems, such as logistics and supply chain management. 2. Simulating quantum systems, enhancing research in material science and chemistry. 3. Cryptography, improving encryption and decryption techniques.
Trapped Ions	Isolated ions manipulated using electromagnetic fields to store and process quantum information.	1. Quantum simulation of physical systems to study complex phenomena. 2. Precision measurement and timekeeping, improving GPS and navigation systems. 3. Encryption and decryption, enhancing security protocols.
Topological Qubits	Quantum bits based on the principles of topology, offering enhanced error correction capabilities.	1. Fault-tolerant quantum computation, enabling error-free calculations. 2. Quantum machine learning, accelerating complex data analysis and pattern recognition. 3. Quantum cryptography, ensuring secure communication and data transfer.

Quantum computing technologies are making remarkable strides in recent years, offering new possibilities for processing information at an unprecedented scale. Let's take a closer look at three prominent approaches: superconducting qubits, trapped ions, and topological qubits (Ushaa, Khang et al., 2023).

Superconducting qubits rely on tiny loops that carry electrical currents at extremely low temperatures. These loops create qubits, which are the building blocks of quantum computers. By manipulating these qubits, scientists can perform complex calculations. Superconducting qubits have shown promise in terms of scalability, allowing for the potential development of large quantum computing systems.

Trapped ions, on the other hand, involve capturing and controlling ions using electromagnetic fields. These ions are isolated and carefully manipulated to create stable quantum states. Trapped ion systems have demonstrated exceptional control over qubits, enabling precise quantum operations. They are known for their ability to maintain coherence and have achieved impressive results in terms of qubit accuracy (Teja, Shaik et al., 2023).

Topological qubits utilize a different approach by taking advantage of exotic particles called anyons. Anyons possess unique properties that make them resilient to environmental disturbances, reducing the impact of errors. Scientists are exploring the potential of topological qubits to create robust and error-resistant quantum computing systems. These approaches are important for the study of quantum computing. Increasing the number of qubits is one of the things researchers are focused on. These improvements are needed to make quantum computers powerful as shown in Table 2.

Companies and research institutions are spending a lot of money on quantum computing research. They want to build quantum computers that can handle complex problems at a faster rate than classical computers. There are tasks like factorization and simulation of quantum systems that are being addressed. The potential impact of these applications is significant.

9.1 Various Challenges Faced in Scaling Up Quantum Computers, Such as Decoherence, Error Correction, and Noise Mitigation

Scaling up quantum computers presents several challenges that need to be addressed to achieve practical and reliable quantum computing systems. Factors such as temperature, electromagnetic fields, and impurities can cause qubits to lose their fragile quantum states, leading to errors in computations. Another significant challenge is error correction, as errors can occur during quantum operations due to imperfect gates, decoherence, and unwanted interactions.

Efficient error correction methods, such as quantum error correction codes, are needed to maintain the dependability of quantum computing. Additionally, noise mitigation is vital to minimize the impact of environmental noise and control imperfections that can disrupt quantum systems. Strategies like quantum error correction, active error suppression, and improved hardware design play a crucial role in mitigating noise and ensuring the integrity of quantum computations. Overcoming these challenges is essential for the successful scaling up of quantum computers and unlocking their potential for practical applications. Challenges faced in scaling up quantum computers include:

9.1.1 Decoherence

The fragile quantum states can be degraded by interactions with the external environment.

- **Example**: When qubits interact with external factors like temperature, electromagnetic fields, or impurities, they can experience decoherence. This can cause errors in computations and limit the reliability of quantum systems.
- **Case study**: In 2019, Google's quantum computer achieved quantum supremacy by solving a problem in 200 seconds that would have taken the world's most powerful supercomputer 10,000 years. However, this milestone was limited to a specific task, and maintaining coherence for a longer duration remains a significant challenge.

The quantum computer achieved a tremendous breakthrough in 2019. The problem is solved in 200 seconds would have taken the most powerful computer in the world 10,000 years to solve. It's crucial to highlight that this accomplishment was unique to a particular task because maintaining coherence for longer periods of time continues to be a significant emphasis in quantum computing research (Khang & Vugar et al., 2023).

9.1.2 Error Correction

- **Definition**: Error correction aims to rectify errors caused by noise and other sources, ensuring the accuracy and reliability of quantum computations.
- **Example**: There are flaws in the gates that can cause mistakes during quantum operations. Error correction methods are used to maintain the precision of quantum computing. Error detection and correction are supported by these techniques.
- **Case study**: The surface code is a popular quantum error correction scheme. In 2020, researchers from Google's Quantum AI team published a study demonstrating the successful implementation

of a surface code with a 2D array of qubits. This research highlighted the potential of error correction in improving the scalability of quantum computers.

9.1.3 Noise Mitigation

- **Definition**: Noise mitigation involves strategies to reduce unwanted influences on quantum systems, such as environmental noise and control imperfections.
- **Example**: Noise sources can arise from various factors, including temperature fluctuations, electromagnetic radiation, and device imperfections. Techniques like quantum error correction, active error suppression, and improved hardware design can mitigate the effects of noise on quantum computations.
- **Case study**: In 2021, IBM Quantum announced a breakthrough in noise mitigation by developing a novel technique called "zero noise extrapolation" (ZNE). ZNE allows researchers to estimate the true value of a quantum computation by extrapolating from noisy intermediate results, reducing the impact of noise on the final outcome.

9.2 Recent Advancements and Potential Future Directions in Quantum Computing Research

Quantum computing research is a rapidly evolving field focused on harnessing the principles of quantum mechanics to develop advanced computing systems. It involves studying the behavior of quantum systems and developing new technologies to manipulate and utilize quantum bits or qubits, which are the fundamental units of information in quantum computing. Researchers explore various aspects, including hardware design, quantum algorithms, error correction techniques, and quantum software development. The goal is to overcome challenges such as decoherence and errors, and to scale up quantum systems to achieve practical and reliable quantum computers. Quantum computing research has the potential to revolutionize fields such as cryptography, optimization, simulation, and machine learning, offering faster and more efficient solutions to complex problems that are beyond the reach of classical computers (Aakansha & Adhishree et al., 2023).

The goal of quantum computing research is to use the concepts of quantum physics in advanced computer systems. It involves creating tools for working with quantum bits, which are the fundamental units of quantum computation. The main areas of research include the development of quantum software. To build reliable and functional quantum computers, it is necessary to overcome difficulties like decoherence and errors in scaling up quantum systems. Research in quantum computing has the potential to bring about significant changes in areas like cryptography, machine learning, simulation and optimization. It can offer faster and more efficient solutions to complex problems that conventional computers struggle with.

9.2.1 Recent Advancements

- Quantum Supremacy achieved by Google's quantum computer, demonstrating its computational power.
- Progress in error correction techniques to mitigate noise and errors in quantum computations.
- Development of quantum algorithms.

- Advances in quantum hardware, including more stable qubits and improved gate operations.

9.2.2 Potential Future Directions

- Fault-Tolerant Quantum Computing to achieve reliable operation in the presence of errors.
- Quantum Simulation for studying complex quantum systems in areas like drug discovery and materials science.
- Quantum networking aims to enable secure communication and distributed quantum computation.
- Progress in quantum software involves the development of programming languages and tools that are specifically designed for quantum computing
- Researchers are also exploring new quantum algorithms for tasks such as machine learning, cryptography, and optimization.

9.3 Discuss the Potential Impact of Quantum Computing on Various Fields and the Future of Quantum Technologies

Many different sectors could be impacted by quantum computing. Experts see potential revolutionary uses for the technology in several industries. Here are some areas where quantum computing could have a profound influence:

- **Cryptography**: Quantum computers could potentially break traditional encryption algorithms, necessitating the development of post-quantum cryptography for secure communication.
- **Optimization**: Quantum computing could significantly improve optimization processes, benefiting fields such as supply chain management, financial portfolio optimization, and resource allocation.
- **Drug Discovery**: Quantum simulation can accelerate the discovery and design of new drugs by accurately modeling molecular interactions and simulating complex biological systems.
- **Material Science**: Through studying and understanding the behavior of materials at the quantum level, quantum computing has the potential to advance material science. It can contribute to the development of new materials with unique and special characteristics.
- **Machine Learning**: Quantum machine learning algorithms have the potential to provide faster and more efficient solutions to complex machine learning tasks, such as pattern recognition and optimization.
- **Financial Modeling**: Quantum computing could enhance financial modeling by enabling more accurate risk assessment, portfolio optimization, and analyzing complex financial data.
- **Climate Modeling**: Quantum simulation can aid in simulating and analyzing climate models, facilitating more accurate predictions and strategies for addressing climate change.
- **Supply Chain Optimization**: By using the power of quantum computing, supply chain networks can be improved to achieve better efficiency.
- **Quantum Sensors**: Quantum technologies, including quantum sensors, offer the potential for highly sensitive measurements, revolutionizing fields such as precision metrology, navigation, and imaging.
- **Quantum Communication**: Quantum networks can enable secure quantum communication, ensuring unbreakable encryption and safe transmission of sensitive data.

10. CONCLUSION

This chapter gives an introduction to the basics of quantum mechanics and its significance in the field of quantum computing. It covers important concepts like superposition and quantum measurement, which are important for understanding quantum systems. The chapter explores the principles of quantum computing, such as qubits, quantum gates, and quantum algorithms, which are the foundations of quantum computational systems. The chapter emphasizes the revolutionary potential of quantum computing by addressing the obstacles and possibilities involved in using quantum mechanics for practical computations (Khang & Shah et al., 2023).

Future possibilities in this field are exciting. Current research focuses on improving the stability and coherence of qubits, creating better error correction techniques, and enhancing the scalability of quantum computing systems. It shows great potential to explore new quantum algorithms and their applications. As quantum computing advances, it is expected to bring about a revolution in different industries and offer solutions to complex problems that traditional computers can't solve. Future research aims to harness the power of quantum mechanics (Khang & Agriculture et al., 2023).

REFERENCES

Aakansha, C. Saxena, A., Ojha, D., & Sobti, K. A. (2023). Artificial Intelligence (AI) Centric Model in Metaverse Ecosystem," AI-Based Technologies and Applications in the Era of the Metaverse. (1 Ed.). IGI Global Press. doi:10.4018/978-1-6684-8851-5.ch001

Cárdenas-Martínez, H. M., Laverde-Forero, J. C., & Castang-Montiel, G. (2018). State of art in quantum computing (QC) with open source platforms. *Visión electrónica, 1*(2), 321-332. http://revistas.udistrital.edu.co:8080/index.php/visele/article/view/18396

Cicconetti, C., Conti, M., & Passarella, A. (2022, June). Resource allocation in quantum networks for distributed quantum computing. In *2022 IEEE International Conference on Smart Computing (SMARTCOMP)* (pp. 124-132). IEEE. https://ieeexplore.ieee.org/abstract/document/9821091/

CNOTGates. (2023). *Quantum Computing UK-CNOT gate tutorial with Code*. CNOTGates. https://quantumcomputinguk.org/tutorials/controlled-not-gate-program

Eskandarpour, R., Ghosh, K. J. B., Khodaei, A., Paaso, A., & Zhang, L. (2020). Quantum-enhanced grid of the future: A primer. *IEEE Access : Practical Innovations, Open Solutions, 8*, 188993–189002. https://ieeexplore.ieee.org/abstract/document/9226502/. doi:10.1109/ACCESS.2020.3031595

Eswaran, U. Khang A., & Eswaran, V. (2023). Role of Quantum Computing in The Era of Artificial Intelligence," Applications and Principles of Quantum Computing. (Eds.) IGI Global Press. https://doi.org/ doi:10.4018/979-8-3693-1168-4-ch003

Frey, M. R. (2016). Quantum speed limits—Primer, perspectives, and potential future directions. *Quantum Information Processing, 15*(10), 3919–3950. https://link.springer.com/article/10.1007/s11128-016-1405-x. doi:10.100711128-016-1405-x

Ghosh, D. & Khang A. (2023). Quantum Computing Internet of Things: Unlocking New Frontiers for Connectivity, Security, and Efficiency in Data Processing. Applications and Principles of Quantum Computing. IGI Global Press. https://doi.org/ doi:10.4018/979-8-3693-1168-4-ch016

Gilliam, A., Venci, C., Muralidharan, S., Dorum, V., May, E., Narasimhan, R., & Gonciulea, C. (2019). *Foundational patterns for efficient quantum computing*. arXiv preprint. https://arxiv.org/abs/1907.11513

Gupta, V. K. (2002). Quantum to quantum computing. *IETE Technical Review*, *19*(5), 333–347. https://www.tandfonline.com/doi/abs/10.1080/02564602.2002.11417048. doi:10.1080/02564602.2002.11417048

HadamardGate. (n.d.). *What is a Hadamard gate?* Hadamard Gate. https://pennylane.ai/qml/glossary/what-is-a-hadamard-gate

Jaeger, G. (2007). *A Short Introduction to Quantum Information and Quantum Computation*. Google Books. https://www.google.com/books?hl=en&lr=&id=eQIP1_fc7gYC&oi=fnd&pg=PR1&dq=A+Short+Introduction+to+Quantum+Information+and+Quantum+Computation&ots=jSEn_N2UMq&sig=tOoLdo_Yzz1nw9fYgP5Z3dXY9M8

Jordan, S. P. (2017). Fast quantum computation at arbitrarily low energy. *Physical Review. A*, *95*(3), 032305. https://journals.aps.org/pra/abstract/10.1103/PhysRevA.95.032305. doi:10.1103/PhysRevA.95.032305

Kak, S. (2000). Active agents, intelligence and quantum computing. *Information Sciences*, *128*(1-2), 1–17. https://www.sciencedirect.com/science/article/pii/S0020025500000438. doi:10.1016/S0020-0255(00)00043-8

Khang, A. (2023). *Advanced Technologies and AI-Equipped IoT Applications in High-Tech Agriculture* (1st ed.). IGI Global Press. doi:10.4018/978-1-6684-9231-4

Khang, A. Rath, K., Satapathy, S., Kumar, A., Das, S., & Panda, M. (2023). Enabling the Future of Manufacturing: Integration of Robotics and IoT to Smart Factory Infrastructure in Industry 4.0. AI-Based Technologies and Applications in the Era of the Metaverse. IGI Global Press. doi:10.4018/978-1-6684-8851-5.ch002

Khang, A. & Alyar, A., Khalilov, M., Murad, B., & Litvinova, E. (2023). Introduction to Quantum Computing and Its Integration Applications. Applications and Principles of Quantum Computing. (Eds.) IGI Global Press. https://doi.org/ doi:10.4018/979-8-3693-1168-4-ch002

Khang A. (2024). The Quantum Evolution: Application of AI and Robotics in the Future of Quantum Technology. CRC Press. https://doi.org/ doi:10.1201/978-1-032-64207-9

Khang, A., & Muthmainnah, M. (2023). AI-Aided Teaching Model for the Education 5.0 Ecosystem" AI-Based Technologies and Applications in the Era of the Metaverse. (1 Ed.) IGI Global Press. doi:10.4018/978-1-6684-8851-5.ch004

Khang, A., Shah, V., & Rani, S. (2023). AI-Based Technologies and Applications in the Era of the Metaverse. (1 Ed.) IGI Global Press. doi:10.4018/978-1-6684-8851-5

Kribs, D. W. (2005). A quantum computing primer for operator theorists. *Linear Algebra and Its Applications, 400*, 147–167. https://www.sciencedirect.com/science/article/pii/S0024379504004793. doi:10.1016/j.laa.2004.11.010

Kribs, D. W., Pasieka, A., Laforest, M., Ryan, C., & da Silva, M. P. (2009). Research problems on numerical ranges in quantum computing. *Linear and Multilinear Algebra, 57*(5), 491–502. https://www.tandfonline.com/doi/abs/10.1080/03081080802677441. doi:10.1080/03081080802677441

Miranda, E. R. (2021). Quantum computer: Hello, music! *Handbook of artificial intelligence for music: Foundations, advanced approaches, and developments for creativity*, (pp. 963-994). Springer. https://link.springer.com/chapter/10.1007/978-3-030-72116-9_34

Mueller, F., Byrd, G., & Dreher, P. (2019, February). Programming quantum computers: a primer with IBM Q and D-Wave exercises. In *PPoPP* (p. 451). https://arcb.csc.ncsu.edu/~mueller/qc/qc-tut/NEW-ASPLOS-4-14-2019.pdf

Nimbe, P., Weyori, B. A., & Adekoya, A. F. (2021). Models in quantum computing: A systematic review. *Quantum Information Processing, 20*(2), 80. https://link.springer.com/article/10.1007/s11128-021-03021-3. doi:10.100711128-021-03021-3

Petschnigg, C., Brandstötter, M., Pichler, H., Hofbaur, M., & Dieber, B. (2019, May). Quantum computation in robotic science and applications. In *2019 International Conference on Robotics and Automation* (ICRA) (pp. 803-810). IEEE. https://ieeexplore.ieee.org/abstract/document/8793768/

Prabu, B., & Khang, A. (2023). Quantum Computing Smart City: Fostering Sustainability and Green Infrastructure with Smart City. Applications and Principles of Quantum Computing. (Eds.) IGI Global Press. https://doi.org/ doi:10.4018/979-8-3693-1168-4-ch009

Shyam, R. (2023). Effects of Quantum Technology on Metaverse," AI-Based Technologies and Applications in the Era of the Metaverse. IGI Global Press. doi:10.4018/978-1-6684-8851-5.ch009

Soljanin, E. (2020). Quantum Information Processing: An Essential Primer. *IEEE Journal on Selected Areas in Information Theory, 1*(2), 351–366. https://ieeexplore.ieee.org/abstract/document/9177097/. doi:10.1109/JSAIT.2020.3019340

Spiller, T. P. (1998). Basic elements of quantum information technology. In *Introduction to Quantum Computation and Information* (pp. 1-28). Wold Scientific. https://www.worldscientific.com/doi/abs/10.1142/9789812385253_0001

Teja, S. B. B., & Saheb, S. H. & Khang A. (2023). Quantum Deep Learning: Detection of Drowsiness in Drivers Using Tensorflow Framework. Applications and Principles of Quantum Computing. IGI Global Press. https://doi.org/ doi:10.4018/979-8-3693-1168-4-ch015

Chapter 2
Introduction to Quantum Computing and Its Integration Applications

Alex Khang
https://orcid.org/0000-0001-8379-4659
Global Research Institute of Technology and Engineering, USA

Vugar Abdullayev
https://orcid.org/0000-0002-3348-2267
Azerbaijan University of Architecture and Construction, Azerbaijan

Abuzarova Vusala Alyar
https://orcid.org/0000-0001-7480-4632
Azerbaijan State Oil and Industry University, Azerbaijan

Matlab Khalilov
Department of Computer Engineering, Azerbaijan State Oil and Industry University, Azerbaijan

Nazila Ali Ragimova
Azerbaijan State Oil and Industry University, Baku, Azerbaijan

Yitong Niu
School of Aeronautical Engineering, AnYang University, Anyang, China

ABSTRACT

As humans continue to evolve, they discover new worlds. One of them is the Quantum World, one of the main discoveries of the last century. The Quantum World, which is the world of subatomic particles, has gone through great development stages since its first discovery. This process, which started with quantum physics, continues with the application and study of quanta in many different fields. One of the main fields that use the possibilities of quantum theory and quantum mechanics is the field of computer technologies. Today is an era in which the quantum computer, which is considered to be the next gen-

DOI: 10.4018/979-8-3693-1168-4.ch002

eration of computers, continues to be improved. This innovation, which can advance the era of classic computers to the next level, will guide the future technology sector. In the chapter, a very short version of the long history of evolution from the creation of the universe to quantum theory and finally to quantum computing is mentioned, and the main parts of this theory that most need to be known are mentioned.

1. INTRODUCTION

Although various theories have been tried to prove how the universe was created, it is still not completely clear how exactly this process happened. One of the main theories about how the universe started is the Big Bang theory. With the Big Bang, the universe began to form, but before that there was a great void. Another theory is trying to explain this big gap. This is the "Inflation" theory first proposed by Alan in 1979.

Will Kinney, author of An Infinity of Worlds: Cosmic Inflation and the Beginning of the Universe, had this to say about inflation in a related discussion (Kinney, 2022). "Inflation tells us that the time before the Big Bang was extremely cold, almost absolute zero, and everything but space was empty, and that empty space carried the energy that brought the universe to this enormous size and initial state before the Big Bang." The key word that stands out in this statement will be our keyword. Which is "Energy". Energy is what existed before and after the universe (Khang & Kali, 2024).

Today, when we think of the main components of the universe, matter and energy come to mind. Energy is one of the properties of matter, but energy can exist without matter. Which is exactly how energy existed in all the empty space before the Big Bang. It is known from physics that "energy is the ability to do work". And it can exist in potential, kinetic, thermal, electrical, chemical, nuclear, or various other forms. On the other hand, energy can be of different sizes in different places. Here again we need to look back to before the Big Bang. There was a huge gap before the explosion.

Belgian Catholic priest, theoretical physicist, mathematician, astronomer and physics professor Georges Lemaitre, who is called the father of the Big Bang theory (Georges, 1949), argued that "the physical universe was in the beginning a single particle - the 'primordial atom' -" and that it "disintegrated as a result of the explosion into space and time and the universe today led to the expansion that continued until. This theory later became known as the Big Bang. Today, when talking about the creation of the universe, scientists put forward the Big Bang theory, but what actually caused it is still unknown.

A statement by Georges Lemaitre simply says that everything that exists today was contained within a tiny "dot" billions of years ago, which then exploded and grew. This "tiny particle", which could be called a "ball of energy", was the beginning of "Everything". But in fact, in the beginning of everything, there was another topic that is being studied more and more today. It was subatomic particles, not the "primordial atom" proposed by Georges Lemaitre. This constituted the quantum world. The atom, a ball of energy and once known as the smallest existence, it was itself actually composed of smaller particles. Subatomic particles - these were called "quarks". This is currently the theory - So quarks are the smallest particles and contain nothing - "Hollow particles". But it is debatable how true the statement is. At one time the smallest particle was the atom, later it was suggested that there were subatomic particles. In other words, energy levels were studied (Khang & Shah et al., 2023).

Energy levels currently studied up to the Quark may indicate that another, smaller particle exists later on. The idea of quarks was proposed in 1964. Evidence of their existence was seen in 1968 in experiments conducted at the Stanford Linear Accelerator Center (SLAC). That is, the existence of quarks is confirmed, they exist but we cannot see them. Quarks are high-level energy points that move quickly. Here again the key word is energy.

In this regard, let's start the topic with Energy, the main component of the Universe. First of all, let's mention such a controversial issue: "According to some, the Universe has no energy. That is equal to zero (Will Kinney, 2022). On the other hand, according to some - including us - the Universe has energy. Energy existed before the universe and still exists, and the universe, everything in it, beings, objects, particles, particles, each has energy." In this regard, the next parts that we will mention will be in this direction. For an article, we can list the table of contents as follows:

1. Energy – Just energy
2. Quantum Physics – Fundamental; and a separate concept from Classical Physics
3. From Quantum Mechanics to Quantum Computing: About the Modern – next-generation computer – Quantum Computer.

2. ENERGY– JUST ENERGY

The concept of "energy" can be found in many different fields, from everyday life to various fields of science. However, the main field of science in which the basic principles of the concept of energy are studied is Physics. Energy has a certain existence in the field of physics, it is one of its fundamental concepts.

Energy comes from the Greek word "Enérgeia: (ἐνέργεια)". This concept is rooted in Aristotle. The concept of energy used today is different from the form of the past. The modern concept of energy was first introduced to physics in 1800 by Thomas Young. At first, this concept was not popular. Young later determined the wave nature of light through interference experiments. The related term "work" was defined in 1828/29 by Gustave Gaspard de Coriolis and Jean-Victor Poncelet (Energy, 2023).

Basically, in physics, energy is basically known as the "capacity to do work". It is simply about using energy to do work by changing it from one form to another. But not all energy is spent on work. This is reflected in certain laws of energy. First, let's mention the forms of energy. There are two main (basic) types of energy, such as kinetic and potential. All energy basically takes its various forms in these two types: heat, chemical, magnetic, electric, nuclear, sound, light, and mechanical energy.

2.1 Kinetic and Potential Energy

According to fundamental physics theory, the type of energy that Kinetic and Potential (2002) energy together form is total mechanical energy: ME = KE + PE. Energy mainly means mechanical energy. It characterizes the ability of objects to work. In the international system of measurement (units), Joule is accepted as the unit of Energy as shown in Formula 1.

$$1J = 1\frac{\text{kg}m^2}{sec^2} \tag{1}$$

The energy gained by an object due to its own motion is Kinetic energy. Derived from the Greek word "kinētikos" meaning "movement" as shown in Formula 2.

$$KE = \frac{mv^2}{2} \tag{2}$$

The energy determined by the mutual state of objects is called Potential energy. Its origin is from the Latin "potentia", which means "power, possibility". Potential energy is determined by the Formula 3.

$$PE = mgh \tag{3}$$

According to the formula, Potential energy is the work required to raise an object of mass m from its initial position to a height h (Physics, 2023), and m is the mass of the object, h is the height, and g is the force of gravity. From this it is known that the potential energy is related to the weight of the earth (gravitational force of the earth).

2.2 Other Forms of Energy and in Simple Terms

- Electric Energy – is a type of energy produced as a result of the movement of electrons.
- Chemical Energy – is the energy stored in the bonds of atoms and molecules. Batteries, biomass, oil, natural gas, and coal are examples of chemical energy (EnergyExplained, 2023).
- Heat Energy – this type of energy, also known as thermal energy, is created as a result of the movement of atoms, molecules, and particles. The faster the movement, the faster the heat energy.
- Magnetic Energy – is a type of energy produced as a result of the strength of the magnetic field. Depending on the direction of the electrons, the interaction in the magnetic field can be repulsive and attractive.
- Nuclear Energy – The most powerful form of energy, nuclear energy is the form of energy obtained from the nucleus of an atom. Electricity is obtained from this type of energy. It is most often associated with Albert Einstein's E=mc2 formula. It is an important formula for understanding how energy is released from the nucleus.
- Light – Light Energy – related to quantum theory - is a form of electromagnetic radiation, a type of kinetic energy created by photons. Light energy is the only visible form of radiant energy. It is possible to mention within the same category.
- Sound Energy – The movement of energy in the form of waves through water or air. It mainly occurs when the object (or substance) vibrates.

Each of the above-mentioned types of energy has quite a wide range of theories. What is mentioned is just a small part of their definitions. Each of these types of energy represents a different division of physics. There are many concepts related to energy, its definitions, measurement units, laws, formulas, theories, forms, types, existence in various fields, transfer, etc.

The law of conservation of energy, which is one of the basic and fundamental concepts of physics, also says an interesting statement related to energy. Energy in 'nature' does not arise from nothing and does not disappear. Its amount is unchanged. This corresponds to the first of the three laws of thermodynamics. According to the first law of thermodynamics (Thermodynamics, 2023), when a system (object) changes from one state to another, its internal energy changes, which occurs in two ways: by applying a work force on it from the outside and by giving off an amount of heat. The law of conservation of energy

Figure 1. The relationship between E and m in thermodynamics

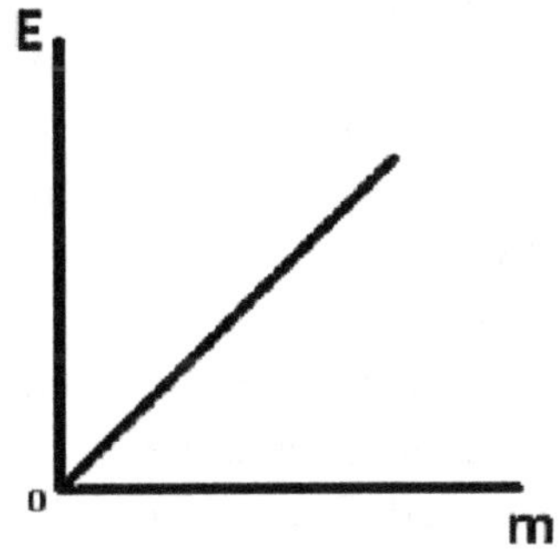

also supports the theory that the amount of energy that enters the system is the same amount of energy that is released from it as shown in Figure 1.

Alt-text: Figure 1 depicts one of the most famous formulas related to energy, E=mc2 defines the main characteristics of an object. Any object has both mass and energy in a vacuum. The relationship between the mass and energy of an object is a straight line, supported by the above formula. Just as an object in motion has energy, it also has a large amount of energy at rest as shown in Formula 4.

$$E0=m0c2 \tag{4}$$

It is the energy of the object at rest. The total energy of an object is equal to the sum of its rest energy and kinetic energy as shown in Formula 5.

$$E = E0 + K \tag{5}$$

There is all energy in the universe. The universe (according to many experts) is a balance of positive and negative energies. In this respect, the "total" energy of the universe is "zero". On the other hand, there are also speculations about the existence of "dark energy" as the largest and most powerful energy in the universe, which is the main energy of the universe. If "dark energy" exists, it would definitely be the most powerful energy in the universe.

- Energy is studied in different branches of Physics.
- One of them is quantum physics.

3. QUANTUM PHYSICS: FUNDAMENTAL, AND A SEPARATE CONCEPT FROM CLASSICAL PHYSICS

Classic shows itself in different areas. Referring to the past, learning from the past and taking it as a foundation. Today, when we turn to many fields, we can see that they have a basic classical theory. Which has greatly helped to shape the modern field: Classical literature, classical art, classical everything.

And also, Classical Physics. Classical physics, which forms the basis of modern physics, is a collection of classical theories that were acquired and studied in previous centuries and allowed the formation of today's modern physics. Classical Physics is a model that studies macroscopic systems, their activities and development, and accepted energy and matter as the basic components. Modern physics, on the other hand, aims to study microscopic systems and still provides a fundamental study of energy and matter. Modern Physics is mainly the result of two important scientific breakthroughs (discoveries): Theory of Relativity (Einstein, 1905) and Quantum Physics (Planck et al., 1900).

There are also differences between classical physics and modern physics, which mainly characterize the contradictions between them. Although Quantum Physics (as modern physics) is considered the most correct physics, it is still not fully studied. Quantum physics can reflect the reality of the universe. Just like how the universe came to be, the big bang theory and the realities of the world of many microsystems can allow us to study the realities of the macroscopic world as well. Considering that the macroscopic world consists of millions of microscopic worlds, different results can be obtained after studying the microscopic worlds than anything studied so far. The study of microscopic systems (or worlds) can change classical physics.

For example, someone who is an observer looks at a mountain. He recognizes it. He is learning. But only as a macroscopic system. He does not study its basic interior, its smallest units, and its microscopic world. According to the observer, this mountain has a perfect structure and is quite strong. However, when the observer goes deeper and studies its microscopic world, he realizes that this mountain is so weak that it can be completely destroyed by a simple (for example, chemical) reaction. Then he realizes that the macroscopic mountain is completely different from the microscopic mountain, which is the basis of the similarities and contrasts between the two systems. The microscopic mountain changes all knowledge and practice of the macroscopic mountain.

Or vice versa. That is, the similarities between the two can be proven (or supported) by the research conducted in the microscopic system, and the theories proposed by the macroscopic system. In classical systems, something is either a pure wave or a pure particle. You can choose one or the other to classify certain behaviors. But in quantum mechanics, these two properties are combined: everything is both a particle and a wave at the same time, and always exhibits some properties of both (Paul, 2019).

The universe is not in a state of absolute stillness (or absolute zero). All the particles that make it up are constantly in motion. Absolute zero is never achieved in the general universe. The main purpose of mentioning this is to show that quanta are not at rest either. These were the simple and basic fundamental concepts to get into Quantum Physics, at least to get to the front door. Now, let's get into quantum physics.

3.1 Definition

Various dictionaries have similar definitions of quantum physics:

- A branch of science that examines the principles of quantum theory to understand the behavior of particles at the atomic and subatomic level (Oxford Advanced Learner's Dictionary)
- The theory of mechanics of atoms, molecules and other physical systems subject to the uncertainty principle1 (Dictionary.com (website)
- According to the correspondence principle2, classical physics is the limiting case of quantum physics as we approach higher quantum numbers, the quantum results merge with the classical ones (Cambridge English Corpus).

- The branch of physics that uses quantum theory to describe and predict the properties of a physical system (American Heritage, 2016)

3.2 History

Quantum Physics also has two historical evolutions within it. Quantum physics, which forms the basis of ancient quantum physics and modern physics. The history of quantum physics first started with M. Planck. This dates back to the 1900s. According to Planck, a particle radiates electromagnetic energy as a quantum, the smallest amount of energy that can be emitted or absorbed as radiation. Energy of radiation - quantum is calculated by the formula: E = hv. h^3 – Planck's constant, v – represents the frequency. The study (initiation) of Quantum Physics is related to giving the model of the atom.

3.3 Atomic and Quantum

Any objects (including atoms) can only emit or absorb a certain amount of energy. Energy is quantized; occurs in constant amounts rather than continuously. Each constant amount of energy is called a quantum. An atom changes its energy state by emitting or absorbing one or more quanta of energy (McGraw, 2013). Quantum Physics characterizes the fundamental components of physics by studying atomic and subatomic particles and their properties. The quantum world (system) characterizes the subatomic world.

Quantum physics is also related to the processes carried out in the direction of giving the model of the atom for the first time. Bohr and Planck are the founders of quantum physics. As a result of their work in this direction, they were awarded the Nobel Prize. At the same time, Albert Einstein is also one of the founders. Their work in this direction can be noted as follows.

- The main reasons for the emergence of Quantum Physics were the flaws in Maxwell's electromagnetic theory4 (1865).
- According to Maxwell's theory, light was an electromagnetic wave. There were a number of events that confirmed the truth of this theory. However, this theory could not explain a number of regularities observed in thermal radiation. So, according to Electrodynamics, when a heated object radiates electromagnetic waves, it must lose its energy and approach zero. But this remained a theory, such a situation was not observed in practice.
- In connection with this, for the first time, M. Planck put forward a hypothesis that laid the foundation of Quantum physics.
- In 1900, Planck proposed the following hypothesis: "Atoms release (radiate) energy in the form of certain portions. These are called quanta. The energy of each quantum is defined as E=hv. Light quanta were called "photons"".

Again, one of the shortcomings of Maxwell's theory was the inability to explain the phenomenon of the photo effect. In this direction, Stoletov conducted research and put forward laws. However, the main leading step for Quantum Physics was taken by Einstein. Albert Einstein gave the "Photoelectric effect" equation as shown in Formula 6.

$$hv = W + \frac{mv^2}{2} \tag{6}$$

According to Einstein, "as a result of the absorption of a photon, its hv energy is entirely used to do the work W to remove the electron from the metal, and the rest is given in the form of kinetic energy of the electron leaving the metal. Einstein was awarded the Nobel Prize (1921) for his work in quantum physics on the Photoelectric Effect. After that, in 1913, Bohr's model of the hydrogen atom started the basic history of quantum physics.

For the first time in 1911, Rutherford presented the planetary model of the atom. However, there have been shortcomings related to the issue of atomic stability. So, at this time, Bohr put forward several postulates in this direction (related to Bohr's atomic model theory). These (AtomicStructure, 2023),

- In an atom, the electrons (negatively charged) revolve around the positively charged nucleus in certain circular paths called orbits or shells. Each orbit or shell has a fixed energy, and these circular orbits are known as orbital shells.
- Energy levels are represented by an integer (n=1, 2, 3…) known as a quantum number. This quantum number range starts from the nuclear side with the lowest energy level n=1. n=1, 2, 3, and 4… orbits are defined as K, L, M, and N…. When the shells and electron reach their lowest energy level, it is said to be in a stable state.
- Electrons in an atom move from a lower energy level to a higher energy level by gaining the required energy, and an electron moves from a higher energy level to a lower energy level by losing energy

When an atom emits a photon, it loses energy, changing from energy level Ei to level Ej. At this time, the photon has energy Ei-Ej and frequency v as shown in Formula 7 and Formula 8.

$$E = E_i - E_j = hv \tag{7}$$

$$v = \frac{E_i - E_j}{h} = \frac{E}{h} \tag{8}$$

Nevertheless, Bohr's postulates contradicted a number of theories. The most important of them was Maxwell's electromagnetic theory. Another was Heisenberg's uncertainty principle.

The period up to this point covered the period of Old Quantum physics. In the 1920s, physicists tried to apply Planck's concept of energy quanta to the atom and its components. At the end of the decade, Erwin Schrödinger and Werner Heisenberg invented a new quantum theory of physics. The Institute of Physics of the University of Zurich published Schrödinger's lectures on wave mechanics (first on January 27, 1926) and in 1930 Heisenberg's book "Physical Principles of Quantum Theory". With this (Erwin, 1926), the era of modern quantum physics had already begun.

4. FROM QUANTUM MECHANICS TO QUANTUM COMPUTING: ABOUT THE MODERN/NEXT-GENERATION COMPUTER – QUANTUM COMPUTER

The process of using computer technologies to perform a certain task is called "Computing". The history of computing is related to the history of the computer. The first computers were used within limited capacities to perform various tasks.

There are different types of calculation. Often, it is possible to find Cloud computing, Edge computing, Grid computing, and Distributed computing, etc. One of them is Quantum computing, which uses the possibilities of Quantum mechanics.

- Quantum computing is a rapidly developing technology that uses the laws of quantum mechanics to solve problems too complex for classical computers (IBM, 2023).
- Quantum computing uses subatomic particles such as photons. Other computers use 0s and 1s, where the response of any process can be either a 0 or a 1.

According to Hard Computing, the answer to tasks should be two: Yes or No, True or False, White or Black, 1 or 0. On the other hand, according to Soft Computing (which is basically one of the main theories for the development of artificial intelligence technology, which allows the technology to perform a decision-making process close to a human), there should also be vague answers: Maybe, Probably, 1.3, 0.4, etc.

Briefly supports the existence of ambiguous answers in ambiguous situations. Quantum Computing, on the other hand, supports a different theory. 1 or 0, or 0-1. That is, the answer can be either simply 1, or simply 0, or both at the same time (0-1). Getting this state called superposition is not such an easy process. With the simplest and possible example, the principle of superposition can be explained in this way. Sometimes when making any decision (to choose one of two alternatives) we can use "decision coin". At this time, we throw the coin in the air and when it lands on the intended side, we make the appropriate decision.

For example, we want to make a decision whether to go to the theater in the evening or not. For this, we use a "decision coin" with "Yes" on one side and "No" on the other. Naturally, the result of this process will be one of two alternatives. But there is an interval of time within the duration of the process – which is the time when this coin is spinning in the air with circles – where both alternatives are valid. That is, "yes-no, yes-no, yes-no". It is repeated many times. This process characterizes the state of superposition. But this is a finite process and the result is one of the alternatives.

One of the most famous experiments related to the superposition state is the "Schrödinger's cat"5 experiment. The work of a quantum computer is based on the use of particles in superposition. Here, instead of representing bits, such particles represent qubits (aka, quantum bits) that can hold 0, 1, or both at the same time. The basic unit of information in quantum computers is qubits.

The first computers were quite large and tried to perform tasks within limited capabilities. Later, the size of computers began to decrease, while their capabilities began to increase. And now, the Quantum computer, which is one of the modern computers, has a large enough size, but still has many possibilities.

4.1 Properties of Quantum Computing

Special properties of quantum objects (atoms, ions, and photons) are superposition, interference, and entanglement. Superposition refers to particles existing simultaneously in all possible states. Interference is the condition in which interference from environmental noise damages a quantum object, as well as the possibility that the wave functions of particles can amplify or attenuate each other. Entanglement means that groups of particles are bound together and can interact in such a way that, even when the particles are separated by great distances, the quantum state of each particle cannot be described independently of the states of the others (AQC, 2019).

4.2 Advantages and Disadvantages of Quantum Computing

4.2.1 Advantages

- **Possibility of Parallel Operation** - In which it is possible to solve the problem that cannot be solved in classic computers with simple steps. This allows the quantum computer to become even faster. Note: A quantum computer performs operations sometimes as fast as classical computers, sometimes much faster, depending on the task at hand.
- **(In the future) to have a wide field of application** - Quantum computing can be applied to decision-making processes in business, drug development and similar operations in healthcare, finance, military, etc. Still a developing technology, quantum technologies are likely to be used in more fields and for a wider range of purposes in the near future (Khang & Medicine, 2023).
- **Quantum computers have "qubits"** - Compared to classical bits, qubits are used to store and compute more information. Qubits save more memory and time than bits.

4.2.2 Disadvantages

Limited number of qubits - A quantum computer must have enough qubits to solve real-world problems. On the other hand, it is difficult to store and manage a large number of qubits.

- **Qubit Decoherence** - In other words, incompatibility or loss of features. Qubits are extremely sensitive to their environment, and even a small process error can cause them to lose their qubit properties. This leads to distortion of information.
- **High probability of error** - One of the major drawbacks of quantum computing is its high error rate. The results of the operation can often be wrong output. This increases especially as the complexity increases as well as the number of qubits.
- **Costs** - A quantum computer can be expensive to operate and upgrade. One of the main reasons for this is that it is a very new technology and has a complex infrastructure. On the other hand, this drawback leads to its application in limited spaces. Currently, quantum computing is being used mainly in research laboratories, as well as in a few possible leading companies.

Since it is a new technology and is still being improved, it has many drawbacks, but this will not prevent the future potential of the quantum computer. Quantum computers will be much more powerful, faster and more productive than classical computers and will have a wide range of applications. Although

Figure 2. Architecture of quantum computer
Alt-text: In the scheme shown in Figure 2, quantum registers represent quantum memory. In this part, quantum data is stored to execute the algorithm intended for the solution of the problem. In many types of quantum hardware, these registers are physically static, and operations are performed directly on the qubit.

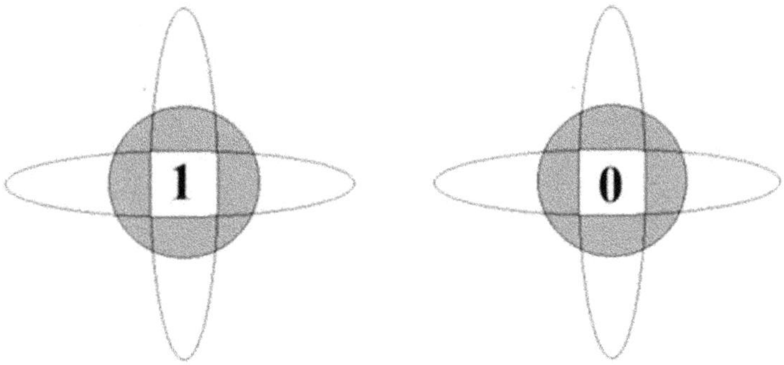

quantum computers are still relatively large, in the near future we may see them as the mainframe computers of the past are the microcomputers of today.

4.3 Architecture of Quantum Computer (Computing)

Here we can consider two parts. Standard structure and layered architecture (QCA, 2023). A standard structure for a Quantum Computer can be stated as shown in Figure 2.

The basic requirement of every computer is "software". Without software, a computer and any automated technology in general cannot work. A quantum computer is no exception. Algorithms intended for the problem to be solved are understood by the quantum computer through the program.

On the other hand, quantum gates are one of the main building blocks of this scheme. They are used to manipulate the quantum state of qubits. A quantum gate is a mathematical operation that affects one or more qubit states, and it can be represented by a matrix. In classical computing, logic gates such as AND, OR, and NOT are used to manipulate the state of bits. The most commonly used gates in quantum computing are Pauli gates, Hadamard gates, CNOT gates and T gates (Mayank, 2023).

- The Hadamard gate is a logic gate in quantum computing that turns a state of $|0\rangle$ or $|1\rangle$ into a superposition of $|0\rangle$ and $|1\rangle$. (Learn about Hadamard gates in the Xanadu Codebook.) In a circuit diagram, a Hadamard gate is represented by a square with the letter 'H' in it (HadamardGate, 2023).
- The Pauli gates are the three Pauli matrices and act on a single qubit. The Pauli X, Y and Z equate, respectively, to a rotation around the x, y and z axes of the Bloch sphere by radians (PauliGates, 2023).
- The CNOT gate is a mulit-qubit gate that consists of two qubits. The first qubit is known as the control qubit and the second is known as the target qubit. If the control qubit is |1>then it will flip the targets qubit state from |0>to |1>or vice versa (CNOTGate, 2023).
- The T gate is a single-qubit operation given by: T = (1 0 0 exp (i π 4)) \ T = \begin{pmatrix} 1 & 0 \ 0 & \exp \left (\frac{i \pi}{4} \right) \end{pmatrix} T=(100exp(4iπ)) The T gate is related to the S gate by the relationship S = T 2 \ S=T^{2} S=T2 (TGate, 2023)

Figure 3. Diagram of five layers of quantum computer
Alt-text: Figure 3 depicts 5 layers of quantum computer

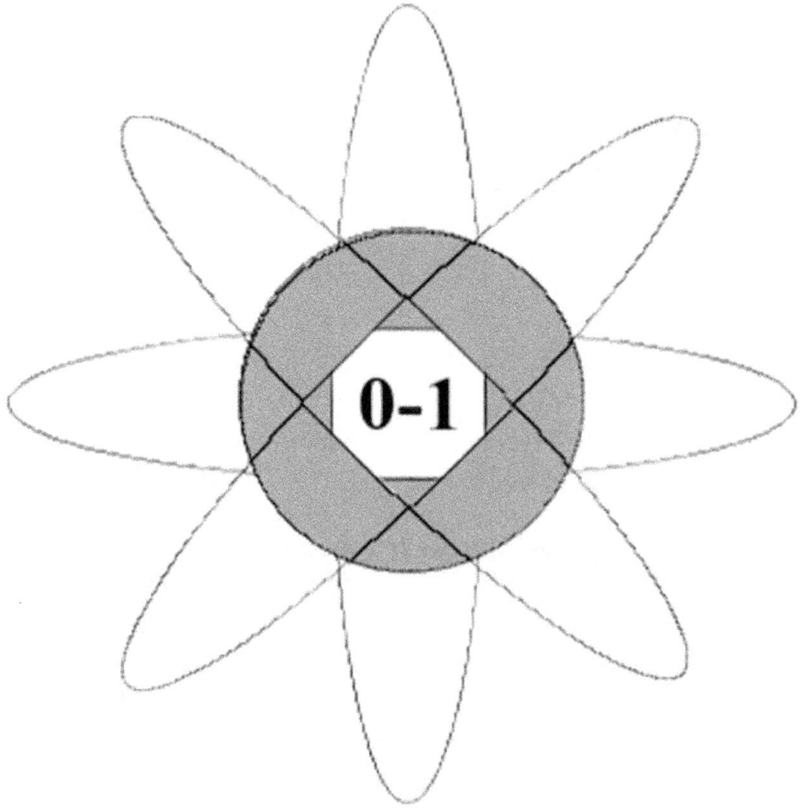

Other blocks are generally similar to a classic computer. It should be noted that the quantum computer is not a completely different model and has similarities with the classical computer. The "layered" architecture intended for a quantum computer mainly consists of 5 layers. The structure of this architecture (Sarkar et al., 2018) as shown in Figure 3.

- Physical layer (layer) – This is the lowest and primary layer, It contains hardware including physical qubits and control operations.
- Virtual layer - This layer characterizes open-loop fault cancellation as dynamic decoupling.
- Quantum error correction (detection) layer - This layer corrects arbitrary system errors if the rates are below the threshold
- Logic layer - This layer builds the substrate (sublayer) that supports universal quantum computing
- Application layer - It includes quantum algorithms and a classical user interface.

Figure 4. At least five fields can integrate with quantum computing

4.4 Integration of Quantum Computing

Today, quantum computing is in the process of development. On the other hand, it continues this process in the environment of integration with various fields and technologies. These fields and technologies form the foundations of next generation development and some of these fields and technologies are Artificial Intelligence, Internet of Things, (Big) Data Analytics as well as Cyber Security. They are also interconnected and integrated in different sectors as shown in Figure 4.

Artificial Intelligence gives many "technological entities" in the world the opportunity to move to a new level: It makes them intelligent. In other words, it enables them to think, learn, make decisions and act accordingly. At the moment, it allows and requires integration in many other areas. These are Internet of Things, Big Data Analytics, as well as the main part Security. These parts form certain branches of Artificial Intelligence. In this regard, first of all, it is necessary to look at the integration of Artificial Intelligence and Quantum technology.

The integration of Artificial Intelligence and Quantum technology has led to the emergence of the concept of Quantum Artificial Intelligence. Artificial Intelligence and Quantum Computing converge at some point to make things faster, more accurate and more productive using qubits unlike the classics as shown in Figure 5.

Quantum Artificial Intelligence is the use of quantum computing to compute machine learning algorithms. Thanks to the computational advantages of quantum computing, quantum artificial intelligence can help achieve results that cannot be achieved with classical computers (Zhang et al., 2023).

The convergence of quantum computing and artificial intelligence opens the door to accelerated training of machine learning models, improved pattern recognition, and the ability to solve previously intractable problems. The combination of quantum computing and Artificial Intelligence ushers in a new era: This is quantum machine Learning (QML). Algorithms in this field harness the power of qubits to open new possibilities and redefine the boundaries of artificial intelligence (David, 2023).

Healthcare, Finance and Banking, Logistics, Transportation, Environment, Research, etc. can be mentioned as local application areas of Quantum Artificial Intelligence (Khang, Vugar and Abuzarova

Figure 5. Classical bits and qubit
Alt-text: Figure 5 depicts the classical bits and qubit in Quantum Computing.

et al., 2023). Health care uses the capabilities of Artificial Intelligence in the implementation of similar processes: organization of tests and laboratories, diagnosis and prediction of medical devices, discovery of drugs, improvement of patient-doctor relations, management of medical data, etc. With the help of Quantum Artificial Intelligence, the same processes can be carried out more accurately, efficiently, productively, quickly, in a short time, at low cost (if the costs of quantum artificial intelligence are not counted).

Logistics also uses the capabilities of artificial intelligence to implement similar processes: warehouse control, resource management, route planning, inventory control, organization of producer-consumer relations, etc. Just like in Healthcare, with the help of Quantum Artificial Intelligence, the same processes can be carried out more accurately, efficiently, productively, quickly, in a short time, and at low cost (Keerthika et al., 2023). In a generalized form, we can note areas of application in the Table 1.

Quantum Artificial Intelligence, Data Analytics, Internet of Things and Security which uses the capabilities of artificial intelligence and ICT in general, converge at a certain point and develop then implementation of processes, more accurately, efficiently, productively, quickly, in a short time, at low cost (if the costs of quantum artificial intelligence are not counted).

4.5 Internet of Things and Quantum Computing

The Internet of Things is used for various purposes in sectors like healthcare, education, daily life, manufacturing (Khang & Kali et al., 2023), military, logistics, finance and banking, entertainment, tourism, etc (Khang & Muthmainnah et al., 2023). The Internet of Things is an environment where smart devices

Table 1. Areas of artificial intelligence use cases

Healthcare	Organization of tests and laboratories, diagnosis and forecasting of medical devices, drug discovery, improvement of patient-physician relations, management of medical information, etc.
Logistics	Warehouse control, resource management, route planning, inventory control, organization of producer-consumer relations, etc.
Transportation	Optimization of route directions, monitoring, control and management of traffic jams, implementation of optimal transport solutions, etc.
Finance and Banking	Risk assessment, anti-fraud, customer relationship improvement, customer data management and security, etc.
Environment	Controlling global environmental problems, forecasting, organizing solutions for efficient use of natural resources, etc.
Research	Studying, testing, observation of the research object, modeling in areas where the capabilities of the human worker are not sufficient, or carrying out direct research, etc.

and the process of data exchange between them are implemented. In terms of data management, Internet of Things devices: sensors, smart devices, Internet of Things components in general are applied together (Nayak and Atmika et al., 2023).

The joint integration of Internet of Things and Quantum Computing has led to the emergence of the concept of Quantum Internet of Things. Here, Internet of Things data merges with quantum data. Attributes of quantum information include the concept of uncertainty, non-clone quantum theory, quantum teleportation, and hidden quantum information properties that can be exploited for resistance attacks during cyberspace transmission (Heisenberg et al., 1985) (Shamshad et al., 2023). The Internet of Things and Quantum Computing can combine to create a new Quantum Internet of Things infrastructure. This may be as in the following illustration (Kelsey, 2023):

- **Application layer -** This layer covers the management of data flows: collection, transmission, processing, etc. Quantum algorithms and qubits are used here.
- **Quantum Teleportation layer -** The quantum teleportation layer contains only quantum replicators. That is, it manages the integrity of the qubits to process and transmit data. Briefly, quantum repeaters play a key role in this layer (Kelsey, 2023).
- **Quantum network layer -** The quantum network layer is responsible for controlling information transfer through quantum and classical systems, it consists of two components: Quantum server and gateway (Kelsey, 2023).
- **Physical layer -** There is the hardware layer, which includes the proprietary Internet of Things devices, or IoT components for short. IoT communication in the future Quantum computing will be essential for secure communication (Diksha et al., 2023).

4.6 Data Analytics and Quantum Computing

The integration of quantum computing with data analytics is characterized by the ability to work with large data sets. Data (whether large or small) is the key to all processes. Being able to work with data correctly, to manage them correctly, also creates conditions for the proper implementation of processes in general. Being able to obtain data in a timely, complete form, without loss as much as possible, is important in data management and implementation of processes.

- **Integration -** Quantum computing provides high-speed detection, analysis, integration, and diagnostics when working with large, dispersed data sets (Rakesh, 2022).
- **Data management -** Data management with quantum computers is faster and more efficient with less risk than classical computers. The main issue in data management with the help of quantum algorithms working with qubits is to be able to solve even complex problems in a faster time than classical computers.
- **Data Analytics -** Quantum Computing also uses the capabilities of Artificial Intelligence (for example, machine learning algorithms) to ensure the proper management of data. At this point, Quantum Artificial Intelligence merges with Data Analytics.

In its generalized form, the application of Quantum computing in Data analytics allows (Corey, 2023):

- **Dispersed data sets**: Quantum computing offers high-speed detection, analysis capabilities, integration and diagnostics when using large, distributed data sets.
- **Patterns**: Quantum computers can quickly find patterns in large, unsorted data sets by looking at every item in a huge database simultaneously.
- **Calculations**: While classical computers take a long time to perform extremely complex calculations, quantum computers can perform them in mere seconds.
- **AI applications**: Current AI applications are often used to manage big data, as well as analyze datasets to find patterns (Aakansha & Adhishree et al., 2023). Three areas of artificial intelligence benefit from the speed and capabilities of quantum computing:
    - Natural language processing
    - Machine Learning
    - Predictive Analytics

4.7 Cyber Security and Quantum Computing

Cyber Security, one of the main "Global" problems, is also one of the areas that will be affected by the possibilities of Quantum Computing. The key word here is "Cryptography"6. With Quantum Computing, more rigorous and robust solutions can be created to protect the privacy of personal, critical and important data. Quantum Computer is still several steps ahead of classical computers for quick detection of certain security holes, as well as faster and more efficient organization of diagnostics and security solutions. According to Forbes (Paul, 2021), Quantum computing will transform cyber security in four areas:

4.7.1 The Generation of Quantum Random Numbers Is Fundamental to Cryptography

A quantum random number generator (QRNG) relies on the intrinsic randomness of quantum mechanics to generate truly random numbers important in information processing tasks. Due to the existence of the superposition state, a quantum computer can be used as a true random number generator. It creates randomness by measuring quantum processes that are non-deterministic in nature (Debosree, Khang et al., 2023).

4.7.2 Quantum-Secure Communication, Especially Quantum Key Distribution (QKD)

Quantum key distribution (QKD) is a secure communication method for exchanging encryption keys known only between shared parties. Encryption allows two parties to generate and share a key used to encrypt and decrypt messages. Concretely, QKD is a method of key distribution between parties.

QKD-based quantum secure communication (QSC), which can be integrated with various cryptographic applications and communication protocols, increasing the security of key generation and update speed, has become one of the important solutions to improve information security.

4.7.3 The Most Controversial Application of the Quantum Computer Is Its Potential to Break Public Key Cryptography: the RSA Algorithm

Millions of physical qubits are needed to break the RSA-2048 scheme. Breaking a 1024-bit or 2048-bit RSA key requires a quantum computer with huge resources. A group of researchers in China reported that they found a factorization method that can break a 2,048-bit RSA key using a 372-qubit quantum system. So, the fact that this is actually happening is proof that Quantum Computing is one of the biggest impacts in the field of Cyber Security.

4.7.4 Machine Learning Has Revolutionized Cyber Security, Enabling New Attacks To Be Detected and Blocked

Quantum Machine Learning can drive faster, more energy-, cost-, and time-saving Machine Learning algorithms that can create highly effective algorithms to detect and solve new cyber-attack methods.

5. CONCLUSION

With the study of the Quantum World, it has become clear that small particles are a more ideal tool for solving larger and more complex problems. This branch of physics, which is studied after classical physics and still retains its mystique for specialists, called Quantum Physics, has an important place in computer science today. The Quantum World is like the still unexplored oceans in a space where Physics and other sciences exist. The quantum world, which carries a large universe within itself, existed before the creation of the universe and will continue to exist (Khang & Quantum, 2023).

Quantum Computing is a type of computing as a subsystem, which gained a special place in Computer Science as a result of the study of Quantum Physics. As mentioned earlier, "Quantum computing is a rapidly developing technology that uses the laws of quantum mechanics to solve problems too complex for classical computers."

Today, Quantum Computers continue to develop, offering higher accuracy, higher speed and correspondingly higher efficiency and productivity in performing operations than classical ones. At present, the number of quantum computers is quite small, and they are mainly used by research or very large enterprises. Quantum computing, in turn, has certain advantages and disadvantages, as well as architecture and infrastructure, which we have mentioned in detail above (Khang & Kali, 2023).

Quantum computing has begun to influence other fields and technologies such as Artificial Intelligence, Internet of Things, Data Analytics, and Cyber Security. It is an inevitable beginning that future

Computer Science as well as development will merge with Quantum Computing – Quantum World at some point. Which will give impetus to the beginning of a new era - a new revolution (Shyam & Khang et al., 2023).

REFERENCES

Aakansha, C. Saxena, A., Sobti, D., & Khang A. (2023). Artificial Intelligence (AI) Centric Model in Metaverse Ecosystem. AI-Based Technologies and Applications in the Era of the Metaverse. IGI Global Press. doi:10.4018/978-1-6684-8851-5.ch001

Agarwal, M. (2023). What Is Quantum gates and circuits. *Advancements in Quantum Computing. Exploring the Feasibility of Quantum Computing for Future Technology Applications.* LinkedIn. https://www.linkedin.com/pulse/what-quantum-gates-circuits-mayank-agarwal

American Heritage. (2016). *The American Heritage® Dictionary of the English Language, Fifth Edition by the Editors of the American Heritage Dictionaries.* Houghton Mifflin Harcourt Publishing Company. https://www.ahdictionary.com/

AQC. (2019). Advanced Quantum Computing: Interference and Entanglement. *Quantum Computing.* World Scientific (Europe). https://ieeexplore.ieee.org/abstract/document/8528677/

AtomicStructure. (2023). *Atomic Structure - Discovery of Subatomic Particles.* JEEIIT, JEE, Study Material. https://byjus.com/jee/atomic-structure/

Brummer, D. (2023). Quantum Computing and Artificial Intelligence. A Paradigm Shift for Investors. *OurCrowd.* https://www.ourcrowd.com/learn/quantum-computing-and-artificial-intelligence

Chawla, D., & Mehra, P. S. (2023). A Survey on Quantum Computing for Internet of Things Security. *Procedia Computer Science, 218,* 2191–2200. https://www.sciencedirect.com/science/article/pii/S1877050923001953. doi:10.1016/j.procs.2023.01.195

CNOTGates. (2023) *Quantum Computing UK-CNOT gate tutorial with Code.* CNOTGates. https://quantumcomputinguk.org/tutorials/controlled-not-gate-program

Einstein. (1905). *Annus Mirabilis of Albert Einstein. The 1905 Papers.* LOC. https://guides.loc.gov/einstein-annus-mirabilis/1905-papers

EnergyExplained. (2023). *Forms of energy, What is energy?* U.S. Energy Information Administration. https://www.eia.gov/energyexplained/what-is-energy/forms-of-energy.php

Ghosh, D. & Khang, A. (2023). Quantum Computing Internet of Things: Unlocking New Frontiers for Connectivity, Security, and Efficiency in Data Processing. Applications and Principles of Quantum Computing. IGI Global Press. https://doi.org/ doi:10.4018/979-8-3693-1168-4-ch016

Ginsberg, C. (2023). *How is Quantum Computing Used in Data Analytics?* Noble Desktop. https://www.nobledesktop.com/classes-near-me/blog/quantum-computing-in-data-analytics

Guth, A. (1979). *Waiting for the Big Bang: Innovators.* National Geographic. https://www.nationalgeographic.com/ science/article/140630-alan-guth-profile-inflation-cosmology-science

HadamardGate. (n.d.). *What is a Hadamard gate?* Penny Lane. https://pennylane.ai/qml/glossary/what-is-a-hadamard-gate

IBM. (2023). *What is quantum computing?* IBM. https://www.ibm.com/topics/quantum-computing

Kenneth, S. (2018). Philosophy of Science. *Physical Chemistry.* Elsevier. https://shop.elsevier.com/books/physical-chemistry/schmitz/978-0-12-800513-2

Khang, A. (2023). Applications and Principles of Quantum Computing. IGI Global Press. https://doi.org/ doi:10.4018/979-8-3693-1168-4

Khang, A. (2023). Quantum Mechanics Primer - Fundamentals and Quantum Computing. Applications and Principles of Quantum Computing. IGI Global Press. https://doi.org/ doi:10.4018/979-8-3693-1168-4-ch001

Khang, A. (2023). Enabling the Future of Manufacturing: Integration of Robotics and IoT to Smart Factory Infrastructure in Industry 4.0. AI-Based Technologies and Applications in the Era of the Metaverse. IGI Global Press. doi:10.4018/978-1-6684-8851-5.ch002

Khang, A. (2023). AI and IoT-Based Technologies for Precision Medicine. IGI Global Press. doi:10.4018/979-8-3693-0876-9

Khang, A. (2023). AI-Aided Data Analytics Tools and Applications for the Healthcare Sector. AI and IoT-Based Technologies for Precision Medicine. IGI Global Press. doi:10.4018/979-8-3693-0876-9.ch018

Khang, A., & Muthmainnah, M. (2023). AI-Aided Teaching Model for the Education 5.0 Ecosystem" AI-Based Technologies and Applications in the Era of the Metaverse. IGI Global Press. doi:10.4018/978-1-6684-8851-5.ch004

Khang, A., Shah, V., & Rani, S. (2023). AI-Based Technologies and Applications in the Era of the Metaverse. IGI Global Press. doi:10.4018/978-1-6684-8851-5

Khang A. & Rath, K. (2024). *The Quantum Evolution: Application of AI and Robotics in the Future of Quantum Technology.* CRC Press. https://doi.org/ doi:10.1201/978-1-032-64207-9

Lemaitre, G. (1949). *Father of the Big Bang.* AMNH. https://www.amnh.org/learn-teach/curriculum-collections/cosmic-horizons-book/georges-lemaitre-big-bang

Lipman, P., & How, Q. C. W. T. C. (2021). *Forbes Technology Council.* Forbes Technology Council. https://www.forbes.com/sites/forbestechcouncil/2021/01/04/how-quantum-computing-will-transform-cybersecurity/?sh=9a638b57d3fb

McGraw-Hill Ryerson Limited. (2013). Quantum Theory and Atomic Structure. McGraw-Hill. https://mysite.science.uottawa.ca/sgambarotta/sites/default/files/CHM%201311F/slide%20show/Ch-6%20atom/7_lecture.pdf

Physics. (2023). *Gravitational Potential Energy, Work, Energy, and Energy Resources.* Lumen Learning. https://courses.lumenlearning.com/suny-physics/chapter/7-3-gravitational-potential-energy/

Potential and Kinetic Energy (Conservation of Energy). 2002. http://www.csun.edu/~psk17793/S9CP/S9%20potential_and_kinetic_energy.htm

QCA. (2023). *Quantum Computer Architecture/量子コンピュータアーキテクチャ*. Keio University. https://www.futurelearn.com/info/courses/intro-to-quantum-computing-j/0/steps/121378

Rakesh, P. (2022). *Data Analytics and Quantum Computing*. LinkedIn. https://www.linkedin.com/pulse/data-analytics-quantum-computing-rakesh-p

Sarkar, S., & Bertels, K. (2018). *The Computer Architecture of Quantum Computing*. IEEE. https://ieeexplore.ieee.org/abstract/document/9076341/

Shamshad, S., Riaz, F., Riaz, R., Rizvi, S. S., & Abdulla, S. (2023). *Quantum Computing Supremacy in the Internet of Things*. Encyclopedia. https://encyclopedia.pub/entry/34723

Sutter, P. (2019). *Physicists are still trying to reconcile two different worlds: the quantum and the macro. Do We Live in a Quantum World?* Space. https://www.space.com/do-we-live-in-quantum-world.html

Taylor, K. (2023). IoT and Quantum Computing: A Futuristic Convergence. *Hitech Nectar.* https://www.hitechnectar.com/blogs/iot-and-quantum-computing-a-futuristic-convergence/

Will, K. (2022). *An Infinity of Worlds: Cosmic Inflation and the Beginning of the Universe*. GoodReads. https://www.goodreads.com/en/book/show/58275961

Zhang, X., Wang, L., Helwig, J., Luo, Y., Fu, C., & Xie, Y. (2023). *Artificial intelligence for science in quantum, atomistic, and continuum systems.* arXiv. https://arxiv.org/abs/2307.08423

ENDNOTES

1. The uncertainty principle was formulated by Heisenberg in 1927 in order to eliminate the contradictions between wave and particle. According to the principle, it is impossible to accurately determine both the momentum and the position of the particle at the same time.
2. The correspondence principle states that in the limit of large quantum numbers, the predictions of quantum physics coincide with the predictions of classical physics (Kenneth et al., 2018). It is a principle proposed by Niels Bohr.
3. $h = 6.63 \times 10^{-34}$ J × sec
4. Electromagnetic Theory: Stated that light is a propagating wave of electric and magnetic fields.
5. Schrödinger's cat is a thought experiment in quantum mechanics that is sometimes described as a "paradox". The experiment was developed in 1935 by the famous Austrian physicist Erwin Schrödinger. The cat is placed in a sealed box with a bottle of poison and a radioactive source. If a sensor inside the box detects radioactivity, a mechanism breaks the vial, and the poison kills the cat. However, the observer does not know whether the cat in the box is dead or alive. A cat in a box is either dead or alive until someone looks at it. That is, according to this principle, in a certain time interval, the cat characterizes both of the two situations.

6 The area of mathematics concerned with secure information encoding is often based on mathematics such as the prime factorization of very large numbers (Cryptography, 2023). [The most obvious and common application of cryptography is to secure communications, thus providing security and privacy.

Chapter 3
Role of Quantum Computing in the Era of Artificial Intelligence (AI)

Ushaa Eswaran
https://orcid.org/0000-0002-5116-3403
Indira Institute of Technology and Sciences, Markapur, India

Alex Khang
https://orcid.org/0000-0001-8379-4659
AI and Data Science, Global Research Institute of Technology and Engineering, USA

Vishal Eswaran
CVS Health, USA

ABSTRACT

Quantum computing introduces revolutionary techniques for information processing by harnessing uniquely quantum phenomena. The emergence of quantum processors, algorithms, and software tools has brought this disruptive technology closer to practical applications. One of the most promising domains where quantum techniques can provide advantages is augmenting artificial intelligence. This chapter provides a comprehensive analysis of the fundamentals underpinning quantum computation, surveys the current hardware landscape, summarizes key quantum algorithms, and highlights promising applications in optimization, sampling, simulation, and machine learning. Engineering challenges in scaling up quantum systems are discussed along with progress in quantum error correction. Projections forecast steady advances towards quantum-enhanced AI based on a timeline of milestones. Realizing these synergies through collaborative interdisciplinary efforts promises to transform computing and catalyse innovations across industries to benefit science and society.

DOI: 10.4018/979-8-3693-1168-4.ch003

1. INTRODUCTION

Artificial intelligence has achieved remarkable success in recent years, enabled by increasing computational power, large datasets, and algorithmic advances (Jordan and Mitchell, 2015). However, fundamental limitations remain in mainstream AI techniques. Deep neural networks rely heavily on vast training data, and suffer performance degradation when applied to distributions differing from the training set, as seen in failures like Microsoft's Tay chatbot which was manipulated into making offensive statements on Twitter (Kiela et al., 2021). Lack of model interpretability creates trust issues for high-stakes applications like healthcare (Tonekaboni et al., 2019), where opaque predictions can potentially cause harm. Combinatorial optimization tasks like protein folding become intractable as problem sizes grow (Perdomo-Ortiz et al., 2015), limiting the ability to model complex biomolecular structures.

Meanwhile, quantum computing has emerged as a revolutionary computational paradigm introducing new physics-inspired approaches that classical systems fundamentally cannot replicate (Dowling and Milburn, 2003). By leveraging uniquely quantum properties like superposition, entanglement, and interference, quantum processors operate in ways drastically different from classical machines. For instance, superposition enables intrinsic parallelism where one operation evaluates multiple states simultaneously. Entanglement can represent correlated datasets unavailable to classical systems. Cleverly designed quantum algorithms like Grover's search (Grover, 1996) and Shor's factoring (Shor, 1997) demonstrate quadratic or exponential speedups over the best known classical techniques for certain problems (Khang & Kali, 2024).

Recent rapid advances in quantum processor engineering, algorithms, and software tools have brought this disruptive technology closer to delivering practical applications (Mohseni et al., 2017). Accessible via cloud services, noisy intermediate-scale quantum (NISQ) processors with 50-100 qubits are already being applied to specialized tasks like quantum chemistry simulations (Cao et al., 2019a), finance (Orús et al., 2019), and optimization (Neukart et al., 2017). The prospect of scaled-up fault-tolerant quantum computers emerging in the next 10-15 years has sparked tremendous excitement about the monumental impact across industries (Prasad et al., 2021).

One of the most promising application domains is augmenting artificial intelligence. The combination of quantum computing and AI offers new avenues for processing massive data sets, achieving speedups on constrained problems, modeling quantum systems, and enhancing deep learning techniques. Realizing this potential requires collaborative expertise spanning physics, computer science, and mathematics, engineering, and domain sciences. While reaching practical quantum-enhanced AI still faces multiple challenges, steady progress on the hardware and algorithm fronts aims to unlock the joint benefits (Shyam & Khang et al., 2023). This chapter provides a comprehensive analysis of the emerging interplay between quantum computing and artificial intelligence. The key topics covered are:

1. Fundamentals of quantum computing formalism and qubit implementations
2. Quantum algorithms and software frameworks
3. Quantum improvements for optimization, sampling, and simulation
4. Quantum neural networks, quantum deep learning, and quantum machine learning
5. Engineering challenges in scaling up quantum systems
6. Promising application areas and real-world demonstrations
7. Projections and timelines forecasting the future of quantum AI

Through accessible explanations of core concepts supplemented by examples, this chapter aims to spur further research and collaboration to unlock the synergies between quantum techniques and AI. Harnessing these joint advantages promises to transform computing and catalyze breakthrough innovations across sectors to benefit science and society. The future outlook is tremendously exciting for the convergence of quantum computing and artificial intelligence – a revolution just beginning! (Khang & Quantum, 2023).

2. FUNDAMENTALS OF QUANTUM COMPUTING

Quantum computing utilizes unique phenomena from quantum physics to perform information processing and computations in ways impossible on classical machines. This section summarizes the mathematical formalism and physical implementations underlying quantum computation.

2.1 Qubits and Quantum States

The basic unit of information in a quantum computer is the quantum bit or qubit. Unlike classical bits representing 0 or 1, a qubit can exist in a superposition of $|0\rangle$ and $|1\rangle$ states as shown in Equation 1:

$$|\psi\rangle = \alpha|0\rangle + \beta|1\rangle \tag{1}$$

Here α and β are probability amplitudes satisfying normalization $|\alpha|2 + |\beta|2 = 1$. This enables a qubit to represent a coherent combination of 0 and 1 simultaneously. Measurement collapses the superposition probabilistically to a single state.

Qubit state manipulations are represented by unitary matrices. For an n-qubit system with 2n dimensions, the state $|\psi\rangle$ is a vector in this Hilbert space. Quantum circuits perform algorithms by applying unitary gates on the input state.

2.2 Quantum Parallelism and Entanglement

A key advantage of qubits is the ability to evaluate operations in parallel on superposition of exponential states. Applying a single gate to n qubits creates a superposition of 2n states on which computation occurs simultaneously.

When multiple qubits become correlated such that the state of one qubit depends on the others, this non-local correlation is called entanglement. It enables representations of data inaccessible to classical systems.

2.3 Diverse Qubit Implementations

There are various physical implementations of qubits being engineered:

- **Superconducting circuits:** Microfabricated Josephson junctions like transmon qubits exhibit quantum behavior at cryogenic temperatures (Kjaergaard et al., 2020). IBM and Rigetti use these.

- **Trapped ions:** Qubits encoded in electronic states of trapped ions manipulated by lasers. IonQ utilizes this approach (Wright et al., 2019).
- **Photonics:** Leverage properties of single photons like polarization. Allow low-noise operations over long distances.
- **Quantum dots:** Isolated electrons trapped in semiconductor nanostructures form spin qubits.
- **Topological qubits:** Store information non-locally making them robust to errors. Still need scaling up.

Each approach has relative trade-offs in quality, control, fabrication, and scaling. Hybrid systems combining multiple qubit types are also explored.

2.4 Quantum Logic Gates and Circuits

Analogous to classical logic gates, quantum gates perform reversible operations on qubits to manipulate their states. Single qubit gates like the Hadamard, Phase and Pauli gates can create superposition, interfere states, and flip qubit states. Multi-qubit gates like CNOT enact conditional logic by entanglement. Composing primitive gates enables realizing complex quantum circuits.

- The Hadamard gate is a logic gate in quantum computing that turns a state of $|0\rangle$ or $|1\rangle$ into a superposition of $|0\rangle$ and $|1\rangle$. (Learn about Hadamard gates in the Xanadu Codebook.) In a circuit diagram, a Hadamard gate is represented by a square with the letter 'H' in it (HadamardGate, 2023).
- The Pauli gates are the three Pauli matrices and act on a single qubit. The Pauli X, Y and Z equate, respectively, to a rotation around the x, y and z axes of the Bloch sphere by radians (PauliGates, 2023).
- The CNOT gate is a mulit-qubit gate that consists of two qubits. The first qubit is known as the control qubit and the second is known as the target qubit. If the control qubit is |1>then it will flip the targets qubit state from |0>to |1>or vice versa (CNOTGate, 2023).
- The T gate is a single-qubit operation given by: T = (1 0 0 exp > (i π 4)) \ T = \begin{pmatrix} 1 & 0 \ 0 & \exp \left (\frac{i \pi}{4} \right) \end{pmatrix} T=(100exp(4iπ)) The T gate is related to the S gate by the relationship S = T 2 \ S=T^{2} S=T2 (TGate, 2023)

2.5 Comparison to Classical Computation

While a classical register of n bits represents one of 2n states, an n qubit quantum state compactly represents all 2n states in superposition. This exponential state space provides quantum parallelism.

Certain problems like simulating quantum systems, factoring large integers, and combinatorial optimization become intractable for classical computers as they grow in size. Quantum algorithms can provide exponential or quadratic speedups by leveraging superposition, entanglement, and interference.

But engineering challenges like noise, decoherence, and algorithm design remain. Robust error correction techniques are required to realize fault tolerant quantum computation.

2.6 Quantum Computing Hardware Landscape

Quantum computing has evolved from a theoretical concept to an emerging computational model spanning diverse qubit technologies, programming frameworks, and cloud access. This section surveys the current landscape and latest achievements.

2.7 Timeline of Key Milestones

- **1980s:** Quantum computing proposed theoretically with seminal algorithms (Nielsen and Chuang, 2002).
- **1990s:** Important algorithms like Shor's and Grover's discovered (Grover, 1996; Shor 1997). Quantum error correction developed (Shor, 1995).
- **2000s:** Small quantum processors with a few qubits built.
- **2010s:** IBM, Google, Rigetti enter with superconducting qubits (Kjaergaard et al., 2020). Quantum cloud services launched.
- **2019:** Google announces quantum supremacy demonstration (Arute et al., 2019).
- **2020s**: Rapid growth with superconducting processors exceeding 100 qubits (IBM Quantum systems). Trapped ion systems reach high fidelity (Wright et al., 2019).

2.8 Diverse Models of Quantum Computation

There are different approaches to designing quantum computers:

- **Gate Model:** Circuits composed of quantum logic gates. Followed by IBM, Rigetti, Google.
- **Adiabatic Model:** Evolve physical system slowly into ground state encoding solution. D-Wave annealers use this (Kadowaki and Nishimori, 1998).
- **Measurement-based:** Adaptive measurements on entangled cluster states drive computation.
- **Topological Model:** Use braiding of non-Abelian anyons allowing inherent fault tolerance.

Each model has relative merits and tailored qubit implementations being engineered to leverage them. Hybrid systems combining multiple approaches also hold promise.

2.9. Leading Qubit Modalities

The table below compares the leading qubit platforms and metrics:

Table 1. Leading qubit platforms and metrics

Qubit Type	Coherence Time	Qubit Count	Gate Fidelity	Connectivity
Superconducting	50-100 μs	>100 qubits	99.90%	Limited
Trapped Ion	>10 sec	~10 qubits	100.00%	All-to-all
Photonics	Excellent	Prototype arrays	Challenging	Very high
Quantum Dot	~200 μs	~5 qubits	99%	Neighboring

2.9.1 Recent Hardware Achievements

- **IBM Osprey Processor:** In 2022, IBM unveiled the Osprey processor, featuring 433 qubits using superconducting transmon qubits. This represents a significant milestone in quantum hardware development, demonstrating progress in scaling up the number of qubits in a superconducting quantum processor.
- **Google Sycamore Processor:** Google achieved a major milestone in 2019 with its 72-qubit Sycamore processor, which demonstrated a two-qubit gate fidelity of 99.9%. This achievement highlighted the potential of superconducting qubits for performing complex quantum computations with high accuracy.
- **IonQ Trapped Ion Qubits:** IonQ demonstrated high-fidelity two-qubit gates in systems consisting of 11-32 trapped ion qubits. Their experiments achieved a gate fidelity of 99.97%, showcasing the potential of trapped ion qubits for reliable and accurate quantum operations.
- **PsiQuantum Photonic Chips:** PsiQuantum has made progress in building reconfigurable photonic chips with over 100 qubits. Photonics-based quantum computing offers advantages in terms of scalability and low error rates, and PsiQuantum's achievement is a significant step towards realizing large-scale quantum photonic processors.
- **Quantum Dot Spin Qubits:** Quantum dot spin qubits have shown promise in achieving high gate fidelities. In 2021, researchers demonstrated single-qubit gate fidelity of 98% in quantum systems based on quantum dot spin qubits. This achievement contributes to the development of solid-state qubits for quantum computing.

2.9.2 Case Studies

- **Quantum Chemistry Simulation:** The use of large-scale quantum processors, such as IBM's Osprey processor, enables advancements in quantum chemistry simulations. Researchers can leverage the increased number of qubits and improved gate fidelities to tackle complex molecular simulations, which can have applications in drug discovery, materials science, and catalyst design.
- **Optimization and Machine Learning:** Quantum processors with higher qubit counts and improved gate fidelities, such as Google's Sycamore processor, offer potential for solving optimization problems and enhancing machine learning algorithms. By leveraging quantum algorithms, researchers can explore applications such as portfolio optimization, logistics planning, and machine learning model training.
- **Quantum Error Correction:** Achieving high-fidelity gates is crucial for implementing error correction codes in quantum computing systems. The advancements in gate fidelity demonstrated by IonQ and other platforms contribute to the development of error correction techniques, which are essential for building fault-tolerant quantum computers capable of performing error-free computations.
- **Quantum Communication and Networking:** Photonic-based quantum processors, such as those developed by PsiQuantum, hold promise for enabling secure quantum communication and building large-scale quantum networks. The scalability of photonic qubits and their low error rates make them suitable for long-distance quantum communication and quantum key distribution applications.

These recent hardware achievements and case studies illustrate the progress made in quantum computing hardware, showcasing advancements in qubit counts, gate fidelities, and different qubit technologies. These developments pave the way for exploring a wide range of applications and driving the field of quantum computing forward.

2.10 Quantum Software Stack

The quantum software stack consists of various tools, libraries, and frameworks that enable the development, execution, and optimization of quantum programs. Here are some key components of the quantum software stack:

- **High-Level SDKs:** High-**level** software development kits (SDKs) like Qiskit and Cirq provide programming interfaces and tools for writing and executing quantum programs. These SDKs offer abstractions and APIs that simplify the process of designing quantum circuits, implementing quantum algorithms, and running simulations. They also provide access to quantum hardware for real-world experimentation.
    - **Qiskit:** Qiskit is an open-source SDK developed by IBM. It offers a comprehensive suite of tools for quantum circuit design, quantum algorithm development, and quantum simulation. Qiskit supports multiple backends, including simulators and real quantum hardware.
    - **Cirq:** Cirq is an open-source SDK developed by Google. It provides a framework for writing quantum circuits and running quantum algorithms on simulators and quantum processors. Cirq focuses on low-level operations, allowing developers to have fine-grained control over their quantum programs.
- **Open Source Simulators:** Open source simulators are crucial for debugging and testing quantum circuits and algorithms. These simulators allow researchers and developers to simulate the behavior of quantum systems, assess the performance of quantum algorithms, and validate their designs before running them on real quantum hardware.
    - **Qiskit Aer:** Qiskit Aer is a high-performance simulator included in the Qiskit SDK. It provides various simulation backends, including statevector simulators, unitary simulators, and noise simulators. Qiskit Aer allows users to run large-scale simulations and analyze the behavior of quantum circuits.
    - **ProjectQ:** ProjectQ is an open-source software framework for quantum computing that includes a simulator capable of simulating quantum circuits on classical hardware. It offers a user-friendly interface for circuit construction and simulation.
- **Cloud Access to Quantum Processors:** Several cloud platforms provide access to real quantum processors, allowing researchers and developers to run their quantum programs on actual hardware. These cloud-based services enable experimentation and collaboration, even if users do not have direct access to physical quantum devices.
    - **IBM Quantum Experience:** IBM Quantum Experience is a cloud-based platform that provides access to IBM's quantum processors and simulators. Users can write quantum programs in Qiskit and execute them on IBM's quantum hardware or simulators.
    - **Rigetti Quantum Cloud Services:** Rigetti Quantum Cloud Services offer cloud access to Rigetti's quantum processors. Developers can use Rigetti's SDKs to design and run quantum programs on their quantum hardware.

- **Algorithm Libraries:** Algorithm libraries provide pre-implemented quantum algorithms and techniques for various applications. These libraries offer high-level functions and tools that developers can utilize to build quantum applications in domains such as optimization, machine learning, finance, and more.
    - **PennyLane:** PennyLane is an open-source library for quantum machine learning and optimization developed by Xanadu. It integrates with popular deep learning frameworks like TensorFlow and PyTorch, allowing seamless integration of quantum circuits into classical machine learning pipelines.
    - **Qiskit Aqua:** Qiskit Aqua is a library within the Qiskit SDK that focuses on quantum algorithms for optimization, machine learning, and finance. It provides a collection of pre-implemented algorithms and tools to address specific problem domains.
- **Case Studies:**
    - **Quantum Chemistry Simulation:** The Quantum Chemistry Module in Qiskit Aqua provides pre-implemented algorithms for simulating molecular systems and solving quantum chemistry problems. Researchers can use these tools to study chemical reactions, optimize molecular structures, and explore the behavior of complex quantum systems in chemistry.
    - **Quantum Machine Learning:** Libraries like PennyLane and Qiskit Aqua enable the development and execution of quantum machine learning algorithms. These algorithms leverage the power of quantum computing to enhance classical machine learning tasks, such as classification, regression, clustering, and dimensionality reduction.
    - **Optimization:** Quantum algorithms and libraries, such as Qiskit Aqua's optimization module, provide tools for solving optimization problems. These algorithms can be applied to various domains, including portfolio optimization, supply chain management, and scheduling, to find optimal solutions faster using quantum optimization techniques.

The rapid progress in the quantum software stack, along with advancements in quantum hardware, expands the possibilities for developing and running quantum applications. Researchers and developers can leverage these tools to explore new algorithms, optimize existing processes, and push the boundaries of quantum computing.

3. QUANTUM ALGORITHMS AND APPLICATIONS

3.1 Grover's Algorithm for Database Search

Grover's algorithm is a quantum search algorithm that offers a quadratic speedup over classical linear search algorithms. It can efficiently search through an unstructured database to find a specific item of interest. Example: Imagine a scenario where you have a large database of phone numbers, and you want to find a specific phone number. Grover's algorithm can search through the database with fewer queries compared to classical methods, significantly reducing the search time.

3.2 Application: Bio Sequence Matching for Drug Discovery

In drug discovery, bio sequence matching plays a vital role in identifying potential drug candidates. Grover's algorithm can be used to speed up the process of matching biosequences, such as DNA or protein sequences, against large databases. This can accelerate the identification of molecules with desired therapeutic properties, aiding in the development of new drugs (Iriyama et al., 2020).

3.3 Shor's Algorithm for Factoring

Shor's algorithm is a quantum algorithm that can efficiently factorize large composite numbers. It demonstrates an exponential speedup compared to classical factoring algorithms, such as the general number field sieve. Example: Factoring large numbers into their prime factors is a computationally intensive task for classical computers. Shor's algorithm can factorize large numbers efficiently, making it a potential threat to the security of cryptographic systems that rely on the difficulty of factoring large numbers.

3.4 Application: RSA Cryptography

The RSA (Rivest-Shamir-Adleman) cryptographic algorithm is widely used for secure communication and data encryption. It relies on the difficulty of factoring large numbers into their primes. Shor's algorithm, when scaled up with a large enough quantum computer, could break RSA cryptography. This has significant implications for the security of digital communications, and it highlights the need for post-quantum cryptographic algorithms to ensure secure communication in the future.

3.5 Case Study: Quantum Machine Learning

Quantum machine learning is an emerging field that combines quantum computing with classical machine learning techniques. Quantum algorithms, such as the quantum support vector machine (QSVM) and quantum neural networks, offer potential advantages for solving machine learning problems.

Quantum algorithms can be applied to tasks such as image classification, pattern recognition, and optimization problems. For instance, quantum machine learning algorithms can improve image recognition capabilities by harnessing the power of quantum superposition and entanglement. This can have applications in fields such as medical imaging, autonomous vehicles, and computer vision (Khang, Vugar and Abuzarova et al., 2023).

These examples and case studies highlight the transformative potential of quantum algorithms in various domains. Quantum computing has the ability to revolutionize industries and solve complex problems more efficiently than classical approaches in certain scenarios. While quantum computers are still in the early stages of development, ongoing research and advancements are paving the way for practical applications of quantum algorithms in the near future.

3.6 Experimental Demonstration

Shor's algorithm has been experimentally demonstrated on a small scale, showcasing its practical feasibility. Researchers at Lucero et al. (2012) implemented a rudimentary version of the algorithm using a quantum processor.

These examples and case studies highlight the transformative impact of quantum algorithms. Grover's algorithm accelerates search tasks, while Shor's algorithm has the potential to revolutionize cryptography. These applications exemplify the prowess of quantum computing in addressing real-world challenges, offering solutions that were previously beyond the reach of classical computation. As quantum technologies continue to advance, these algorithms serve as inspiration for harnessing the unique capabilities of quantum systems to unlock new frontiers of innovation.

3.7 Quantum Machine Learning Algorithms

The convergence of quantum computing and machine learning has led to transformative advancements in data analysis and pattern recognition. Quantum machine learning techniques leverage quantum properties to enhance classical algorithms, enabling unparalleled computational efficiency (Biamonte et al., 2017).

3.8 Quantum Neural Networks

Quantum neural networks utilize qubits and quantum circuits to represent and process data. These networks can perform complex operations such as quantum data compression and quantum feature extraction. By exploiting quantum properties like entanglement and superposition, quantum neural networks have the potential to outperform classical counterparts.

3.9 Quantum Support Vector Machines

Quantum support vector machines (QSVMs) offer significant improvements in classification tasks. These algorithms leverage the quadratic speedup provided by quantum computing to enhance the efficiency of classification operations. QSVMs excel in scenarios where large datasets require rapid classification.

3.10 Quantum Principal Component Analysis

Quantum principal component analysis (PCA) introduces exponential speedup to the process of dimensionality reduction. By extracting key features from high-dimensional datasets more efficiently, quantum PCA enhances the accuracy and efficiency of data analysis tasks.

3.11 Quantum Reinforcement Learning

Quantum reinforcement learning exploits quantum computing to accelerate subroutines such as policy evaluation. Quantum-enhanced reinforcement learning has the potential to expedite decision-making processes in dynamic environments, making it particularly valuable for applications in robotics and optimization.

3.12 Gate-Model and Quantum Annealing

The field of quantum machine learning encompasses two promising paradigms: gate-model quantum computing and quantum annealing. Gate-model quantum computing involves manipulating qubits through quantum gates, while quantum annealing leverages quantum fluctuations to identify optimal solutions in optimization problems. Both approaches offer unique advantages for enhancing machine learning tasks.

As quantum machine learning algorithms continue to evolve, they hold the promise of revolutionizing various fields, including finance, healthcare, and artificial intelligence. By combining quantum properties with machine learning techniques, researchers are unlocking unprecedented computational power, marking a significant leap toward addressing complex challenges that were previously out of reach for classical computers.

3.13 Quantum Simulation Algorithms

Quantum computers excel at simulating quantum systems, making them valuable for applications in chemistry, materials science, and fundamental physics (Cao et al., 2019b). Several quantum simulation algorithms have been developed:

3.13.1 Trotterization Techniques

Trotterization techniques decompose the evolution of complex quantum systems into sequential gate operations. This enables the simulation of quantum systems by approximating their behavior through a series of simpler quantum operations.

3.13.2 Phase Estimation

Phase estimation is a quantum algorithm that allows the extraction of properties from molecular Hamiltonians. It can be used to estimate energy levels, molecular spectra, and other characteristics of quantum systems, providing valuable insights for chemical and materials research.

3.13.3 Variational Quantum Eigensolvers (VQE)

Variational quantum eigensolvers combine classical and quantum algorithms to find the ground state energy of a given quantum system. They employ a hybrid approach, utilizing a classical optimizer to iteratively refine quantum circuits and approximate the lowest energy state of the system.

Small proofs-of-concept have been demonstrated in quantum simulations of simple molecules like hydrogen and lithium hydride (Cao et al., 2019b). As quantum computers continue to improve in scalability and coherence, quantum simulation algorithms hold the potential to revolutionize fields that rely on understanding and manipulating quantum systems, enabling advancements in materials design, drug discovery, and fundamental physics research.

4. ENHANCING MACHINE LEARNING WITH QUANTUM

While the realization of large fault-tolerant quantum computers remains a challenge, the potential for advancement is already manifest in today's noisy intermediate-scale quantum (NISQ) processors, accessible via cloud platforms. These NISQ systems can enhance specific machine learning techniques, including optimization, sampling, regularization, and quantum neural networks, through the integration of quantum and classical algorithms.

4.1 Quantum Optimization

Combinatorial optimization problems, ubiquitous in domains like vehicle routing, scheduling, protein folding, and quantitative finance, often become intractable as they scale up. Quantum optimization algorithms offer a novel approach to tackling these challenges by exploring extensive search spaces more efficiently:

- **Quantum Annealing:** Quantum annealing leverages quantum fluctuations to navigate complex landscapes (Kadowaki and Nishimori, 1998). Notably, D-Wave systems implement this approach, showcasing its practicality.
- **Quantum Approximate Optimization Algorithm (QAOA):** QAOA combines the power of superposition with classical optimization techniques (Farhi et al., 2014), offering promising results in solving complex optimization problems.
- **Grover Adaptive Search:** Grover adaptive search introduces multi-objective optimization through iterative Grover rotations (Dalzell et al., 2020). This innovative technique holds potential for tackling problems with multiple objectives.
- **Variational Quantum Optimization:** Variational quantum optimization hybridizes quantum evaluation with classical optimization, creating a flexible approach for solving intricate optimization challenges (Peruzzo et al., 2014).

4.2 Quantum Sampling

Quantum computers offer advantages in generating samples from complex probability distributions, a task critical for Monte Carlo simulations, machine learning, and optimization:

- **Boson Sampling:** Utilizing photonic interference, boson sampling enables the efficient sampling of output photon distributions (Zhong et al., 2020). This approach holds promise for tackling challenging sampling problems.
- **Adiabatic Quantum Computing:** Adiabatic quantum computing samples low-energy states of physical models (Kadowaki and Nishimori, 1998), contributing to efficient sampling in various applications.
- **Universal Quantum Circuits:** Universal quantum circuits have demonstrated the capability to sample useful probability distributions efficiently (Arute et al., 2019), showcasing their potential for diverse applications.

4.3 Quantum Regularization

Leveraging high-dimensional quantum states for data encoding enhances separability, providing advantages for certain machine learning problems. Quantum regularization techniques offer advantages over classical implementations:

- **Compact Representation:** Using n qubits effectively encodes 2n dimensional feature spaces, enabling efficient data representation.
- **Parameterized Gates:** Parameterized gates facilitate the exploration of superpositions of features, offering a flexible approach to data manipulation.
- **Experimental Benefits:** Experimental studies have demonstrated the benefits of quantum regularization for classification tasks with limited training data (Havlíček et al., 2019), highlighting its potential in practical scenarios.

These advancements underscore the evolving synergy between quantum and classical computing paradigms, presenting a transformative trajectory for the field of machine learning. As NISQ processors continue to mature, hybrid quantum-classical algorithms are set to reshape the landscape of machine learning, offering unprecedented computational power and innovative solutions to complex problems.

5. QUANTUM NEURAL NETWORKS

Quantum neural networks (QNNs) are models that replace classical neurons and activations with qubits and unitary gates. The following characteristics are associated with quantum neural networks (Biamonte et al., 2017):

- **Trainable Parameters:** In QNNs, trainable parameters are encoded in qubit rotation gates. These gates allow for the manipulation and adjustment of the qubit states during training.
- **Feedback and Recurrence:** QNNs can incorporate feedback, enabling the utilization of recurrent quantum neural networks. This feedback mechanism allows information to be propagated through the network over multiple iterations, enabling more complex computations.
- **Hybrid Quantum-Classical Networks:** Hybrid quantum-classical networks combine both quantum and classical processing. Quantum components can handle certain parts of the computation, while classical components are responsible for tasks such as data preprocessing and optimization.
- **Noise Mitigation Techniques:** Due to the presence of noise in quantum systems, noise mitigation techniques have been developed to reduce the impact of noise on QNNs. These techniques aim to improve the overall performance and reliability of QNNs.

5.1 Applications and Advantages

Quantum neural networks show promise in various areas, including:

- **Pattern Recognition:** QNNs have the potential to excel in pattern recognition tasks by leveraging the computational power of quantum systems and exploiting quantum phenomena such as superposition and entanglement.
- **Generative Modeling:** QNNs can be utilized for generative modeling, enabling the generation of new samples that exhibit similar characteristics to the training data. This has applications in areas such as image synthesis and text generation.

- **Nonlinear Classification:** QNNs can perform nonlinear classification tasks more efficiently compared to classical neural networks. The utilization of quantum gates and qubits allows for complex computations that can capture intricate decision boundaries.

5.2 Achieving Quantum Advantage

Quantum advantage refers to the ability of quantum algorithms to outperform classical algorithms for specific tasks. Several routes and algorithms have been identified to demonstrate quantum advantage:

Grover's Algorithm: Grover's algorithm provides a speedup for searching unsorted databases, offering an $O(\sqrt{N})$ search complexity compared to the classical O(N) search complexity (Grover, 1996).

Shor's Algorithm: Shor's algorithm offers an exponential speedup for integer factoring, which is a computationally intensive task in classical computing (Shor, 1997).

Quantum Machine Learning: Quantum machine learning algorithms have the potential for polynomial or exponential speedups compared to classical counterparts (Biamonte et al., 2017). This can significantly enhance the efficiency of machine learning tasks.

Quantum Simulation: Quantum simulation algorithms can provide a polynomial speedup over classical techniques for simulating physical systems (Cao et al., 2019b). This has applications in fields such as quantum chemistry and materials science.

6. REAL-WORLD QUANTUM ADVANTAGE DEMONSTRATIONS

Several real-world demonstrations have showcased the potential of quantum advantage in specific domains:

Optimization: Google demonstrated that a quantum processor could sample Ising model states that are unachievable classically (Arute et al., 2019). Additionally, Volkswagen utilized a D-Wave annealer for traffic optimization, surpassing classical methods (Neukart et al., 2017).

Machine Learning: Quantum classifiers have been developed that outperform classical counterparts, especially when dealing with limited training data (Havlíček et al., 2019). Furthermore, a quantum reinforcement learning algorithm demonstrated a speedup in learning robot behaviors (Lockwood and Si, 2020).

Quantum Chemistry: Quantum processors have been utilized for simulations of molecular energetics, matching theoretical predictions (Hempel et al., 2018). Proof-of-concept simulations of water molecule hydration energy using four qubits have also been achieved (Cao et al., 2019b).

While these examples provide evidence of emerging quantum advantage in specific contexts, achieving fault-tolerant quantum computation that is broadly applicable remains a significant challenge.

7. ENGINEERING CHALLENGES IN SCALING UP QUANTUM SYSTEMS

While quantum computing holds great potential, there are significant engineering challenges that need to be addressed when scaling up quantum systems while maintaining qubit coherence. The following are key hardware limitations and recent advancements in quantum error correction:

7.1 Sources of Noise and Decoherence

- **Cosmic Rays, Vibrations, and Defects:** Superconducting processors are susceptible to various sources of noise and decoherence, including cosmic rays, vibrations, and defects. These factors can adversely affect the stability and reliability of qubits.
- **Quantum Error Correction**: To combat errors in quantum systems, quantum error correction techniques are employed. These techniques involve encoding quantum states in error correction codes that utilize redundant qubits to detect and correct errors. Some notable error correction codes include:
    - **Shor Code:** The Shor code uses nine qubits to correct one qubit errors, providing a level of error resilience.
    - **Surface Codes:** Surface codes are two-dimensional lattice configurations that exhibit high error tolerance. They can efficiently detect and correct errors.
    - **Topological Codes:** Topological codes, such as 3D color codes, aim for around a 3% error threshold. These codes leverage the topological properties of qubits to enhance error resilience.

Executing logical qubit operations fault-tolerantly within these error correction codes is a crucial aspect, although managing the resource overheads associated with error correction remains a significant challenge.

7.2 Recent Experimental Progress

Recent experimental advancements have demonstrated progress in implementing error correction and logical qubits:

- **Logical Qubits in Oscillator Grid States:** A logical qubit encoded in grid states of an oscillator has been demonstrated with a low error rate (Campagne-Ibarcq et al., 2020).
- **Successful Error Correction in Small Surface Codes:** Small surface codes have been employed to successfully correct induced errors (Egan et al., 2021).
- **Sustained Entanglement in Logical Bell Pair:** Logical Bell pairs encoded in trapped ion qubits have been shown to sustain entanglement (Wright et al., 2019).

Verification and validation play crucial roles in assessing the performance and capabilities of quantum systems. However, the inherent complexity and sensitivity of quantum systems pose challenges in these processes. Here are some key aspects of verification and validation in quantum computing:

- **Exponential Scaling in Full Quantum State Tomography:** Full quantum state tomography aims to reconstruct the complete quantum state of a system by characterizing the density matrix. As the number of qubits increases, the number of parameters required to describe the quantum state grows exponentially, making full tomography infeasible for large-scale systems. Therefore, alternative techniques that provide meaningful information while reducing the number of measurements are being explored.
- **Gate Fidelity Estimates with Randomized Benchmarking:** Randomized benchmarking is a technique used to estimate the average gate fidelity of a quantum system. It involves the applica-

tion of random sequences of gates and measuring the resulting error rates. While randomized benchmarking provides valuable insights into gate performance, it does not provide detailed information about the specific noise sources affecting the system. Additional methods are needed to gain a comprehensive understanding of noise and error characteristics.

- **Scalable Validation with Partial Tomographic Methods:** Partial tomographic methods, such as shadow tomography, have been developed to address the scalability challenges associated with full quantum state tomography. These methods aim to reconstruct a subset of the density matrix or extract specific information about the quantum system while reducing the measurement requirements. By focusing on relevant aspects of the system, partial tomography techniques enable more scalable validation procedures.
- **Statistical Comparative Benchmarking:** Statistical comparative benchmarking is an approach to compare the performance of different quantum systems, algorithms, or protocols. It involves running experiments on multiple systems and statistically analyzing their outcomes to identify differences in performance. This technique provides a comparative assessment, allowing researchers to evaluate the advantages and limitations of different approaches.

Addressing the challenges in verification and validation is crucial for advancing quantum computing. As quantum systems continue to scale up, developing efficient and scalable techniques for characterizing and validating their performance will be essential for ensuring their reliability and enabling the realization of quantum advantage.

8. RESPONSIBLE DEVELOPMENT OF QUANTUM COMPUTING

The advancement of quantum computing brings with it transformative potential, but it also raises ethical considerations and risks that need to be addressed.

Here are some key considerations and strategies for the responsible development of quantum technologies:

- **Threats to Cryptographic Security:** Quantum computers have the potential to break widely used cryptographic schemes, such as RSA, which rely on the difficulty of factoring large numbers. To ensure future-proof security, a transition to post-quantum cryptography is recommended. Some post-quantum cryptographic schemes include:
- **Lattice-based schemes:** These schemes rely on the hardness of lattice problems for security.
    - Code-based cryptography: Error-correcting codes are used to provide security in this approach.
    - Multivariate schemes: These schemes utilize systems of polynomial equations for encryption.
    - Fortunately, many post-quantum cryptographic techniques have efficient key sizes, which makes them suitable for practical implementation.
- **Financial System Stability:** Quantum optimization algorithms have the potential to find arbitrage opportunities in complex financial markets. To ensure the safe deployment of quantum technologies in finance, governance mechanisms are necessary to prevent market instability. Some strategies to consider include:

- **Research into systemic implications:** Understanding the potential impact of quantum optimization on financial systems is crucial for developing appropriate safeguards.
- Mechanisms like circuit breakers: Implementing circuit breakers or other mechanisms can help mitigate disruptions and prevent extreme market reactions.
- **International coordination:** Given the global nature of financial markets, international cooperation and coordination are essential for regulating transnational markets and addressing potential risks.
- **Ethics for AI Applications:** Quantum machine learning introduces new paradigms for data representation, model training, and inference. Ethical considerations must be taken into account when deploying quantum machine learning in sensitive domains like healthcare (Khang & Medicine, 2023). Some important points to consider include:
- **Lack of interpretability:** Quantum machine learning models may lack interpretability, making it challenging to understand the reasoning behind their predictions. Caution should be exercised when making high-stakes decisions based on such models.
- **Contextual understanding:** Blindly applying quantum machine learning models without considering the specific context and potential biases can lead to harmful outcomes. Careful consideration of the ethical implications is necessary.
- **Extending best practices:** Many principles and best practices from classical AI ethics can be extended to quantum machine learning. Transparency, fairness, and accountability should be core considerations in the development and deployment of quantum AI applications.

9. CASE STUDIES

9.1 Post-Quantum Cryptography Standardization

The National Institute of Standards and Technology (NIST) launched a competition for standardizing post-quantum cryptographic algorithms. The goal is to identify and develop robust cryptographic algorithms that can resist attacks from both classical and quantum computers (Debosree, Khang et al., 2023).

9.2 Quantum-Safe Blockchain

Researchers and developers are exploring the integration of quantum-resistant cryptographic techniques into blockchain technology. This ensures that blockchain systems remain secure even in the presence of powerful quantum computers (Khang & Chowdhury et al., 2022).

9.3 Ethical Guidelines for Quantum AI

Organizations and researchers are working on developing ethical guidelines specific to quantum machine learning and AI applications. These guidelines address issues such as fairness, transparency, privacy, and accountability to ensure the responsible development and deployment of quantum AI technologies.

By proactively addressing ethical considerations, engaging in responsible research practices, and involving stakeholders in governance discussions, the development of quantum technologies can be guided towards maximizing societal benefits while minimizing potential risks.

10. OUTLOOK ON QUANTUM COMPUTING AND AI

The combination of quantum computing and artificial intelligence represents one of the most impactful technology opportunities on the horizon. The following timelines present a projection of key milestones forecasting the progress toward quantum-enhanced AI (Khang & Shah et al., 2023).

10.1 Near Term Outlook (0-5 years)

- Expanded cloud access to NISQ processors with ~100 qubits.
- Growth of hybrid algorithms, software stacks, and application libraries.
- Small demonstrations of quantum advantage for specialized tasks.
- Improved quantum programming frameworks and compilers.
- New qubit modalities and encoding techniques.
- Basic components of fault tolerance demonstrated.

10.2 Medium Term (5-10 years)

- Increased qubit numbers and gate fidelities surpassing fault tolerance thresholds.
- Realization of quantum neural networks and deep learning algorithms.
- Machine learning techniques adapted for quantum training.
- Real-world quantum advantage shown for some industry problems (Khang & Kali, 2023).
- Components combined into fault tolerant quantum processors.

10.3 Long Term (10+ years)

- Early availability of fully fault tolerant quantum computers.
- Quantum processors widely accessible via cloud services.
- Significant quantum speedups and capabilities demonstrated.
- Quantum techniques integrated into mainstream computing.
- Broad exploration of quantum AI applications across sectors.

The future is tremendously exciting for quantum computing and AI. Sustaining investment in research and engineering focused on responsibly realizing their joint potential promises to transform computing and catalyze breakthrough innovations across industries (Khang & Vugar et al., 2023).

11. CONCLUSION

In conclusion, quantum computing represents a revolutionary computational paradigm introducing new physics-inspired capabilities unattainable for classical systems. Harnessing properties like superposition, entanglement, and interference unlocks exponential state spaces, intrinsic parallelism, and quantum speedups for suitable problems. Engineering challenges remain in scaling up qubit numbers, gate operations, connectivity and error correction to realize reliable quantum processors. However, rapid progress is being made on both the hardware and software fronts to advance towards this goal (Khang & Muthmainnah et al., 2023).

The emergence of quantum techniques opens up new prospects for augmenting artificial intelligence. Quantum improvements for optimization, sampling, simulation and machine learning workflows hold immense promise for tackling intractable big data challenges, modeling quantum systems, and enhancing deep learning (Khanh & Khang et al., 2021). Real-world applications in domains ranging from quantum chemistry to optimization to finance are beginning to demonstrate pathways to quantum advantage. Further collaborative interdisciplinary efforts focused on responsibly realizing scalable quantum-enhanced AI promise to transform computing and catalyze innovations across sectors to benefit science and society (Khang & Kali et al., 2023).

REFERENCES

AlexanderR.KanadeG.LiW.YangL.GanzhornM.PetrasG.CandelS. (2022). Qiskit: An open-source framework for quantum computing. Zenodo. https://joss.theoj.org/papers/10.21105/joss.05329.pdf

Arute, F., Arya, K., Babbush, R., Bacon, D., Bardin, J. C., Barends, R., & Chen, Y. (2019). Quantum supremacy using a programmable superconducting processor. *Nature*, *574*(7779), 505–510. https://www.nature.com/articles/s41586%20019%201666%205. doi:10.103841586-019-1666-5 PMID:31645734

Bharti, K., Cervera-Lierta, A., Kyaw, T. H., Haug, T., Alperin-Lea, S., Anand, A., & Guerreschi, G. G. (2022). Noisy intermediate-scale quantum algorithms. *Reviews of Modern Physics*, *94*(1), 015004. https://journals.aps.org/rmp/abstract/10.1103/RevModPhys.94.015004. doi:10.1103/RevModPhys.94.015004

Biamonte, J., Wittek, P., Pancotti, N., Rebentrost, P., Wiebe, N., & Lloyd, S. (2017). Quantum machine learning. *Nature*, *549*(7671), 195–202. https://www.nature.com/articles/nature23474. doi:10.1038/nature23474 PMID:28905917

Campagne-Ibarcq, P., Eickbusch, A., Touzard, S., Zalys-Geller, E., Frattini, N. E., Sivak, V., & Schoelkopf, R. J. (2020). Quantum error correction of a qubit encoded in grid states of an oscillator. *Nature*, *584*(7821), 368–372. https://www.nature.com/articles/s41586-020-2603-3. doi:10.103841586-020-2603-3 PMID:32814889

Cao, Y., Romero, J., Olson, J. P., Degroote, M., Johnson, P. D., Kieferová, M., & Aspuru-Guzik, A. (2019). Quantum chemistry in the age of quantum computing. *Chemical Reviews*, *119*(19), 10856–10915. https://pubs.acs.org/doi/abs/10.1021/acs.chemrev.8b00803. doi:10.1021/acs.chemrev.8b00803 PMID:31469277

Chen, L., Jordan, S. P., Liu, Y. K., Moody, D., Peralta, R., Perlner, R., & Smith-Tone, D. (2016). *Report on post-quantum cryptography*. US Department of Commerce, National Institute of Standards and Technology. https://nvlpubs.nist.gov/nistpubs/ir/2016/nist.ir.8105.pdf

CNOTGates. (2023). *Quantum Computing UK-CNOT gate tutorial with Code*. CNOTGates. https://quantumcomputinguk.org/tutorials/controlled-not-gate-program

Dalzell, A. M., Yoder, T. J., & Chong, F. T. (2020). How quantum annealing can fail. *Quantum*, *4*, 317. doi:10.4018/979-8-3693-1168-4

DowlingJ. P.MilburnG. J. (2003). Quantum technology: the second quantum revolution. Philosophical Transactions of the Royal Society of London. Series A: Mathematical, Physical and Engineering Sciences, 361(1809), 1655-1674. https://arxiv.org/abs/quant-ph/0206091

Egan, L., Deb, B., Noel, C., Risinger, A., Zhu, D., Biswas, D., & Monroe, C. (2021). Fault-tolerant operation of a quantum error-correction code. *Nature*, *598*(7882), 281–286. https://arxiv.org/abs/2009.11482. doi:10.103841586-021-03928-y PMID:34608286

Erhard, A., Wallman, J. J., Postler, L., Meth, M., Stricker, R., Martinez, E. A., & Blatt, R. (2019). Characterizing large-scale quantum computers via cycle benchmarking. *Nature Communications*, *10*(1), 1–7. https://www.nature.com/articles/s41467-019-13068-7. doi:10.103841467-019-13068-7 PMID:31767840

Farhi, E., Goldstone, J., & Gutmann, S. (2014). A quantum approximate optimization algorithm. arXiv preprint arXiv:1411.4028.

Ghosh, D. & Khang A. (2023). Quantum Computing Internet of Things: Unlocking New Frontiers for Connectivity, Security, and Efficiency in Data Processing. Applications and Principles of Quantum Computing. IGI Global Press. https://doi.org/ doi:10.4018/979-8-3693-1168-4-ch016

Grover, L. K. (1996). A fast quantum mechanical algorithm for database search. *Proceedings of the twenty-eighth annual ACM symposium on Theory of computing*, (pp. 212-219). ACM. https://dl.acm.org/doi/pdf/10.1145/237814.237866

Guerreschi, G. G., & Matsuura, A. Y. (2022). QX: A high-performance quantum computer simulation platform. *Quantum Science and Technology*, *7*(4), 045003. https://ieeexplore.ieee.org/ abstract/document/7927034/. doi:10.1088/2058-9565/aacf0b

HadamardGate. (2023). *What is a Hadamard gate?* Hadmard Gate. https://pennylane.ai/qml/ glossary/what-is-a-hadamard-gate

Havlíček, V., Córcoles, A. D., Temme, K., Harrow, A. W., Kandala, A., Chow, J. M., & Gambetta, J. M. (2019). Supervised learning with quantum-enhanced feature spaces. *Nature*, *567*(7747), 209–212. https://www.nature.com/articles/s41586-019-0980-2. doi:10.103841586-019-0980-2 PMID:30867609

Hempel, C., Maier, C., Romero, J., McClean, J., Monz, T., Shen, H., & Aspuru-Guzik, A. (2018). Quantum chemistry calculations on a trapped-ion quantum simulator. *Physical Review X*, *8*(3), 031022. https://journals.aps.org/prx/abstract/10.1103/PhysRevX.8.031022. doi:10.1103/PhysRevX.8.031022

Hendrickx, N. W., Franke, D. P., Sammak, A., Kouwenhoven, M., Sabbagh, D., Yeoh, L., & Tagliaferri, M. L. V. (2021). A four-qubit germanium quantum processor. *Nature*, *591*(7849), 580–585. https://www.nature.com/articles/s41586-021-03332-6. doi:10.103841586-021-03332-6 PMID:33762771

Iriyama, S., Inoue, Y., Tamura, Y., Asaka, M., Nagano, S., & Mikami, H. (2020). A variational quantum algorithm for molecular sequence alignments. *iScience*, *23*(8), 101256. doi:10.4018/979-8-3693-1168-4

Jurcevic, P., Javadi-Abhari, A., Bishop, L. S., Lauer, I., Borgorin, D., Brink, M., & Blais, A. (2021). Demonstration of quantum volume 64 on a superconducting quantum computing system. *Quantum Science and Technology*, *7*(1), 015020. https://iopscience.iop.org/article/10.1088/2058-9565/abe519/meta. doi:10.1088/2058-9565/abe519

Kadowaki, T., & Nishimori, H. (1998). Quantum annealing in the transverse Ising model. *Physical Review E: Statistical Physics, Plasmas, Fluids, and Related Interdisciplinary Topics, 58*(5), 5355–5363. https://journals.aps.org/pre/abstract/10.1103/PhysRevE.58.5355. doi:10.1103/PhysRevE.58.5355

Khang, A. (2023). Applications and Principles of Quantum Computing. IGI Global Press. https://doi.org/ doi:10.4018/979-8-3693-1168-4

Khang, A. (2023). Quantum Mechanics Primer - Fundamentals and Quantum Computing. Applications and Principles of Quantum Computing. IGI Global Press. https://doi.org/ doi:10.4018/979-8-3693-1168-4-ch001

Khang, A. (2023). Introduction to Quantum Computing and Its Integration Applications. Applications and Principles of Quantum Computing. IGI Global Press. https://doi.org/ doi:10.4018/979-8-3693-1168-4-ch002

Khang, A. (2023). AI and IoT-Based Technologies for Precision Medicine. IGI Global Press. doi:10.4018/979-8-3693-0876-9

Khang, A. (2023). AI-Aided Data Analytics Tools and Applications for the Healthcare Sector. AI and IoT-Based Technologies for Precision Medicine. IGI Global Press. doi:10.4018/979-8-3693-0876-9.ch018

Khang, A. (2023). Enabling the Future of Manufacturing: Integration of Robotics and IoT to Smart Factory Infrastructure in Industry 4.0. AI-Based Technologies and Applications in the Era of the Metaverse. IGI Global Press. doi:10.4018/978-1-6684-8851-5.ch002

Khang A. (2024). *The Quantum Evolution: Application of AI and Robotics in the Future of Quantum Technology.* CRC Press. https://doi.org/ doi:10.1201/978-1-032-64207-9

Khang, A., Chowdhury, S., & Sharma, S. (2022). *The Data-Driven Blockchain Ecosystem: Fundamentals, Applications, and Emerging Technologies* (1st ed.). CRC Press. doi:10.1201/9781003269281

Khang, A., & Muthmainnah, M. (2023). AI-Aided Teaching Model for the Education 5.0 Ecosystem" AI-Based Technologies and Applications in the Era of the Metaverse. IGI Global Press. doi:10.4018/978-1-6684-8851-5.ch004

Khang, A., Shah, V., & Rani, S. (2023). AI-Based Technologies and Applications in the Era of the Metaverse. IGI Global Press. doi:10.4018/978-1-6684-8851-5

Khanh, H. H. & Khang, A. (2021). The Role of Artificial Intelligence in Blockchain Applications. *Reinventing Manufacturing and Business Processes through Artificial Intelligence*, 2(20-40). CRC Press. doi:10.1201/9781003145011-2

Kjaergaard, M., Schwartz, M. E., Braumüller, J., Krantz, P., Wang, J. I. J., Gustavsson, S., & Oliver, W. D. (2020). Superconducting qubits: Current state of play. *Annual Review of Condensed Matter Physics, 11*(1), 369–395. https://www.annualreviews.org/doi/abs/10.1146/annurev-conmatphys-031119-050605. doi:10.1146/annurev-conmatphys-031119-050605

Lockwood, O., & Si, M. (2020). *Quantum algorithm for reinforcement learning in infinite MDPs with policy advice*. arXiv preprint arXiv:2008.01082.

Lucero, E., Barends, R., Chen, Y., Kelly, J., Mariantoni, M., Megrant, A., & Nigg, D. (2012). Computing prime factors with a Josephson phase qubit quantum processor. *Nature Physics*, *8*(10), 719–723. https://www.nature.com/articles/nphys2385. doi:10.1038/nphys2385

Mitarai, K., Negoro, M., Kitagawa, M., & Fujii, K. (2018). Quantum circuit learning. *Physical Review. A*, *98*(3), 032309. https://journals.aps.org/pra/abstract/10.1103/PhysRevA.98.032309. doi:10.1103/PhysRevA.98.032309

Mohseni, M., Read, P., Neven, H., Boixo, S., Denchev, V., Babbush, R., & Smelyanskiy, V. (2017). Commercialize quantum technologies in five years. *NATNews*, *543*(7644), 171. https://www.nature.com/articles/543171a PMID:28277529

Neukart, F., Compostella, G., Seidel, C., Von Dollen, D., Yarkoni, S., & Parney, B. (2017, January). Traffic flow optimization using a quantum annealer. In *Frontiers in ICT* (Vol. 4, p. 29). Frontiers. https://www.frontiersin.org/articles/10.3389/fict.2017.00029/full

Nielsen, M. A., & Chuang, I. (2002). *Quantum computation and quantum information*. MIG. http://mjg.global.s3.amazonaws.com/static/projects/refs/Quantum_Computation_and_Quantum_Information.pdf

Orús, R., Mugel, S., & Lizaso, E. (2019). Quantum computing for finance: Overview and prospects. *Reviews in Physics*, *4*, 100028. https://www.sciencedirect.com/science/ article/pii/S2405428318300571. doi:10.1016/j.revip.2019.100028

PauliGates. (2023). *Knowledge base*. Quantum Inspire. https://www.quantum-inspire.com/kbase/pauli-x/

Pechen, A., Eliot, L., Dutton, Z., Carrasco, J. A., Hunt, T., Kalay, A. E., & Dunkel, J. (2022). *Ethical considerations for quantum machine learning research and development.* arXiv preprint arXiv:2209.05204.

Perdomo-Ortiz, A., Fluegemann, J., Narasimhan, S., Biswas, R., & Smelyanskiy, V. (2015). *A quantum-inspired classical algorithm for recommendation systems*. arXiv preprint arXiv:1503.01334.

Peruzzo, A., McClean, J., Shadbolt, P., Yung, M. H., Zhou, X. Q., Love, P. J., & O'brien, J. L. (2014). A variational eigenvalue solver on a photonic quantum processor. *Nature Communications*, *5*(1), 1–7. https://www.nature.com/articles/ncomms5213. doi:10.1038/ncomms5213 PMID:25055053

Prasad, N. K., Wille, R., Broekema, P. C., Rahman, N., Sung, J., Bishop, L. S., & Amin, M. H. (2021). *The future of quantum information processing*. arXiv preprint arXiv:2103.06269.

Preskill, J. (2018). Quantum Computing in the NISQ era and beyond. *Quantum, 2*, 79. https://quantum-journal.org/papers/q-2018-08-06-79/?fbclid=IwAR0FRb9N2fas7ETWu2M40OS6prXB5QvFME_WRELpm2CAUcccVIEzA_UmLn4

Shor, P. W. (1995). Scheme for reducing decoherence in quantum computer memory. *Physical Review*, *52*(4), R2493–R2496. https://journals.aps.org/pra/abstract/10.1103/PhysRevA.52.R2493. doi:10.1103/PhysRevA.52.R2493 PMID:9912632

Shor, P. W. (1997). Polynomial-time algorithms for prime factorization and discrete logarithms on a quantum computer. *SIAM Review*, *41*(2), 303–332. https://search.proquest.com/openview/9335e070823d763bdbdf578ed5271127/1?pq-origsite=gscholar&cbl=47353. doi:10.1137/S0036144598347011

Terhal, B. M. (2015). Quantum error correction for quantum memories. *Reviews of Modern Physics*, *87*(2), 307–346. https://journals.aps.org/rmp/abstract/10.1103/RevModPhys.87.307. doi:10.1103/RevModPhys.87.307

Tonekaboni, S., Joshi, S., McCradden, M. D., & Goldenberg, A. (2019). What clinicians want: contextualizing explainable machine learning for clinical end use. In *Machine Learning for Healthcare Conference* (pp. 359-380). PMLR. https://proceedings.mlr.press/v106/tonekaboni19a.html

Wright, K., Beck, K. M., Debnath, S., Amini, J. M., Nam, Y., Grzesiak, N., & Figgatt, C. (2019). Benchmarking an 11-qubit quantum computer. *Nature Communications*, *10*(1), 1–6. https://www.nature.com/articles/s41467-019-13534-2. doi:10.103841467-019-13534-2 PMID:31784527

Zhong, H. S., Wang, H., Deng, Y. H., Lu, M., Peng, L. C., Luo, Y. H., & Pan, J. W. (2020). Quantum computational advantage using photons. *Science*, *370*(6523), 1460–1463. doi:10.1126cience.abe8770 PMID:33273064

Zhu, D., Linke, N. M., Benedetti, M., Landsman, K. A., Nguyen, M., Alderete, C. H., & Monroe, C. (2019). Training of quantum circuits on a hybrid quantum computer. *Science Advances*, *5*(10), eaaw9918. https://www.science.org/doi/abs/10.1126/sciadv.aaw9918. doi:10.1126ciadv.aaw9918 PMID:31667342

Chapter 4
The Impact of Quantum Computing on Artificial Intelligence

K. Sakthi Srinivasan
Vellore Institute of Technology, India

Varun Kesavan
https://orcid.org/0000-0002-2373-2613
Vellore Institute of Technology, India

ABSTRACT

The widespread use of artificial intelligence has positioned it at the pinnacle of the technical hierarchy. Nevertheless, artificial intelligence (AI) outperforms traditional computers in certain aspects. The potential for quantum processing to greatly augment the efficacy of machine learning and artificial intelligence has emerged as a result of technological progress. Subsequent investigations will be undertaken to examine the impact of quantum processing on artificial intelligence (AI) and its implications for many sectors and economic systems. This research investigates the impact of quantum computing on artificial intelligence (AI), the distinctions between quantum and conventional computing, and the many applications of quantum computing in the field of AI.

1. INTRODUCTION

According to Sahota (2023), the integration of quantum computing with artificial intelligence (AI) has ushered in a new epoch of scientific advancement. The emergence of quantum computing, which utilizes concepts derived from quantum mechanics, is anticipated to provide a substantial augmentation in computational capabilities and problem-solving proficiency. According to projections, the worldwide quantum computing industry is anticipated to have a valuation of $4,375,000,000 by the year 2028, with a compound annual growth rate (CAGR) of 38.3%.

DOI: 10.4018/979-8-3693-1168-4.ch004

2. REVIEW OF LITERATURE

Quantum computing models include several approaches, one of which is quantum machine learning (QML). QML is an emerging subject that combines quantum information algorithms with machine learning techniques (Ayoade et al., 2022; Schuld & Petruccione, 2018). One such approach is the use of conventional machine learning methods on quantum computers. According to complexity theory, the use of this blending zone is expected to enable us to address difficulties with improved performance, while simultaneously reducing both the operating time and memory space requirements.

Quantum algorithms refer to a series of sequential operations performed on a quantum computer, utilizing principles from quantum physics, such as superposition and entanglement, to achieve computational tasks with enhanced speed, efficiency, and precision compared to classical computing systems (Ayoade et al., 2022; Montanaro, 2016). Although the study in this particular domain is still at an early stage, notable advancements have been achieved in the pursuit of quantum-compatible algorithms.

2.1 The Objectives of the Study

- This paper aims to elucidate the distinctions between quantum computing and traditional computing.
- Further research is required to examine the influence of quantum computing on the field of artificial intelligence.
- This paper aims to analyze the potential impact of quantum processing on the advancement of artificial intelligence.

2.2 The implementation of Quantum Computing

Quantum computing is an emerging and interdisciplinary area that combines computer technology, physics, and mathematics to efficiently process and manipulate data by using the laws of quantum mechanics. The topic under consideration covers both research on hardware and development of applications, and is pushing the boundaries of computational possibilities (Sahota, 2023). IBM Quantum is considered a prominent player in the field due to their successful commercialization of quantum hardware, hence enabling a large number of programmers to access and use this technology. The attainment of true quantum computers, which seemed as a far dream three decades ago, signifies a noteworthy milestone. Quantum computers exhibit a higher level of efficacy compared to conventional computers in addressing certain problem domains, owing to their utilization of quantum mechanical principles such as superposition and quantum interference.

2.3 Quantum Computers Are Differentiated From Conventional Computers by a Number of Key Characteristics

Qubits, which are used in quantum computers, serve as the distinguishing feature that sets conventional computing apart from quantum computing. According to Kanwar (2023), the ability of qubits to represent both zero and one enables them to do several calculations concurrently. Moreover, quantum computers have a distinct advantage over conventional computers in terms of error susceptibility when used for intricate applications like artificial intelligence. Hence, they exhibit greater suitability for integration

Figure 1. Quantum computing vs. classical computing
Alt-text: Figure 1 depicts features of Quantum Computing Vs Classical Computing.
Source: https://www.cbinsights.com/research/quantum-computing-classical-computing-comparison-infographic/

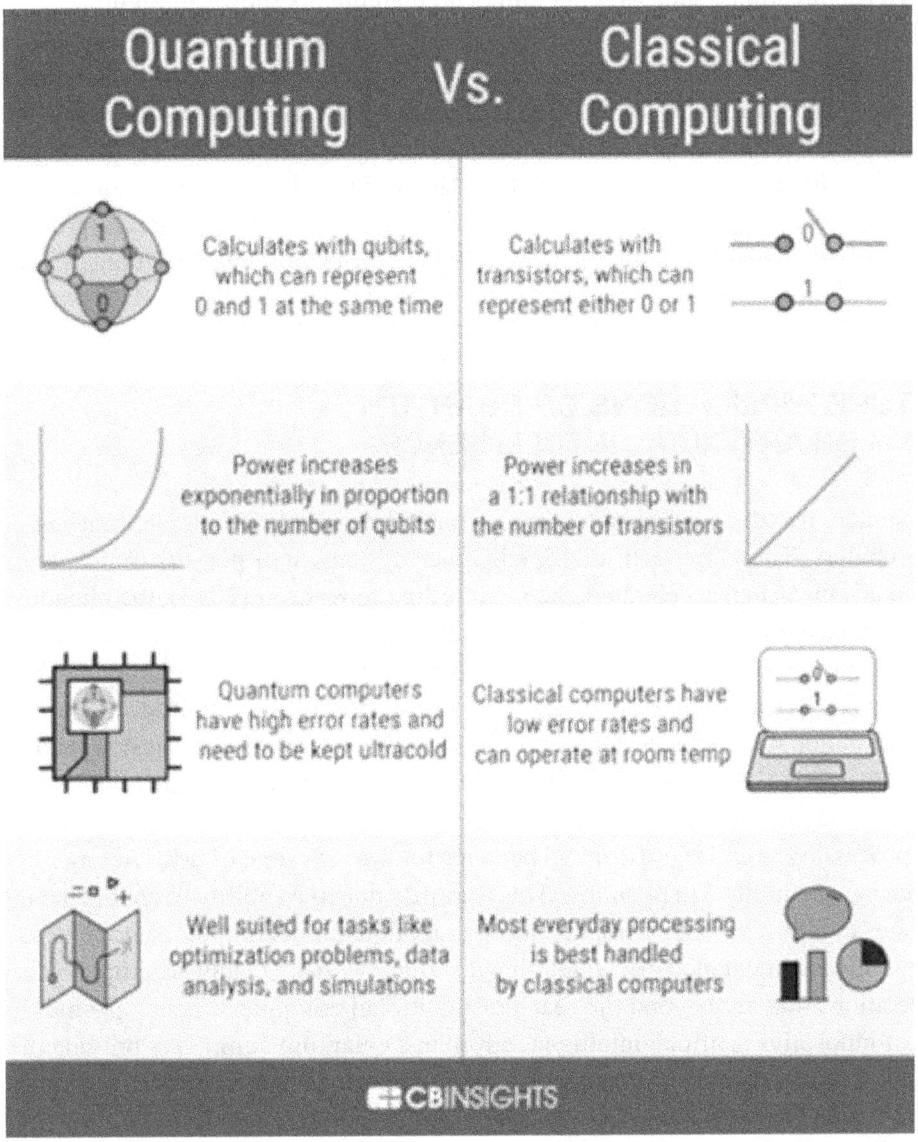

into artificial intelligence (AI) systems. The primary goal of the quantum computing community is to supplement and maybe augment the existing powers of conventional computing. It is expected that quantum computers would operate in conjunction with classical computers, aiding with tasks such as optimizing the efficiency of a preexisting system. Specialized jobs may be executed with more efficiency and effectiveness by these computers, offering programmers an alternative choice.

Quantum computing has superior performance compared to traditional computation in several aspects, such as heightened accuracy and efficiency. According to Sahota (2023), the increased data processing speeds of these systems facilitate accelerated learning and development of AI. Quantum entanglement enhances the strength of correlations among variables. Quantum computers have a strong affinity for intricate optimisation issues, hence enabling the development of artificial intelligence systems that surpass the constraints imposed by traditional physics. They play a significant role in several artificial intelligence (AI) applications, such as scheduling and planning as shown in Figure 1.

The acceleration of AI architecture development is facilitated via the execution of tasks that are beyond the capabilities of conventional computers, such as complex calculations and algorithmic creation. According to Sahota (2023), these issues are to matters that are not amenable to resolution by traditional methods. By using quantum computers, we can enhance the speed of learning and improve the precision of simulations for real-world events, while mitigating the risk of significant data loss. Quantum computers has the capability to address the issue of cybercrime, with its parallel processing and data security functionalities. Quantum computers contribute to the development of more efficient algorithms by virtue of their ability to live in several states concurrently.

3. THE FUTURE IMPLICATIONS OF QUANTUM COMPUTING ON ARTIFICIAL INTELLIGENCE

Quantum computers has the capability to process data at a notably accelerated rate in comparison to traditional computers. Therefore, artificial intelligence systems will possess the capability to acquire knowledge and advance at an accelerated pace. According to Kanwar (2023), the phenomenon of quantum entanglement has the potential to facilitate the use of correlations between variables by algorithms.

Furthermore, the use of quantum computers enhances the performance of algorithms employed in artificial intelligence by enabling the resolution of complex optimisation issues that are beyond the capabilities of conventional computers. Due to its departure from traditional physics principles, quantum computing has the potential to provide artificial intelligence (AI) systems that possess unprecedented levels of potency and intelligence.

Quantum processing offers significant advantages for a wide range of artificial intelligence (AI) applications, such as scheduling and planning. This is mostly due to its ability to enable the investigation of alternative issue solutions. Furthermore, the utilization of quantum computers facilitates the substantial and expeditious advancement of artificial intelligence frameworks. Quantum computers has the capability to do calculations that are beyond the reach of traditional computers, hence paving the path for the development of innovative artificial intelligence systems. Quantum computers provide the capability to replicate quantum processes with higher precision in comparison to conventional computers. Additionally, Shor's algorithm has shown its efficacy in factoring very large numbers.

Quantum annealing facilitates the resolution of computational problems that are beyond the capabilities of traditional computing methods. Quantum computers has the capability to verify the precision of AI algorithms' outputs, hence eliminating the potential for errors. AI systems have the potential to acquire knowledge at an accelerated pace and enhance their readiness for real-world situations as a result of the robust simulation environments facilitated by quantum computers. In contrast to conventional neural networks, quantum computers do not experience the phenomenon of catastrophic amnesia. This attribute enhances their aptitude for lifetime learning since it enables individuals to acquire new abilities without the need to revisit foundational knowledge.

Artificial intelligence systems have the capability to use quantum computers for the purpose of safeguarding confidential information. Furthermore, the mitigation of cybercrime may be achieved by the use of parallel processing techniques. Quantum computers, characterized by their ability to exist in superposition states, have shown superior performance in the realm of algorithm discovery when compared to conventional computers.

4. INTRODUCTION TO THE FUNDAMENTALS OF QUANTUM COMPUTING

The concept of superposition refers to the ability of a physical system to exist in multiple states. In the realm of quantum physics, the superposition principle allows for the amalgamation of states, resulting in the emergence of novel and valid states. This bears resemblance to the manner in which classical physics supports the summation of waves. According to Sahota (2023), it is possible to express any quantum state as a superposition of several other states. The inherent characteristic of superposition confers onto quantum computers an inherent parallelism. Through the use of parallelism, these computers have the capability to execute several operations concurrently, up to millions in total.

The concept of entanglement refers to a phenomenon in quantum mechanics where two or more particles. When two distinct systems achieve quantum entanglement, although being physically separated, they establish an inseparable connection. Despite the spatial disconnection, the acquisition of data pertaining to one entangled particle instantaneously discloses information about the other. Quantum algorithms, which are built around this unique correlation, serve as the fundamental basis for addressing intricate issues. Furthermore, the computational capacity experiences exponential growth with the incremental inclusion of qubits.

4.1 The Absence of Coherence

Decoherence refers to the occurrence of quantum systems, specifically qubits, losing their quantum behavior due to external factors such as vibrations or temperature fluctuations. These disruptions have the potential to induce the loss of superposition in qubits, leading to computational mistakes. Viable protective methods for qubits include insulation, supercooled freezers, and vacuum chambers.

The applications of Artificial Intelligence (AI) and Quantum Computers have garnered significant attention in several fields. AI, a branch of computer science, focuses on the development of intelligent machines that can do tasks that typically require human intelligence. Quantum Computers, on the other hand, use principles of quantum mechanics to process information and solve complex problems more efficiently.

4.2 Effectively and Efficiently Navigate Challenging Circumstances

The increasing intricacy and volume of datasets surpasses the capacities of contemporary computers, hence imposing a substantial burden on our computing infrastructure. According to Kanwar (2023), the existing computational powers of contemporary computers are insufficient to answer complex issues that quantum computers are anticipated to address within a matter of seconds. IBM challenges the assertion made by Google in 2019 on the achievement of Quantum Supremacy, which refers to the ability of a quantum computer to do calculations that were previously estimated to take thousands of years in a mere 200 seconds.

4.3 The Management of Large-Scale Datasets

The daily creation of data amounts to an estimated 2.5 exabytes. Quantum computers are specifically designed to efficiently detect abnormalities and patterns within extensive datasets, a task that would overwhelm traditional central processing units (CPUs) and graphics processing units (GPUs) due to the sheer volume of data involved.

4.4 The Identification and Prevention of Fraudulent Activities

The integration of quantum computing and artificial intelligence (AI) in the banking and financial sectors is anticipated to bring forth advancements in the realm of fraud detection. Moreover, quantum computer-based models has the capability to effectively address the substantial volumes of data that may be processed by these devices, while also being adept at identifying intricate patterns that are difficult to discern. Furthermore, the implementation of algorithmic improvements would contribute to the achievement of this objective.

5. THE TASK AT HAND INVOLVES THE DEVELOPMENT OF MORE ACCURATE AND REFINED MODELS

Modern companies have moved away from outdated computer systems and now rely on advanced technology to do comprehensive assessments of current data sets. These enterprises require very intricate models capable of evaluating all possible scenarios. By the year 2025, it is projected that the healthcare sector would produce a much larger volume of data compared to other businesses such as manufacturing, banking, logistics, and e-commerce. This increase is estimated to be around 36%, and is expected to occur at a compound annual growth rate of 6%.

The use of quantum computing has the potential to facilitate the development of improved models, which in turn might result in progress in the fields of illness detection and treatment, financial stability, and cooperation. Quantum computers have shown the capability to enhance DNA sequencing in the medical domain, while also demonstrating proficiency in precisely forecasting traffic levels within the transportation industry. Quantum computing has great potential for advancing our comprehension of biology and evolution, facilitating the creation of enhanced therapeutic interventions for ailments like cancer, and mitigating the impact of global warming.

6. THE IMPACT OF ARTIFICIAL INTELLIGENCE AND QUANTUM COMPUTING

By using the capabilities of quantum processing, artificial intelligence has the potential to achieve unparalleled levels of speed, efficiency, and accuracy. By using qubits and non-linear processes, this technology is capable of overcoming the constraints imposed by traditional computers (Sahota, 2023). As a result of this technological progress, quantum computing may now be used in a wide range of artificial intelligence applications. Industries such as shipping and logistics, semiconductors, electric cars, lights, and power are increasingly using the remarkable problem-solving capabilities offered by quantum computing.

6.1 The Provision of Medical Treatment

A medical board refers to a collective of doctors from several specialties that work together to diagnose and provide treatment for intricate medical issues in patients who are severely sick. The typical method is problematic since it hinders the efficiency of the decision-making process (Sahota, 2023). The resolution is inside the realm of quantum computing. Through the use of diverse situations and pertinent historical data, healthcare professionals have the capability to employ quantum computing in order to expeditiously assess the effectiveness of a therapeutic intervention. As a result of the quantum advantage, service providers are able to get crucial information in a much-reduced timeframe.

Artificial Intelligence (AI) refers to the field of computer science that focuses on the development of intelligent machines capable of doing tasks that typically require human intelligence.

The field of machine learning (ML) operates on the fundamental assumption that computers possess the ability to acquire knowledge from data and then generate or identify patterns in a manner akin to human cognition. According to Sahota (2023), the efficacy of machine learning may be diminished due to inadequate data quality and delayed data availability. The capacity of quantum computing to efficiently handle vast quantities of data has the potential to provide similar benefits to the field of machine learning.

6.2 The Topic of Security and Encryption is of Paramount Importance in the Field of Information Technology

Quantum computing has the potential to enhance data security by enabling the implementation of more robust security procedures via the use of qubits to compute a wide range of possible breach strategies, hence increasing the difficulty for unauthorized persons to compromise data. According to Sahota (2023), there are critics who contend that the use of quantum computing's speed and computational capabilities might potentially enable hackers to bypass existing security mechanisms, therefore obtaining unauthorized access to crucial data kept on servers.

Due to this rationale, the National Institute of Standards and Technology (NIST) is dedicating significant resources towards the development of encryption techniques that possess resistance against quantum computing. The objective of the National Institute of Standards and Technology's (NIST) Post-Quantum Cryptography Standardization Program is to facilitate the creation of an advanced cryptographic system that is resilient against quantum computing. The rivalry between the United States and China in the development of quantum technology is influenced by the geopolitical landscape, as the United States is motivated by China's pursuit of enhanced security measures.

The integration of quantum computing has the potential to enhance market forecasting and risk management within the financial sector. A proof-of-concept research was conducted by Sahota (2023), including Multiverse computing, Pasqal, and Crédit Agricole, to explore the potential use of quantum computing in the assessment of financial products and credit risk. The project spanned a duration of 1.5 years. Substantial advancements have facilitated the reduction of memory use and the enhancement of computational speed. The successful generation of precise outcomes by a 50-qubit quantum processor, which are equivalent to those obtained using conventional techniques, signifies a significant milestone towards the practical implementation of quantum computing. This achievement has particular promise for the field of derivatives pricing, as it opens up possibilities for real-world applications in this domain. The commercial availability of a 300-qubit quantum processor in 2024 is expected to mark the start of a transformative period characterized by rapid technical advancements. This pioneering research introduces the first quantum-enhanced machine learning framework specifically developed for the purpose of forecasting credit rating downgrades.

6.3 The Field of Study Often Referred to as "Natural Language Processing"

The use of quantum computing has the potential to boost the effectiveness and accuracy of communication in the domains of natural language processing (NLP) and voice recognition. In order to achieve this objective, Sahota (2023) proposes the use of quantum natural language processing (QNLP) to encode linguistic information for subsequent decoding by quantum computers. Quantum natural language processing (QNLP) utilizes current progress in quantum computing to enhance efficiency, expedite processing, diminish power usage, and minimize environmental impact. Quantum Natural Language Processing (QNLP) utilizes quantum computing techniques to enhance the efficiency of training Artificial Intelligence (AI) models that need substantial data volumes.

The University College London (UCL) and the British Broadcasting Corporation (BBC) have established a partnership with Quantinuum, a company specializing in the integration of software and hardware solutions for quantum computing. The objective of this collaboration is to conduct research on the commercial potential of content discovery and archive retrieval using Quantum Natural Language Processing (QNLP) techniques. By using string diagrams and the process of translating words into logical representations, Quantum Natural Language Processing (QNLP) simplifies the intricacy of Natural Language Processing (NLP) design via the use of quantum technologies. The storage of quantum circuits in this way has a high degree of compatibility with the application of artificial intelligence.

7. THE ADVANTAGES OF QUANTUM COMPUTING IN THE CONTEXT OF ARTIFICIAL INTELLIGENCE

The current infrastructure has challenges in keeping pace with the rapid expansion of intricate data sets, beyond the computational capabilities of traditional computers. According to Sahota (2023), quantum computing emerges as a prospective answer to these issues within a matter of seconds. The rapid computational capabilities of quantum computing play a pivotal role in addressing complex issues by efficiently handling vast quantities of data. Quantum simulations have significant potential in addressing intricate challenges, such as climate change and healthcare. Quantum computing offers many advantages for artificial intelligence (AI), including its exceptional computational speed and its capacity to effectively manage large datasets. An exploration of the subject matter will be provided within this context.

7.1 Improvement of the Solution

The accuracy and speed offered by quantum computing might potentially provide advantages in the domains of portfolio analysis, distribution channels, and customized treatment techniques. The alteration of computer architecture by big data necessitates the development of a novel approach for its efficient management (Sahota, 2023). Quantum computers have exceptional proficiency in resolving sequential issues, so conferring a competitive edge to both enterprises and consumers in the realm of decision-making. As a result, this unique technology is granted further financial support.

7.2 Uncovering Patterns in Vast Data Sets

When quantum computing is used, it exhibits the ability to efficiently analyze extensive and disorganized datasets, hence facilitating the identification of patterns or anomalies within them. According to Sahota (2023), the system has the capability to rapidly conduct searches inside a database, efficiently identifying corresponding entries within a short time frame. Current solutions are dependent on sequential analysis, a method that is not efficient and consumes a significant amount of time when dealing with large volumes of data.

7.3 Facilitating the Integration of Disparate Data Sets

The transformative potential of quantum computers stems from their ability to integrate vast datasets. At the outset, it may be necessary for human intervention to facilitate the integration process. In the field of artificial intelligence, scholars will encounter the formidable task of augmenting their capacity to analyze natural language (Sahota, 2023). The successful incorporation of the system's integration training is heavily contingent upon human involvement.

8. THE LIMITATIONS OF QUANTUM COMPUTING IN THE CONTEXT OF ARTIFICIAL INTELLIGENCE

Qubit decoherence, which refers to the loss or degradation of quantum information in a qubit, is a significant obstacle in the field of quantum computing. According to Sahota (2023), the qubits exhibit a high level of environmental sensitivity, making them susceptible to the loss of their quantum features even in the presence of minimal perturbations. In order to overcome this challenge, it will be necessary for researchers to explore non-traditional quantum approaches, use novel materials, and advance computational methodologies. It is important to acknowledge that the obstacles associated with quantum computing include a scope that surpasses the intricacies of hardware. The intricacy of quantum algorithms necessitates programmers to use innovative thinking when faced with computational obstacles. The advent of quantum computing has introduced novel complexities that need the engagement of scientists, programmers, and entrepreneurs. This analysis will examine eight unresolved difficulties.

8.1 Rectifying an Error

Undoubtedly, the most significant problem encountered by academics and practitioners is error correction. According to Sahota (2023), the quantum computing system has been seen to exhibit a high degree of vulnerability to both environmental disturbances and human errors.

These inaccuracies have the potential to accumulate and result in computations that are not correct. The refinement of established error correcting algorithms is of utmost importance in the development of functional quantum computers.

8.2 The Capacity to Expand or Adjust in Size or Scope

Quantum computers have shown notable capabilities in some sectors; yet, they are unable to rival regular computers in terms of their magnitude and range. According to Sahota (2023), a significant obstacle in the field of quantum computing lies in the task of scaling up quantum computers to support a larger number of qubits, ranging from hundreds to thousands. This endeavor necessitates the preservation of coherence and the reduction of errors. The process of scaling quantum computers necessitates augmenting their ability to effectively handle a significant number of qubits, potentially ranging from hundreds to thousands, all the while preserving coherence and minimizing mistakes.

8.3 The Development of Novel Hardware Technologies

The development of reliable quantum hardware, including components like qubits and control circuits, poses a significant challenge. According to Sahota (2023), there exists a wide range of qubit technologies, each with distinct merits and demerits. The research has placed major emphasis on the development of a qubit technology that is both scalable and capable of tolerating errors. The field of software development encompasses the process of designing, coding, testing, and maintaining computer software. It involves the use of engineering.

The current state of quantum algorithms and software development tools is characterized by their early stages of development. Consequently, there is a need to advance the field by creating novel programming languages, compilers, and optimisation tools. These advancements should optimize the capabilities of quantum computers.

8.4 The Topic of Discussion Pertains to Interfaces Designed for Legacy Computers

The primary goal in the development of quantum computers is not to make conventional computers obsolete, but rather to provide a complementary technological framework. In order to facilitate the implementation of real use cases, it is essential to provide efficient and dependable techniques for the exchange of data between traditional and quantum computing systems.

Protocols and guidelines are established frameworks that provide systematic instructions and recommendations for doing various activities or processes. These frameworks serve as standardized procedures to ensure consistency, efficiency, and quality. In order to ensure quantum computing systems are compatible, interoperable, and effectively evaluated for performance, it is imperative to build standardized frameworks, protocols, and benchmarks. As the field progresses, these aspects have more significance in the domains of hardware, software, and communication interfaces.

8.5 Acquired Expertise

Finding suitable professionals in the field of quantum physics might be a significant challenge. The situation might be characterized as a catch-22, since the progress of the field is contingent upon the presence of a larger number of persons who possess a heightened level of enthusiasm. However, the realization of this need is impeded by the limited accessibility to quantum processors. Quantum computing encounters a substantial impediment in the shape of considerable initial expenses (Sahota, 2023). The acquisition of quantum knowledge, equipment, and the establishment of complicated supply networks entail significant financial costs. In the next years, it is anticipated that scientists and company owners would be required to overcome these expenses and actively seek funding.

9. THE ANTICIPATED IMPACT OF QUANTUM COMPUTERS IN THE COMMERCIAL SECTOR

Based on the analysis of specialists in the area, it seems improbable that the competition to commercialize quantum computing would reach a conclusion prior to the year 2027. According to Sahota (2023), as the area of quantum computing progresses, there is an anticipation of a reduction in hardware costs, hence enhancing the feasibility of this technology for enterprises across all scales. The pricing dynamics of quantum computers are anticipated to be significantly impacted by technical progress, user preferences, and the degree of competition among quantum computing enterprises.

Several prominent companies, including IBM, Amazon, and Google, have made significant advancements in the field of quantum computing by successfully developing cutting-edge quantum processors boasting hundreds of qubits. In order to have a substantial influence on commercial applications, it may be necessary to increase the quantity of qubits to the magnitude of hundreds or even millions. Notwithstanding these technological progressions, there are impediments that hinder the future improvement.

An example of a technological application that necessitates advanced refrigeration systems is superconducting quantum computing. These systems are essential for preserving the optimum performance of qubits inside the ultra-cold environment necessary for their operation, while also implementing reliable error-correction techniques. The presence of these challenges renders the commercial availability of quantum computers difficult to forecast. However, professionals across several fields have a strong belief that this groundbreaking technology will provide favorable results in the next decade or two.

10. CONCLUSION

It can be inferred that the aforementioned points support the notion that, the field of quantum computing has a profound influence on the development of artificial intelligence (AI) and has great potential for a wide range of businesses and applications. Quantum computing has superior capabilities compared to conventional computers, since it enhances the velocity, efficacy, and accuracy of artificial intelligence systems. As a result, several domains like healthcare, machine learning, encryption and cybersecurity, financial applications, and natural language processing are able to use the advantageous impacts of quantum computing.

REFERENCES

Ayoade, O., Rivas, P., & Orduz, J. (2022). Artificial intelligence computing at the quantum level. *Data, 7*(3), 28. doi:10.3390/data7030028

Kanwar, P. (2023). *Quantum computing in artificial intelligence around the corner*. eInfo Chips. https://www.einfochips.com/blog/quantum-computing-in-artificial-intelligence-around-the-corner/

Montanaro, A. (2016). Quantum algorithms: An overview. *npj Quantum Information, 2*. doi:10.1038/npjqi.2015.23

Sahota, N. (2023, June 7). *Quantum computing and its impact on AI*. Neil Sahota. https://www.neilsahota.com/quantum-computing-and-its-impact-on-ai-unlocking-limitless-potential/

Schuld, M., & Petruccione, F. (2018). Supervised learning with quantum computers. *Quantum Science and Technology*. doi:10.1007/978-3-319-96424-9

Chapter 5
Quantum Computing Principles:
Harnessing the Potential of Quantum Mechanics

M. Gayathri
Department of Computing Technologies, SRM Institute of Science and Technology, Kattankulathur, India

P. C. Karthik
Department of Data Science and Business Systems, SRM Institute of Science and Technology, Kattankulathur, India

G. K. Sandhia
Department of Computing Technologies, SRM Institute of Science and Technology, Kattankulathur, India

R. Thilagavathy
Department of Computing Technologies, SRM Institute of Science and Technology, Kattankulathur, India

M. Pushpalatha
Department of Computing Technologies, SRM Institute of Science and Technology, Kattankulathur, India

ABSTRACT

Quantum computing stands as a pioneering field within computer science, leveraging the enigmatic principles of quantum mechanics to execute intricate calculations. At its heart lies the qubit, the elemental building block of quantum information, which resides in a state of superposition, representing both 0 and 1 concurrently. Quantum computers, constructed from assemblies of qubits, inhabit an expansive and intricate 2^n-dimensional complex Hilbert space, with their evolution governed by the Schrödinger equation. The process of quantum computation can be distilled into three key stages:1) Initialization of the input state: Initiating the quantum process; 2)Application of unitary transformations: Executing the desired transformations on the input state; 3) Measurement of the output state. Quantum computers

DOI: 10.4018/979-8-3693-1168-4.ch005

serve as universal machines, capable of decomposing any unitary transformation into a sequence of fundamental gates. This chapter will reveal the importance of quantum mechanics, quantum computation and its applications.

1. INTRODUCTION

The science of quantum computing is expanding quickly and has the potential to completely change how we approach difficult challenges. Its foundation is the theory of quantum mechanics, which explains how matter and energy behave at the atomic and subatomic scales. Data is represented by quantum bits, or qubits, in quantum computers. Qubits have the ability to exist in a superposition of states, which allows them to be both 0 and 1 simultaneously. Because of this, calculations on quantum computers can be completed tenfold quicker than on traditional computers. Cryptography is one of the most potential uses of quantum computing (Khang & Quantum, 2023).

Cybersecurity would be significantly impacted if present encryption techniques were cracked by quantum computers. Chemical, materials science, and drug development difficulties could all be resolved by quantum computers. They might also be applied to the creation of fresh artificial intelligence and machine learning algorithms. Although quantum computing is still in its infancy, there are a lot of potential advantages. Quantum computing has the ability to transform both our personal and professional lives and to provide solutions for some of the most difficult issues facing humanity. Practical applications domains exemplifying how quantum computers could reshape our world:

- **Encryption Decryption**: Quantum computers hold the capability to potentially break current encryption standards, which rely on the difficulty of factoring large numbers, thereby posing a significant cybersecurity challenge.
- **Molecular Simulation**: Quantum computers offer the capacity to simulate molecular behaviours, promising discoveries of novel drugs and materials with unprecedented efficiency.
- **Database Searches:** Quantum computers excel in database searches, outperforming classical computers significantly. This can accelerate data pattern recognition and the detection of fraudulent transactions.

The realm of quantum computing represents a dynamic arena poised to revolutionize various industries as it matures. As quantum computers gain in computational power, we anticipate even more transformative applications in the near future. Quantum computers serve as universal machines, capable of decomposing any unitary transformation into a sequence of fundamental gates. These gates facilitate the implementation of core Boolean functions, arithmetic operations, and intricate quantum algorithms (Khang & Kali, 2023).

These algorithms harness the fundamental attributes of quantum mechanics—superposition, entanglement, and interference—to address specific computational challenges with unparalleled efficiency. These tasks include: Grover's Algorithm: Efficiently locating a marked item within an unstructured database. Shor's Algorithm: Rapidly factoring large integers, a significant threat to classical cryptography. Physical System Simulation: Quantum computers excel in simulating intricate physical systems, unlocking possibilities in fields such as materials science and chemistry. Quantum computing is poised to revolutionize numerous sectors, encompassing cybersecurity, drug discovery, and materials science. Despite its early stage of development, this field has already achieved remarkable milestones:

- In 1998, Shor's algorithm was successfully executed on a quantum computer comprising only 7 qubits. By 2012, Grover's algorithm had been demonstrated on a 14-qubit quantum computer.
- In 2019, Google's research team announced a historic milestone: quantum supremacy, signifying their ability to perform a calculation on a quantum computer that would be infeasible for a classical computer. The quantum computing landscape is evolving at a rapid pace, promising further groundbreaking advancements in the years ahead.

2. GLIMPSE: HISTORY OF QUANTUM COMPUTING

Quantum computing is a rapidly developing field with a rich history. The basic concepts of quantum computing were first developed in the early 20th century, but it was not until the 1980s that the first quantum computers were built (Khang & Vugar et al., 2023).

2.1 The Early Years

In the early 1900s, physicists Werner Heisenberg, Erwin Schrödinger, and Niels Bohr lay the groundwork for quantum computing. These scientists created the quantum mechanics hypothesis, which explains how matter and energy behave at the atomic and subatomic levels. Quantum mechanics is a very different way of thinking about the world than classical physics. In classical physics, objects have well-defined properties such as position and momentum. Nonetheless, objects can exist in a superposition of states in quantum mechanics, which allows them to exhibit several attributes at once. Quantum computing requires this state superposition. It enables calculations on quantum computers that are not feasible on classical computers (Khang & Kali, 2024).

2.2 The First Quantum Computers

The first quantum computers were built in the 1980s. These computers were very small and simple, and they could only perform very basic calculations. However, they were a proof of concept that quantum computers were possible. Significant advancements were made in the realm of quantum computing during the 1990s. The quantity of qubits—the fundamental information unit in quantum computing—increased along with the development of new kinds of quantum computers.

2.3 The Current State of Quantum Computing

Quantum computing is still in its early stages, but there has been a lot of progress in recent years. Today, there are many different quantum computing projects underway, and the field is advancing rapidly. Some of the most promising quantum computing projects are being developed by companies such as Google, IBM, and Microsoft. These companies are working on developing quantum computers that can solve real-world problems. The challenges of quantum computing are significant, but there is a lot of optimism about the future of this technology. Many experts believe that quantum computers will have a major impact on society in the years to come (Khang & Shah et al., 2023).

2.4 The Future of Quantum Computing

The future of quantum computing is unknown, there are a lot of possible advantages. Numerous industries, including materials science, medicine development, and encryption, stand to benefit greatly from the advent of quantum computing. They might also be applied to address some of the most urgent issues facing the globe, like illness and climate change. It is an exciting time to be involved in quantum computing, and I believe that we will see many breakthroughs in the years to come.

3. QUANTUM MECHANICS

The foundational theory of physics known as quantum mechanics describes the physical characteristics of matter at the size of atoms and subatomic particles. It serves as the cornerstone for all branches of quantum physics, such as quantum information science, quantum field theory, quantum chemistry, and quantum technology. Quantum mechanics differs from classical physics in many ways, most notably in its description of energy, matter, and light. In classical physics, energy, matter, and light are all considered to be continuous quantities (Khang & Kali et al., 2023).

However, in quantum mechanics, energy, matter, and light are all considered to be quantized, meaning that they can only exist in discrete units. Superposition is one of the key ideas of quantum physics. According to superposition, a particle can have more than one state at once. For example, an electron can be both a wave and a particle. This is a very counterintuitive concept, and it is one of the things that makes quantum mechanics so difficult to understand. Another important concept in quantum mechanics is entanglement. When two particles are entangled, they are connected in a way that determines their fate together, regardless of how far apart they are. This is another counterintuitive concept, and it has led to some of the most interesting and important discoveries in quantum physics (Khang & Muthmainnah et al., 2023).

Quantum tunnelling is a phenomenon in which a particle can pass through a barrier that is classically impossible to cross. This is because particles in quantum mechanics can behave like waves, and waves can pass through barriers. Quantum tunnelling is important in many different applications, including nuclear fusion and the operation of transistors. These are just a few of the basic concepts of quantum mechanics. Quantum mechanics is a vast and complex subject, and there is much more to learn about it. However, the concepts outlined above provide a good foundation for understanding the basics of this fascinating field of physics (Nielsen and Chuang, 2010).

3.1 Superposition

A key idea in quantum mechanics is superposition, which asserts that a particle can exist in more than one state at once. A particle can only exist in one state at a time in classical physics, in contrast to this. The best way to understand superposition is to think of a wave. A wave can have a wide range of possible amplitudes and frequencies. However, at any given moment, the wave can only have one specific amplitude and frequency. A particle can be thought of as a wave in the context of quantum mechanics. A particle's wave function lists every potential state that the particle could be in.

The likelihood that a particle is in a specific condition is represented by the wave function's amplitude. For example, the wave function of an electron in an atom can describe all the possible states that the electron can be in. The amplitude of the wave function will be highest for the states that are most likely, and it will be lowest for the states that are least likely. When an electron is measured, it will always be found in one specific state. However, before it is measured, it can be in any of the possible states. This is because the electron is in a superposition of states (Ciaran et al., 2021).

3.2 Entanglement

When two particles are entangled, they are connected in a way that determines their fate together, regardless of how far apart they are. This is a very counterintuitive concept, and it is one of the most important concepts in quantum mechanics. Entanglement can be thought of as a special type of superposition. In entanglement, the wave functions of the two particles are linked together. This means that the particles cannot be described independently of each other. The best way to understand entanglement is to think of two particles that are created together. The wave function of these particles will be entangled from the moment they are created. This means that the particles will always be found in the same state, no matter how far apart they are separated. Entanglement has many important implications for quantum physics. It is used in quantum teleportation, quantum cryptography, and other quantum technologies.

3.3 Quantum Tunnelling

Quantum tunnelling is a phenomenon in which a particle can pass through a barrier that is classically impossible to cross. This is because particles in quantum mechanics can behave like waves, and waves can pass through barriers. Quantum tunnelling is important in many different applications, including nuclear fusion and the operation of transistors. The best way to understand quantum tunnelling is to think of a wave that is incident on a barrier. The wave will be partially reflected by the barrier, and partially transmitted through the barrier. The amount of wave that is transmitted through the barrier depends on the height and width of the barrier. If the barrier is too high or too wide, the wave will not be able to pass through. However, if the barrier is not too high or too wide, the wave will be able to tunnel through the barrier.

4. DIFFERENT TYPES OF QUANTUM COMPUTERS

At present, multiple varieties of quantum computers are under development, each possessing unique advantages and disadvantages. We shall talk about the five most promising kinds of quantum computers in this article: quantum dots, photonic, neutral atoms, trapped ions, and superconducting (Vishal, 2007).

4.1 Superconducting Quantum Computers

Superconducting quantum computers are the most developed type of quantum compu ter. They use superconducting circuits to create qubits. Superconducting circuits are made of materials that lose all electrical resistance at very low temperatures. This allows the qubits to be very stable and easy to control. Superconducting quantum computers have been used to achieve some impressive results, such as factoring a 15-digit number. However, they are still limited by the number of qubits that can be reliably created and controlled (Dominique, 2018).

4.2 Photonic Quantum Computers

Photonic quantum computers use photons, or particles of light, to create qubits. Photons are very well-suited for quantum computing because they can be manipulated with high precision. It is difficult to create and maintain large-scale photonic quantum computers.

4.3 Trapped Ions Quantum Computers

Trapped ions quantum computers use ions, or charged atoms, to create qubits. The ions are trapped in a vacuum and manipulated with lasers. Trapped ions quantum computers are very stable and can be scaled up to a large number of qubits. Trapped ions quantum computers are also very expensive to build and operate.

4.4 Neutral Atom Quantum Computers

Neutral atom quantum computers use neutral atoms to create qubits. The atoms are cooled to very low temperatures and trapped in a vacuum. Neutral atom quantum computers are very promising because they are relatively easy to create and control. They are also very sensitive to noise and interference.

4.5 Quantum Dot Quantum Computers

Quantum dot quantum computers use quantum dots, or artificial atoms, to create qubits. Quantum dots are small semiconductor particles that can be manipulated with lasers. Quantu m dot quantum computers are relatively new, but they have the potential to be very powerful. However, they are also very difficult to create and control.

In addition to the five types of quantum computers discussed in this article, there are also several other promising approaches being explored. These include adiabatic quantum computers, topological quantum computers, and liquid-state quantum computers. The future of quantum computing is bright, and it is exciting to see how these different technologies will develop in the years to come.

5. QUANTUM BITS (QUBITS)

In quantum computing, a qubit is the fundamental unit of information. It is a quantum incarnation of the traditional binary bit, which is capable of holding two values: 0 and 1. In contrast, a qubit can simultaneously be in a superposition of 0 and 1. This explains why quantum computers are so powerful: a qubit may represent an endless number of possibilities. To understand how qubits work, it is important to understand the concept of superposition. In classical physics, a particle can only be in one state at a time. However, in quantum mechanics, a particle can be in a superposition of states. This means that the particle can be in multiple states at the same time.

The superposition of states is a very counterintuitive concept, but it is one of the fundamental principles of quantum mechanics. It is what allows qubits to be so powerful. A qubit can be represented by a wave function. The wave function of a qubit is a mathematical object that describes all the possible states that the qubit can be in. The amplitude of the wave function represents the probability of the qubit being in a particular state (Mikio and Tetsuo, 2008).For example, the wave function of a qubit that is in a superposition of 0 and 1 can be represented by the equation 1

$$\psi = a|0\rangle + b|1\rangle \tag{1}$$

where a and b represent complex numbers that depicts the probability amplitudes of the qubit being in the zero and one states, respectively.

Quantum gates allow one to control a qubit's wave function. Qubits can undergo operations called quantum gates to alter their states. Although there are numerous varieties of quantum gates, the Hadamard, CNOT, and T gates are among the most widely used ones. The Hadamard gate is a quantum gate that can be used to put a qubit into a superposition of states. The CNOT gate is a quantum gate that can be used to entangle two qubits. The T gate is a quantum gate that can be used to rotate the state of a qubit by 90 degrees. By manipulating the wave functions of qubits using quantum gates, it is possible to perform a wide variety of quantum computations. These computations can be used to solve problems that are intractable on classical computers. (David, 2007)

6. QUANTUM ALGORITHMS

A computing process that can be carried out on a quantum computer is called a quantum algorithm. The special abilities of quantum mechanics, such as superposition and entanglement, are used by quantum algorithms. (Tharrmashastha & Maitra, 2020). Quantum algorithms can be used to solve problems that are intractable on classical computers. Some of the most well-known quantum algorithms include Shor, Grover, Simon and QFT.These are just a few of the many quantum algorithms that have been developed.

As quantum computers become more powerful, more and more quantum algorithms will be developed. The field of developing quantum algorithms is one that is fast developing. Before quantum algorithms are applied to real-world situations, a number of obstacles must be removed. (Stephen, 2011). The four quantum algorithms mentioned above are

- Shor's algorithm was developed by Peter Shor in 1994. It can be used to factor large integers in polynomial time, which is exponentially faster than any known classical algorithm. This has implications for breaking many of the most widely used encryption algorithms, such as RSA (Menezes, Van Oorschot and Vanstone, 2018)
- Grover's algorithm was developed by Lov Grover in 1996. It can be used to search an unstructured database in quadratic time, which is exponentially faster than any known classical algorithm. This has implications for many different applications, such as finding patterns in data and medical diagnosis.

- Simon's algorithm was developed by Daniel in 1997. It can be used to determine whether a function is one-to-one or two-to-one. This has implications for many different applications, such as cryptography and error correction.
- Quantum Fourier transform is a fundamental operation in quantum computing. It can be used to perform many different quantum algorithms, such as Shor's algorithm and Grover's algorithm.

7. QUANTUM GATES

One fundamental action that can be carried out on a qubit is a quantum gate. The fundamental components of quantum circuits—the analogous of computer programs in quantum computing—are known as quantum gates. (Kaye, Laflamme, and Mosca. 2006). Although there are numerous varieties of quantum gates, the following are some of the more widely used ones,

- **The Hadamard gate**: The Hadamard gate is a quantum gate that can be used to put a qubit into a superposition of states.
- **The C NOT gate**: The CNOT gate is a quantum gate that can be used to entangle two qubits.
- **The T gate**: The T gate is a quantum gate that can be used to rotate the state of a qubit by 90 degrees.
- **The Phase gate**: The Phase gate is a quantum gate that can be used to add a phase shift to the state of a qubit.

Combining quantum gates allows for the execution of more intricate computations. For instance, Shor's method, which factors big integers, can be implemented using a series of Hadamard gates and CNOT gates. The design of quantum gates is a complex and challenging task. The goal is to design gates that are as efficient as possible, while also being reliable and robust to noise. There are many different ways to implement quantum gates. Some of the most common methods include:

- **Optical gates**: Optical gates can be implemented using photons.
- **Electrical gates**: Electrical gates can be implemented using superconducting circuits.
- **Ionic gates**: Ionic gates can be implemented using trapped ions.

The particular application will determine which implementation approach is used. For instance, applications requiring vast distances are better served by optical gates, whereas those requiring rapid speeds are better served by electrical gates. Quantum gates are a fundamental building block of quantum computing. The development of efficient and reliable quantum gates is essential for the realization of practical quantum computers. Additional details about the four quantum gates mentioned above:

The Hadamard gate is a unitary operator that can be represented by the following truth table. The Hadamard gate has the effect of putting a qubit into a superposition of the 0 and 1 states. Hadamard gate truth table is represented by Table1.

The CNOT gate has the effect of flipping the state of the second qubit if the first qubit is in the 1 state. CNOT gate truth table is represented by Table 2.

The T gate is a unitary operator that can be represented by the following truth as shown in Table 3. The T gate has the effect of rotating the state of a qubit by 45 degrees.

Table 1. Hadamard gate truth table

Input	Output
$\vert 1\rangle$	$\vert -\rangle$
$\vert 0\rangle$	$\vert +\rangle$
$\vert +\rangle$	$\vert 0\rangle$
$\vert -\rangle$	$\vert 1\rangle$

Table 2. CNOT gate truth table

Input	Output
$\vert 00\rangle$	$\vert 00\rangle$
$\vert 01\rangle$	$\vert 01\rangle$
$\vert 10\rangle$	$\vert 11\rangle$
$\vert 11\rangle$	$\vert 10\rangle$

Table 3. T gate truth table

Truth Table for $P\left(\frac{\pi}{4}\right)$	
Input	Output
$\vert 0\rangle$	$\vert 0\rangle$
$\vert 1\rangle$	$e^{i\frac{\pi}{4}}\vert 1\rangle$

Table 4. Phase gate truth table

Truth Table for $P(\phi)$	
Input	Output
$\vert 0\rangle$	$\vert 0\rangle$
$\vert 1\rangle$	$e^{i\phi}\vert 1\rangle$

The Phase gate is a unitary operator that can be represented by the following truth as shown in table 4. The Phase gate has the effect of adding a phase shift of $\pi/2$ to the state of a qubit.

8. QUANTUM ERROR CORRECTION

A method for shielding quantum data from computing faults is known as quantum error correction. Due to the fact that quantum bits, or qubits, are subject to the laws of quantum physics, quantum computers are prone to errors. There are many different types of errors that can occur in quantum computers, including:

- **Bit flips**: A bit flip is an error that flips the state of a qubit from 0 to 1 or vice versa.
- **Phase flips**: A phase flip is an error that changes the phase of a qubit.
- **Loss of coherence**: Loss of coherence is an error that causes the qubit to lose its superposition of states.

Quantum error correction techniques can be used to correct these errors and ensure that the quantum information remains accurate. Stabilizer code is one of the most often used quantum error correcting methods. The way stabilizer codes function is by using more qubits to encode the quantum information. Errors in the other qubits are detected and corrected by the additional qubits. Another common quantum error correction technique is called the surface code. The surface code is a more complex technique than the stabilizer code, but it is also more efficient. The surface code can be used to correct a wider range of errors than the stabilizer code. In the field of quantum computing, developing quantum error correcting methods is a crucial task. Building trustworthy quantum computers would be impossible without quantum error correction. The details about the two quantum error correction techniques mentioned above are,

- Stabilizer codes are a class of quantum error correction codes that can correct a limited number of errors. They are relatively simple to implement, but they are not very efficient.
- Surface codes are a class of quantum error correction codes that can correct a larger number of errors. They are more complex to implement, but they are more efficient.

The particular application determines which quantum error correcting technique is best. Applications requiring a high degree of error correction, for instance, can benefit from surface codes, but simpler and more effective applications might benefit from stabilizer codes. Quantum error correction is a rapidly developing field. There are many new techniques being developed, and the field is constantly evolving. The development of efficient and reliable quantum error correction techniques is essential for the realization of practical quantum computers.

9. QUANTUM APPLICATIONS

Some of the potential applications of quantum computing include, breaking the RSA encryption algorithm, which is used to secure many of the internet's most important applications, Finding new drugs and treatments for diseases, designing new materials and devices and simulating complex physical systems

- **Materials science:** By using quantum computing, new materials with desirable characteristics like low conductivity or great strength could be designed.
- **Artificial intelligence:** New algorithms for machine learning and artificial intelligence may be created using quantum computers. This could result in advancements in fields like robotics, natural language processing, and facial recognition (Maria and Francesco, 2021).
- **Finance:** The development of new trading algorithms and financial models using quantum computers may increase the effectiveness of the financial markets. By analysing financial data, quantum computers may help investors make wiser choices.
- **Cybersecurity:** Current encryption techniques could be broken by quantum computers, which would have a significant impact on cybersecurity.
- **Optimization:** Issues with optimization, like the traveling salesman issue, that are too complex for conventional computers may be resolved by quantum computers.
- **Drug discovery:** By simulating the behaviour of molecules, quantum computers may be able to assist scientists in creating novel medications and medical treatments.
- **Cryptography:** New encryption techniques that are impenetrable by conventional computers may be created using quantum computers.
- **Machine learning**: More precise and effective machine learning models could be trained with quantum computers than with conventional (Maria and Francesco, 2018), (Peter, 2014).
- **Weather forecasting**: By simulating how weather systems behave, quantum computers may be able to increase the accuracy of weather forecasts.
- **Traffic optimization**: By maximizing traffic flow, quantum computers may be able to lessen traffic and enhance air quality.
- **Aerospace engineering**: More secure and efficient new spaceships and aircraft might be designed using quantum computing.
- **Robotics:** By using quantum computing, it may be possible to create more intelligent robots that can function in intricate surroundings.

10. CHALLENGES OF QUANTUM COMPUTING

The field of quantum computing is developing quickly in spite of the difficulties. Many businesses and academic institutions are working on creating quantum computers, and interest in the possible uses of this technology is rising. Quantum computing is an exciting new technology that could fundamentally alter the way we live and work. The field of quantum computing is rapidly growing and has the power to drastically change a variety of industries.

However, before building a working quantum computer, a number of obstacles still need to be solved. Developing reliable qubits that can maintain their quantum state for long periods of time. Developing quantum algorithms that can take advantage of the power of quantum computers. Overcoming the noise and errors that occur in quantum systems. Some of the biggest challenges of quantum computing include:

- **Qubit coherence:** Qubits have the potential to lose their quantum state due to mistakes and decoherence. Maintaining the precision of quantum computing becomes challenging as a result.
- **Quantum error correction:** Quantum error correction is essential for protecting quantum information from errors. However, it is a complex and challenging task.

- **Scaling:** Quantum computers need to be scaled up to a large number of qubits in order to be useful for practical applications. However, this is a difficult and expensive task.
- **Control:** Quantum computers need to be precisely controlled in order to perform accurate computations. This is a challenging task, as quantum systems are very sensitive to noise and interference.
- **Applications:** There are many potential applications for quantum computing, but most of them are still in the early stages of development. It is not yet clear which applications will be the most successful.

Despite these challenges, there is a lot of progress being made in the field of quantum computing. Many companies and research institutions are investing heavily in this technology, and there is a growing community of scientists and engineers working on developing practical quantum computers. It is still too early to say when practical quantum computers will be available. However, the potential benefits of this technology are immense, and the field is advancing rapidly. It is an exciting time to be involved in quantum computing, and we believe that we will see many breakthroughs in the years to come.

Although quantum computing presents several obstacles, the science is making great strides. An increasing number of businesses and academic organizations are making significant investments in this technology, and scientists and engineers are collaborating to create usable quantum computers. Think that in the years to come, quantum computing will see a lot of advancements. This technology has the power to transform a wide range of sectors and provide solutions for some of the most important issues facing humanity.

11. THE FUTURE OF QUANTUM COMPUTING

Quantum computing is a rapidly developing field with the potential to revolutionize many different industries. However, it is still in its early stages, and it is not yet clear what the future holds for this technology. There are many different opinions about the future of quantum computing. Some experts believe that it will have a major impact on society, while others believe that it will be limited to a few niche applications.

12. CONCLUSION

There are many potential advantages, the future of quantum computing remains unclear. Quantum computers have the potential to transform numerous industries and address some of the most critical global issues if the obstacles are successfully surmounted. The future of quantum computing holds a lot of great discoveries, and it's an exciting moment to be involved in the field. Although it is still in its infancy, the field of quantum computing is growing quickly. It is currently unclear which variety of quantum computer will ultimately be the most effective because each type has pros and cons of its own.

All of the many kinds of quantum computers, however, will undoubtedly be very important to computing in the future and have the power to completely transform a wide range of businesses. These are but a handful of the numerous possible uses for quantum computing. With the increasing power of quantum computers, there's a good chance we'll find even more applications for them to tackle issues that classical computers are presently unable to handle. Although there is still much to learn about the development of quantum applications, the promise of this technology is very exciting (Khang & Medicine, 2023).

A growing community of scientists and engineers is working on creating new applications for quantum computing, and numerous businesses and academic institutes are making significant investments in this field. Although it will probably take many years for quantum computing to become widely used, there are a lot of potential advantages. They might also result in the development of previously unimaginable new technologies. Being in the realm of quantum computing is fascinating because it has a bright future. Many online and in-person resources are available if you're interested in learning more about quantum computing. Additionally, there are many of chances for you to participate in research and development (Khang & Agriculture et al., 2023).

REFERENCES

David Mermin, N. (2007). *Quantum Computer Science*. Cambridge Press. https://www.google.com/books?hl=en&lr=&id=q2S9APxFdUQC

Hughes, C., Isaacson, J., Perry, A., Sun, R. F., & Turner, J. (2021). *Quantum Computing for the Quantum Curious*. Springer. https://library.oapen.org/handle/20.500.12657/48236

Jordan, S. (2011). *Home*. Quantum Algorithm Zoo. https://quantumalgorithmzoo.org/

Kaye, A. P., Laflamme, R., & Mosca, M. (2006). *Introduction to Quantum Computing*. Oxford University Press., https://www.google.com/books?hl=en&lr=&id=18_-rgkdQGIC doi:10.1093/oso/9780198570004.001.0001

Khang, A. (2023). Enabling the Future of Manufacturing: Integration of Robotics and IoT to Smart Factory Infrastructure in Industry 4.0. AI-Based Technologies and Applications in the Era of the Metaverse. IGI Global Press. doi:10.4018/978-1-6684-8851-5.ch002

Khang, A. (2023). *Advanced Technologies and AI-Equipped IoT Applications in High-Tech Agriculture* (1st ed.). IGI Global Press. doi:10.4018/978-1-6684-9231-4

Khang, A. (2023). Applications and Principles of Quantum Computing. IGI Global Press. https://doi.org/ doi:10.4018/979-8-3693-1168-4

Khang, A. (2023). Quantum Mechanics Primer - Fundamentals and Quantum Computing," Applications and Principles of Quantum Computing. IGI Global Press. https://doi.org/ doi:10.4018/979-8-3693-1168-4-ch001

Khang, A. (2023). Introduction to Quantum Computing and Its Integration Applications. Applications and Principles of Quantum Computing. IGI Global Press. https://doi.org/ doi:10.4018/979-8-3693-1168-4-ch002

Khang, A. (2023). AI and IoT-Based Technologies for Precision Medicine. IGI Global Press. doi:10.4018/979-8-3693-0876-9

Khang A. & Rath, K. (2024). *The Quantum Evolution: Application of AI and Robotics in the Future of Quantum Technology*. CRC Press. https://doi.org/ doi:10.1201/978-1-032-64207-9

Khang, A., & Muthmainnah, M. (2023). AI-Aided Teaching Model for the Education 5.0 Ecosystem. AI-Based Technologies and Applications in the Era of the Metaverse. IGI Global Press. doi:10.4018/978-1-6684-8851-5.ch004

Khang, A., Shah, V., & Rani, S. (2023). AI-Based Technologies and Applications in the Era of the Metaverse. IGI Global Press. doi:10.4018/978-1-6684-8851-5

Schuld, M. & Petruccione, F. (2018). *Supervised Learning with Quantum Computers.* Springer. https://link.springer.com/content/pdf/10.1007/978-3-319-96424-9.pdf

Menezes, P. van Oorschot, C., & Vanstone, S. (2018). *Handbook of Applied Cryptography*. CRC Press. http://galois.azc.uam.mx/mate/propaganda/Menezes.pdf

Nakahara M., & Ohmi, T. (2008). *Quantum Computing*. CRC Press. https://www.google.com/books?hl=en&lr=&id=VdHJsTdoyAMC

Nielsen, M. A., & Chuang, I. L. (2010). *Quantum Computation and Quantum Information*. Cambridge University Press., doi:10.1017/CBO9780511976667

Sahni, V. (2007), *Quantum computing*. Tata Mc Graw Hill Publishing Company Limited. https://www.mitindia.edu/images/mit/Departments/EI/syllabus/BE_EIE-R2019-_A-I.pdf#page=24

Schuld, M., & Petruccione, F. (2021). *Machine Learning with Quantum Computers* (2nd ed.). Springer. https://link.springer.com/content/pdf/10.1007/978-3-030-83098-4.pdf doi:10.1007/978-3-030-83098-4

Tharrmashastha, S. A. (2020). *Quantum Algorithms for Cryptographically Significant Boolean Functions - An IBMQ Experience.* Springer. https://link.springer.com/content/pdf/10.1007/978-981-16-3061-3.pdf

Unruh, D. (2018) *Quantum Cryptography*. UT. https://courses.cs.ut.ee/all/MTAT.07.024/2017_fall/uploads/

Wittek, P. (2014). *Quantum Machine Learning – What Quantum Computing Means to Data Mining*. Elsevier. https://www.google.com/books?hl=en&lr=&id=92hzAwAAQBAJ

Chapter 6
Quantum Computing: Unveiling the Paradigm Shift and Diverse Applications

R. Siva Subramanian
https://orcid.org/0000-0002-7509-9223
R.M.K. College of Engineering and Technology, India

B. Maheswari
R.M.K. College of Engineering and Technology, India

T. Nithya
Rajalakshmi Institute of Technology, India

P. Girija
https://orcid.org/0009-0000-1691-8180
Vel Tech Rangarajan Dr.Sagunthala R&D Institute of Science and Technology, India

M. Karthikeyan
R.M.K. College of Engineering and Technology, India

T. Saraswathi
https://orcid.org/0000-0003-1095-9015
Easwari Engineering College, India

ABSTRACT

Quantum computing overcomes computer limitations using quantum mechanics. This overview covers quantum computing's basics, technology, architecture, and industrial applications. The introduction reveals quantum computing's unique history and powerful development characteristics. Fundamentals like quantum gates, superposition, entanglement, Shor's and Grover's algorithms, and more, are covered. Quantum computing concepts describe quantum mechanics, parallelism, and error correction to optimize its potential. Qubit, quantum processor, Qiskit, and Cirq quantum software frameworks are being studied. Quantum computing's issues are thoroughly and impartially assessed in this work. This

DOI: 10.4018/979-8-3693-1168-4.ch006

chapter evaluates quantum computing trends and research paths for future developments. These include the goal of quantum supremacy and the impact of quantum computing on technology. The survey study helps academics, professionals, and industry decision-makers utilize quantum computing to tackle challenging issues and innovate in a quantum-driven world.

1. INTRODUCTION

Utilizing the concepts of quantum mechanics to get beyond the limits of conventional computing, quantum computing offers a revolutionary paradigm in the field of computation. A cutting-edge computing paradigm known as quantum computing uses the ideas of quantum mechanics to process data in ways that are radically different from those of conventional computers (Herman et al., 2022). Quantum bits, also known as qubits, are the basic building blocks of quantum computing that may exist in numerous states at once owing to a phenomena called superposition. Quantum computers can execute certain sorts of computations tenfold quicker than conventional computers because to this special characteristic (Rietsche et al., 2022).

Richard Feynman, a physicist, first proposed the idea of quantum computing as an effective way to model quantum systems in the early 1980s. David Deutsch developed the idea of a quantum universal Turing computer a few years later, establishing the foundation for quantum computing. When Peter Shor developed the well-known Shor's algorithm in the 1990s (Ahnefeld et al., 2022), it marked a key turning point by proving that quantum computers could factor big numbers exponentially faster than traditional machines.

Grover's method, created at the same time by Lov Grover, demonstrated the possibility of quantum speedup in search tasks. With the advent of several quantum hardware platforms, research and development in quantum computing accelerated in the early 2000s. Superconducting qubit research was started by firms including IBM, Google, and Rigetti, while others looked at trapped ions and photon-based qubits. These initiatives represented a big step forward in the development of quantum computers that can handle more difficult computations. 2019 saw one of the pivotal events in the history of quantum computing when Google said it had achieved "quantum supremacy." Their quantum processor, Sycamore, demonstrated the potential for real-world quantum advantage by completing a particular task quicker than the most sophisticated conventional supercomputers (Khang & Quantum, 2023).

Though still in its infancy, quantum computing faces significant difficulties such qubit stability, error correction, and scalability. Researchers are always investigating fresh strategies for creating more reliable and potent quantum computers. These innovations spurred scientists all around the globe to start working on building usable quantum computers. Different quantum computing techniques have developed throughout time, each with its own benefits and difficulties, such as superconducting qubits, trapped ions, and topological qubits (Khang & Kali, 2023).

1.1 Motivation for Quantum Computing Studies

- **Unprecedented Computational Power:** Quantum computers have the potential to solve difficult problems that were previously beyond the computational capabilities of conventional computers. This covers quantum system simulation, optimization, and cryptography.

- **Revolutionizing Industries:** Quantum computing has the potential to revolutionize a variety of sectors, including banking, healthcare, materials research, and artificial intelligence. It can speed up the development of new drugs, improve supply systems, and transform machine learning techniques.
- **Quantum-Safe Cryptography**: As quantum computers grow more prevalent, traditional encryption techniques will be more open to attack. For the future development of quantum-safe cryptography methods to secure digital communication, studying quantum computing is essential.
- **Scientific Discovery:** Researchers can model and simulate quantum systems with unheard-of accuracy thanks to quantum computing. The knowledge of basic physics, chemistry, and materials science is profoundly affected by this.
- **Changing Technological Environment:** As technology develops quickly, there is an increasing chance that quantum computing may upend established computer paradigms. It's essential to comprehend quantum computing if you want to remain on the cutting edge of technological development.
- **Quantum Internet:** The creation of a quantum internet, which might allow secure communication utilizing quantum teleportation and quantum key distribution (QKD), is closely related to quantum computing. The future of communication networks will be shaped by research into quantum computing.
- **Environmental Impact:** Quantum computing has the ability to streamline resource-intensive procedures and provide calculations that use less energy. This may help lessen the impact of computer technology on the environment.

1.2 Goal of the Study

The survey's objective is to provide a thorough and in-depth examination of quantum computing, including its fundamental ideas, historical development, important technologies, architecture, and a broad range of applications in many fields. It seeks to provide readers with a comprehensive grasp of quantum computing's importance, possible applications, difficulties, and limitations. In addition, the survey aims to be a useful tool for academics, professionals, and decision-makers who desire to understand the intricate and quickly developing subject of quantum computing, supporting informed choices and encouraging innovation in a quantum-driven future (Khang & Vugar et al., 2023).

In conclusion, the drive to research quantum computing goes beyond simple pique at the subject. It includes efforts to develop cutting-edge computing capabilities, revolutionize industries, protect digital security, and enhance scientific understanding. The quest for innovation in the constantly changing technological environment is shown by the exploration of quantum computing. The study of quantum computing is an expedition into unknown computational space, propelled by the alluring potential of unleashing unmatched processing power and transforming business and scientific research. This study, which straddles the fields of mechanics, computer science, and mathematics, has the potential to completely alter the technological landscape in the years to come (Ushaa, Khang et al., 2023).

2. FUNDAMENTALS AND CONCEPTS OF QUANTUM COMPUTING

In contrast to conventional computing, quantum computing is based on a number of basic ideas from quantum mechanics. Let's explore these fundamental ideas in more depth.

2.1 Quantum Bits (Qubits)

The fundamental units of information in quantum computing are quantum bits, or qubits. Qubits have a noteworthy characteristic known as superposition, which sets them apart from traditional bits, which can only exist in one of two states (0 or 1) (Gyongyosi & Imre, 2019). A qubit may be in both a linear combination of 0 and 1 states and superposition at the same time. A qubit in superposition is represented mathematically as shown in Formula 1:

$$|\psi\rangle = \alpha|0\rangle + \beta|1\rangle \tag{1}$$

Here, the probabilities of seeing a measurement of 0 or 1 are represented by the complex probability amplitudes and respectively. The odds must adhere to the normalization condition as shown in Formula 2

$$|\alpha|^2 + |\beta|^2 = 1. \tag{2}$$

Where

- **Superposition:** Superposition allows quantum computers to analyze many scenarios concurrently, providing exponential speedup for certain algorithms. It is a basic characteristic that underpins quantum computation's power.
- **Quantum Entanglement**: Qubits may also be entangled in addition to superposition. When two or more qubits are entangled, a rare quantum occurrence, their states become inextricably connected even if the qubits are physically isolated from one another. This entangled state is a valuable resource for quantum algorithms since it cannot be independently explained (Shannon et al., 2020).
- **Measurement:** A qubit collapses into one of its base states (0 or 1) when it is measured, with probabilities defined by the squared magnitudes of α and β.

2.2 Quantum Gates and Circuits

The quantum analogue of conventional logic gates, quantum gates change the states of qubits. The representation of these gates as unitary matrices ensures the reversibility of quantum processes. Common examples of quantum gates include. The "X Gate (Pauli-X)" switches a qubit's state from "0" to "1," and vice versa.

- **Hadamard Gate:** The Hadamard gate prepares a qubit for quantum computations by spinning its state along several axes.
- **CNOT Gate:** If and only if a control qubit is in state |1, the CNOT gate executes an X gate operation on a target qubit. It is essential in producing entangled states.

By placing a series of quantum gates that operate on qubits, quantum circuits are built. The positioning, sequencing, and kinds of gates in a circuit determine its operation. In essence, quantum algorithms are quantum circuits that describe how to prepare, operate, and measure qubits in order to solve certain tasks.

2.3 Superposition and Entanglements

- **Characteristic**: Superposition, a fundamental concept in quantum computing, enables qubits to exist in several states concurrently. This characteristic is used by quantum algorithms to do calculations simultaneously on a huge number of possibilities. It is a key characteristic that distinguishes quantum computing from traditional computing.
- **Entanglement:** Entanglement is a phenomena in which qubits are intertwined to the point that their states are immediately connected even when they are separated by great distances. Enabling safe communication and data transmission, entanglement is essential to quantum teleportation and quantum cryptography.

2.4 Quantum Algorithms, Such as Grover's and Shor's Algorithms

- **Shor's Algorithm:** A quantum technique called Shor's algorithm was developed to effectively factor huge numbers into their prime parts. This capacity presents a serious danger to traditional public-key encryption schemes, such as RSA, since it may allow for the speedy discovery of huge composite number's prime factors, which would allow for their decryption. Shor's algorithm has significant ramifications for data security and cryptography as a result (Ugwuishiwu et al., 2020).
- **The Grover algorithm:** Grover's algorithm, a quantum search technique, outperforms conventional search algorithms. It is especially useful for jobs like unstructured database searches, where it may drastically cut down on the amount of processing power needed to discover the desired outcome (Ramezani et al., 2020).

In essence, the principles of quantum computing rely upon the use of quantum gates to control qubit states and the special characteristics of qubits, such as superposition and entanglement. These concepts serve as the cornerstone of quantum algorithms, allowing them to solve certain problems exceptionally well and possibly revolutionizing areas like encryption, optimization, and scientific simulation (Prabu, Khang et al., 2023).

3. PRINCIPLES OF QUANTUM COMPUTING

Several important ideas are necessary for understanding how quantum algorithms function and why quantum computers have the potential to surpass conventional computers in the area of quantum computing. Let's take a closer look at these principles:

3.1 Quantum Mechanics Fundamentals

Quantum mechanics is the discipline of mechanics that regulates the behavior of quantum particles such as atoms and subatomic particles. Quantum computing is based on the following quantum mechanics fundamental concepts:

- **Wave-Particle Duality:** According to quantum mechanics, particles such as electrons and photons have both particle-like and wave-like qualities. The idea of wave functions, which represent the probabilistic character of a particle's location and momentum, captures this duality.
- **Uncertainty Principle:** According to Werner Heisenberg, it is impossible to know both the exact location and momentum of a quantum particle at the same time. This intrinsic uncertainty has far-reaching consequences for quantum computing and cryptography.
- **Quantum Superposition:** Superposition is a basic concept of quantum mechanics. It enables quantum bits (qubits) to exist in several states at the same time. In other words, a qubit may be in a superposition of 0 and 1, denoted by |0 + |1. Because of this characteristic, quantum computers can do parallel computations.
- **Quantum Entanglement:** Quantum entanglement is a phenomena that occurs when the states of two or more particles become entangled while being separated by huge distances. Changes in the state of one particle have an immediate impact on the state of the entangled particles. Entanglement is an important resource in quantum computing and cryptography.

3.2 Quantum Parallelism

Quantum parallelism is an important idea derived from superposition. It allows quantum computers to investigate several computing possibilities at the same time. A bit may be in one of two states (0 or 1) in classical computing, and each operation is executed sequentially. Quantum bits (qubits) in a superposition state, on the other hand, may represent both 0 and 1.

Quantum algorithms use parallelism to analyze huge datasets and execute calculations more effectively. Grover's approach, for example, may search an unsorted database of N items in around N steps, providing a quadratic speedup over conventional search algorithms (Shah & Khang et al., 2023).

3.3 Quantum Error Correction

Because qubits are subject to ambient noise and decoherence, quantum error correction is critical. Quantum error correction codes are similar to classical error correction codes, but they are meant to safeguard quantum information against mistakes induced by environmental interactions.

- **Codes for Quantum Error Correction:** To identify and repair faults, these algorithms use entangled qubits. The surface code, which organizes qubits in a two-dimensional lattice, is the most well-known quantum error correcting code. By monitoring syndrome qubits in the lattice, errors may be found and remedied.
- **Fault-Tolerant Quantum Computing:** Quantum error correction allows for the development of fault-tolerant quantum computers in which qubits are redundantly encoded and mistakes are con-

tinually checked and fixed. This is critical for the development of effective and scalable quantum computing systems.

In summary, the foundations of quantum computing are strongly entrenched in quantum mechanics, including ideas such as superposition, entanglement, and error correction. These concepts underlie quantum computers' unique processing capacity, enabling them to execute concurrent computations and possibly transform sectors such as encryption, optimization, and scientific modelling.

4. TECHNOLOGIES FOR QUANTUM COMPUTING

The hardware, software, and algorithms that make up quantum computing technologies allow for the operation of quantum computers. Together, these elements enable the use of quantum mechanics for computing. Let's look more closely at these technologies:

4.1 Quantum Hardware, Such as Quantum Bits and Processors

- **Quantum Processors:** A quantum computer's brain is its quantum processor. They are actual objects that operate on qubits and carry out quantum calculations. Quantum processors are constructed using a variety of methods, such as topological qubits, trapped ions, and superconducting circuits.
- **Quantum bits (Qubits):** The quantum equivalent of conventional bits are called qubits. They may exist in a superposition of states and are the fundamental building blocks of quantum information. Superconducting qubits, ion trap qubits, and photon-based qubits are a few examples of different kinds of qubits. Qubits need error correction to maintain their coherence since they are sensitive to outside noise.
- **Quantum Gates:** Implemented on quantum computers, quantum gates operate on qubits to change their states. These gates are in charge of carrying out quantum algorithms. X gates, Hadamard gates, and CNOT gates are typical quantum gates (Gill et al, 2022).

4.2 Quantum Programming Languages (Such as Qiskit and Cirq)

- **Quantum Software Frameworks:** These frameworks provide libraries and tools for creating and simulating quantum algorithms. The two well-known instances are Qiskit and Cirq.
- **Qiskit:** Created by IBM, Qiskit is an open-source quantum computing framework that enables users to either mimic quantum circuits on conventional computers or develop and operate them on IBM's quantum hardware. It interacts with Python and offers a user-friendly interface for quantum programming.
- **Cirq:** Another open-source quantum framework for creating, enhancing, and operating quantum circuits was created by Google. It offers support for several quantum hardware platforms and is designed for academics and developers that are interested in quantum computing.

4.3 Quantum Simulators and Algorithms

- **Quantum Algorithms:** Quantum algorithms make use of quantum concepts like superposition and entanglement and are particularly created to be run on quantum computers. Examples include Grover's method for quantum search and Shor's algorithm for factoring big numbers.
- **Quantum simulators:** These computer programs mimic the behavior of quantum systems on classical machines. Before putting quantum algorithms to use on actual quantum hardware, these simulators are essential for testing and troubleshooting them. They assist researchers in comprehending the operation of quantum algorithms and locating areas for advancement.

In conclusion, the hardware, software frameworks, and algorithms needed for quantum processing are all included in quantum computing technologies. Quantum software frameworks enable quantum programming, quantum processors use quantum gates to handle qubits, and simulators support the creation and testing of algorithms. Together, these technologies help quantum computing evolve, providing the promise for exponentially faster processing and game-changing applications across a range of industries (Khang & Kali, 2024).

5. ARCHITECTURE FOR QUANTUM COMPUTING

The structural and organizational elements that make it possible for quantum computers to function are included in the quantum computing architecture. The hardware, software stack, and cutting-edge concepts like quantum cloud computing are all included in this design. Let's examine these features in further detail:

5.1 Quantum Hardware Architecture

- **Quantum Processors:** At the core of the hardware architecture are quantum processors. They are made up of qubits (quantum bits) and the equipment required to manipulate them, such as cooling and control systems. Superconducting qubits, trapped ions, and photon-based qubits are only a few of the technologies used in quantum computers; each has benefits and drawbacks.
- **Control Electronics:** Control electronics are in charge of producing and directing the microwave or laser pulses required for qubit manipulation. These circuits guarantee accurate timing and synchronization of qubit operations.
- **Cryogenic Systems:** To sustain qubit coherence, many quantum computers need very low temperatures (near absolute zero). The design of quantum hardware includes dilution refrigerators and other cryogenic devices as essential elements.
- **Qubit Connectivity:** Quantum computers are often made up of a number of qubits organized in a certain architecture. Qubit connectivity, which controls which qubits may communicate with one another, is essential for effectively running quantum algorithms.

5.2 Quantum Software Stack

- **Quantum Software Frameworks:** The quantum software stack includes quantum software frameworks like Qiskit and Cirq. For the creation of algorithms, simulation, and quantum programming, these frameworks provide tools and libraries. Users may create quantum circuits, use quantum hardware to execute them, or mimic those using conventional computers (Pandey, R et al 2023).

- **Quantum Compilers:** Quantum compilers convert high-level quantum programs into low-level instructions that quantum processors can understand. They enhance compilation to reduce mistakes and boost the efficiency of quantum circuits.
- **Quantum Simulators:** Researchers may simulate the behavior of quantum systems on classical computers using quantum simulators, which are often included in software stacks. Simulators support algorithm design, problem fixing, and performance evaluation.

5.3 Quantum Cloud Computing

- **Quantum Cloud Services:** Quantum cloud computing is a new paradigm that allows cloud services to provide remote access to quantum computers. Leading providers of quantum technology, like IBM and Google, provide cloud-based quantum computing services that let customers execute quantum circuits on quantum hardware that is housed in data centers. For companies and researchers without direct access to quantum hardware, this accessibility is essential.
- **Hybrid quantum-cloud systems:** These systems integrate the capabilities of both conventional and quantum computing. They enable scalable and useful quantum applications by allowing users to offload certain quantum workloads to quantum processors while using conventional computing for other computational activities.

In conclusion, the quantum computing architecture comprises the quantum software stack, which consists of frameworks, compilers, and simulators, as well as the quantum hardware, which includes quantum processors and related components. By enabling remote access through cloud services, quantum cloud computing increases the accessibility of quantum computers. The development of quantum computing research and applications depends on this entire architecture.

6. APPLICATIONS OF QUANTUM COMPUTING

By providing extraordinary processing power and the capacity to tackle complicated issues that have long been a hurdle to conventional computers, quantum computing is set to transform a number of fields. Let's look at some of the areas where quantum computing is making considerable progress:

6.1 In Artificial Intelligence (AI), Quantum Computing

The potential for improving AI capabilities is enormous. Machine learning processes like training deep neural networks may be expedited by quantum algorithms by effectively navigating complicated parameter spaces. This may result in improvements to AI optimization issues, image recognition, and natural language processing (Acampora, 2019).

Figure 1. Applications of quantum computing

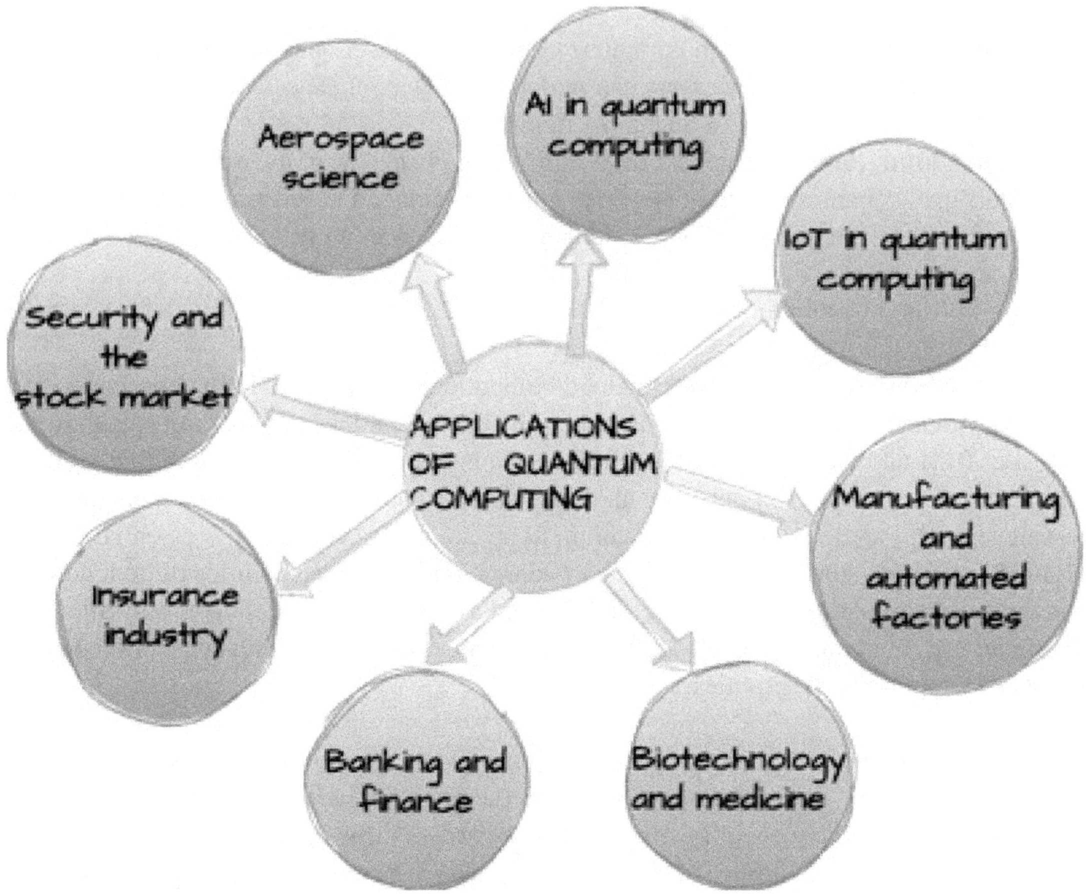

6.2 In the Internet of Things (IoT), Quantum Computing

Large networks of linked devices are essential to the IoT. In order to improve IoT infrastructure, quantum computing must solve problems with data processing, communication, and security. Quantum-resistant encryption techniques help safeguard IoT data from future dangers (Abd El-Aziz et al., 2022).

6.3 Quantum Computing Applications in Manufacturing and Automated Factories

Manufacturing may be revolutionized by quantum computing by streamlining production schedules, logistics for the supply chain, and quality assurance procedures. This improves industrial automation efficiency and lowers costs.

6.4 Quantum Computing in Biotechnology and Medicine

By modelling molecular interactions and drug binding, quantum computing may speed up the drug development process in the fields of medicine and biotechnology. Additionally, complex biological systems may be modeled using quantum simulations, assisting in the study of genetics and the creation of specialized treatments (Rasool et al., 2023).

6.5 Computational Chemistry and Quantum Computing

The behavior of molecules and materials at the quantum level can be accurately simulated by quantum computers. This capacity is crucial for computational chemistry because it makes it possible to simulate chemical processes precisely and forecast material characteristics and develop catalysts.

6.6 Quantum Computing in Banking and Finance

By enhancing portfolio optimization, fraud detection, and risk assessment, quantum computing has the potential to revolutionize the banking industry. More accurate investing judgments may be made as a result of quantum algorithms' superior speed at solving complicated financial equations compared to traditional computers.

6.7 Quantum Computing in the Insurance Industry

By enhancing actuarial calculations and risk modeling, quantum computing helps the insurance sector. In the end, quantum algorithms may help insurance firms save money by improving price accuracy and underwriting procedures.

6.8 Quantum Computing Applications in Security and the Stock Market

By offering quantum-resistant encryption methods that protect sensitive data from future quantum assaults, quantum computing may support cybersecurity. Quantum algorithms may enhance trading methods on the stock market, resulting in better financial results (Canabarro et al., 2022).

6.9 Quantum Computing Applications in Aerospace Science

Quantum simulations that precisely simulate aerodynamics, optimize flight routes, and improve spacecraft design are advantageous for aerospace research. This may result in the creation of spacecraft and airplanes that use less fuel.

6.10 Quantum Computing in the Army's National Defense

In order to better defend against quantum attacks, cryptographic techniques are strengthened by quantum computing. Additionally, it can improve resource allocation during defensive operations and simulate intricate military situations, increasing overall military effectiveness.

6.11 Quantum Computing Applications in Big Data and Analytics

Large-scale data processing and analysis are areas where quantum computing thrives. Data analytics has advanced due to the efficiency with which quantum algorithms can extract insights and patterns from massive data compared to their conventional equivalents (Egger at al., 2020).

6.12 Cloud Computing and Quantum Computing

Businesses may use quantum power without having to spend money on specialized hardware thanks to quantum cloud services that provide scalable access to quantum computers. This accessibility hastens research and development across a range of fields.

6.13 Cybersecurity and Quantum Computing

Cryptography that can withstand quantum assaults is essential to protecting digital communication against them in the future. In a world that is becoming more and more linked, quantum computing may be crucial in protecting the integrity and privacy of data (Faruk et al., 2022).

These uses provide a sneak preview of quantum computing's enormous potential in a variety of industries. These applications are anticipated to expand and transform sectors as quantum technology develops and quantum computers become more affordable, ushering in a new age of computing power.

7. CHALLENGES AND CONSTRAINTS OF QUANTUM COMPUTING

Although quantum computing is a promising topic with the potential to change many sectors, it also has a number of difficulties and restrictions. We'll look at the existing limitations and challenges facing quantum computing technology here:

Figure 2. Challenges of quantum computing

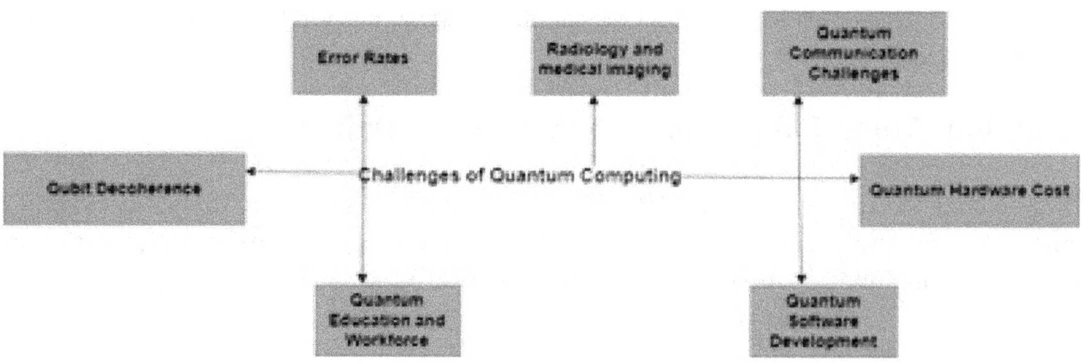

- **Qubit Decoherence:** Due to the great sensitivity of qubits to their surroundings, decoherence occurs. Qubits gradually lose their quantum characteristics, such as superposition and entanglement, due to decoherence. Long-term qubit coherence maintenance is a difficult task.
- **Error Rates:** Thermal noise and flaws in qubit gates are two reasons why quantum technology is prone to mistakes. The reliability and scalability of quantum calculations are constrained by high error rates. To overcome this restriction, error-correction procedures must be developed.
- **Limited Qubit connection:** Quantum computers often feature limited qubit connection, which prevents all qubits from being able to communicate with one another directly. This limits the use of certain quantum algorithms and necessitates the creation of methods for efficient qubit routing.
- **Qubit Quality and Fidelity:** A key difficulty is ensuring low error rates and high-quality qubits. To successfully carry out intricate and error-sensitive quantum calculations, qubit fidelity must be improved.
- **Scalability of quantum hardware:** It is difficult to create quantum computers that are more powerful and bigger. It is a difficult undertaking to increase the amount of qubits while keeping low error rates. Scalable quantum hardware is required to support practical applications.
- **Quantum Hardware Variability:** Qubit characteristics may vary, even within the same processor, which makes it difficult to provide stable and predictable quantum behavior. For reliable quantum computing, this variability must be eliminated.
- **Quantum error correction:** Quantum computing faults must be minimized by using efficient quantum error correction codes. Error-correcting codes, on the other hand, may considerably increase the quantity of qubits needed, making massive quantum calculations more difficult.

The absence of "killer applications" Since quantum computing is still in its infancy, finding real-world, useful applications that can fully utilize quantum capabilities is still difficult. Many quantum algorithms are theoretical and have not yet been shown to be useful in real-world applications.

- **Quantum Hardware Cost:** The cost of developing and maintaining quantum hardware might be prohibitive. For widespread deployment, the price of quantum computers and related infrastructure must be brought down.
- **Quantum Communication Challenges**: Quantum key distribution, for example, has difficulties due to distance restrictions and the need for quantum repeaters to increase the range of safe quantum communication.
- **Quantum Software Development:** Developing quantum software demands specific skills and equipment. It is still difficult to connect real-world software applications with quantum algorithms.
- **Quantum Education and Workforce:** It's crucial to have a knowledgeable staff with experience in quantum computing. The expansion of the sector depends on creating educational programs and training possibilities for quantum scientists, engineers, and developers.

Despite these difficulties and restrictions, quantum computing is still developing quickly. It is the goal of ongoing research and development to remove these barriers and realize the full potential of quantum computing, opening the door for game-changing applications in industries like materials science, cryptography, and optimization.

8. FUTURE TRENDS AND DIRECTIONS FOR QUANTUM COMPUTING

The future of quantum computing is full with fascinating innovations and trends that promise to revolutionize science and technology. Here is a detailed analysis of the newest developments and trends in quantum computing:

- **Advances in Quantum Hardware:** Quantum hardware will continue to develop, with a focus on raising the number of qubits, enhancing their quality, and reducing error rates. Trapped ions, superconducting qubits, and other qubit technologies will coexist and compete while each providing distinct benefits.
- **Quantum Software Ecosystem:** As quantum software frameworks like Qiskit, Cirq, and others emerge, researchers and developers will be able to more quickly and simply design quantum algorithms and applications. Tools and libraries for developing quantum software will become more comprehensible and available.
- **Quantum Cloud Services:** With the expansion of quantum cloud services, developers, enterprises, and academics will be able to access quantum computers from a distance. Innovation and cooperation will be stimulated by this democratization of quantum computing.

Figure 3. Future trends and directions for quantum computing

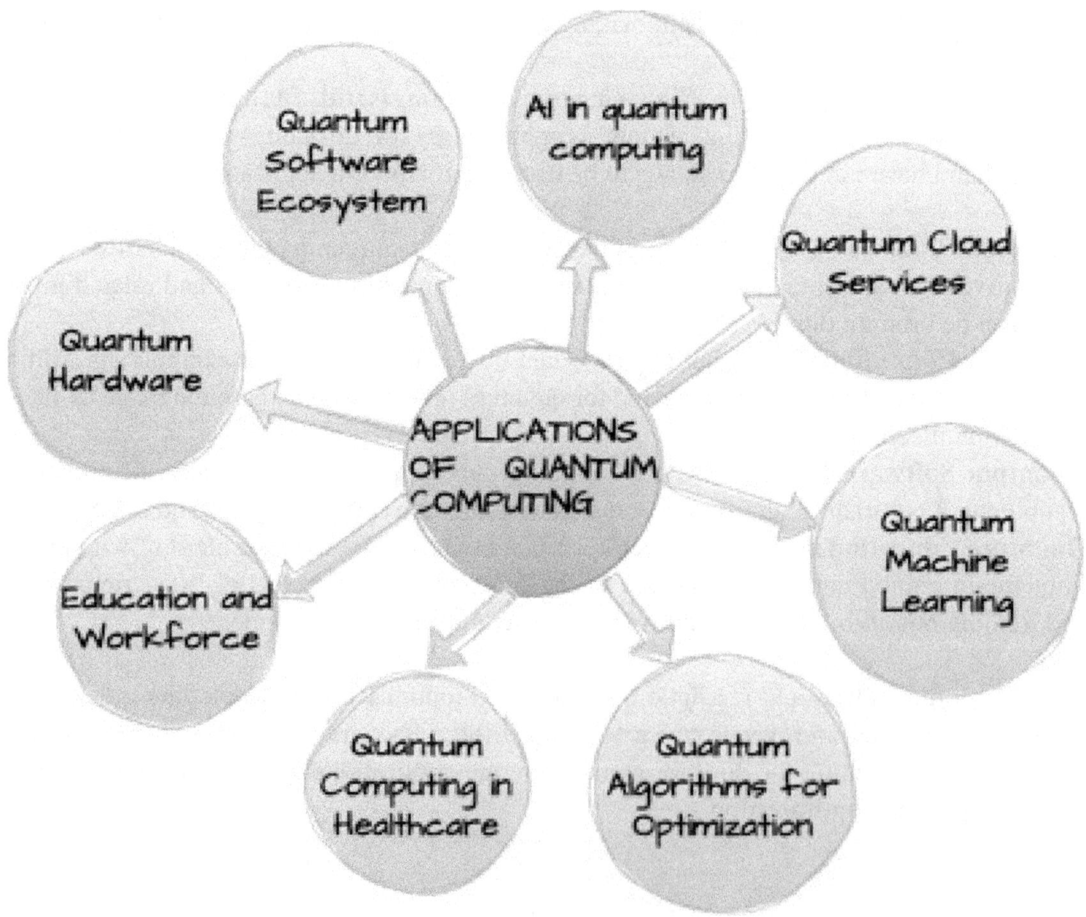

- **Dimension and Beyond:** Quantum Supremacy The quest for quantum supremacy, which occurs when quantum systems outperform classical ones in certain tasks, will go on. Beyond dominance, quantum computing will open up new avenues for quantum simulations, optimization, and cryptography.
- **Quantum-resistant encryption:** Quantum-resistant cryptography methods will become more important as quantum computers develop. The goal of research and development will be to protect digital communication against future quantum assaults.
- **Quantum Simulation:** Quantum computers will be used to model intricate quantum materials, systems, and chemical processes. This will have significant effects on drug development, quantum chemistry, and materials research.
- **Quantum Machine Learning:** Quantum machine learning algorithms will become more popular, providing deep neural network training that is quicker and more effective optimization issues. Applications for quantum-enhanced AI will develop in many fields.
- **Quantum Internet and Communication:** Through the development of quantum key distribution (QKD) networks, secure long-distance communication will be made possible. These networks will find use in delicate industries like banking and government.
- **Quantum Sensors**: Quantum sensors, such as gravimeters and magnetometers, will proliferate. They will improve accuracy in a variety of sectors, including geology, environmental monitoring, and navigation.
- **Quantum Algorithms for Optimization:** Quantum algorithms developed for complicated optimization problems will find extensive use in finance, supply chain management, portfolio optimization, and logistics.
- **Quantum Computing in Materials Discovery:** Quantum computing will quicken the process of discovering new materials by modelling and forecasting their characteristics. This will result in the creation of cutting-edge materials for a variety of applications.
- **The Use of Quantum Computing in Healthcare:** Personalized medicine, genetics research, and drug development will all benefit greatly from quantum computing. It will make it possible for simulations and analyses to be more precise, eventually improving patient outcomes.
- **Quantum Computing Education and Workforce:** Educational endeavors to prepare a workforce for quantum computing will develop. The future generation of quantum scientists, engineers, and developers will be developed via the programs and resources provided by universities and organizations (Khang & Muthmainnah et al., 2023).
- **Interdisciplinary Research:** Quantum computing will encourage more interaction between researchers from many fields. Innovative applications and solutions will result from this idea-cross-pollination across a variety of industries.
- **Ethical and Societal Considerations:** Quantum technologies' ethical, legal, and societal repercussions will be highlighted. As quantum technologies develop, concerns about data privacy, security, and ethical usage will be addressed.

Quantum computing has the potential to completely change how technology is used in the future. Quantum computing has the potential to transform businesses, solve difficult issues, and spur creativity in a variety of fields as it continues to improve. Quantum technologies will have a major influence on science, business, and society as they develop and get beyond their present obstacles (Khang & Kali et al., 2023).

9. CONCLUSION

A meticulous examination of the complicated area of quantum computing emerged in this comprehensive review, showing its basic foundations, new technologies, and far-reaching possibilities. The study began with the revelation of the fundamental principles of quantum computing. These fundamental principles, which include quantum bits (qubits), quantum gates, superposition, and entanglement, are at the heart of quantum computation and hold the key to breaking free from traditional computer constraints.

The exploration descended even further into the core of quantum computing principles, illuminating the critical roles of quantum physics, quantum parallelism, and quantum error correction. These concepts are critical for realizing quantum computing's full potential and overcoming the long-standing issues of qubit decoherence and error rates that throw doubt on its practicality. Further continued our investigation by traversing the growing quantum technology environment, where we saw continuous advances in quantum hardware. This involves the race to increase qubit counts, improve qubit quality, and reduce error rates.

From superconducting qubits to trapped ions, several qubit technologies cooperate and compete to shape the future of quantum computing. Beyond hardware, our path took us into the developing quantum software environment, where we were able to simplify the once-complex chore of quantum programming. Quantum software frameworks like Qiskit and Cirq are fast evolving, allowing developers and researchers to more easily explore quantum algorithms and applications. In addition, the democratization of quantum computing via quantum cloud services has opened up new possibilities. These services broaden access to quantum computing, boosting innovation and worldwide cooperation.

The future of quantum computing, as shown during our journey, is full of promise and trends. The pursuit of quantum supremacy, in which quantum computers beat conventional equivalents in specified tasks, remains a lofty goal. The transformational promise of quantum computing extends to a wide range of sectors, including banking, healthcare, materials research, and artificial intelligence. Quantum-resistant encryption is developing as a defender of digital communication security against prospective quantum assaults. As quantum computing overcomes conventional boundaries, sparking cross-disciplinary cooperation, interdisciplinary research is booming (Khang, Vugar and Abuzarova et al., 2023).

Materials discovery, healthcare, and education are all positioned to profit significantly from the quantum revolution, sparking innovation and research. To summarize, the exploration into the quantum domain has revealed a technology that constitutes a revolution in computing rather than an evolution. Despite severe difficulties like as qubit decoherence and error rates, continuous research and development efforts are paving the route toward realizing the transformational promise of quantum computing (Khang & Medicine, 2023).

Quantum computing is a seismic change set to transform the future of technology and science, not merely a boost in processing capability. Its far-reaching implications for industry, scientific research, and society are set to be nothing short of revolutionary. The quantum revolution has begun, and its potential to transform our world is both thrilling and terrifying (Khang & Shah et al., 2023).

REFERENCES

Abd El-Aziz, R. M., Taloba, A. I., & Alghamdi, F. A. (2022). Quantum computing optimization technique for IoT platform using modified deep residual approach. *Alexandria Engineering Journal*, *61*(12), 12497–12509. https://www.sciencedirect.com/science/article/pii/S1110016822004094. doi:10.1016/j.aej.2022.06.029

Acampora, G. (2019). Quantum machine intelligence: Launching the first journal in the area of quantum artificial intelligence. *Quantum Machine Intelligence*, *1*(1-2), 1–3. https://link.springer.com/article/10.1007/s42484-019-00006-5. doi:10.100742484-019-00006-5

Ahnefeld, F., Theurer, T., Egloff, D., Matera, J. M., & Plenio, M. B. (2022). *On the Role of Coherence in Shor's Algorithm.* arXiv preprint arXiv:2203.10632.

Canabarro, A., Mendonc, T. M., Nery, R., Moreno, G., Albino, A. S., de Jesus, G. F., & Chaves, R. (2022). *Quantum Finance: a tutorial on quantum computing applied to the financial market.* arXiv. org Papers, (2208.04382). https://ideas.repec.org/p/arx/papers/2208.04382.html

Egger, D. J., Gutiérrez, R. G., Mestre, J. C., & Woerner, S. (2020). Credit risk analysis using quantum computers. *IEEE Transactions on Computers*, *70*(12), 2136–2145. https://ieeexplore.ieee.org/abstract/document/9259208/. doi:10.1109/TC.2020.3038063

Eswaran, U., Khang, A., & Eswaran, V. (2023). Role of Quantum Computing in The Era of Artificial Intelligence. Applications and Principles of Quantum Computing. IGI Global Press. https://doi.org/doi:10.4018/979-8-3693-1168-4-ch003

Faruk, M. J. H., Tahora, S., Tasnim, M., Shahriar, H., & Sakib, N. (2022). A review of quantum cybersecurity: threats, risks and opportunities. In *2022 1st International Conference on AI in Cybersecurity (ICAIC)* (pp. 1-8). IEEE. https://ieeexplore.ieee.org/abstract/document/9896970/

Gill, S. S., Kumar, A., Singh, H., Singh, M., Kaur, K., Usman, M., & Buyya, R. (2022). Quantum computing: A taxonomy, systematic review and future directions. *Software, Practice & Experience*, *52*(1), 66–114. https://onlinelibrary.wiley.com/doi/abs/10.1002/spe.3039. doi:10.1002pe.3039

Gyongyosi, L., & Imre, S. (2019). A survey on quantum computing technology. *Computer Science Review*, *31*, 51–71. https://www.sciencedirect.com/science/article/pii/S1574013718301709. doi:10.1016/j.cosrev.2018.11.002

Herman, D., Shaydulin, R., Sun, Y., Chakrabarti, S., Hu, S., Minssen, P., & Pistoia, M. (2022). Portfolio optimization via quantum zeno dynamics on a quantum processor. arXiv preprint arXiv:2209.15024.

Khang, A. (2023). AI and IoT-Based Technologies for Precision Medicine. IGI Global Press. doi:10.4018/979-8-3693-0876-9

Khang, A. (2023). AI-Aided Data Analytics Tools and Applications for the Healthcare Sector. AI and IoT-Based Technologies for Precision Medicine. IGI Global Press. doi:10.4018/979-8-3693-0876-9.ch018

Khang, A. (2023). Enabling the Future of Manufacturing: Integration of Robotics and IoT to Smart Factory Infrastructure in Industry 4.0. AI-Based Technologies and Applications in the Era of the Metaverse. IGI Global Press. doi:10.4018/978-1-6684-8851-5.ch002

Khang, A. (2023). AI and IoT-Based Technologies for Precision Medicine. IGI Global Press. doi:10.4018/979-8-3693-0876-9

Khang, A. (2023). AI-Aided Data Analytics Tools and Applications for the Healthcare Sector. AI and IoT-Based Technologies for Precision Medicine. IGI Global Press. doi:10.4018/979-8-3693-0876-9.ch018

Khang, A. (2023). Applications and Principles of Quantum Computing. IGI Global Press. https://doi.org/ doi:10.4018/979-8-3693-1168-4

Khang, A. (2023). Quantum Mechanics Primer - Fundamentals and Quantum Computing. Applications and Principles of Quantum Computing. IGI Global Press. https://doi.org/ doi:10.4018/979-8-3693-1168-4-ch001

Khang, A. (2023). Introduction to Quantum Computing and Its Integration Applications. Applications and Principles of Quantum Computing. IGI Global Press. https://doi.org/ doi:10.4018/979-8-3693-1168-4-ch002

Khang A. (2024). The Quantum Evolution: Application of AI and Robotics in the Future of Quantum Technology. CRC Press. https://doi.org/ doi:10.1201/978-1-032-64207-9

Khang, A., & Muthmainnah, M. (2023). AI-Aided Teaching Model for the Education 5.0 Ecosystem. AI-Based Technologies and Applications in the Era of the Metaverse. IGI Global Press. doi:10.4018/978-1-6684-8851-5.ch004

Khang, A., Shah, V., & Rani, S. (2023). AI-Based Technologies and Applications in the Era of the Metaverse. IGI Global Press. doi:10.4018/978-1-6684-8851-5

Pandey, R., Maurya, P., Singh, G. D., & Faiyaz, M. S. (2023). *Simulating Quantum Principles: Qiskit Versus Cirq. In Quantum Computing: A Shift from Bits to Qubits* (pp. 333-348). Springer Nature Singapore. https://link.springer.com/chapter/10.1007/978-981-19-9530-9_18

Prabu, B., & Khang, A. (2023). Quantum Computing Smart City: Fostering Sustainability and Green Infrastructure with Smart City. Applications and Principles of Quantum Computing. IGI Global Press. https://doi.org/ doi:10.4018/979-8-3693-1168-4-ch009

Ramezani, S. B., Sommers, A., Manchukonda, H. K., Rahimi, S., & Amirlatifi, A. (2020, July). Machine learning algorithms in quantum computing: A survey. In *2020 International joint conference on neural networks (IJCNN)* (pp. 1-8). IEEE. https://ieeexplore.ieee.org/abstract/document/9207714/

Rietsche, R., Dremel, C., Bosch, S., Steinacker, L., Meckel, M., & Leimeister, J. M. (2022). Quantum computing. *Electronic Markets, 32*(4), 2525-36. https://link.springer.com/article/10.1007/s12525-022-00570-y

Shah, V., & Khang, A. (2023). Metaverse-Enabling IoT Technology for a Futuristic Healthcare System. AI-Based Technologies and Applications in the Era of the Metaverse. IGI Global Press. doi:10.4018/978-1-6684-8851-5.ch008

Shannon, K., Towe, E., & Tonguz, O. K. (2020). *On the use of quantum entanglement in secure communications: a survey.* arXiv preprint arXiv:2003.07907.

Ugwuishiwu, C. H., Orji, U. E., Ugwu, C. I., & Asogwa, C. N. (2020). An overview of quantum cryptography and shor's algorithm. *Int. J. Adv. Trends Comput. Sci. Eng, 9*(5). https://www.academia.edu/download/84737888/ijatcse82952020.pdf

Ur Rasool, R., Ahmad, H. F., Rafique, W., Qayyum, A., Qadir, J., & Anwar, Z. (2023). Quantum computing for healthcare: A review. *Future Internet, 15*(3), 94. https://www.mdpi.com/1999-5903/15/3/94. doi:10.3390/fi15030094

Chapter 7
Quantum Computing AI:
Application of Artificial Intelligence in the Era of Quantum Computing

Ankita Nayak
Transtrack Aeroservices Pvt. Ltd., Bhubaneswar, India

Atmika Patnaik
University of Cambridge, UK

Ipseeta Satpathy
Kalinga Institute of Industrial Technology, India

Alex Khang
https://orcid.org/0000-0001-8379-4659
Global Research Institute of Technology and Engineering, USA

B. C. M. Patnaik
KIIT School of Management, KIIT University, India

ABSTRACT

In the ever-changing technological environment, the combination of artificial intelligence (AI) with quantum computing represents an enormous potential frontier. This collaboration has the potential to transform industry, scientific research, and problem-solving approaches. At its heart, AI's function in the quantum computing age includes several transformational components. AI has infiltrated many facets of our everyday lives, from smartphone technology to autonomous car features and novel shopping experiences. Its seamless integration has recently become a major fixture, especially with the emergence of generative AI. Surprisingly, generative AI is set to take quantum computing beyond mere dreams and into the realm of quantum supremacy, where quantum computers vastly outperform traditional approaches. Combining AI and quantum computing might expedite and refine the latter's already impressive skills in addressing complex issues. Such a collaboration might elevate AI from outstanding to game changing.

DOI: 10.4018/979-8-3693-1168-4.ch007

1. INTRODUCTION

Quantum computing is defined as computing based on quantum physics. Data is traditionally represented as bits that can be either 1 or 0. Because of the feature of superposition, qubits in quantum computing can be both 1 and 0. It is an area of research that performs computer tasks using quantum mechanics ideas. It makes use of quantum bits, or qubits, which may be in several states at the same time, as opposed to classical bits, which can only be in a 0 or 1 state. Because of its capacity to deal with several states simultaneously, quantum computers have the potential to be far more powerful than conventional computers (Shyam & Khang et al., 2023).

Various approaches, such as Ion Trap and Quantum Dot techniques, have been used to show the notion of quantum computing. When you read the word "quantum," you may envision an atom or molecule. Quantum computers operate on the same principles. Processing takes place at the bit level in a conventional computer. Quantum computers are governed by a certain behavior known as quantum physics. There are several ways in quantum physics for characterizing the interaction between different atoms. These atoms are referred to as "qubits" in the context of Quantum Computers. (Chatterjee & Chakraborty, 2020). Quantum computing evolved from the notion of quantum mechanics, which was created by physicists such as Max Planck and Albert Einstein in the early twentieth century.

In 1982, scientist Richard Feynman advocated employing quantum mechanics for computation, claiming that quantum computers might handle issues that were difficult or impossible for conventional computers. Peter Shor, a mathematician, created a quantum method in 1994 that could effectively factor enormous numbers, which has important implications for encryption. This finding heightened interest in quantum computing. Experimental demonstrations of fundamental quantum computing processes were obtained in the late 1990s and early 2000s utilizing diverse approaches such as ion traps and quantum dots. These investigations lay the groundwork for the practical application of quantum computers. Since then, advances in qubit technology, error correction methods, and quantum algorithms have fueled continued research and development in quantum computing (Hidary & Hidary, 2020).

Quantum computing offers the ability to tackle complicated problems that traditional computers are incapable of solving, notably in the realm of quantum mechanics. Quantum computers may do certain computations significantly quicker than traditional computers, which has important ramifications for a variety of sectors like cryptography, optimization, and drug development. It can improve machine learning algorithms by more effectively processing and analyzing massive volumes of data, resulting in greater pattern detection and optimization. These are capable of simulating quantum systems, allowing scientists to examine and comprehend events at the atomic and molecular levels, which has applications in materials science, chemistry, and physics.

By tackling complicated optimization problems more effectively, quantum computing can help to breakthroughs in disciplines such as weather forecasting, financial modeling, and logistics optimization (Gill et al., 2022). The quantum computing field is quickly developing and offers several opportunities in virtually every subject, but the most essential aspect is its capacity to aid farmers in enhancing agricultural efficiency and productivity (Khang & Agriculture et al., 2023). A quantum computing platform might also aid in the development of chemicals used in the production of energy-efficient fertilizers. These considerations have contributed to the increasing use of quantum computing technology in agriculture (Khang & Santosh et al., 2023).

Furthermore, governments all across the world are aiding and funding quantum computing specialists. (Leon et al., 2021). According to earlier research, the worldwide quantum computer industry was

Figure 1. Market size of quantum computing 2021 to 2030
Source - https://www.precedenceresearch.com/quantum-computing-market
Alt-text: Figure 1 depicts statistic of Market Size of Quantum Computing, 2021 to 2030 (USD Billion).

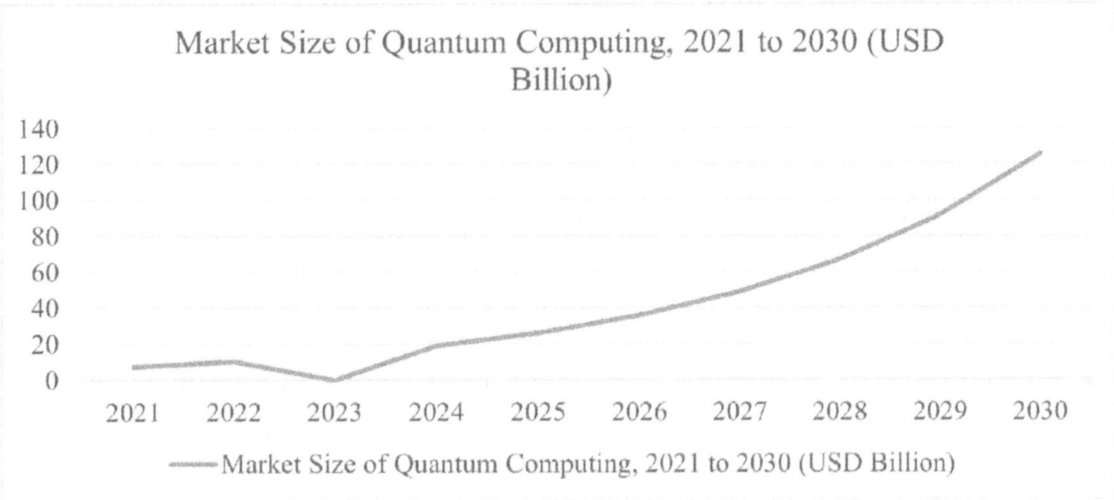

valued at US$10.13 billion in 2022 and is expected to reach US$125 billion by 2030, with a projected CAGR of 36.89% from 2022 to 2030 as shown in Figure 1.

Quantum computing is a complicated mathematics that is based on both quantum mechanics and quantum theory. Notwithstanding the fact that it just employs zeros and one bit, it uses every state among zero and one, resulting in incredible speed. Sales are likely to rise as the technologies utilized in quantum parallelism, algorithms, encoding, and quantum simulation improve.

The integration of AI with quantum computing has the potential to significantly boost computational power and speed in processing complex data and datasets. The combination of AI with quantum computing is designed to address pressing difficulties that standard computers and algorithms are incapable of addressing. AI systems can employ quantum computers to safeguard sensitive data. Furthermore, parallel computing can be used to combat cybercrime. Unlike ordinary computers, which are limited to functioning in one mode at a time, quantum computers are able to exist in several states at the same time, allowing them to create better algorithms (Rietsche et al., 2022).

AI offers a unique chance to create efficient numerical approaches for solving quantum many-body issues, which are critical in condensed matter physics. AI in quantum computing can help allow us to comprehend the mechanics of new physics like high Tc superconductivity by seeking optimal solutions in the ever-expanding Hilbert space. AI may improve the precision of solutions in quantum many-body simulations by increasing the number of variational parameters, resulting in exceptional accuracy and time-to-solution (Zhao et al., 2022).

Reinforcement neural networks and evolutionary algorithms are two AI technologies that may be used to optimize and improve the performance of quantum algorithms. It can help with the creation of error correction codes and methods for reducing the impact of noise and errors in quantum systems. AI algorithms may be used to rebuild a system's quantum state from measurement data, allowing for improved characterization and understanding of quantum systems. It can also help optimize control settings for quantum systems, resulting in more efficient and precise quantum state manipulation.

AI methods may be used to extract significant information and patterns from quantum data, allowing the creation of quantum machine learning algorithms (Liang et al., 2023). AI plays a critical role in quantum computing for trust-based applications by improving system performance and dependability algorithms can be utilized to improve the efficiency and accuracy of quantum algorithms in trust-based AI applications. Machine learning and other AI approaches may be applied to quantum computing to analyze and understand complicated data, allowing for better decision-making in trust-based AI systems. It can help in the development of quantum algorithms for trust-based AI applications, allowing for quicker and more accurate calculations that are beyond the capability of traditional computing.

AI can also aid in the design and optimization of quantum circuits, hence increasing the overall efficacy of quantum computing systems for trust-based applications (Kaur et al., 2023). Quantum computing, on the other hand, is frequently heralded as the next great thing in artificial intelligence. Quantum computers can process a large number of possibilities at the same time. This has the potential to accelerate AI algorithms and allow them to analyze more datasets more effectively, resulting in more powerful AI models. According to a recent Boston Consulting Group report, quantum prospects in generative, foundational, and horizontal AI have a market potential of $50 billion to $100 billion, influencing virtually all sectors. According to BCG, there are other multibillion-dollar potentials in fraud and money laundering prevention, as well as automotive AI algorithms (Boger, 2023).

The befitting example to prove the same are Google AI is utilizing artificial intelligence to create new quantum algorithms for machine learning and materials research. IBM Research is utilizing artificial intelligence to create new error-correcting algorithms for quantum computers. Microsoft Quantum uses artificial intelligence to mimic quantum computers and create novel quantum algorithms for cryptography and drug discovery.

Figure 2. Various applications of AI and quantum computing
Source – Compiled by the Authors
Alt-text: Figure 2 depicts components of Solve Complex Issues in a Short Time, Management of Large Data Sets, and Detection and Repression of Fraud.

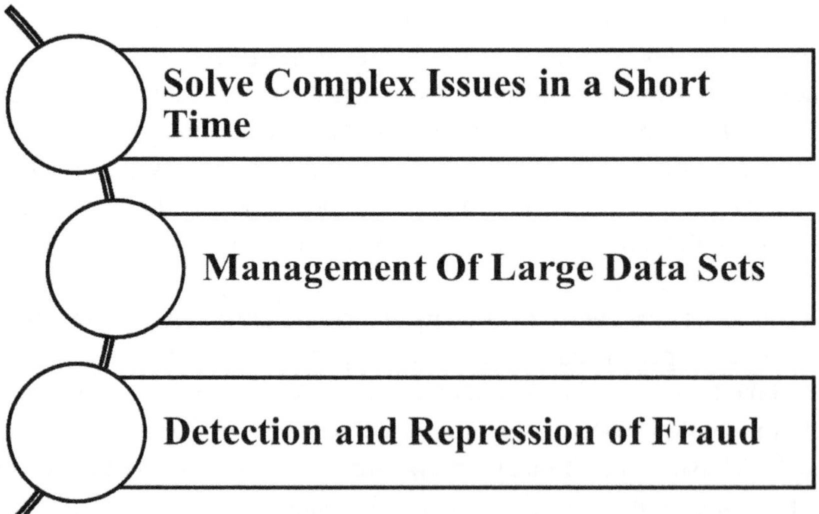

A quantum computer outperforms a regular computer or supercomputer by orders of magnitude. Sycamore, Google's quantum computer, which is still in development, is believed to have performed a computation in 200 seconds, contrasted to the 10,000 years it would take on of the world's fastest computers. The various applications of AI and Quantum Computing are shown in Figure 2 and described below (Kanwar, 2023).

- **Solve Complex Issues in a Short Time**: Data sets are getting increasingly complex and big, beyond the capabilities of our present computers, placing tremendous strain on our computing infrastructure. Today's computers are incapable of addressing complicated issues that quantum computing, which is projected to solve similar problems in seconds, can. Computations that would normally take thousands of years may now be completed in 200 seconds according to Google's claim of Quantum Supremacy (the ability of a quantum computer), which it claimed in 2019 (a claim that IBM disputes).
- **Management of Large Data Sets**: We create around 2.5 exabytes of data every day. Ordinary CPUs and GPUs cannot manage such vast amounts of data, but quantum computers are built to swiftly find patterns and abnormalities in such massive amounts of data.
- **Detection and Repression of Fraud**: Fraud detection will be strengthened and strengthened when quantum computing and artificial intelligence are used in the banking and finance industries. Models trained on quantum computers would be able to recognize patterns that would be difficult to identify with standard equipment, as well as manage the vast amounts of data that these machines could handle. Algorithm advancements would also help in reaching this aim.

AI has an opportunity to play a major role in the development and deployment of quantum computing. AI can help make quantum computers more efficient, accurate, and dependable by making them more efficient, accurate, and consistent.

2. ARTIFICIAL INTELLIGENCE FOR QUANTUM DATA PROCESSING

The use of quantum computers to perform computational tasks such as optimisation and ML at potentially quicker rates than standard approaches is known as quantum data analysis. Quantum data processing is critical in quantum computers, allowing them to do complicated computations and solve problems that conventional computers cannot. It entails manipulating and processing quantum information, which is represented by quantum bits or qubits, using quantum computers. At its heart, quantum data processing deals with massive amounts of data that is quantum, which, due to the quantum nature of combination and entanglement, can be rapidly complex AI algorithms, particularly the ML and deep learning models that generate them.

Quantum computers use quantum physics properties like as superposition and entanglement to execute calculations on massive volumes of data at the same time, resulting in exponential speedup in certain jobs. Quantum data processing enables the construction of quantum algorithms, which are more effective than conventional algorithms in solving issues in cryptography, optimization, and simulation. Quantum computers have an opportunity to revolutionize a wide number of fields by using quantum information-processing capabilities (Rose et al., 2019). Because of quantum algorithms that use superposition and entanglement, quantum computers can solve certain sorts of problems quicker than classical digital computers.

Quantum data processing enables the manipulation and storage of information in quantum states, resulting in more efficient and powerful calculations. Quantum algorithms, such as the Shor algorithm, use superposition and entanglement to more effectively handle issues like prime factorization and encryption. Quantum data processing allows for the investigation of quantum states as well as the use of quantum features like interference and coherence to accomplish complicated calculations. Quantum computers have the potential to revolutionize industries such as encryption, optimization, and simulation by exploiting the capabilities of quantum data processing (Balynsky et al., 2021). AI aids in the processing of quantum data by enabling the development of transdisciplinary technologies such as quantum neural networks.

Quantum reservoir processing, a type of artificial neural network, is a quantum information processing platform that can do both qualitative and quantitative tasks such as recognizing entangled quantum states and estimating nonlinear functions of input quantum states. On a trained quantum reservoir processor, this lowers experimental approaches that need numerous observable observations to only one observable. A hybrid quantum-classical neural network model combining a variational quantum circuit into a deep neural network has also been described for biosignal processing applications, offering state-of-the-art performance with a minimum number of trainable parameters (Tacchino et al., 2021). This has the potential to significantly improve, because quantum computers are used to handle complicated information,

AI is vital in quantum data processing. Natural language processing (NLP) and semantic networking are complex/complex NP challenges that typical computer methods cannot tackle. By inventing quantum algorithms, quantum computing has the ability to overcome these complicated issues. For discrete optimisation issues, D-Wave commercial quantum annealing technologies can reach a global optimum. A unique technique for converting symbolic AI issues in NLP to quadratic unconstrained binary optimisation (QUBO) has been created, which can partition text documents as QUBO function optimizations and then apply D-Wave quantum annealer to energy output.

The lowest point in the data is to share fresh results through statistical analysis (Ayanzadeh, 2018). AI aids the processing of quantum data by enabling the development of cross-disciplinary technologies such as quantum neural networks. Quantum reservoir processing (QRP) is a quantum information processing platform that is built on the reservoir computing idea, which is similar to an artificial neural network. QRP is capable of recognizing entangled quantum states and estimating nonlinear properties of input quantum states.

On a trained quantum reservoir processor, QRP reduces multiple observable observations in experimental procedures to a single observable measurement. QRP may be used as programmable quantum hardware or as a quantum machine learning software architecture that maximizes its output weight matrix employing a training dataset to best suit available training data for a particular purpose. It has been proven to recognize non-Gaussian entangled states effectively, implying that it can recognize entanglement patterns from Gaussian input states. Entanglement classification with QRP needs quantum coherence across reservoir nodes, showing that it processes quantum information rather than binary information (Ghosh et al., 2019).

AI aids in quantum data processing by using a quantum processor to enable unsupervised or semi-supervised feature learning. The quantum processor is programmed to perform Hierarchical Deep Learning (HDL) on one or more data sets. Searching for, analyzing, and finding maximum recurring patterns inside or across data sets. Detecting maximum repeated patterns in or across data using sparse coding, such as L0 and L1 sparse coding. Labels are appended, incorporated, or attached to dictionary items or constituent parts of one or more dictionaries. Creating a logical link between labels and labeled items to aid in the process of unsupervised or semi-supervised feature learning (Hamze et al., 2017; Rose et al., 2017).

By aiding in the creation of new and more efficient quantum algorithms, AI has the potential to revolutionize the field of quantum computing. One example is optimizing the layout of qubits in a quantum circuit, which can result in quicker and more accurate results. Furthermore, AI may be utilized to create novel error correction strategies for quantum computers, which are susceptible to faults. The accuracy and reliability of quantum data processing may be enhanced by identifying and fixing these defects in real-time.

Artificial intelligence (AI) may also be used to mimic quantum computers, allowing researchers to test novel algorithms and error correction approaches before implementing them on genuine quantum hardware (Henderson et al., 2021). Using a quantum processor, AI provides unsupervised or semi-supervised feature learning, allowing for quicker data processing and analysis algorithms that may be built into quantum computers to perform Hierarchical Deep Learning (HDL) across one or more data sets, increasing quantum data processing capability (Rose et al., 2019). As quantum bits or qubits are intrinsically error-taken from the environment, the major problem in quantum data processing is quantum error correction.

AI-driven algorithms are error-correcting rules, and error-tolerant quanta -, as a result, using gates can assist in discovering and reducing mistakes, boosting the dependability of quantum processing. In addition, quantum data analytics can analyze vast volumes of data quicker than traditional methods. Algorithms for quantum machine learning, such as quantum support vector machines and quantum neural networks, leverage AI approaches to expedite data processing processes, making them particularly valuable for drug discovery, optimization challenges, and cryptography (Dukatz et al., 2019). AI can also help to hasten the development of quantum data applications in general.

AI, for example, may be used to develop and test quantum devices automatically. This can aid researchers in developing novel quantum algorithms and applications more quickly. The integration of AI with quantum data processing leverages quantum computing's computational capacity while combining AI's ability to handle, analyze, and optimize quantum data. This parallel relationship has considerable potential since it will transition from materials science to cryptography and then to quantum technology, information processing that is powered has the ability to motivate.

3. THE IMPACT OF ARTIFICIAL INTELLIGENCE (AI) FOR QUANTUM SAFE CRYPTOGRAPHY

Quantum-secure cryptography, often known as post-quantum cryptography, refers to cryptographic techniques and algorithms designed to protect against quantum computer assaults. Quantum computers may execute calculations much quicker than conventional computers such as RSA and ECC (Elliptic). (Curve Cryptography) It has the potential to compromise conventional encryption approaches. Future quantum computers may render common encryption techniques insecure, posing a severe danger to network security. Quantum-safe encryption techniques seek to address this issue by creating algorithms that are immune to quantum computer crimes.

Quantum key distribution (QKD) protocols are one of the quantum-safe techniques that allow secure key exchange using quantum mechanics principles. Another quantum-safe option is post-quantum cryptography (PQC) algorithms, which attempt to offer secure encryption and digital signatures even in the presence of quantum computers. The National Institute of Standards and Technology (NIST) has revealed the examination and comparison of four post-quantum cryptography algorithms (Xu et al., 2023).

Quantum-safe Cryptography reconstructs the cryptographic vault, making it resistant to quantum and conventional assaults. Cryptography has two primary applications: encryption and authentication. Encryption shields data from prying eyes, while authentication stops malicious actors from impersonating others. These methods are intended to keep sensitive information and secure communication networks safe from quantum computing advances. The study of quantum secure cryptography entails the standardization of asymmetric encryption systems that are difficult for quantum computers to break. (Strand, 2021).

Quantum computers are significantly quicker than classical computers in performing complicated computations, which implies they can readily crack encryption systems that are now deemed safe. To solve this issue, researchers are creating quantum cryptography, which provides a theoretically safe solution to standard encryption's flaws in the face of quantum computing.

Quantum cryptography employs quantum mechanics principles such as qubit superposition and entanglement to enable secure communication and data encryption. Quantum cryptography can allow safe key distribution and protect against eavesdropping attacks by using the unique aspects of quantum physics, making it important for sustaining security in the era of quantum computing (Mohamed et al., 2022).

In a quantum computing system, quantum cryptography provides a safe technique for sending information. It guarantees that private information is neither intercepted nor interfered with by unauthorized persons. Quantum computing is based on concepts of quantum physics, such as the usage of quantum bits, or qubits. These qubits may exist in several states at the same time, allowing for parallel processing and the potential to solve complex problems faster than conventional computer systems. This, however, exposes quantum computers to certain sorts of assaults, such as quantum attacks on classical encryption methods.

Quantum cryptography addresses this weakness by offering quantum-safe cryptographic algorithms that are immune to quantum computer assaults. These algorithms are built to resist the processing power of quantum computers while still ensuring the confidentiality of sensitive data. Researchers can improve the security of quantum computing systems and guard against possible quantum assaults by implementing quantum-safe lightweight cryptography, such as AES-QPP, into these systems (Khang and Vladimir, 2022).

AES-QPP combines the features of the quantum permutation pad (QPP) with the AES algorithm to provide a high degree of security while lowering the number of encryption rounds necessary (Kuang et al., 2021) as shown in Figure 3.

To protect sensitive data, AI systems can use quantum computers. Parallel computing may also be utilized to tackle cybercrime. Unlike conventional computers, which are able to reside in one state at a time, quantum computers can exist in several states at once, allowing them to create superior algorithms. AI can optimize the distribution of cryptographic keys in quantum cryptography systems, guaranteeing efficient and secure communication between operators. It can analyze system vulnerabilities and discover possible flaws in quantum cryptography protocols, enabling for proactive security solutions to be implemented (Tosh et al., 2020).

AI may be used to design and refine quantum cryptography encryption algorithms, making them less susceptible to assaults and maintaining the security of sensitive data. It may be used in tandem with quantum computing to improve the speed and effectiveness of quantum cryptography systems, allowing for quicker and more secure communication (Badhwar, 2021). AI-powered systems can monitor quantum

Figure 3. Quantum cryptography market global forecast to 2028 (USD BN)
Source: https://www.marketsandmarkets.com/Market-Reports/quantum-cryptography-market-45857130.html
Alt-text: Figure 3 depicts statistic of the Quantum between 2023 and 2028, the cryptography market will rise at a compound annual growth rate (CAGR) of 41.2%, from USD 0.5 billion to USD 3.0 billion. As seen in Figure X.3, increased cybersecurity investment and upcoming next generation wireless network technologies are driving market development).

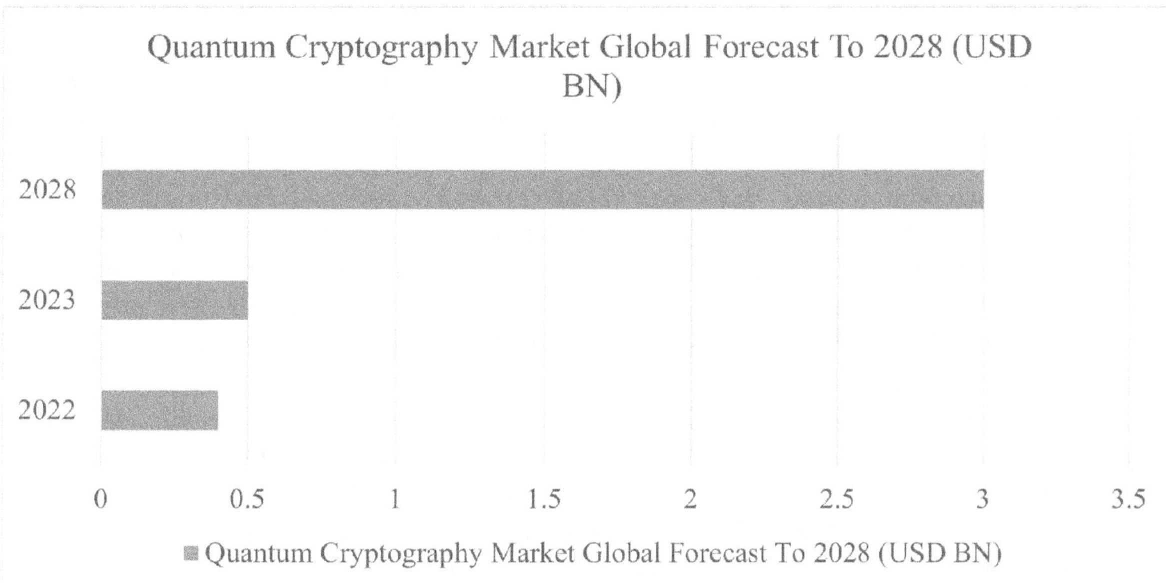

cryptography systems for abnormalities and potential security breaches in real-time. AI can recognize patterns that indicate assaults or attacks, allowing them to respond more quickly.

AI can simulate possible quantum assaults and weaknesses to assess the security of quantum cryptography systems, assisting researchers and organizations in developing more strong defenses (Akdemir, 2021). The role of AI in quantum cryptography has been diversified and its impact is on the banking sector as well, by merging AI with quantum cryptography, AI can improve the security of e-banking by protecting the privacy and legitimacy of data in transactions.AI approaches such as fuzzy logic and knowledge bases can be used to construct a resilient and responsive e-banking security architecture (Hassan, 2019).

AI-powered systems may continuously monitor quantum cryptography systems for anomalies and potential security breaches. AI can recognize patterns that signal an assault or attack, allowing it to respond faster and may simulate potential quantum attacks and vulnerabilities in order to test the security of quantum cryptography systems, supporting researchers and organizations in designing more robust defenses.

4. ARTIFICIAL INTELLIGENCE IMPACT ON QUANTUM MACHINE LEARNING (QML)

Quantum machine learning is a discipline that applies quantum computing concepts to machine learning approaches to address challenging issues. It entails enhancing the capabilities of classical machine learning algorithms with quantum algorithms and quantum systems. Quantum machine learning seeks to use quantum systems' unique qualities, such as superposition and entanglement, to execute computations more efficiently and precisely.

Quantum machine learning has the potential to revolutionize a variety of disciplines by leveraging the power of quantum mechanics, including optimization, pattern identification, and data analysis (Thomas, 2023). QML has the possibility of improve conventional machine learning approaches by utilizing the unique properties of quantum systems such as superposition and entanglement. It may be applied to optimization situations where quantum techniques can give an exponential speedup over traditional algorithms.

Quantum machine learning may also be used to improve the accuracy and efficiency of data categorization and clustering operations. Another use is in data analysis, where quantum machine learning algorithms may more effectively manage big and complicated datasets, resulting in better insights and decision-making. Quantum machine learning can give unique answers and breakthroughs in a variety of disciplines, including finance, healthcare (Khang, Vugar and Abuzarova et al., 2023), materials research, and encryption (Biamonte et al., 2017).

Quantum machine learning has emerged as a result of quantum computing breakthroughs and its ability to tackle certain problems more effectively than classical computing. Quantum parallelism, a distinguishing property of quantum computing, enables exponentially faster solutions of factorization and unordered search tasks. When integrating conventional and quantum computing to accelerate processing speed, challenges occur. Experiments with three datasets for binary classification were carried out, using the Support Vector Machine (SVM) and Quantum SVM (QSVM) methods. This has the potential to significantly improve the QSVM approach outperform basic SVM on complex datasets, depending to the findings, and the performance gap between quantum and classical models expands as dataset complexity increases (Tychola et al., 2023).

Quantum machine learning uses quantum computing to tackle problems more efficiently than traditional computing. Quantum parallelism, a distinguishing property of quantum computing, enables exponentially faster solutions of factorization and unordered search tasks. Quantum machine learning has enormous promise in fields like unsupervised learning and generative models. By utilizing quantum physics concepts, quantum machine learning can overcome the constraints of conventional machine learning (Cerezo et al., 2022). QML takes advantage of quantum computing's ability to accelerate and improve machine learning on the "classical" computers we use every day.

Quantum computers, which are built utilizing the sometimes conflicting principles of quantum physics, can store and process ten times the amount of data as the tablets, smartphones, and supercomputers that power the bulk of the modern world. According to McKinsey & Company, just 2,000 to 5,000 quantum computers will be operating by 2030, and those capable of handling the most complex problems may not exist until 2035 or later (Ushaa, Khang et al., 2023).

Both quantum computing and machine learning have the potential to make waves in the future, from assessing massive amounts of data to powering transformative technological advancements. While QML is still in its infancy, researchers and professionals have begun putting it to use in a range of applications. These are some of the applications: Create new machine learning algorithms, accelerate existing machine learning algorithms, employ quantum-enhanced reinforcement learning, which entails a machine learning algorithm learning based on its interactions with each other in a quantum environment, and Make quantum neural networks that can operate in fewer steps and at a quicker rate. (Ciliberto et al., 2018).

AI algorithms can be deployed on quantum computers, which may outperform traditional hardware in certain applications. Certain challenges in machine learning and AI that are difficult for traditional hardware can be handled naturally by quantum computers. Using ensemble learning methods and optimization methodologies, quantum machine learning may be applied for pattern recognition tasks such as person identification based on face scans (Malov, 2020). PQCs provide a fresh way for machine learning, but they pose a challenge to existing eXplainable AI (xAI) methods due to probabilistic mistakes and the exponential expansion of a quantum circuit's phase space with the number of qubits.

To analyze the performance of PQCs and investigate strategies to speed up their calculation, established xAI methods such as Baseline SHAP and Integrated Gradients are employed (Steinmüller et al., 2022). Quantum machine learning combines quantum computing with artificial intelligence principles to develop algorithms and models that can operate on quantum computers. Machine learning algorithms, for example, might be implemented on quantum computers to make use of their unique properties like superposition and entanglement.

QML algorithms have an opportunity to outperform regular machine learning algorithms in specific tasks, such as pattern identification and optimization. QML has a chance to improve AI systems' skills in a variety of domains, notably image recognition, NLP, and data analysis (Ciliberto, 2018). For processing industrial information in which there are of near-term quantum devices, quantum-assisted deep learning systems like as quantum-assisted Helmholtz machines have been developed.

AI algorithms may be transferred onto quantum platforms like quantum annealers to take use of quantum phenomena like tunneling, superposition, and entangled. Quantum machine learning has the potential to substantially expedite a wide range of activities, leading to new technologies and discoveries that have the potential to revolutionize problem-solving in a variety of sectors. To deal with hardware limits and make the most of available quantum resources, hybrid quantum-classical techniques are being investigated (Biswas, 2020).

Quantum breakthroughs are possible in all three core areas of machine learning: supervised, unsupervised, and reinforcement learning. While supervised and unsupervised learning have received more attention in QML, reinforcement learning has received less. Reward learning (RL) is a broader field that represents an interactive style of learning in which the learning agent learns how to behave appropriately through reward signals. There have been detailed constructions of quantum-enhanced reinforcement learning agents in the setting of QML that surpass their classical counterparts in terms of learning effectiveness and efficiency (Dunjko, 2016).

Deep learning algorithms have demonstrated promising outcomes for image processing applications such as remote sensing and picture recognition. Quantum processing techniques have been shown to tackle problems that traditional algorithms and processors cannot identify. AI can use the capabilities of both techniques to handle challenging challenges in image recognition and other applications by merging deep learning algorithms with quantum processing algorithms. To solve data analysis and machine learning issues, matrix operations are performed to vectors rather than higher order vectors. In quantum mathematics, qubit quantum states are vectors in a 2a-dimensional complex vector space. This region is teeming with shapeshifters. Quantum computers can perform simple linear algebra problems like Fourier transformations, eigenvectors and eigenvalues, and linear sets of equations in 2a-dimensional vector spaces in polynomial time (and considerably quicker than classical computers owing to quantum speed).

Harrow is one example. Hassidim, Lloyd (HHL) Application (Remanan, 2021). AI can have an influence on quantum machine learning by offering a deeper and more complete framework for deep learning and better objective function optimisation. When compared to conventional computing, quantum computing has been proven to reduce the time necessary to train a deep-restricted Boltzmann machine. In comparison to classical computing, quantum computing provides a deeper and more extensive foundation for deep learning (Khang & Vugar et al., 2023).

Quantum computing leads to considerable gains in deep learning optimization of the underlying goal function. Quantum approaches enable the effective training of multilayer and fully linked deep learning models (Weibe et al., 2016). AI has the potential to improve quantum machine learning by harnessing quantum computing capabilities and inventing unique techniques to more effectively solve complicated issues.

5. CONCLUSION

In the age of quantum computing, artificial intelligence's task is to build efficient algorithms that mitigate the effects of quantum decoherence. Quantum computers, which use quantum physics phenomena such as superposition and entanglement to solve complex problems, can provide remarkable performance. This has the potential to significantly improve AI by enabling the creation of efficient algorithms for tasks like machine learning and optimization (Khang & Kali, 2023).

AI methods, on the other hand, might be used to mitigate the effects of quantum decoherence, rendering quantum computing more reliable. The amalgamation of quantum computing with AI has an opportunity to exceed current methods and pave the way for new computer architecture. AI is advancing thanks to quantum computing, and the possibilities are endless. This collaboration has the prospect of revolutionary discoveries that will change the future of computers and our understanding of the universe, from addressing hard scientific and mathematical issues to expanding AI applications in healthcare (Khang & Medicine, 2023), finance, and beyond. It is an expedition into unexplored terrain, where invention collides with understanding, and where the limits of what is possible are continuously redefined (Khang & Quantum, 2023).

REFERENCES

Akdemir, Z., Menekse, M., Hosseini, M., Nandi, A., & Furuya, K. (2021). Introducing quantum key distribution to high school students. *Science Teacher (Normal, Ill.)*, *88*, 44–51. https://www.researchgate.net/profile/Muhsin-Menekse-2/publication/349042469_Introducing_Quantum_Key_Distribution_to_High_School_Students/links/601c3b8392851c4ed54bc973/Introducing-Quantum-Key-Distribution-to-High-School-Students.pdf

Ayanzadeh, R. (2018, February). Quantum artificial intelligence for natural language processing applications. In *Proceedings of the 49th ACM Technical Symposium on Computer Science Education* (pp. 273-273). ACM. https://dl.acm.org/doi/abs/10.1145/3159450.3162338

Badhwar, R. (2021). *The CISO's Next Frontier: AI, Post-Quantum Cryptography and Advanced Security Paradigms*. Springer. https://link.springer.com/chapter/10.1007/978-3-031-22032-6_7 doi:10.1007/978-3-030-75354-2

Balynsky, M., Chiang, H., Gutierrez, D., Kozhevnikov, A., Filimonov, Y., & Khitun, A. (2021). Quantum computing without quantum computers: Database search and data processing using classical wave superposition. *Journal of Applied Physics*, *130*(16), 164903. https://pubs.aip.org/aip/jap/article/130/16/164903/568206. doi:10.1063/5.0068316

Biamonte, J., Wittek, P., Pancotti, N., Rebentrost, P., Wiebe, N., & Lloyd, S. (2017). Quantum machine learning. *Nature*, *549*(7671), 195–202. https://www.nature.com/articles/nature23474. doi:10.1038/nature23474 PMID:28905917

Boger, Y. (2023, July 24). *Quantum computing and ai: A leap forward or a distant dream?* Datanami. https://www.datanami.com/2023/07/24/quantum-computing-and-ai-a-leap-forward-or-a-distant-dream/

Cerezo, M., Verdon, G., Huang, H. Y., Cincio, L., & Coles, P. J. (2022). Challenges and opportunities in quantum machine learning. *Nature Computational Science*, *2*(9), 567–576. https://www.nature.com/articles/s43588-022-00311-3. doi:10.103843588-022-00311-3

Chatterjee, P., & Chakraborty, R. (2020). A Brief Study on Quantum Computing. *International Journal of Innovative Research in Physics*, *1*(4), 58–63. https://www.ingentaconnect.com/content/smart/ijiip/2020/00000001/00000004/art00007. doi:10.15864/ijiip.1407

Ciliberto, C., Herbster, M., Ialongo, A. D., Pontil, M., Rocchetto, A., Severini, S., & Wossnig, L. (2018). Quantum machine learning: A classical perspective. *Proceedings - Royal Society. Mathematical, Physical and Engineering Sciences*, *474*(2209), 20170551. https://royalsocietypublishing.org/doi/abs/10.1098/rspa.2017.0551. doi:10.1098/rspa.2017.0551 PMID:29434508

Ciliberto, C., Herbster, M., Ialongo, A. D., Pontil, M., Rocchetto, A., Severini, S., & Wossnig, L. (2018). Quantum machine learning: A classical perspective. *Proceedings - Royal Society. Mathematical, Physical and Engineering Sciences*, *474*(2209), 20170551. https://royalsocietypublishing.org/doi/abs/10.1098/rspa.2017.0551. doi:10.1098/rspa.2017.0551 PMID:29434508

De Leon, N. P., Itoh, K. M., Kim, D., Mehta, K. K., Northup, T. E., Paik, H., Palmer, B. S., Samarth, N., Sangtawesin, S., & Steuerman, D. W. (2021). Materials challenges and opportunities for quantum computing hardware. *Science*, *372*(6539), eabb2823. https://www.science.org/doi/abs/10.1126/science.abb2823. doi:10.1126cience.abb2823 PMID:33859004

Dukatz, C. M., Garrison, D., Forrester, L., & Hollenbeck, C. (2019). U.S. Patent No. 10,275,721. Washington, DC: U.S. Patent and Trademark Office. https://doi.org/ doi:10.4018/979-8-3693-1168-4

Dunjko, V., Taylor, J. M., & Briegel, H. J. (2016). Quantum-enhanced machine learning. *Physical Review Letters*, *117*(13), 130501. https://journals.aps.org/prl/abstract/10.1103/PhysRevLett.117.130501. doi:10.1103/PhysRevLett.117.130501 PMID:27715099

Easttom, W. C., II. (2021). *Quantum computing fundamentals*. Addison-Wesley Professional. https://www.google.com/books?hl=en&lr=&id=otnPEAAAQBAJ&oi=fnd&pg=PT19&dq=Quantum+computing+fundamentals.+Addison-Wesley+Professional&ots=EUw6AzwVIS&sig=nPx4BVkAUuM2PSiqjhAHfvUuB4M

Eswaran, U., Khang A., & Eswaran, V. (2023). Role of Quantum Computing in The Era of Artificial Intelligence," Applications and Principles of Quantum Computing. IGI Global Press. https://doi.org/ doi:10.4018/979-8-3693-1168-4-ch003

Ghosh, S., Opala, A., Matuszewski, M., Paterek, T., & Liew, T. C. (2019). Quantum reservoir processing. *npj Quantum Information, 5*(1), 35. https://www.nature.com/articles/s41534-019-0149-8

Gill, S. S., Kumar, A., Singh, H., Singh, M., Kaur, K., Usman, M., & Buyya, R. (2022). Quantum computing: A taxonomy, systematic review and future directions. *Software, Practice & Experience, 52*(1), 66–114. https://onlinelibrary.wiley.com/doi/abs/10.1002/spe.3039. doi:10.1002pe.3039

Hamze, F., King, A. D., Raymond, J., Roy, A. P., Israel, R., Andriyash, E., & Ranjbar, M. (2018). *U.S. Patent No. 9,881,256*. U.S. Patent and Trademark Office. https://heinonline.org/hol-cgi-bin/get_pdf.cgi?handle=hein.journals/jstl27&section=5

Hassan, T., & Ahmed, F. (2019). Transaction and Identity Authentication Security Model for E-Banking: Confluence of Quantum Cryptography and AI. In *Intelligent Technologies and Applications: First International Conference*. Springer Singapore. https://link.springer.com/chapter/10.1007/978-981-13-6052-7_29

Henderson, M., Gallina, J., & Brett, M. (2021). Methods for accelerating geospatial data processing using quantum computers. *Quantum Machine Intelligence, 3*(1), 4. https://link.springer.com/article/10.1007/s42484-020-00034-6. doi:10.100742484-020-00034-6

Hidary, J. D., & Hidary, J. D. (2021). A brief history of quantum computing. *Quantum Computing: An Applied Approach*, 15-21. Springer. https://link.springer.com/content/pdf/10.1007/978-3-030-83274-2.pdf

Kanwar, P. (2023, May 8). *Quantum Computing in artificial intelligence around the corner*. Product Engineering Services. https://www.einfochips.com/blog/quantum-computing-in-artificial-intelligence-around-the-corner

Kaur, D., Uslu, S., & Durresi, A. (2023, June). Quantum Algorithms for Trust-Based AI Applications. In *Conference on Complex, Intelligent, and Software Intensive Systems* (pp. 1-12). Cham: Springer Nature. https://link.springer.com/chapter/10.1007/978-3-031-35734-3_1

Khang, A. (2022). Cyber-Physical-Social System and İncident Management. AI-Centric Smart City Ecosystems: Technologies, Design and Implementation. CRC Press. doi:10.1201/9781003252542-2

Khang, A. (2023). *Advanced Technologies and AI-Equipped IoT Applications in High-Tech Agriculture* (1st ed.). IGI Global Press. doi:10.4018/978-1-6684-9231-4

Khang, A. (2023). Revolutionizing Agriculture: Exploring Advanced Technologies for Plant Protection in the Agriculture Sector. Handbook of Research on AI-Equipped IoT Applications in High-Tech Agriculture. IGI Global. doi:10.4018/978-1-6684-9231-4.ch001

Khang, A. (2023). Enabling the Future of Manufacturing: Integration of Robotics and IoT to Smart Factory Infrastructure in Industry 4.0. AI-Based Technologies and Applications in the Era of the Metaverse. IGI Global Press. doi:10.4018/978-1-6684-8851-5.ch002

Khang, A. (2023). Applications and Principles of Quantum Computing. IGI Global Press. https://doi.org/ doi:10.4018/979-8-3693-1168-4

Khang, A. (2023). Quantum Mechanics Primer - Fundamentals and Quantum Computing. Applications and Principles of Quantum Computing. IGI Global Press. https://doi.org/ doi:10.4018/979-8-3693-1168-4-ch001

Khang, A. (2023). Introduction to Quantum Computing and Its Integration Applications. Applications and Principles of Quantum Computing. IGI Global Press. https://doi.org/ doi:10.4018/979-8-3693-1168-4-ch002

Khang, A. (2023). AI and IoT-Based Technologies for Precision Medicine. IGI Global Press. doi:10.4018/979-8-3693-0876-9

Khang, A. (2023). AI-Aided Data Analytics Tools and Applications for the Healthcare Sector. AI and IoT-Based Technologies for Precision Medicine. IGI Global Press. doi:10.4018/979-8-3693-0876-9.ch018

Khang, A., & Muthmainnah, M. (2023). AI-Aided Teaching Model for the Education 5.0 Ecosystem" AI-Based Technologies and Applications in the Era of the Metaverse. IGI Global Press. doi:10.4018/978-1-6684-8851-5.ch004

Khang, A., Shah, V., & Rani, S. (2023). AI-Based Technologies and Applications in the Era of the Metaverse. IGI Global Press. doi:10.4018/978-1-6684-8851-5

Kuang, R., Lou, D., He, A., & Conlon, A. (2021, April). Quantum safe lightweight cryptography with quantum permutation pad. In *2021 IEEE 6th international conference on computer and communication systems (ICCCS)* (pp. 790-795). IEEE. https://ieeexplore.ieee.org/abstract/document/9449247/

Liang, X., Li, M., Xiao, Q., Chen, J., Yang, C., An, H., & He, L. (2023). Deep learning representations for quantum many-body systems on heterogeneous hardware. *Machine Learning: Science and Technology, 4*(1), 015035. https://iopscience.iop.org/article/10.1088/2632-2153/acc56a/meta. doi:10.1088/2632-2153/acc56a

Malov, D. (2020, August). Quantum Algebraic Machine Learning. In *2020 IEEE 10th International Conference on Intelligent Systems (IS)* (pp. 426-430). IEEE. https://ieeexplore.ieee.org/abstract/document/9199982/

Mohamed, B. S., Kumar, M. S., & Srinivasagan, K. G. (2022). Quantum Computing and Quantum Cryptography. In Privacy and Security Challenges in Cloud Computing (pp. 171-189). CRC Press. doi:10.1201/9781003219880-9

Remanan, S. (2021, April 9). *Beginner's Guide to Quantum Machine Learning.* Paperspace Blog. https://blog.paperspace.com/beginners-guide-to-quantum-machine-learning/

Rietsche, R., Dremel, C., Bosch, S., Steinacker, L., Meckel, M., & Leimeister, J. M. (2022). Quantum computing. *Electronic Markets, 32*(4), 2525–2536. https://link.springer.com/article/10.1007/s12525-022-00570-y. doi:10.100712525-022-00570-y

Rose, G., Gildert, S., Macready, W. G., & Walliman, D. C. (2019). *U.S. Patent No. 10,318,881*. U.S. Patent and Trademark Office. https://heinonline.org/hol-cgi-bin/get_pdf.cgi?handle=hein.journals/jstl27&section=5

Shah, V., & Khang, A. (2023). Metaverse-Enabling IoT Technology for a Futuristic Healthcare System. AI-Based Technologies and Applications in the Era of the Metaverse. IGI Global Press. doi:10.4018/978-1-6684-8851-5.ch008

Steinmüller, P., Schulz, T., Graf, F., & Herr, D. (2022). *eXplainable AI for Quantum Machine Learning*. arXiv preprint arXiv:2211.01441.

Strand, M. (2021, May). A Status Update on Quantum Safe Cryptography. In *2021 International Conference on Military Communication and Information Systems (ICMCIS)* (pp. 1-7). IEEE. https://ieeexplore.ieee.org/abstract/document/9486413/

Tacchino, F., Mangini, S., Barkoutsos, P. K., Macchiavello, C., Gerace, D., Tavernelli, I., & Bajoni, D. (2021). Variational learning for quantum artificial neural networks. *IEEE Transactions on Quantum Engineering*, *2*, 1–10. https://ieeexplore.ieee.org/abstract/document/9364892/. doi:10.1109/TQE.2021.3062494

Thomas, R. K. (2023). Quantum Neural Networks: Bridging Quantum Computing and Machine Learning. *Asian Journal of Research in Computer Science*, *16*(3), 1–10. https://www.tandfonline.com/doi/pdf/10.1080/00018735700101156. doi:10.9734/ajrcos/2023/v16i3340

Tosh, D., Galindo, O., Kreinovich, V., & Kosheleva, O. (2020). *How Quantum Cryptography and Quantum Computing Can Make Cyber-Physical Systems More Secure*. Scholar Works. https://scholarworks.utep.edu/cs_techrep/1397/

Tychola, K. A., Kalampokas, T., & Papakostas, G. A. (2023). Quantum Machine Learning—An Overview. *Electronics (Basel)*, *12*(11), 2379. https://www.researchgate.net/profile/Kyriaki-Tychola/publication/371001049_Quantum_Machine_Learning-An_Overview/links/646e1f426a0082273fa86d0f/Quantum-Machine-Learning-An-Overview.pdf. doi:10.3390/electronics12112379

Wiebe, N., Kapoor, A., & Svore, K. M. (2014). *Quantum deep learning*. arXiv preprint arXiv:1412.3489.

Xu, G., Mao, J., Sakk, E., & Wang, S. P. (2023, March). An Overview of Quantum-Safe Approaches: Quantum Key Distribution and Post-Quantum Cryptography. In *2023 57th Annual Conference on Information Sciences and Systems* (CISS) (pp. 1-6). IEEE. https://ieeexplore.ieee.org/abstract/document/10089619/

Zhao, X., Li, M., Xiao, Q., Chen, J., Wang, F., Shen, L., & Liang, X. (2022, November). AI for quantum mechanics: high performance quantum many-body simulations via deep learning. In *SC22: International Conference for High Performance Computing, Networking, Storage and Analysis* (pp. 1-15). IEEE. https://ieeexplore.ieee.org/abstract/document/10046063/

Chapter 8
Quantum Machine Learning: Enhancing AI With Quantum Computing

Priyanga Subbiah
https://orcid.org/0000-0002-2395-7492
Department of Networking and Communications, School of Computing, Faculty of Engineering and Technology, SRM Institute of Science and Technology, Kattankullar, India

N. Krishnaraj
Department of Networking and Communications, School of Computing, Faculty of Engineering and Technology, SRM Institute of Science and Technology, Kattankullar, India

Kiran Bellam
Department of Computer Science, Prairie View A&M University, USA

ABSTRACT

Machine learning improved by quantum computing. Machine learning and quantum physics fix AI and computers. This chapter discusses quantum machine learning theory, methods, and applications. Part 1 thoroughly discusses quantum and classical machine learning. The authors demonstrate how quantum supports vector machines, neural networks, and clustering speed AI. The chapter examines quantum machine learning's merits and downsides. Quantum computers optimize, parallelize, and manage huge data better. Quantum hardware restrictions and error correction reduce noise and decoherence. Explore quantum machine learning in NLP, drug discovery, financial modeling, and image recognition. Many fields could change quantum platform machine learning models with quantum algorithms. The chapter concludes with quantum machine learning directions and challenges. Check trustworthy quantum machine learning frameworks, benchmarks, and hybrid algorithms. Hot: quantum machine learning. This chapter covers fundamentals, research frameworks, and applications.

1. INTRODUCTION

In the fast-developing subject of quantum computing, computation is done using the concepts of quantum mechanics. It makes use of quantum bits, also known as qubits, which can be in several states at once

DOI: 10.4018/979-8-3693-1168-4.ch008

due to the concept of superposition. As a result, quantum computers can handle enormous volumes of information at once and might even be able to solve some problems more quickly than traditional computers. Entanglement, in which qubits are correlated, and quantum gates, which change qubit states, are important ideas in quantum computing (Adebiyi & Fatinikun et al., 2023).

The goal of artificial intelligence (AI) is to create intelligent machines with capabilities equivalent to those of human beings in terms of perception, learning, reasoning, and decision-making. In order to teach algorithms, which are a subset of AI, how to recognize patterns and make predictions or judgements, massive datasets are used in machine learning. Deep learning, a kind of machine learning, processes complex data and extracts useful features using neural networks with numerous layers. Natural language processing, computer vision, robotics, recommendation systems, and other technologies are just a few of the many uses for AI (Hajimahmud & Khang et al., 2022).

The potential for quantum computing and AI to improve and complement one another is what binds them together. The enormous computational power and simultaneous data processing capacity of quantum computing has the potential to speed up and improve AI algorithms. Quantum algorithms created expressly for AI activities to address optimization issues, better pattern recognition, and improve data processing can use the special characteristics of quantum systems. Moreover, conventional AI algorithms may be improved and quantum-inspired algorithms, which are inspired by quantum principles but can be implemented on classical computers, may overcome some of its drawbacks (Khang & Kali, 2024).

A tool for AI research, quantum computing enables researchers to explore novel architectures, simulate and analyze sophisticated AI models, and create cutting-edge methods that push the field's limits. The relationship between quantum computing and AI is still not fully understood, but it shows promise for addressing difficult issues, enhancing computational effectiveness, and expanding the capabilities of intelligent systems. Unlocking the transformational potential of this connection depends on ongoing research and cooperation between the communities of quantum computing and AI (Khang & Shah et al., 2023).

The domains of machine learning and quantum computing have each made outstanding advancements recently, changing the technological landscape and creating new opportunities for innovation. Using the laws of quantum physics, quantum computing has the potential to tackle complicated problems tenfold more quickly than traditional computers. A revolution in artificial intelligence has been brought about by machine learning, which enables computers to learn from data and make wise predictions or judgements (Khang & Vugar et al., 2023). Now that these two domains have converged, a new field called quantum machine learning (QML) has emerged. QML aims to use quantum computers to improve AI skills and realize the quantum advantage. QML has a lot of potential to revolutionize a number of AI-related fields, including pattern identification, optimization, and data analysis (Metawei et al., 2022). QML algorithms have the potential to surpass their classical counterparts in terms of computational correctness and efficiency by taking advantage of the special characteristics of quantum systems, such as superposition and entanglement. Researchers and industry professionals alike have paid close attention to the interaction between quantum computing and machine learning, which has led to fascinating developments and discoveries (Ushaa, Khang et al., 2023).

In this article, we investigate the field of quantum machine learning and examine how the computational power of quantum computing might be used to improve AI. We will go over the fundamental ideas, procedures, and methods that underlie QML in order to highlight its revolutionary potential for a range of applications (Lin et al., 2023). We will also look at the difficulties and factors that come with combining machine learning with quantum computing, such as algorithm design, hardware limitations, and the creation of specialized quantum software frameworks. We hope to give a thorough overview of this developing topic and highlight its significance for the development of artificial intelligence through this investigation of quantum machine learning. We can unleash the full potential of machine learning and expand the frontiers of artificial intelligence by utilizing the power of quantum computing (Khang & Kali, 2023).

2. CONTRASTING CLASSICAL MACHINE LEARNING WITH QUANTUM MACHINE LEARNING: UNDERSTANDING THE FUNDAMENTAL DIFFERENCES

The primary distinction between classical machine learning (ML) and quantum machine learning (QML) is found in the computational models that underlie each approach, as well as the methods for processing and analyzing data. Classical machine learning employs a deterministic and sequential execution model and runs on traditional computers. Information is represented and processed using traditional bits. Traditional machine learning (ML) methods are made to operate with data that is often represented as numerical values or categorical labels (Zeguendry et al., 2023). These algorithms make use of statistical techniques, mathematical algorithms like gradient descent, decision trees, and support vector machines, as well as traditional optimization techniques. Since classical ML has been thoroughly researched and optimized for classical computer architectures, it can be used to solve a variety of real-world issues.

Figure 1. Classical machine learning with quantum machine learning

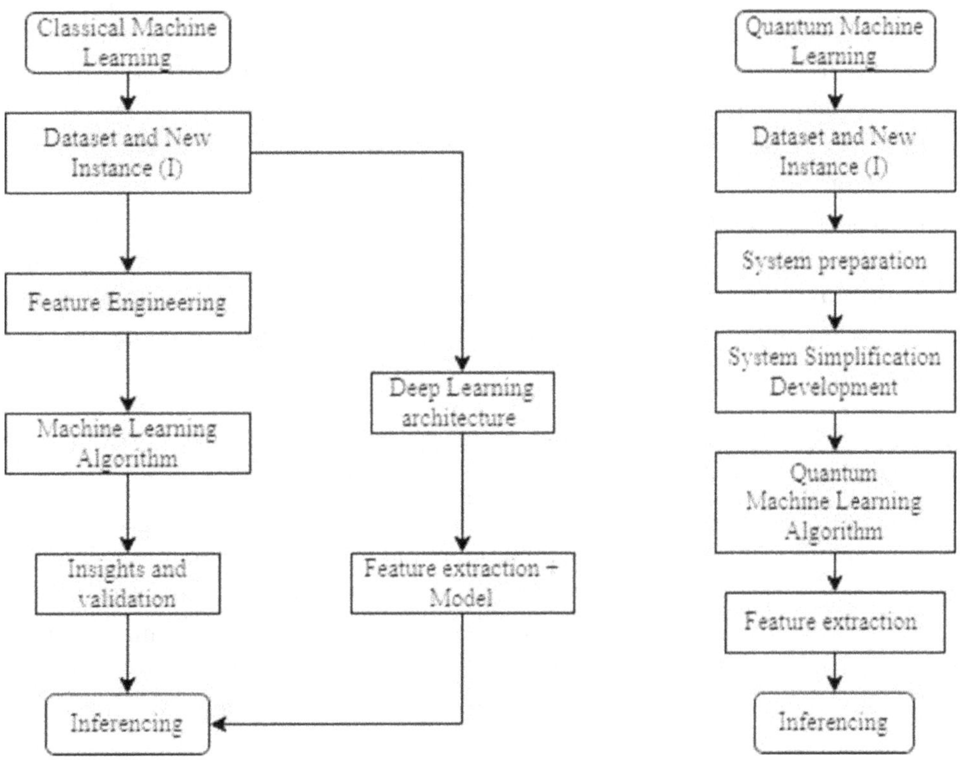

On the other hand, QML makes use of quantum computing principles to improve machine-learning capabilities. Qubits, which can represent and process information in quantum states, are the building blocks of quantum computers. Qubits, which differ from traditional bits in that they can reside in a superposition of 0 and 1, allow for parallel processing. It is the purpose of QML algorithms to take use of the special qualities of quantum systems. To compute on quantum states, they use quantum gates and quantum circuits. Quantum versions of conventional machine learning algorithms as well as original algorithms created especially for quantum systems are both included in QML algorithms. These algorithms may be able to exponentially speed up specific tasks like data classification, optimization, and pattern recognition by taking advantage of quantum phenomena like superposition and entanglement. Both classical and quantum data can be included in the data representation in QML (Cruz-Benito & Martín-Fernández et al., 2018). Quantum states can be used to encode classical data, enabling parallel processing and possible speedup. Contrarily, quantum data is information that can be processed directly by quantum algorithms since it is fundamentally quantum in nature. It's crucial to remember that real-world QML applications are still developing since it's difficult to create viable quantum computers with enough qubits and low error rates as shown in Figure 1.

Alt-text: Figure 1 depicts the contrasting Classical Machine Learning with Quantum Machine Learning.

3. QUANTUM POWER HARNESSED FOR ADVANCED MACHINE LEARNING

In many domains, including machine learning, new frontiers have been made possible by the quick development of quantum computing. The use of quantum computing allows for the performance of intricate calculations and the resolution of issues that are beyond the capabilities of conventional computers. Researchers have looked into the possibility of using quantum computing for sophisticated machine learning tasks in recent years. In addition to highlighting the key ideas and methods used in utilizing quantum computing for improved machine learning capabilities, this paper provides an overview of the developing topic of quantum machine learning. We explore quantum algorithms that have the potential to enhance pattern recognition, optimization, and data processing tasks, such as quantum support vector machines and quantum neural networks. Also covered are the difficulties and possibilities of quantum machine learning, such as hardware limitations, algorithmic design, and the requirement for specialized quantum software development (Fernández et al., 2018). We hope to stimulate additional research and development in this fascinating field and open the door for revolutionary developments in artificial intelligence by investigating the synergy between quantum computing and machine learning.

4. COMBINING MACHINE LEARNING AND QUANTUM COMPUTING TO UNLEASH THE QUANTUM ADVANTAGE IN AI

By utilizing the quantum advantage, machine learning and quantum computing have the potential to revolutionize the field of AI. The capabilities of classical computers are surpassed by quantum computing, which applies the ideas of quantum mechanics to computation. Machine learning algorithms provide the ability for computers to learn from data and make predictions or judgements. Combining these two potent technologies allows us to take use of the special qualities of quantum systems and improve the efficacy and efficiency of AI applications.

In this chapter, we investigate the relationship between machine learning and quantum computing, emphasizing crucial applications of quantum advantage in AI, Maria, Ilya, et al., 2015. We examine quantum-inspired algorithms, quantum neural networks, and quantum-enhanced optimization methods, demonstrating their capability to speed up training procedures, boost pattern recognition, and resolve challenging optimization issues. We also explore the difficulties and factors involved in combining machine learning and quantum computing, including algorithm design, hardware limitations, and the requirement for specialized quantum software frameworks. We can advance the field of artificial intelligence by embracing the merger of machine learning with quantum computing, which will result in innovations in a number of industries, including banking, healthcare, and natural language processing (Khang & Medicine, 2023).

5. INVESTIGATING THE RELATIONSHIP BETWEEN QUANTUM COMPUTING AND AI

Both quantum computing and artificial intelligence (AI) have the potential to have significant effects on one another. While quantum computing and AI are distinct fields, there are certain areas of intersection and possible synergy (Metawei et al., 2022). Let's investigate their partnership in more detail.

5.1. Enhanced Processing Speed

Quantum computers may be able to carry out some operations significantly more quickly than conventional ones. Numerous AI tasks, including optimization, machine learning, and data analysis, can benefit from this speedup. Grover's algorithm and quantum support vector machines are two examples of quantum algorithms that have the potential to increase computational efficiency for particular AI issues.

5.2. Quantum Algorithms for Machine Learning

Traditional machine learning methods are intended to be improved using quantum computing resources. Pattern recognition, classification, and prediction tasks may be enhanced by QML techniques such as quantum neural networks and feature selection inspired by quantum mechanics.

5.3. Data Analysis and Optimisation

Using quantum computers, massive datasets may be processed quickly, and challenging optimisation tasks can be completed. Applications of AI that need to manage enormous volumes of data or optimize complex models, such recommendation systems, financial modelling, and supply chain optimisation, can profit from this capabilities.

5.4. AI Algorithms Influenced by Quantum Mechanics

Even in the absence of quantum hardware, researchers have looked into applying quantum-inspired AI techniques to enhance more traditional AI methods. These algorithms improve optimisation, sampling, and search techniques and can be used for AI tasks including reinforcement learning, computer vision, and natural language processing. They are inspired by quantum principles.

5.5. AI Research Using Quantum Computing

Research and development in AI can also make use of quantum computing. In order to better understand complicated AI models and algorithms and identify areas for improvement, quantum simulators can be used. Researchers can investigate unique AI systems and create new methodologies using quantum computing.

5.6. Challenges and Restrictions

Practical quantum computers with suitable qubit counts and low error rates are not yet commonly available. Quantum computing is still in its early phases of development. Additionally, there are difficulties in creating quantum algorithms for AI tasks and integrating them with current AI frameworks. Due to the inherent probabilistic nature of quantum algorithms, it can also be difficult to comprehend and validate the results. It's crucial to highlight that much research is currently being done to determine how quantum computing and AI might complement one another, and that the full depth of this relationship is still being studied. As quantum technology develops, it has the ability to transform different facets of AI, opening up new possibilities and capabilities (Jerbi et al., 2023). Artificial intelligence (AI) and quantum computing are two quickly developing disciplines that have the potential to completely change a variety of industries. This study intends to investigate the connections between quantum computing and AI, stressing the potential advantages, difficulties, and prospective areas of cooperation between both fields. This examination provides light on the complex interactions between quantum computing and AI by reviewing the current research and developments as well as the future prospects.

6. ATOMIC ALGORITHMS MACHINE LEARNING REVOLUTION: FROM THEORY TO PRACTICE

Machine learning has made incredible strides in recent years, revolutionizing sectors and changing how we approach difficult challenges. Among these developments, the idea of atomic algorithms has become recognized as a potent accelerator of machine learning development. Atomic algorithms are small, effective, and specialized algorithms created to handle particular jobs with great accuracy and efficiency (Ramezani et al., 2020). The goal of this paper is to study the potential of atomic algorithms in quickening the machine learning revolution and bridging the gap between theoretical advancements and real-world applications. It does this by delving into the theory and practice of atomic algorithms. The theoretical underpinnings of atomic algorithms are investigated, including their definition, properties, and guiding theories and approaches.

A full discussion of the benefits and difficulties of atomic algorithms is presented. The article discusses the advantages of atomic algorithms, including their improved performance for particular applications, increased computational efficiency, and decreased complexity. However, difficulties with generalizability, interpretability, and adaptability to other datasets are also discussed, highlighting areas in need of additional study and advancement. A variety of machine learning topics are described with real-world applications of atomic algorithms. These applications show in detail how atomic algorithms overcame particular difficulties and outperformed conventional methods in terms of performance (Saini et al., 2020). The study illustrates the transformational potential of atomic algorithms in real-world scenarios by looking at their impact on particular applications and sectors.

Focusing on methods for creating and perfecting these algorithms, the exploration of atomic algorithm development and optimization is undertaken. To increase the efficiency and effectiveness of atomic algorithms, strategies for feature selection, algorithm specialization, and optimization are presented. Researchers and practitioners may now investigate and use these potent tools thanks to the overview of algorithmic frameworks and libraries facilitating the construction of atomic algorithms. One of the most important components of successfully implementing atomic algorithms is bridging the theory-practice divide. The use of atomic algorithms in real-world contexts is examined, as well as methods for converting theoretical developments into useful implementations. It is investigated how atomic algorithms can be adopted and integrated seamlessly with current machine learning workflows and frameworks.

The ramifications for society and ethics are rigorously analyzed, addressing issues with fairness and bias in the context of atomic algorithms. Atomic algorithms' societal effects on various stakeholders are taken into account, coupled with responsible development and deployment procedures to assure moral and just use. Future directions and difficulties in atomic algorithm research and development are covered in the paper's conclusion. In this dynamic sector, new developments and potential research directions are emphasized to encourage further study and cooperation, Mishra, N. *et al.* 2021. The possible implications of atomic algorithms for machine learning in the future are discussed, highlighting the necessity for ongoing developments to fully realize their disruptive potential. This paper as a whole provides a thorough examination of atomic algorithms within the framework of the machine learning revolution. It provides insights into how these small, focused algorithms can revolutionize numerous industries, advancing machine learning and influencing the development of intelligent solutions by bridging the theoretical and practical divide.

7. A QUANTUM LEAP FOR SMARTER AI SYSTEMS: QUANTUM MACHINE LEARNING

Artificial intelligence is advancing thanks to the confluence of quantum computing and AI, which has also revolutionized the capabilities of intelligent systems. This article focuses on the idea of QML, a cutting-edge strategy that uses quantum-computing concepts to improve conventional machine learning techniques. The article begins by exposing readers to the interesting nexus between quantum computing and artificial intelligence while highlighting QML's revolutionary potential. It draws attention to the importance of this developing topic and the advantages it might provide in terms of allowing smarter and more effective AI systems (Essam et al., 2022).

The chapter introduces key quantum computing ideas to create a strong basis. It defines qubits, the fundamental units of quantum systems, and explains important characteristics like superposition and entanglement. It also discusses fundamental quantum operations and quantum gates, which serve as the building blocks of quantum computers. To familiarize readers with the fundamentals of quantum computing, an overview of the quantum circuit model and quantum computational complexity is also presented. The article then digs into machine learning's foundations, introducing traditional machine learning algorithms and approaches. It covers the supervised learning, unsupervised learning, and reinforcement learning paradigms, which are the three primary types of machine learning. Additionally, the difficulties and restrictions of conventional machine learning are examined, laying the groundwork for the introduction of QML as a potential remedy.

In-depth research is done on the fundamental ideas behind quantum machine learning. The article demonstrates how quantum computers can be more advantageous for machine learning applications than classical computing. It focuses on how quantum interference and quantum parallelism might increase computer efficiency and make it possible for learning algorithms to become more potent. In order to show how quantum systems can offer new representations and transformations of data for machine learning, the concepts of quantum feature space and quantum state preparation are also covered, Hu, Wang, et al., 2019. The article's conclusion discusses quantum machine learning's prospective uses and implications.

In particular, it demonstrates how QML may improve a number of AI tasks, including pattern identification, optimization, and data analysis. To demonstrate the revolutionary power of QML in practical situations, examples of prospective use cases in sectors including healthcare (Khang, Vugar and Abuzarova et al., 2023), banking, and cybersecurity are given (Khang and Vladimir, 2022). This article provides a thorough examination of quantum machine learning, demonstrating its potential to significantly increase the power of AI systems. QML opens up new vistas for resolving complicated issues and achieving previously unheard-of levels of intelligence in AI systems by fusing the strength of quantum computing with the principles of machine learning.

8. AI QUANTUM OPTIMIZATION: THE FUTURE OF MACHINE LEARNING

Machine learning has advanced significantly in recent years, but new strategies are required to push the limits of AI further as complicated issues keep cropping up. AI Quantum Optimization is one strategy that uses the capabilities of quantum computing to transform machine learning. In this article, the idea of AI quantum optimization is examined, emphasizing its potential to shape the direction of machine learning (Chen and Shinjae, 2021). Setting the stage early on, the article emphasizes the necessity of advanced optimization methods in machine learning. It examines the difficulties encountered by conventional optimization algorithms when tackling complex, high-dimensional issues as well as the difficulties they have in quickly locating the best solutions.

The next section of the paper covers the idea of AI Quantum Optimization, which makes use of quantum computing principles to improve machine learning optimization processes. It describes how quantum computers have the potential to accelerate optimization and find optimal solutions more quickly due to their capacity to analyses enormous volumes of information concurrently and explore several options at once. It then delves into ideas like quantum annealing, quantum gates, and quantum states to examine the theoretical underpinnings of AI quantum optimization. It demonstrates how these ideas can be used to address issues with parameter tuning, feature selection, and model optimization in machine learning.

The article also illustrates the potential applications of AI quantum optimization in different fields. It emphasizes how this method can, among other things, optimize neural network topologies, hyper parameter optimization, recommendation systems, and financial portfolio management. The revolutionary effects of AI Quantum Optimization are demonstrated through examples drawn from everyday life and success stories. The paper concludes by examining AI quantum optimization's potential and how it may affect machine learning in the future. It highlights current research initiatives, developments in quantum hardware, and the possibilities for hybrid strategies that mix traditional and cutting-edge quantum computing methods (Mandaar & Preeti, 2020). The transformational potential of AI Quantum Optimization in enhancing the capacities of machine learning systems and guiding the future of AI is highlighted in the article's conclusion. In decision, this article offers a thorough examination of AI Quantum Optimization as the upcoming development in machine learning. It demonstrates the benefits of applying quantum-computing ideas to optimization problems and offers examples of possible uses for AI quantum optimization in a range of fields. AI Quantum Optimization lays the way for tackling difficult optimization problems and opening up new horizons in machine learning by fusing the strength of quantum computing with the intelligence of machine learning.

9. ENHANCED QUANTUM LEARNING DEFINING NEW AI BOUNDARIES

In recent years, artificial intelligence (AI) has made incredible strides, allowing machines to carry out difficult tasks and make wise decisions. Traditional AI algorithms still have some limitations, though, when it comes to handling some challenging issues. By fusing the capabilities of quantum computing with machine learning strategies, enhanced quantum learning is an exciting new frontier that aims to push the limits of artificial intelligence (David & Juan et al., 2020). This article examines the idea of enhanced quantum learning and how it can open up new AI horizons. The article starts out by outlining the current difficulties and constraints faced by conventional AI algorithms. It emphasizes the requirement for novel strategies that can get over these restrictions and more successfully deal with difficult issues. This prepares the ground for Enhanced Quantum Learning to be discussed as a potential solution.

The paper then explores the fundamental ideas behind machine learning and quantum computing. It gives a succinct explanation to quantum computing, outlining the potential for exponential computing power and defining essential concepts like superposition and entanglement, Zhang, et al., 2020. To give readers a thorough knowledge of conventional AI methods, it also covers the principles of machine learning, such as supervised and unsupervised learning, reinforcement learning, and deep learning.

The fundamental ideas behind Enhanced Quantum Learning are then investigated. The article describes how, by taking use of the characteristics of quantum systems, quantum computing might improve machine learning methods. It emphasizes how quantum entanglement, quantum interference, and quantum parallelism have the potential to make learning algorithms faster and more effective. It also examines how more accurate and reliable data representation and processing made possible by quantum computing can enhance AI task performance. The paper then goes on to highlight Enhanced Quantum Learning's potential uses in a variety of fields (Fastovets & Bogdanov et al., 2019). It describes how this method can be used to address complex optimization issues, improve natural language processing, increase the accuracy of picture and speech recognition, and, among other things, optimize resource allocation.

Examples from the real world and case studies are used to demonstrate how Enhanced Quantum Learning has completely changed the way that real-world AI applications operate. The article also examines the difficulties and factors to be taken into account while applying enhanced quantum learning. It discusses the current constraints of quantum hardware, the requirement for specialized knowledge in both machine learning and quantum computing, and the integration of quantum algorithms with current AI frameworks. It highlights how crucial interdisciplinary cooperation and ongoing study are to resolving these problems.

The short piece ends by examining the potential applications of enhanced quantum learning and how they can progress artificial intelligence. It examines current research initiatives, developments in quantum hardware, and the possibilities for hybrid strategies that mix traditional and cutting-edge quantum computing methods. The article emphasizes how Enhanced Quantum Learning is pushing the frontiers of AI and opening up new avenues for the solution of challenging issues. In this regard, this study offers a thorough investigation of enhanced quantum learning as a potent strategy to push the limits of artificial intelligence (Uehara et al., 2021). It emphasizes how machine learning methods could be improved by quantum computing, offering more effective and precise answers to challenging issues. Enhanced Quantum Learning expands the boundaries of AI by fusing the strengths of quantum computing and machine learning, paving the path for previously unimaginable developments in intelligent systems and applications.

10. UNRAVEL NEW DIMENSIONS OF INTELLIGENCE WITH QUANTUM-ASSISTED MACHINE LEARNING

Numerous sectors have been altered by machine learning's quick advancement, but as long as difficult challenges exist, researchers are always looking for fresh ideas to expand the capabilities of AI. Quantum-Assisted Machine Learning (QAML), which combines the strength of quantum computing with conventional machine learning methods to uncover new dimensions of intelligence, is one such potential path (Cerezo et al., 2022). This article examines the idea of QAML and how it might revolutionize the field of artificial intelligence. The first section of the article provides readers with an overview of the state of machine learning and the difficulties that conventional algorithms have when attempting to solve complex issues. It highlights the requirement for cutting-edge computational tools capable of handling massive datasets and high-dimensional data representations. The article then delves into the fundamentals of quantum computation, giving a background knowledge of qubits, quantum gates, and quantum operations. It clarifies crucial quantum phenomena like superposition and entanglement, which allow quantum computers to process information in a fundamentally different manner from conventional computers.

The article introduces the idea of quantum-aided machine learning by building on this foundation. It discusses how quantum computing can be used to improve a number of machine learning processes, such as feature selection, model training, and data preprocessing. It describes how qubits can simultaneously encode and process information, potentially accelerating computation exponentially (Chalumuri, 2021). The article then looks at the potential uses of QAML in several fields. It demonstrates how QAML may be applied to a variety of tasks, including complicated pattern recognition, large-scale system optimization, natural language processing, drug development, and financial modelling. Examples from the real world and case studies are used to illustrate how QAML has significantly changed the way that real-world AI applications are developed.

The article also covers the difficulties and factors to be taken into account when using QAML. It discusses the constraints of present quantum hardware, the requirement for specialized knowledge in both quantum computing and machine learning, and the incorporation of quantum algorithms with current machine learning frameworks. It emphasizes how critical it is to create scalable, reliable quantum algorithms and to maximize the fusion of classical and quantum systems. The future prospects of QAML and their consequences for the development of AI are covered in the article's conclusion. It examines current research initiatives, developments in quantum hardware, and the possibilities for hybrid strategies that mix traditional and cutting-edge quantum computing methods (Francesco et al., 2021).

To realize the full potential of QAML, the article emphasizes the necessity for ongoing interdisciplinary cooperation and financial support for quantum computing research. In decision, this study offers a thorough investigation of quantum-assisted machine learning (QAML) as a game-changing strategy to uncover fresh levels of intelligence in AI systems. It demonstrates the multiple ways in which machine learning can be improved by quantum computing and the wide range of disciplines in which QAML is being applied. QAML expands the boundaries of AI, allowing for breakthroughs in the solution of complicated problems and paving the way for the next generation of intelligent systems by fusing the strengths of quantum computing with machine learning (Kushal et al., 2021).

11. UNLOCKING UNPRECEDENTED INSIGHTS AND CAPABILITIES WITH QUANTUM ALGORITHMS FOR AI

Artificial intelligence is not an exception when it comes to the potential influence of quantum computing, which has emerged as a groundbreaking technology. The idea of using quantum algorithms for AI to achieve previously unheard-of insights and capabilities is explored in this article. Researchers are able to get around conventional constraints and open up new opportunities for developing AI by utilizing the potential of quantum computing. The article starts off by giving a general summary of the current obstacles in AI, including managing intricate and sizable datasets, improving machine learning models, and resolving computationally demanding issues. It highlights the requirement for creative solutions to meet these difficulties and expand the capabilities of AI. The paper then presents the basics of quantum computing to the audience. It explains the concepts of superposition and entanglement, which allow for parallel computation and exponential computing capacity in quantum systems.

In order to comprehend quantum algorithms, it is also important to understand key ideas like qubits, quantum gates, and the quantum circuit model (Raschka et al., 2020). The article investigates the possibilities of quantum algorithms for AI by building on this foundation. It explores particular quantum algorithms that could revolutionize many parts of AI activities, such as quantum support vector machines, quantum neural networks, and quantum clustering algorithms. It shows how these algorithms take advantage of the special features of quantum systems to produce previously unheard-of insights and talents.

The article also emphasizes the potential uses of quantum algorithms in AI. It talks about how jobs like data analysis, pattern identification, optimization, and natural language processing can be improved by quantum algorithms, Huang, Broughton, et al., 2021. Examples from the real world and case studies are given to illustrate how quantum algorithms may boost AI applications. The article also discusses the difficulties and factors to be taken into account while applying quantum algorithms to AI. It talks about how to integrate quantum algorithms with conventional machine learning frameworks, the current constraints of quantum hardware, and the necessity for expertise in both quantum computing and AI. It highlights the significance of creating quantum error correction methods, improving quantum circuitry, and encouraging interaction between the communities of quantum computing and AI O'Driscoll, Nichols, et al., 2019.

The article also discusses the ramifications for the development of AI systems as well as the chances for quantum algorithms in the future. It highlights current research initiatives, developments in quantum hardware, and the possibilities for hybrid strategies that mix traditional and cutting-edge quantum computing methods. The article's conclusion emphasizes the revolutionary potential of quantum algorithms in revealing hitherto unheard-of insights and capabilities, leading to substantial developments in the field of AI. In finalization, this article offers a thorough examination of how quantum algorithms for AI can uncover hitherto unheard-of insights and capabilities (Peters and Caldeira et al., 2021). It emphasizes how quantum computing has the potential to surpass preexisting constraints and revolutionize numerous elements of AI activities. Researchers can open new doors and achieve advances in data analysis, optimization, and pattern identification by utilizing the capabilities of quantum algorithms. To fully realize the potential of quantum algorithms for AI, the article emphasizes the significance of ongoing research, collaboration, and quantum computing breakthroughs (Beng et al., 2022).

12. ALGORITHMS FOR QUANTUM MACHINE LEARNING: LAYING THE FOUNDATION FOR THE QUANTUM AI REVOLUTION

It is quite likely that a revolution in intelligent systems will be sparked by the convergence of quantum computing with AI. The revolutionary discipline of quantum machine learning (QML), which combines the capabilities of quantum computing with the fundamentals of machine learning, has developed. In this article, the significance of algorithms in quantum machine learning and how they create the framework for the quantum AI revolution are examined (Nath et al., 2021).

The importance of quantum machine learning in overcoming the drawbacks of traditional machine learning algorithms is covered in the article's opening paragraphs. It emphasizes the quantum computers' exponential processing capacity and how this might fundamentally alter the capabilities of AI systems. In order to comprehend quantum machine learning algorithms, the paper then digs into the fundamental ideas of quantum computing. It clarifies the fundamental concepts of superposition, entanglement, and quantum interference that serve as the foundation for quantum algorithms. Qubits and quantum gates are also introduced, emphasizing their significance in quantum calculations (Quiroga et al., 2021).

On top of this basis, the article investigates various quantum machine learning methods. The article examines quantum variations of conventional machine learning methods, including quantum SVMs, quantum neural networks, and quantum clustering algorithms. It describes how these algorithms take advantage of quantum features to strengthen learning tasks, advance optimization procedures, and manage complex data representations. The article also emphasizes the significance of algorithmic creation and improvement in quantum machine learning. It examines difficulties in using quantum algorithms, including quantum error correction, circuit depth optimization, and utilization of quantum resources. It emphasizes the necessity of algorithmic improvements to fully utilize quantum computing's potential in AI (Khang & Kali et al., 2023).

The article also looks into possible sectors where quantum machine learning techniques could be used. In areas including drug development, financial optimization, recommendation systems, and quantum chemistry simulations, it explains how quantum algorithms can aid in the solution of challenging issues. The revolutionary influence of quantum machine learning algorithms is demonstrated using case studies and real-world situations.

The article's conclusion discusses the potential applications and ramifications of quantum machine learning methods. It examines current research initiatives, improvements in quantum technology, and the possibility for hybrid strategies that blend conventional and novel quantum algorithms. The article highlights the crucial role of algorithms in creating the groundwork for the quantum AI revolution, advancing intelligent systems, and opening up new avenues for resolving challenging issues (Michael, Guillaume, et al., 2022).

In final analysis, this study offers a thorough investigation of quantum machine learning algorithms and their function in setting the groundwork for the quantum AI revolution. It examines numerous quantum machine-learning techniques and emphasizes how quantum computing might improve machine-learning problems. Researchers are laying the groundwork for revolutionary developments in AI systems by creating and improving these algorithms. In order to fully utilize the capabilities of quantum computing in AI, the article emphasizes the significance of ongoing research and collaboration in algorithm development (Shyam & Khang et al., 2023).

13. CONCLUSION

In order to improve AI systems, QML combines the strength of quantum computing with the concepts of machine learning. This article investigates QML's potential to improve AI and fuel revolutionary developments. The paper opens by pointing out the shortcomings of traditional machine learning algorithms when it comes to solving complicated issues. It introduces the idea of quantum computing and describes how special features like superposition and entanglement might fundamentally alter the power of AI systems. The paper then delves into the foundations of QML. It examines qubits, quantum gates, and quantum circuits as well as the theoretical underpinnings of quantum computing. In particular, it highlights how quantum parallelism and quantum interference could speed up machine learning procedures. This foundation is built upon as the article covers QML's uses in AI.

Quantum iterations of well-known machine learning algorithms are covered, including quantum neural networks, quantum SVMs, and quantum clustering methods. It serves as an illustration of how QML can enhance processes like pattern recognition, optimization, and data analysis. The article also discusses the difficulties and factors to take into account when using QML. It talks about how to integrate QML algorithms with current AI frameworks, the current constraints of quantum hardware, and the demand for specialized knowledge in both quantum computing and machine learning. It emphasizes how critical it is to create scalable, reliable quantum algorithms and to maximize the fusion of classical and quantum systems. The article also addresses QML's potential for development and how it might advance AI. It examines current research initiatives, developments in quantum hardware, and the possibilities for hybrid strategies that mix traditional and cutting-edge quantum computing methods (Khang & Muthmainnah et al., 2023).

The transformative potential of QML in enhancing AI system capabilities and opening up new horizons in the field of AI is highlighted in the article's conclusion. In conclusion, this article gives a general overview of quantum machine learning (QML) and how it can improve artificial intelligence (AI) using quantum computers. It draws attention to the potential for quantum phenomena like superposition and entanglement to fundamentally alter machine learning techniques. Researchers can develop AI fundamentally by utilizing QML algorithms, enhancing operations like pattern recognition, optimization, and data analysis. The article emphasizes the need for continued research, collaboration, and advancements in quantum computing to fully harness the potential of QML in enhancing AI (Khang & Shah et al., 2023).

REFERENCES

Adebiyi, M. O., Fatinikun-Olaniyan, D., Osang, F., & Adebiyi, A. A. (2023). Quantum Theory Approach to Performance Enhancement in Machine Learning. *2023 International Conference on Science, Engineering and Business for Sustainable Development Goals (SEB-SDG)*, Omu-Aran, Nigeria. 10.1109/SEB-SDG57117.2023.10124582

Batra, K., Zorn, K. M., Foil, D. H., Minerali, E., Gawriljuk, V. O., Lane, T. R., & Ekins, S. (2021). Quantum Machine Learning Algorithms for Drug Discovery Applications. *Journal of Chemical Information and Modeling*, *61*(6), 2641–2647. doi:10.1021/acs.jcim.1c00166 PMID:34032436

Bova, F., Goldfarb, A., & Melko, R. (2021). Commercial applications of quantum computing. *EPJ Quantum Technol.* . doi:10.1140/epjqt/s40507-021-00091-1

BroughtonM.VerdonG.McCourtT.MartinezA. J.YooJ. H.IsakovS. V. (2003). TensorFlow Quantum: A Software Framework for Quantum Machine Learning. arXiv. https://arxiv.org/abs/2003.02989

Cerezo, M., Verdon, G., Huang, H. Y., Cincio, L., & Coles, P. J. (2022). Challenges and opportunities in quantum machine learning. *Nature Computational Science*, *2*(9), 567–576. doi:10.103843588-022-00311-3

Chalumuri, A., Kune, R., & Manoj, B. S. (2021). A hybrid classical-quantum approach for multi-class classification. *Quantum Information Processing*, *20*(3), 119. doi:10.100711128-021-03029-9

Chen, S. Y.-C., & Yoo, S. (2021). Federated Quantum Machine Learning. *Entropy (Basel, Switzerland)*, *23*(4), 460. doi:10.3390/e23040460 PMID:33924721

Cruz-Benito, J., Faro, I., Martín-Fernández, F., Therón, R., & García-Peñalvo, F. J. (2018). A Deep-Learning-Based Proposal to Aid Users in Quantum Computing Programming. In P. Zaphiris & A. Ioannou (Eds.), Lecture Notes in Computer Science. *Learning and Collaboration Technologies. Learning and Teaching. LCT 2018*. Springer. doi:10.1007/978-3-319-91152-6_32

Essam, H., Abohashima, Z., Elhoseny, M., & Mohamed, W. (2022). Machine learning in the quantum realm: The state-of-the-art, challenges, and future vision. *Expert Systems with Applications, 194*. doi:10.1016/j.eswa.2022.116512

Eswaran, U., Khang, A., & Eswaran, V. (2023). Role of Quantum Computing in the Era of Artificial Intelligence. Applications and Principles of Quantum Computing. IGI Global Press. https://doi.org/doi:10.4018/979-8-3693-1168-4-ch003

Fastovets, D. V., Bogdanov, Yu. I., Bantysh, B. I., & Lukichev, V. F. (2018). Machine learning methods in quantum computing theory. *Proc. SPIE 11022, International Conference on Micro- and Nano-Electronics*. SPIE. 10.1117/12.2522427

Fernández Pérez, I., Prieta, F. l., Rodríguez-González, S., Corchado, J. M., & Prieto, J. (2023). Quantum AI: Achievements and Challenges in the Interplay of Quantum Computing and Artificial Intelligence. In: Julián, V., Carneiro, J., Alonso, R.S., Chamoso, P., Novais, P. (eds) *Ambient Intelligence—Software and Applications—13th International Symposium on Ambient Intelligence*. Springer, Cham. 10.1007/978-3-031-22356-3_15

Gan, B. Y., Leykam, D., & Angelakis, D. G. (2022). Fock state-enhanced expressivity of quantum machine learning models. *EPJ Quantum Technology*, *9*(1), 16. doi:10.1140/epjqt40507-022-00135-0

Hajimahmud, V. A., Khang, A., Hahanov, V., Litvinova, E., Chumachenko, S., & Alyar, V. A. (2022). Autonomous Robots for Smart City: Closer to Augmented Humanity. AI-Centric Smart City Ecosystems: Technologies, Design and Implementation (1st Ed.), 7 (12), CRC Press. doi:10.1201/9781003252542-7

Hu, F., Wang, B.-N., Wang, N., & Wang, C. (2019). Quantum machine learning with D-wave quantum computer. *Quantum Engineering*, *1*(2), e12. doi:10.1002/que2.12

Huang, H. Y., Broughton, M., Mohseni, M., Babbush, R., Boixo, S., Neven, H., & McClean, J. R. (2021). Power of data in quantum machine learning. *Nature Communications*, *12*(1), 2631. doi:10.103841467-021-22539-9 PMID:33976136

Jerbi, S., Fiderer, L. J., Poulsen Nautrup, H., Kübler, J. M., Briegel, H. J., & Dunjko, V. (2023). Quantum machine learning beyond kernel methods. *Nature Communications*, *14*(1), 517. doi:10.103841467-023-36159-y PMID:36720861

Khang, A. (2022). Cyber-Physical-Social System and İncident Management. AI-Centric Smart City Ecosystems: Technologies, Design and Implementation, 7(12), CRC Press. doi:10.1201/9781003252542-2

Khang, A. (2023). AI and IoT-Based Technologies for Precision Medicine. IGI Global Press. doi:10.4018/979-8-3693-0876-9

Khang, A. (2023). AI-Aided Data Analytics Tools and Applications for the Healthcare Sector. AI and IoT-Based Technologies for Precision Medicine. IGI Global Press. doi:10.4018/979-8-3693-0876-9.ch018

Khang, A. (2023). Enabling the Future of Manufacturing: Integration of Robotics and IoT to Smart Factory Infrastructure in Industry 4.0. AI-Based Technologies and Applications in the Era of the Metaverse. IGI Global Press. doi:10.4018/978-1-6684-8851-5.ch002

Khang, A. (2023). Applications and Principles of Quantum Computing. IGI Global Press. https://doi.org/ doi:10.4018/979-8-3693-1168-4

Khang, A. (2023). Quantum Mechanics Primer - Fundamentals and Quantum Computing. Applications and Principles of Quantum Computing. IGI Global Press. https://doi.org/ doi:10.4018/979-8-3693-1168-4-ch001

Khang, A. (2023). Introduction to Quantum Computing and Its Integration Applications. Applications and Principles of Quantum Computing. IGI Global Press. https://doi.org/ doi:10.4018/979-8-3693-1168-4-ch002

Khang A. (2024). *The Quantum Evolution: Application of AI and Robotics in the Future of Quantum Technology*. CRC Press. https://doi.org/ doi:10.1201/978-1-032-64207-9

Khang, A., & Muthmainnah, M. (2023). AI-Aided Teaching Model for the Education 5.0 Ecosystem. AI-Based Technologies and Applications in the Era of the Metaverse. IGI Global Press. doi:10.4018/978-1-6684-8851-5.ch004

Khang, A., Shah, V., & Rani, S. (2023). AI-Based Technologies and Applications in the Era of the Metaverse. (1 Ed.) IGI Global Press. doi:10.4018/978-1-6684-8851-5

Metawei, M. A., Eldeeb, H., Nassar, S. M., & Taher, M. (2022). Quantum Computing Meets Artificial Intelligence: Innovations and Challenges. In M. Virvou, G. A. Tsihrintzis, N. G. Bourbakis, & L. C. Jain (Eds.), *Handbook on Artificial Intelligence-Empowered Applied Software Engineering. Artificial Intelligence-Enhanced Software and Systems Engineering* (Vol. 2). Springer. doi:10.1007/978-3-031-08202-3_12

Metawei, M. A., Eldeeb, H., Nassar, S. M., & Taher, M. (2022). Quantum Computing Meets Artificial Intelligence: Innovations and Challenges. In M. Virvou, G. A. Tsihrintzis, N. G. Bourbakis, & L. C. Jain (Eds.), *Handbook on Artificial Intelligence-Empowered Applied Software Engineering. Artificial Intelligence-Enhanced Software and Systems Engineering* (Vol. 2). Springer. doi:10.1007/978-3-031-08202-3_12

Mishra, N. (2021). Quantum Machine Learning: A Review and Current Status. In N. Sharma, A. Chakrabarti, V. E. Balas, & J. Martinovic (Eds.), *Data Management, Analytics and Innovation. Advances in Intelligent Systems and Computing* (Vol. 1175). Springer. doi:10.1007/978-981-15-5619-7_8

Nath, R. K., Thapliyal, H., & Humble, T. S. (2021). A Review of Machine Learning Classification Using Quantum Annealing for Real-World Applications. *SN Computer Science*, *2*(5), 365. doi:10.100742979-021-00751-0

O'Driscoll, L., Nichols, R., & Knott, P. A. (2019). A hybrid machine learning algorithm for designing quantum experiments. *Quantum Machine Intelligence*, *1*(1-2), 5–15. doi:10.100742484-019-00003-8

Pande, M., & Mulay, P. (2020). Bibliometric Survey of Quantum Machine Learning. *Science & Technology Libraries*, *39*(4), 369–382. doi:10.1080/0194262X.2020.1776193

Peters, E., Caldeira, J., & Ho, A. (2021). Machine learning of high dimensional data on a noisy quantum processor. *npj Quantum Inf.*, *7*, 161. . doi:10.1038/s41534-021-00498-9

Quiroga, D., Date, P., & Pooser, R. (2021). Discriminating Quantum States with Quantum Machine Learning. *2021 IEEE International Conference on Quantum Computing and Engineering (QCE)*, Broomfield, CO, USA. 10.1109/QCE52317.2021.00088

Ramezani, S. B., Sommers, A., Manchukonda, H. K., Rahimi, S., & Amirlatifi, A. (2020). Machine Learning Algorithms in Quantum Computing: A Survey. *2020 International Joint Conference on Neural Networks (IJCNN)*, Glasgow, UK. 10.1109/IJCNN48605.2020.9207714

Raschka, S., Patterson, J., & Nolet, C. (2020). Machine Learning in Python: Main Developments and Technology Trends in Data Science, Machine Learning, and Artificial Intelligence. *Information (Basel)*, *11*(4), 193. doi:10.3390/info11040193

Saini, S., Khosla, P., Kaur, M., & Singh, G. (2020). Quantum Driven Machine Learning. *International Journal of Theoretical Physics*, *59*(12), 4013–4024. doi:10.100710773-020-04656-1

Schuld, M., Sinayskiy, I., & Petruccione, F. (2015). An introduction to quantum machine learning. *Contemporary Physics*, *56*(2), 172–185. doi:10.1080/00107514.2014.964942

Shah, V., & Khang, A. (2023). Metaverse-Enabling IoT Technology for a Futuristic Healthcare System. AI-Based Technologies and Applications in the Era of the Metaverse. IGI Global Press. doi:10.4018/978-1-6684-8851-5.ch008

Shyam, R. (2023). Effects of Quantum Technology on Metaverse. AI-Based Technologies and Applications in the Era of the Metaverse. IGI Global Press. doi:10.4018/978-1-6684-8851-5.ch009

Uehara, G. S., Spanias, A., & Clark, W. (2021). *Quantum Information Processing Algorithms with Emphasis on Machine Learning*. 2021 12th International Conference on Information, Intelligence, Systems & Applications (IISA), Chania Crete, Greece. 10.1109/IISA52424.2021.9555570

Wei, L., Liu, H., Xu, J., Shi, L., Shan, Z., Zhao, B., & Gao, Y. (2023). Quantum machine learning in medical image analysis: A survey. *Neurocomputing, 525*. doi:10.1016/j.neucom.2023.01.049

Zeguendry, A., Jarir, Z., & Quafafou, M. (2023). Quantum Machine Learning: A Review and Case Studies. *Entropy (Basel, Switzerland), 25*(2), 287. doi:10.3390/e25020287 PMID:36832654

Zhang, Y., & Ni, Q. (2020). Recent advances in quantum machine learning. Quantum. *Quantum Engineering*, *2*(1), e34. doi:10.1002/que2.34

Chapter 9
Quantum Computing AI:
Artificial Intelligence and Quantum Computing Applications

Roheen Qamar
https://orcid.org/0000-0003-4169-9455
Quaid-e-Awam University of Engineering, Sciences, and Technology, Sindh, Pakistan

Baqar Ali Zardari
Quaid-e-Awam University of Engineering, Sciences, and Technology, Sindh, Pakistan

Alex Khang
https://orcid.org/0000-0001-8379-4659
Global Research Institute of Technology and Engineering, USA

ABSTRACT

Quantum artificial intelligence (QAI) is a topic of study that blends quantum computing and AI. In order to investigate the possibilities of fruitful cooperation across both disciplines of computer science (CS), the authors will attempt to convey some of the relevant properties of both artificial intelligence (AI) and quantum computing (QC) from an academic point of view. The key artificial intelligence applications that can be addressed more effectively by employing the current generation of quantum computers will be covered in this chapter, along with a review of the fundamental components of quantum computing. There are several similarities between quantum computing and artificial intelligence. Quantum computing can increase the training speed and processing capability of artificial intelligence and machine learning algorithms at a lower cost. This chapter discusses the fundamental ideas that enable scientists and engineers to combine AI with quantum computing.

1. INTRODUCTION

A great technology that has developed over time, quantum computing is 100 trillion times faster than a supercomputer. It is based on quantum theory, which has the ability to perform computations utilizing

DOI: 10.4018/979-8-3693-1168-4.ch009

Figure 1. Artificial intelligence

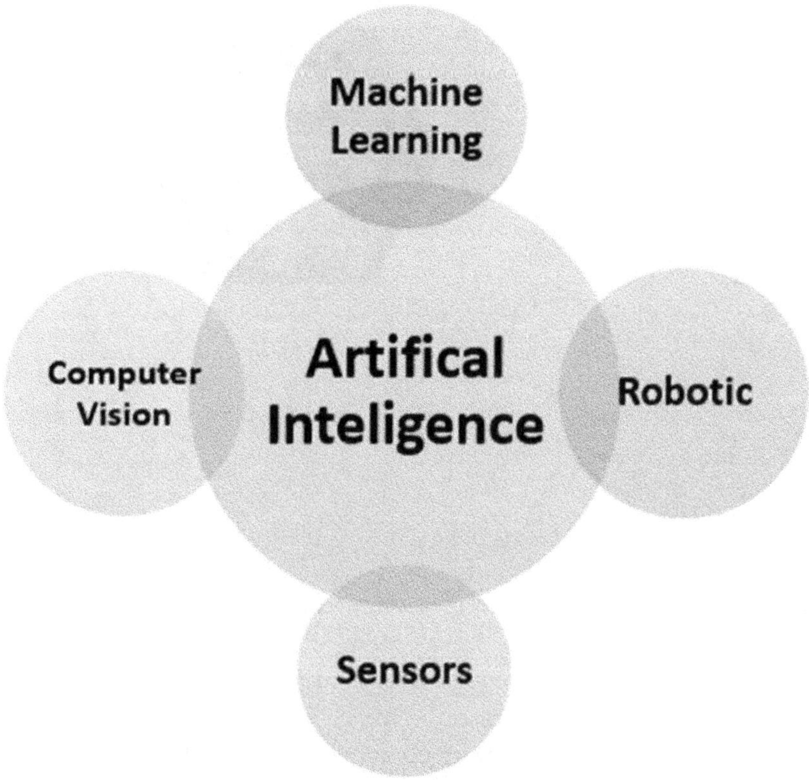

both linear algebra and probability theory. The concept of Quantum Computation (QC) was originally presented in 1982 by Richard Feynman the scientific discipline of quantum computation (QC) investigates how the quantum behavior of specific subatomic particles (such as photons, electrons, etc.) might be used to carry out computation and eventually massive amounts of information processing.

Two crucial quantum phenomena, superposition and entanglement, offer a considerably more effective approach to carry out specific types of computations than traditional algorithmic techniques. Information is kept in quantum registers, which are made up of a series of quantum bits in QC. To conduct straightforward qubit-range operations on quantum registers, QC defines a collection of operators called quantum gates. Computations. Quantum algorithms are applications of numerous quantum techniques in succession do more complex computations using gates on a quantum register as shown in Figure 1 (Gyongyosi et al., 2021).

The field of artificial intelligence (AI) and quantum computing are combined in the ground-breaking technology known as quantum artificial intelligence (QAI). By facilitating quicker and more precise problem-solving, forecasts, and optimizations, this innovative technique has the potential to completely transform the field of artificial intelligence. The whole field of information processing will benefit greatly from QC's capacity to carry out parallel information processing and quick searches over unordered sets of data. The advantages of QC in the field of artificial intelligence (AI) are the main topic of this essay. The current essay makes an effort to take a more comprehensive approach to quantum approaches for

Figure 2. Artificial intelligence applications

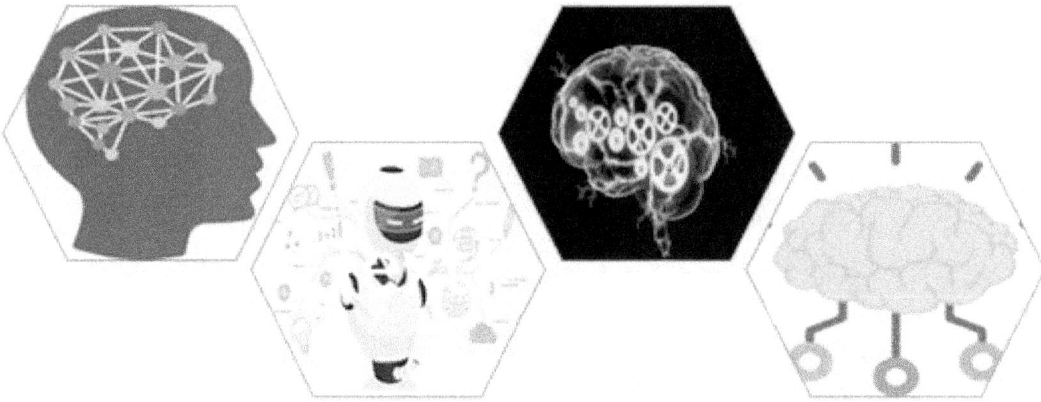

AI applications, taking into account both past research and some big-picture concepts for future study as shown in Figure 2 (Mangini et al., 2021).

But Due to space constraints, it first provides a very quick overview of the fundamentals of quality control Algorithms, simply the basics to grasp the topic. Since several successful applications of AI have demonstrated that we can rely on quantum computers for calculations, interest in using them to compute artificially intelligent algorithms has grown tremendously. Section 1 of this study will explain the foundations of quantum computing from the standpoint of the quantum computing programming paradigm (Wichert et al., 2020).

Understanding quantum computing, QAI's underlying technology, is crucial for a complete understanding. Quantum computing uses qubits, which can be in several states at once, as opposed to classical computing, which processes data using binary bits that can either be 0 or 1.

Due to this special characteristic, quantum computers are able to process complex calculations significantly faster than conventional computers. QAI has the potential to revolutionize our approach to computing and problem-solving by fusing the enormous computational power of quantum computing with the suppleness and adaptability of AI.

According to reports, quantum computing can support artificial intelligence in a variety of ways, including the processing of enormous complicated datasets and the evolution of algorithms to facilitate improved learning, reasoning, and comprehension. Quantum computing opens up a number of opportunities for machine learning algorithms and natural language processing. For instance, a "meaning aware" algorithm was recently created by running a natural language processing algorithm on quantum computing (Dyakonov et al., 2019).

Figure 3. Machine learning

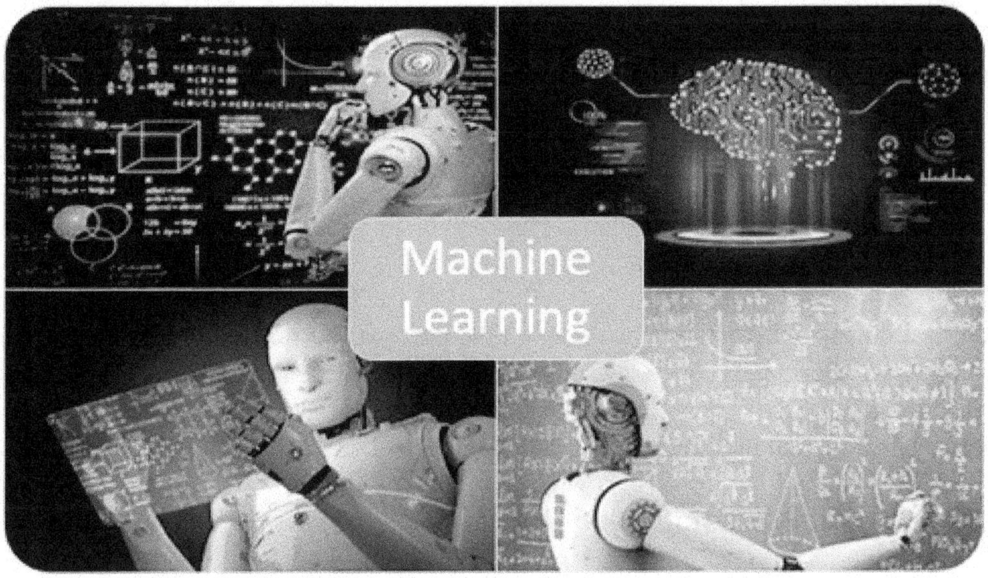

2. MACHINE LEARNING

Machine learning is a subfield of artificial intelligence (AI) and computer science that uses data and algorithms to mimic how humans learn, gradually improving its accuracy. The term "artificial intelligence" refers to systems or machines that mimic human intelligence. Machine learning and AI are commonly discussed together, and the phrases are sometimes used interchangeably, despite the fact that they do not mean the same thing. While all machine learning is AI, not all AI is machine learning (Khang & Quantum, 2023).

Machine Learning is the branch of science that studies how computers can learn without being explicitly programmed. ML is one of the most intriguing technologies that I have ever encountered. As the name implies, it offers the computer the ability to learn, which makes it more human-like. Machine learning is being employed actively now, possibly in more places than one would imagine as shown in Figure 3 (Zhou et al., 2022).

3. QUANTUM COMPUTING BASIC BUILDING BLOCKS

In digital computers, all information is converted into a series of bits, and logic gates are used for all computations. A computer with n states can exist in any one of the 2n possible states because the bit state of a digital computer determines its state.

A quantum computer likewise has bits, but instead of representing 1s and 0s, the quantum bit can simultaneously represent a linear combination of 1 and 0; this is referred to as superposition. A universal model known as quantum mechanics is built on ideas that are not found in everyday experience. For quantum computing to process data, a quantum model of the data is required. For error correction and the proper operation of the quantum computer, hybrid quantum-classical models are also required (Khang & Kali, 2023).

Quantum information serves as the foundation for quantum computers. Information is represented as quantum states in these computers, and quantum computations can be carried out in a quantum computer by taking advantage of the quantum effects provided by quantum mechanics (such as quantum superposition, quantum entanglement, quantum interference, the no cloning theorem, decoherence, etc.) (Cao et al., 2017).

3.1 Quantum Data

For computerization, quantum data can be thought of as data packets housed in qubits. The qualities that make quantum data important, such as superposition and entanglement, also make it difficult to observe and store. Additionally, because quantum data is noisy, machine learning must be used in the process of appropriately evaluating and interpreting this data (Dejpasand et al., 2023).

3.2 Hybrid Quantum-Classical Models

Generating quantum data alone using quantum processors greatly increases the likelihood of obtaining nonsensical data. Because of this, a hybrid model develops when it is driven by quick data processing components like the CPU and GPU, which are typically employed in traditional computers (Dejpasand et al., 2023).

3.3 Algorithms for Quantum Systems

An algorithm is a series of actions that solves a problem. One needs to employ particular instruction sets that the device is built to handle in order to carry out these processes on a device. In contrast to classical computing, quantum computing introduces new instruction sets that are based on a totally different conception of execution. Quantum algorithms seek to speed up problem solving by utilizing quantum phenomena like superposition and entanglement (Dejpasand et al., 2023).

4. AI AND QUANTUM COMPUTING APPLICATIONS

Although artificial intelligence (AI) has advanced quickly over the past ten years, it has not yet overcome technological constraints. Obstacles to obtaining AGI (Artificial General Intelligence) can be removed thanks to quantum computing's special properties. Machine learning models can be trained quickly with quantum computing, and the process can also be utilized to develop improved algorithms. Quantum computing's optimized and reliable AI may speed up years of investigation and provide technological advancements. Some of the core issues facing modern AI include neuro morphic cognitive models, adaptive machine learning, and reasoning under uncertainty. One of the most likely options for future AI is quantum AI (Rawat et al., 2022).

The open-source quantum machine learning library Tensor Flow Quantum (TFQ) from Google is an illustration of a set of tools that blends quantum modeling and machine learning methods. The goal of TFQ is to offer the resources needed to model and control real or created quantum systems (Khang & Vugar et al., 2023).

The following are some of the applications of this super combination of quantum computing and AI, keeping in mind that the term "quantum AI" means the use of quantum computing for computation of machine learning algorithms, which takes advantage of the computational superiority of quantum computing to achieve results that are not possible to achieve with classical computers (Rodríguez-González et al., 2023).

4.1 Encoding of Information

According to projections, quantum computers will use quantum data to process information in a manner akin to that of classical computers. The issue of converting classical data to quantum data has garnered a lot of interest in recent years, leading to numerous breakthroughs. This is because the established conceptuality of QC outperforming CC, the capacity to do classical operations, and the limitations in generating quantum data. Data encoding is the process that is being used. Since data come in a variety of formats, there are numerous ways to convert conventional data to quantum data (Rodríguez-González et al., 2023).

4.2 Processing Big Data Sets

Every day, we generate 2.5 exabytes of data. That is equal to the contents of 250,000 Library of Congresses or 5 million laptops. With 9,722 pins on Pinterest, 347,222 tweets, 4.2 million likes on Facebook, and all the other data we make by taking photos and videos, saving documents, opening accounts, and more, 3.2 billion people utilize the internet every minute of every day.

Quantum computers are made to handle the enormous volume of data as well as fast find patterns and identify anomalies. Developers are now better able to control the potential of quantum bits with each newly released iteration of quantum computer design and the new advancements achieved on the quantum error-correction code. Additionally, the same is optimized for resolving various business issues so that better decisions may be made (Miller et al., 2019).

4.3 Machine Learning Quantum

Innovation in the field of quantum machine learning is ready, and the possibilities are endless. We explore the field of quantum-enhanced classical algorithms, which combines the best aspects of the classical and quantum worlds. Quantum neural networks have the potential to revolutionize deep learning and pattern recognition, while quantum support vector machines offer superior classification capabilities. Imagine a day in the future when medical research has advanced dramatically thanks to quantum machine learning (Khang & Medicine, 2023).

Researchers can evaluate sizable genetic study datasets, comprehend the complexities of protein folding, and speed up medication discovery by utilizing the capabilities of quantum computing. As medical therapies are carefully adapted to each patient's genetic profile, personalized medicine becomes a reality. Medical robots powered by quantum AI help surgeons perform complex procedures with unmatched accuracy and efficiency, saving lives and revolutionizing healthcare (Khan et al., 2020).

4.4 Handling Large Datasets

New technologies like machine learning and AI consume large amounts of data, making traditional computers difficult to evaluate. Quantum computers, designed to manage large datasets and quickly identify patterns, are being improved to better manage potential. Quantum computers offer immense power for businesses and consumers, enabling better decision-making and a revolution in data analysis (Miller et al., 2019).

4.5 Building Better Models

Quantum computers are playing a crucial role in building better models for industries like pharmaceuticals, finance, and life science, enabling complex scenarios and improved data frameworks, potentially leading to better healthcare treatments, decreased financial implosion, and improved manufacturing logistics (Pooranam et al., 2023).

4.6 Integration of Multiple Datasets

Quantum computers can help organizations manage and integrate multiple datasets, making the process quicker and easier. This allows for quick analysis and integration of large datasets, improving machine learning and artificial intelligence capabilities. Quantum computing's ability to handle multiple stakes makes it suitable for solving business problems (Humble et al., 2021).

4.7 Medical Care

In order to simulate the size and complexity of molecules, which are crucial components of drug research, current generation computers are constrained. For instance, if a medication development exercise involves dealing with N atoms, the interactions that must be carried out among these atoms are exponentially complicated. It has been tested and discovered that quantum computers would make it possible to simulate far larger molecules, facilitating interactions between medications and the enormous number of proteins (on the order of 20K+) that are encoded in the human genome as shown in Figure 4. This demonstrates incredibly ambitious advances in pharmacology (Flöther et al., 2023).

4.8 Accounting and Marketing

The use of automated high frequency trading will be obsolete once quantum computers are available. Quantum computers are capable of running sophisticated algorithms that can automate trading and efficiently manage share dealings while taking into account all potential market factors. Quantum computing has the potential to make high-volume trading excessively quickly. The collection and analysis of enormous amounts of customer data might be completed quickly in terms of marketing. Big data analytics will also enable government and e-commerce organizations to target customers and voters with communications tailored to their preferences (Taylor et al., 2020).

Figure 4. Medical care

4.9 Assessment and Therapy

With the use of quantum technology, diagnostics are now more rapid and precise. Quantum computations combined with AI and machine learning can significantly improve pattern detection in MRI-driven medical imaging (Khang, Vugar and Abuzarova et al., 2023). The same is true for radiotherapy treatments, which rely on the capability to quickly model and simulate complicated scenarios, which can be easily handled with quantum computing focused on providing the best possible treatments. This is due to the fact that the therapists can quickly run a large number of simulations, limiting the potential radiation damage to healthy tissues (Shan et al., 2021).

4.10 Identifying Fraud

It is commonly recognized that pattern recognition plays a key role in animal identification. It has been demonstrated that quantum computing speeds up machine learning, substantially lowering the amount of time needed for training and, consequently, for detection. The tools and techniques utilized for fraud management, control, and prevention serve as an illustration of ongoing applied innovation. Systems and processes are able to thwart almost all efforts at fraud in any financial business. However, these remarkable outcomes don't leave much room for

Complacency. Fraudsters are always creating new schemes. Methods and increasing the intensity of their assaults. To identify data patterns that are indicative of fraud, we need new technologies, approaches, and procedures. The principles and phenomena at the heart of quantum physics are used by quantum computers to perform computations in a fundamentally new way. Most significantly, using superposition enables quantum computers to encode 2n values, where n is the quantity of qubits. This allows for the simultaneous evaluation of a large number of potential solutions, combined with entanglement. The efficiency of the computation is increased by this parallelization (Boucher et al., 2019).

4.11 Meteorology

Forecasting the weather accurately is challenging since there are so many variables that are unclear. Improved pattern identification brought about by machine learning on quantum computers would make it simpler to predict extreme weather events and could result in the annual saving of thousands of lives.

In order to gain a better understanding of climate change and how to lessen its harmful effects, climatologists will also be able to create and analyze more intricate climate models. In order to accurately predict the weather, it is necessary to analyze huge amounts of data, including a variety of dynamic elements like air temperature, pressure, and density that interact in complex ways (Nivelkar et al., 2023).

5. RELATED WORK

Wang et al. (2020) In this paper, various aspects of quantum technology are introduced, including quantum teleportation, quantum experiment satellite, quantum cryptography, quantum computers, quantum devices, quantum processing units (QPU), quantum artificial intelligence (QAI), quantum machine learning (QML), quantum cloud computing, and quantum network and communication. Additionally, there are certain difficulties.

This study examines all advised information security and privacy controls from the NIST SP 800-53 Rev. 5 catalog of controls to further address the worry that quantum computers will adversely affect information security. The report also provides a brief overview of current efforts to develop quantum-resistant standards.

Njorbuenwu et al. (2019) Unlike conventional computers, which use and process information in bits (0 or 1), quantum computers utilize and process information in quantum bits, or qubits, which can simultaneously represent other states in addition to ones and zeros. The extra states of superimposition and entanglement enable quantum computers' astonishingly fast computing. Information security risks have increased significantly as a result of the development of quantum computers. In order to clarify both the beneficial and detrimental effects of quantum computers on information security, this research uses a literature review methodology.

Kuppusamy and Nnvs et al. (2022) Machine learning technology is quickly advancing, enabling autonomous operation of devices in several industries. However, the proliferation of sensors in various industries generates a tremendous amount of data that machine learning algorithms use to enhance the performance of autonomous devices. However, real-time problems involving enormous amounts of data cannot be solved using conventional ML techniques and hardware. In order to accelerate the process, researchers have created quantum computing hardware systems and quantum machine learning algorithms. This research effort reviewed the QML methods used to classify the photos as well as quantum computing techniques. The performance comparison of various QML algorithms was shown in this paper. It shown that several QML methods are faster than traditional ML algorithms at classifying photos.

Zhang et al. (2022) this article attempts to provide a thorough evaluation of the most current developments in AI algorithms and applications in RS data processing in light of this age of astonishing evolution. More than 270 research papers, including machine learning, computational intelligence, AI explicability, data mining, natural language processing (NLP), and AI security, are included in the review. We highlight possible future research directions as we wrap up this review.

In remote sensing (RS), artificial intelligence (AI) is becoming more and more important. AI, and machine learning algorithms in particular, are used for a variety of tasks, from basic image processing to advance data analysis and knowledge discovery. All RS domains have experienced notable advancements as a result of the powerful strategy for RS data analysis that AI techniques have developed as.

Faruk et al. (2022) this paper attempts to consolidate basic and foundational research on quantum cyber security, which may be used to address both threats and potential solutions to pressing cyber security problems. We discuss the current proposed methodologies and give a thorough, graphical account of the state-of-the-art quantum computing and cyber security. According to research on cyber security and quantum computing, quantum computing can be used to reduce cyber security concerns while also posing some unanticipated risks. The depth and emphasis of this systematic survey serve as a foundation for future research in this field and offer quantum and cyber security practitioners and scholars a collected body of knowledge about current trends in the field. Quantum computing's promise is not to speed up traditional computing, but rather to provide an exponential advantage for some kinds of problems, with significant consequences for, for example, cyber security. Cyberspace security may soon emerge as the Internet's most pressing issue with the introduction and development of quantum computers.

Choi et al. (2020) The Harrow-Hassidim-Lloyd (HHL) algorithm, Variational Quantum Eigensolver (VQE), Quantum Approximate Optimization Algorithm (QAOA), and other helpful quantum computing approaches are introduced in this work for AI researchers. Artificial Intelligence (AI) and machine learning have been the trendiest research areas for many scientists over the last five years. Numerous AI engineers are still developing as a result of numerous machine learning researches. If AI and machine learning are the focus of current research trends, then quantum computing will be the focus of research trends in the not-too-distant future. Quantum computers now do exist thanks to the qubit implementation using superconductivity, diamond Nitrogen Vacancy (NV) centers, ion-traps, and other techniques. Additionally, the usage of quantum computers remotely for research has been made available via cloud computing for researchers all around the world. Even more so for many AI engineers, the universalization of quantum computing approaches is no longer a science fiction concept.

Torgler et al. (2020) this chapter explores the fascinating potentials that emergent technologies like big data, artificial intelligence, and quantum computing hold for the world of sports. The technologies' ability for more accurate data collecting and analysis, both together and independently, can improve sports-related decision-making and boost organizational performance in many different ways. Torgler also highlights the limitations of technology—as well as issues like inefficiency and privacy concerns—by considering the nature of sport. Finally, it investigates non-technological elements that affect people's intense interest in and emotional commitment to sports and sporting events (Khang & Agriculture et al., 2023).

Lee, Raymond ST, (Lee et al., 2020) this chapter examines two areas of quantum computing and how they interact to influence the development of the modern era of intelligent computing. The examination of the nature and dynamics of financial market operations in the quantum world is the most important question in finance engineering, as was discussed in earlier chapters. Quantum finance is a cross-disciplinary field. One of the main fields of research at the quantum finance forecast center (QFFC.org) is quantum finance. The chapter also explores and introduces additional QFFC aspects and ongoing research in the areas of AI-Fintech, which combines AI technology with cutting-edge financial technology like block chains and digital ledgers; and quantum computing, which is being researched and developed for potential applications in a variety of fields like quantum entanglements in quantum finance, quantum cryptography, and more.

Schuld et al. (2015) this article. It addresses the possibility of a future theory of quantum learning and gives the techniques as well as technical facts in an understandable manner. The developing topic of quantum machine learning is provided in this article. It addresses the possibility of a future theory of quantum learning and gives the techniques as well as technical facts in an understandable manner.

The present study provides a comprehensive review of the burgeoning topic of quantum machine learning. It addresses the possibilities of a future theory of quantum learning and gives techniques as well as technical facts in an understandable manner.

Alvarez-Rodriguez et al. (2018) the first experimental realization of a quantum artificial life algorithm in a quantum computer is presented. The quantum biomimetic protocol encodes specific quantum behaviors of living systems, such as self-replication, mutation, inter-individual interaction, and death, into the IBM ibmqx4 cloud quantum computer. Entanglement extends through generations of individuals in this experiment, where true quantum information properties are inherited via genealogical networks. Experiment data accurately fits the ideal model as a pioneering proof-of-principle. Following that, these and other quantum artificial life models, for which no classical device can foresee their quantum supremacy evolution, can be investigated further in new generations of quantum computers. Quantum biomimetic, quantum machine learning, and quantum artificial intelligence will progress in tandem as quantum complexity increases (Khang & Kali et al., 2023).

Rietsche et al. (2022) discuss potential applications of quantum computing as well as future research objectives in the realm of information systems. Quantum computing has the potential to be the next disruptive technology, with myriad applications and ramifications for businesses and markets. Quantum computers use quantum physics principles such as superposition and entanglement to represent data and perform operations on it. Both of these ideas allow quantum computers to address highly precise, difficult tasks much faster than conventional computers. Against this backdrop, this section provides a high-level overview of a quantum computer's three layers: hardware, system software, and application layer.

Gill et al. (2022) this article provides a comprehensive analysis of the QC literature and offers a QC taxonomy. The proposed taxonomy is used to map different related studies in order to identify research gaps. The present state-of-the-art in quantum software tools and technologies, post-quantum cryptography, and quantum computer hardware research is captured in a complete overview. The essay examines and discusses a number of open concerns as well as possible future avenues for research and innovation in Quebec.

Massive progress in both quantum hardware development and quantum software/algorithm development in recent years has brought QC much closer to reality. Indeed, the discovery of quantum supremacy is a watershed moment in the Noisy Intermediate Scale Quantum (NISQ) era, with the next logical step being the quantum advantage, in which quantum computers handle real-world problems far more effectively than classical computers. Quantum decoherence and qubit interconnection are two of the primary difficulties to achieving quantum advantage in the NISQ era as quantum devices are likely to continuously scale up in the next years. QC is a hotly debated and rapidly evolving subject of study, with significant continuous progress in all areas.

6. QUANTUM ARTIFICIAL INTELLIGENCE AND QUANTUM MACHINE LEARNING

The definition of a quantum neural network is a set of building blocks with a particular input and output. The biological neurons in our brain are the source of these neural structures. There are many layers in a quantum network, such as the input layer, output layer, decision layer, etc. Quantum artificial neural networks (QuANNs) are incredibly adaptable when it comes to processing massive amounts of input data and complex functions to forecast results. The ability of a quantum learning algorithm to respond and adapt in a quantum environment and a classical environment separately is one of its key benefits (Gupta et al., 2017).

A method to quantum machine learning based on networked quantum computation is provided by QuANNs. In QuANNs, the processing of the networks occurs in two stages: the learning stage (during which the networks converge to a particular quantum circuit) and the back propagation stage (during which the networks essentially act as self-programmers). Quantum tunneling is a technique used in one type of quantum machine learning (ML). Using both methods may result in a better solution than either one alone, regardless of whether the tunneling strategy is superior to simulated annealing (Ushaa, Khang et al., 2023).

A collection of theoretically positioned heuristic approaches are referred to as Meta heuristics. Because they will be used with quantum computers, quantum Meta heuristics are approximate algorithms that are designed to affect running time, performance, and metrics. However, classical computers can now run quantum-inspired meta heuristics, and quantum programming can be created by modeling quantum physics phenomena like entanglement and superposition (Khang & Shah et al., 2023). This allows us to forecast future outcomes or behaviors. A hybrid programming approach integrating a quantum computer (handling the quantum chromosome), quantum operators (modifying the register), and a classical computer (evaluating the fitness function and manipulating the algorithm flow) is needed to implement a meta heuristic on a quantum computer based on a circuit model (Gonçalves et al., 2016).

7. QUANTUM FUTURE OF AI

View the intriguing quantum future of artificial intelligence, where research projects and entrepreneurs bravely embrace this quantum revolution. The future entails not only scientific progress but also the democratization of access to quantum computing. We can collectively create a quantum-powered AI future that benefits humanity by encouraging responsible innovation and moral behavior (Prabu, Khang et al., 2023).

Imagine a world in which AI enthusiasts and quantum experts may collaborate because quantum computing is as accessible as traditional computing. Platforms for quantum cloud computing democratize access to quantum resources, fostering a thriving community for quantum artificial intelligence. Open-source quantum AI frameworks enable programmers to create applications powered by quantum technology with ease, sparking a surge of discoveries across industries (Jayaram et al., 2019) (Lu et al., 2019).

8. CONCLUSION

Quantum artificial intelligence (QAI) is a field of study that combines quantum computing and artificial intelligence. We will attempt to explain some of the relevant aspects of both artificial intelligence (AI) and quantum computing (QC) from an academic standpoint in order to study the prospects of beneficial collaboration across both disciplines of computer science (CS) (Khang & Kali, 2024).

This chapter will discuss the important artificial intelligence applications that can be addressed more effectively by using the current generation of quantum computers, as well as an overview of the essential components of quantum computing. Quantum computing and artificial intelligence share significant commonalities. Quantum computing has the potential to improve the training speed and processing capabilities of artificial intelligence and machine learning algorithms while also lowering their cost. The key concepts that allow scientists and engineers to merge AI with quantum computing are covered in this chapter (Khang & Muthmainnah et al., 2023).

REFERENCES

Alvarez-Rodriguez, U., Sanz, M., Lamata, L., & Solano, E. (2018). Quantum artificial life in an IBM quantum computer. *Scientific Reports, 8*(1), 14793. doi:10.103841598-018-33125-3 PMID:30287854

Boucher, P. (2019). How artificial intelligence works. EPRS–European Parliamentary Research Service, In: europarl. Europa, 2.

Cao, Y., Guerreschi, G. G., & Aspuru-Guzik, A. (2017). *Quantum neuron: an elementary building block for machine learning on quantum computers*. arXiv preprint arXiv:1711.11240

Choi, J., Oh, S., & Kim, J. (2020). The useful quantum computing techniques for artificial intelligence engineers. In *2020 International Conference on Information Networking (ICOIN),* (pp. 1-3). IEEE. 10.1109/ICOIN48656.2020.9016555

Dejpasand, M. T., & Ghamsari, M. S. (2023). Research Trends in Quantum Computers by Focusing on Qubits as Their Building Blocks. *Quantum Reports, 5*(3), 597–608. doi:10.3390/quantum5030039

Dyakonov, M. (2019). When will useful quantum computers be constructed? Not in the foreseeable future, this physicist argues. Here's why: The case against: Quantum computing. *IEEE Spectrum, 56*(3), 24–29. doi:10.1109/MSPEC.2019.8651931

Eswaran, U., Khang, A., & Eswaran, V. (2023). Role of Quantum Computing in The Era of Artificial Intelligence. Applications and Principles of Quantum Computing. IGI Global Press. https://doi.org/ doi:10.4018/979-8-3693-1168-4-ch003

Faruk, M. J. H., Tahora, S., Tasnim, M., Shahriar, H., & Sakib, N. (2022). A review of quantum cybersecurity: threats, risks and opportunities. In *2022 1st International Conference on AI in Cybersecurity (ICAIC),* (pp. 1-8). IEEE.

Flöther, F. (2023). *The state of quantum computing applications in health and medicine.* arXiv preprint arXiv:2301.09106

Gill, S. S., Kumar, A., Singh, H., Singh, M., Kaur, K., Usman, M., & Buyya, R. (2022). Quantum computing: A taxonomy, systematic review and future directions. *Software, Practice & Experience*, *52*(1), 66–114. doi:10.1002pe.3039

Gonçalves, C. P. (2016). *Quantum Neural Machine Learning Back propagation and Dynamics.* arXiv preprint arXiv:1609.06935.

Gupta, S., Mohanta, S., Chakraborty, M., & Ghosh, S. (2017). Quantum machine learning-using quantum computation in artificial intelligence and deep neural networks: Quantum computation and machine learning in artificial intelligence. In *2017 8th Annual Industrial Automation and Electromechanical Engineering Conference* (pp. 268-274). IEEE.

Gyongyosi, L., & Imre, S. (2019). A survey on quantum computing technology. *Computer Science Review*, *31*, 51–71. doi:10.1016/j.cosrev.2018.11.002

Humble, T. S., McCaskey, A., Lyakh, D. I., Gowrishankar, M., Frisch, A., & Monz, T. (2021). Quantum computers for high-performance computing. *IEEE Micro*, *41*(5), 15–23. doi:10.1109/MM.2021.3099140

Jayaram, M. A. (2018). Quantum Computing: Some Percepts and Realms of Applications. *International Journal of Computer Applications, 975.*

Khan, T. M., & Robles-Kelly, A. (2020). Machine learning: Quantum vs classical. *IEEE Access : Practical Innovations, Open Solutions*, *8*, 219275–219294. doi:10.1109/ACCESS.2020.3041719

Khang, A. (2023). *Advanced Technologies and AI-Equipped IoT Applications in High-Tech Agriculture* (1st ed.). IGI Global Press. doi:10.4018/978-1-6684-9231-4

Khang, A. (2023). AI and IoT-Based Technologies for Precision Medicine. IGI Global Press. doi:10.4018/979-8-3693-0876-9

Khang, A. (2023). AI-Aided Data Analytics Tools and Applications for the Healthcare Sector. AI and IoT-Based Technologies for Precision Medicine. IGI Global Press. doi:10.4018/979-8-3693-0876-9.ch018

Khang, A. (2023). Enabling the Future of Manufacturing: Integration of Robotics and IoT to Smart Factory Infrastructure in Industry 4.0. AI-Based Technologies and Applications in the Era of the Metaverse. IGI Global Press. doi:10.4018/978-1-6684-8851-5.ch002

Khang, A. (2023). Applications and Principles of Quantum Computing. IGI Global Press. https://doi. org/ doi:10.4018/979-8-3693-1168-4

Khang, A. (2023). Quantum Mechanics Primer - Fundamentals and Quantum Computing. Applications and Principles of Quantum Computing. IGI Global Press. https://doi.org/ doi:10.4018/979-8-3693-1168-4-ch001

Khang, A. (2023). Introduction to Quantum Computing and Its Integration Applications. Applications and Principles of Quantum Computing. IGI Global Press. https://doi.org/ doi:10.4018/979-8-3693-1168-4-ch002

Khang A. & Rath, K. (2024). *The Quantum Evolution: Application of AI and Robotics in the Future of Quantum Technology.* CRC Press. https://doi.org/ doi:10.1201/978-1-032-64207-9

Khang, A., & Muthmainnah, M. (2023). AI-Aided Teaching Model for the Education 5.0 Ecosystem" AI-Based Technologies and Applications in the Era of the Metaverse. IGI Global Press. doi:10.4018/978-1-6684-8851-5.ch004

Khang, A., Shah, V., & Rani, S. (2023). AI-Based Technologies and Applications in the Era of the Metaverse. IGI Global Press. doi:10.4018/978-1-6684-8851-5

Kuppusamy, P. (2022). Quantum Computing and Quantum Machine Learning Classification–A Survey. In *2022 IEEE 4th International Conference on Cybernetics, Cognition and Machine Learning Applications (ICCCMLA),* (pp. 200-204). IEEE.

Lu, Y. (2019). Artificial intelligence: A survey on evolution, models, applications and future trends. *Journal of Management Analytics*, *6*(1), 1–29. doi:10.1080/23270012.2019.1570365

Mangini, S., Tacchino, F., Gerace, D., Bajoni, D., & Macchiavello, C. (2021). Quantum computing models for artificial neural networks. *Europhysics Letters*, *134*(1), 10002. doi:10.1209/0295-5075/134/10002

Miller, A. (2019). The intrinsically linked future for human and Artificial Intelligence interaction. *Journal of Big Data*, *6*(1), 38. doi:10.118640537-019-0202-7

Miller, A. (2019). The intrinsically linked future for human and Artificial Intelligence interaction. *Journal of Big Data*, *6*(1), 38. doi:10.118640537-019-0202-7

Nivelkar, M., Bhirud, S., Singh, M., Ranjan, R., & Kumar, B. (2023). Quantum Computing to Study Cloud Turbulence Properties. *IEEE Access : Practical Innovations, Open Solutions*, *11*, 70679–70690. doi:10.1109/ACCESS.2023.3289924

Njorbuenwu, M., Swar, B., & Zavarsky, P. (2019). A survey on the impacts of quantum computers on information security. In *2019 2nd International conference on data intelligence and security (ICDIS),* (pp. 212-218). IEEE. 10.1109/ICDIS.2019.00039

Pooranam, N. (2023). Quantum computing: future of artificial intelligence and its applications. *Quantum Computing and Artificial Intelligence: Training Machine and Deep Learning Algorithms on Quantum Computers, 163.*

Prabu, B., & Khang, A. (2023). Quantum Computing Smart City: Fostering Sustainability and Green Infrastructure with Smart City. Applications and Principles of Quantum Computing. IGI Global Press. https://doi.org/ doi:10.4018/979-8-3693-1168-4-ch009

Rawat, B., Mehra, N., Bist, A. S., Yusup, M., & Yulia, P. A. S. (2022). Quantum computing and ai: Impacts & possibilities. *ADI Journal on Recent Innovation*, *3*(2), 202–207.

Rietsche, R., Dremel, C., Bosch, S., Steinacker, L., Meckel, M., & Leimeister, J.-M. (2022). Quantum computing. *Electronic Markets*, *32*(4), 2525–2536. doi:10.100712525-022-00570-y

Rodríguez-González. S. (2023). Quantum AI: Achievements and Challenges in the Interplay of Quantum Computing and Artificial Intelligence. In *Ambient Intelligence—Software and Applications—13th International Symposium on Ambient Intelligence, 603*. Springer Nature.

Schuld, M., Sinayskiy, I., & Petruccione, F. (2015). An introduction to quantum machine learning. *Contemporary Physics*, *56*(2), 172–185. doi:10.1080/00107514.2014.964942

Shan, T., Tay, F. R., & Gu, L. (2021). Application of artificial intelligence in dentistry. *Journal of Dental Research*, *100*(3), 232–244. doi:10.1177/0022034520969115 PMID:33118431

Taylor, R. D. (2020). Quantum artificial intelligence: A "Precautionary" US approach? *Telecommunications Policy*, *44*(6), 101909. doi:10.1016/j.telpol.2020.101909

Torgler, B. (2020). Big data, artificial intelligence, and quantum computing in sports. *21st Century Sports: How Technologies Will Change Sports in the Digital Age.*

Wang, L., & Alexander, C. A. (2020). Quantum science and quantum technology: Progress and challenges. *Am. J. Electr. Electron. Eng.*, *8*(2), 43–50.

Zhang, L., & Zhang, L. (2022). Artificial intelligence for remote sensing data analysis: A review of challenges and opportunities. *IEEE Geoscience and Remote Sensing Magazine*, *10*(2), 270–294. doi:10.1109/MGRS.2022.3145854

Zhou, Z.-H. (2022). Open-environment machine learning. *National Science Review*, *9*(8), nwac123. doi:10.1093/nsr/nwac123 PMID:35992239

Chapter 10
Quantum Computing Smart City:
Fostering Sustainability and Green Infrastructure With Smart City

Prabu Baskar
https://orcid.org/0000-0003-2030-7862
Department of Civil Engineering, Sona College of Technology, India

Alex Khang
https://orcid.org/0000-0001-8379-4659
Global Research Institute of Technology and Engineering, USA

Shalini Annadurai
https://orcid.org/0000-0002-2097-2718
Department of Civil Engineering, Sona College of Technology, India

S. S. Amritha
Sona College of Technology, India

Pavithra Annadurai
Vellore Institute of Technology, Vellore, India

ABSTRACT

Nowadays, Quantum computing can impact various aspects by enabling more efficient, optimized, and sustainable solutions to complex issues. Additionally, quantum-inspired algorithms can bring innovative approaches to optimize processes, speed up transfers, innovate product developments, and lessen the overall time durations. Quantum computing also is best suited for resolving complex problems. Particularly, it resolves optimization-related issues from security to financial trading. It can perform algorithms and calculations at an unprecedented speed, allowing it to fix problems that classical computers cannot handle and create green infrastructure for not only industries but also smart cities in the world. It enhances machine learning abilities, simulates quantum systems, optimizes challenges, and

DOI: 10.4018/979-8-3693-1168-4.ch010

contributes to the growth of scientific research as well as provides advanced data protection methods and secure communication.

1. INTRODUCTION

A "smart city" is a city that has undergone a process or a set of stages to make it more livable, resilient, and able to react more quickly to new issues. With these concepts in mind, it is obvious that a "smart city" supports metropolitan areas that offer necessary infrastructure, equitable living conditions for its citizens, a clean and sustainable environment, and the use of "smart" solutions. Compact areas are being researched with an eye towards equitable and sustainable development in order to create a model that would serve as a benchmark for other aspirational cities.

1.1 Features of Smart City

A smart city is one that has the following attributes, according to the Ministry of Urban Development, Government of India:

- Smart homes
- Sufficient water supply
- Security of energy supply
- Solid waste management
- Smart urban transportation
- Efficient internet connection and digitalization
- Health and education
- Safety and security of citizens
- E-government and citizen participation
- A sustainable environment
- Inclusive development

As a result, several traits of fully developed Smart Cities are detailed here. These traits were taken from the Ministry of Urban Development, Government of India

1.1.1 Area-Based Developments Encouraging Mixed Land Use

Planning for "unplanned regions" that include a variety of suitable activities and land uses close to one another is advised in order to increase the effectiveness of land use. The States will offer some freedom in land use and building byelaws in order to adapt to change.

1.1.2 Housing and Diversity

It encourages expanding everyone's access to housing.

1.1.3 Developing Walkable Communities

This is carried out to boost local economies, promote social interaction, maintain safety, and reduce traffic jams, air pollution, and resource depletion. All crucial administrative tasks are located nearby, and roads are constructed or refurbished not only for automobiles and public transportation but also for pedestrians and cyclists.

1.1.4 Protecting and Improving Open Areas

Here, the focus is on constructing parks, playgrounds, and recreational places in order to increase citizen quality of life, minimize the effects of urban heat in some locales, and overall support eco-balance.

1.1.5 Encouraging a Diversity of Travel Choices

It denotes last mile para-transport connectivity, public transportation, and transit oriented development (TOD).

1.1.6 Making Government More Cost- and Citizen-Friendly

The use of mobile devices to reduce service costs and provide services without the need to visit municipal offices, the creation of e-groups to hear from the public and gather feedback, and the use of online monitoring of programs and activities through the use of virtual site visits are all encouraged by this feature.

1.1.7 Providing an Identity to the City

This is based on the main economic activities, such as the production of local food, health care, education, crafts, sporting goods, furniture, hosiery, textiles, dairy products, and other commodities (Khang & Muthmainnah et al., 2023).

1.1.8 Applying Intelligent Solutions to Services and Infrastructure

They suggest that to improve them, area-based development is required. A few examples include designing areas that are less vulnerable to disasters, using fewer resources, and providing services at a lower cost. The following section covers a number of approaches and components that have been created based on the concept of smart cities.

1.2 Components of Smart City

As the idea of a smart city gains popularity, academics and researchers have started to describe the components in their own unique innovative ways. Here, based on the already known research in this field, three approaches to the elements of a smart city are identified. Three different methods exist for learning more specifically (Khang & Rani et al., 2022). There are three different approaches:

- Four components
- Six components
- Eight components

1.2.1 Four Components Approach

Researchers, such as Rumi Aijaz (2016), Raghupathi (2015), and Jain (2015) each identified four elements of a smart city. The following are these four elements:

- Retrofitting
- Redevelopment
- Green field development
- Pan city development

It is clear from reading these elements that they are the abbreviated version of the smart city concept discussed above.

- **Retrofitting:** This aspect of the city improvement process focuses on making existing built-up areas smarter, more livable, and efficient. The city will identify a region of more than 500 acres for this use after consulting with people.
- **Redevelopment** – This part focuses on urban renewal. Decrepit constructed environments must be replaced, and new layouts with improved infrastructure must be created. Urban local bodies (ULBs) will choose a region larger than 50 acres in this regard after consulting with residents.
- **Green Field Development** — this section addresses city growth. It will consider the requirements of the growing and underprivileged population. It will bring smart solutions to unoccupied regions larger than 250 acres.
- **The Pan City Development**-component seeks to enhance services by utilizing data, information, and technology to apply intelligent fixes to the current infrastructure.

Yet, these four components make it clear that the smart city strategy is area-based rather than whole-city-based, where smart solutions for diverse urban infrastructure and services will be planned and executed in an integrated manner using the appropriate technology (Khang & Gupta et al., 2023).

1.2.2 Six components approach

Cohen listed six elements that make up a smart city:

- **Smart Living:** Better access to city amenities and services, like housing and utilities, ensures a higher standard of living. This is what it means to live a "smart" life.
- **Intelligent People**: Individuals who are smart are those who have higher levels of education and skill development, good health, a dedication to lifelong learning, and greater community involvement.
- **Smart Mobility** refers to efficient movement of people, products, and information as well as creative, safe, and sustainable transportation systems.

Figure 1. Six component approach of smart city

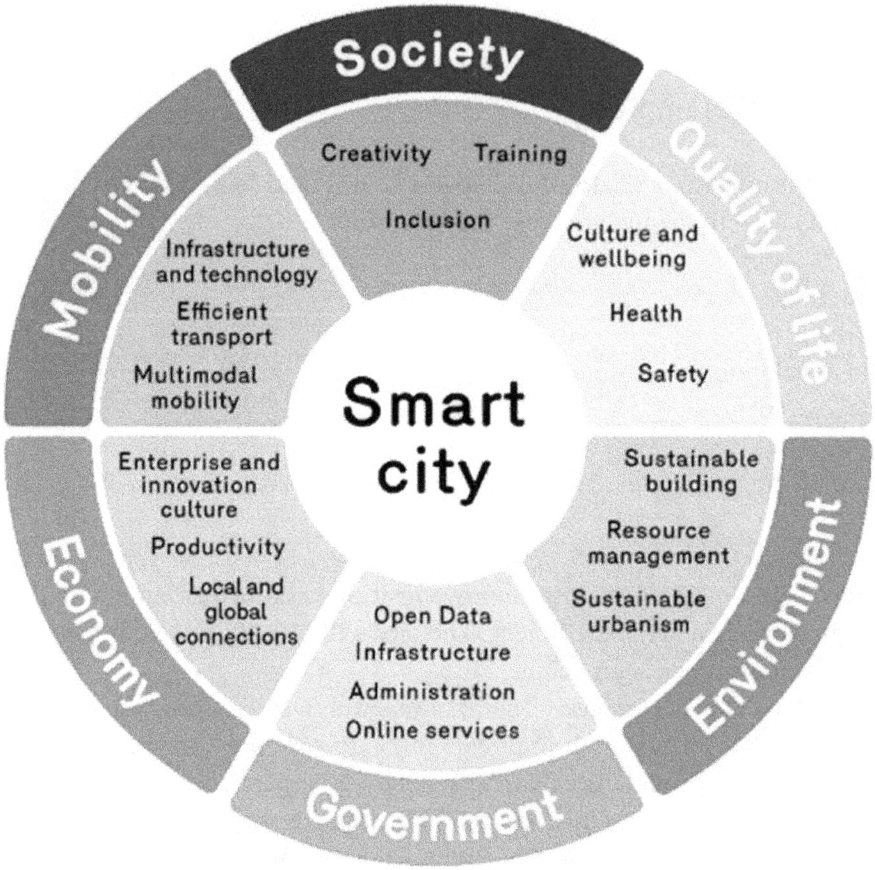

- **Smart Environment** - This word refers to a balance between developed and natural regions as well as sustainable development, resource management, and energy and water use.
- **Smart Economy**: Entrepreneurship, a creative spirit, interconnectedness on a local and global scale, productivity, competitiveness, and the capacity to adapt or change are requirements for the smart economy.
- **Smart Governance**- It refers to the use of ICT and e-government, as well as stakeholder participation in policy development and execution. It also refers to enhanced accountability, transparency, and responsiveness. But for a government to offer smart governance to its citizens, it must first be SMART, which stands for straightforward, moral, practical, receptive, and transparent (Goel, 2011)

1.2.3 Eight Component Approach

The following parameters for the eight smart city components can be further demonstrated using the study by Khaund (2013).

Figure 2. Smart diamond approach

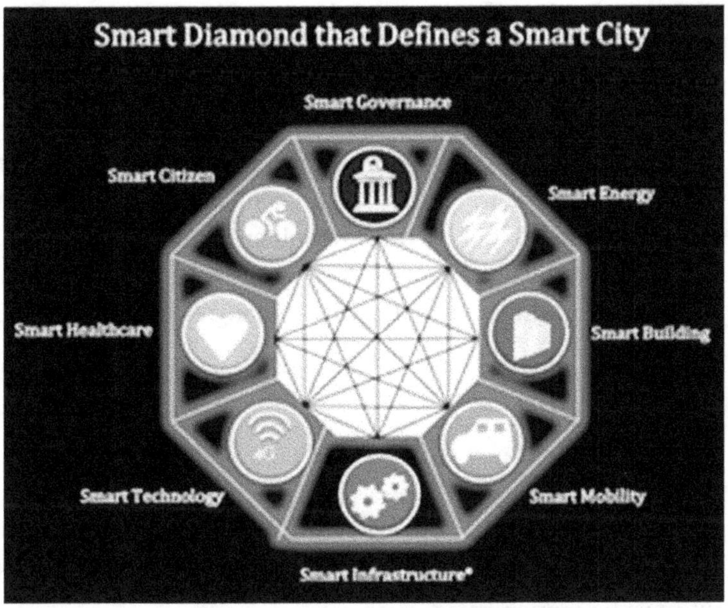

- **Smart Energy:** Digital energy management, Intelligent Grids, Smart Meters, Smart Energy Storage.
- **Smart Structures:** Intelligent Automated Buildings, Integration of Renewable Energies, Building-Integrated Photovoltaic Systems.
- **Smart Mobility:** System for Advanced Traffic Management, Parking Administration, and Transportation Pricing System with ITS Capability.
- **Seamless Connectivity with Smart Technology:** A 4G connection, Super Broadband, Free Wi-Fi, Download rates of 1 Gbps.
- **Smart Infrastructure:** Digital Infrastructure Management, Sensor Networks, Waste and Water Management in the Digital Age.
- **Smart Government:** Mobile government, E-government, E-education, Disaster Management solutions.
- **Smart Healthcare:** Technology for smart Healthcare, Utilization of mHealth and eHealth platforms, Connected and intelligent medical equipment
- **Smart Citizen:** Urban Digital Natives, Using green transportation methods, Sensible lifestyle decisions.

3. FOSTERING SUSTAINABILITY WITH SMART CITY

The term "smart city" can also refer to an urban planning strategy that emphasizes the expanding contribution of ICT infrastructure to enhancing the triple bottom line of urban sustainability. Contrarily, sustainability examines the intricate relationships that exist between an urban environment's various physical, social, economic, and environmental subsystems. By efficiently utilizing big data visualization, virtual reality and ICT have helped urban planners navigate these challenges. In fact, visualizing through virtual reality has allowed for a fresh approach to a variety of sustainable development-related concerns (Khang & Kali, 2023).

As an interactive platform, virtual reality enables sustainable and smart cities to uncover their potentials by becoming more sustainable, smarter, and contributing to urbanization at a faster rate. The main reason that smart cities have emerged, which also emphasizes the significance of urban management systems, is because of the broken link between virtual reality and smart and sustainable cities. As a result, recent research projects have begun to look at ways to enhance smart and sustainable city approaches and enhance sustainability in urban environments. A dynamic virtual environment is presented using web-based visualization technologies like virtual reality (Khang & Quantum, 2023).

4. PROBLEM STATEMENT

4.1 Requirements of Smart City

To achieve the goals of a smart and sustainable city, approaches that are flexible, dependable, scalable, accessible, and resilient must be implemented, such as:

- Enhance the standard of living for its people;
- Encourage economic expansion by creating more job possibilities;
- Ensure that its residents have access to social and community services to improve their well-being
- Create a development strategy that is sustainable and environmentally friendly.
- Make sure that fundamental infrastructure and services including public transit, water supply and drainage, telephone, and other utilities are delivered effectively.
- The capacity to address environmental and climate change challenges
- Establish a strong regulatory and local governance framework to ensure fair policies.

4.2 Smart City Framework

4.2.1 The Architecture of Smart Cities

Researchers are adamant about creating a distinct smart city architecture in order to facilitate the smart city's real deployment. Although theoretically viable, the viability of a basic, common smart city architecture for implementation in the real world is quite far from reality. Generic architectural presumptions are limited by considerable changes in required practicality, yet they are of no use. The four tiers that make up this design are the detection layer, also known as the sensor layer, the transmission layer, the data management layer, and the application layer. Protecting sensitive data is a major concern in every smart city. As a result, security components are included in each tier. Data gathering from physical devices is one of the key duties of the sensing layer at the bottom of the design. The transmit layer sends information to the top levels using a number of communication protocols. In order to provide high levels of service, numerous applications depend on the information management layer to handle and store massive amounts of data (Khang & Vugar et al., 2023).

Figure 3. Layered architechture of smart city

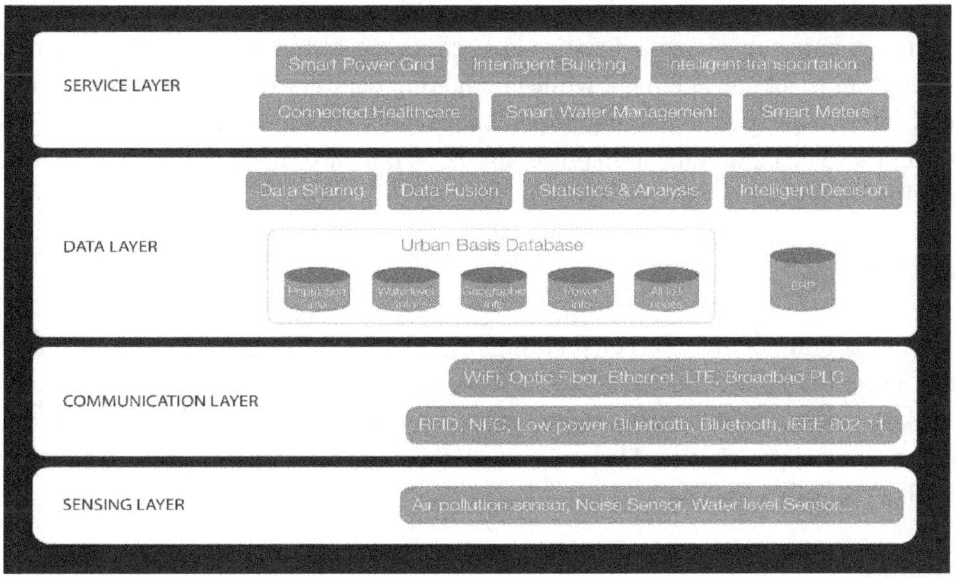

4.2.2 Sensing Layer

Real-world smart cities have a lot of data, composite architectures, data storing, and creative executive skills. Experts claim that using a smart city is so essential to making decisions that it trusts all mathematics and information-gathering procedures. Information collecting is crucial for managing another smart city. Data gathering is the most difficult task because of the data's great heterogeneity. Information gathering techniques and equipment are closely related to the information's context and type. Data from Smart City includes information on how to control personal health epidemics, manage community rubbish for disaster readiness, operate smart home devices to balance smart grid loads, and more. As a result, these discrepancies are greatly altered by the data produced by the city's operations. Large data production also makes it more difficult to collect data. The recognition/information collecting layer is expressed at the lower layer. WSNs, smart devices, and other tools for information gathering are made up of the detecting layer. It gathers information from a variety of sensors and devices.

4.2.3 Transmission Layer

When connecting a data source to a management station, the transport layer serves as the primary support structure of the smart city architecture. The transport layer is composed of many relationship structures. As a result, several gadgets linked to a private company help with individual addressability. The vehicle layer includes various wired, remote, and satellite innovations. By considering this addition, this is separated into two sublayers, access transport and organization transport. Bluetooth, Zigbee, Near Field Communication (NFC), M2M, RFID, and Zwave are just a few examples of what are referred to as short-extend access network enhancements.

Wider inclusion has been improved, especially with long-distance flexible 3G, 4G (LTE), and 5G networks. Transmission network technology also features minimal force consumption (LP-WAN). RFID technology makes use of the radio frequency (RF) region of the electromagnetic spectrum to distinguish objects, people, cars, and other living things. RFID is a scanner that, in contrast to a scanning tag framework, can recognize objects at a distance but cannot tag frameworks. NFC is a wireless network technology that permits acknowledgment while conducting data sharing between NFC-enabled devices at a distance of 10 cm. Bluetooth is another popular access network innovation that employs short-range radio waves to cut save on communication power usage. Within a 10 meter range, low force communication is offered between ZigBee-capable devices.

4.2.4 Data Management Layer

This layer acts as the brain of the smart city, sitting between the sensing and the application layers. This layer performs a variety of jobs, such as connection, evaluation, storing, and dynamic duties. Information management skills are required for a functional smart city because it is essential for the administration and execution of smart city activities. The main objective at the information level is to keep the information active with a focus on cleaning.

4.2.5 Application Layer

The top layer of the smart city design, the application layer, acts as a link between data management and city residents. Target audience, operator perspective, and satisfaction with smart city operations are all directly impacted by application performance. Concern among the people is being raised by the smart behavior of cities that provide smart services (such as smart weather forecasts). The application layer consists of a number of components from several fields (Ushaa, Khang et al., 2023).

Network development, network provisioning, intelligent transportation, climate assessment, etc. are important application layer administrations. The application layer improves city administration by handling and storing the data. Four key technologies that are used in the smart city software platform were identified throughout the literature analysis. These are cloud computing, big data, the internet of things (IoT), and cyber physical systems (CPS). The tech used in smart cities must meet five requirements: Detection, Processing, Networks, Interfaces, and Security (Khang and Vladimir, 2022).

5. ENABLING TECHNOLOGIES OF SMART CITY

5.1 IoT in Smart City

Several IoT applications can be found in smart cities. Identifying energy use, for example, when determining the condition of historic structures, detecting garbage container loads, identifying noise in the city center, and monitoring traffic lights. Studying smart houses, signs, and other devices.

Figure 4 depicts the technologies of smart cities.

Figure 4. Enabling technologies of smart cities

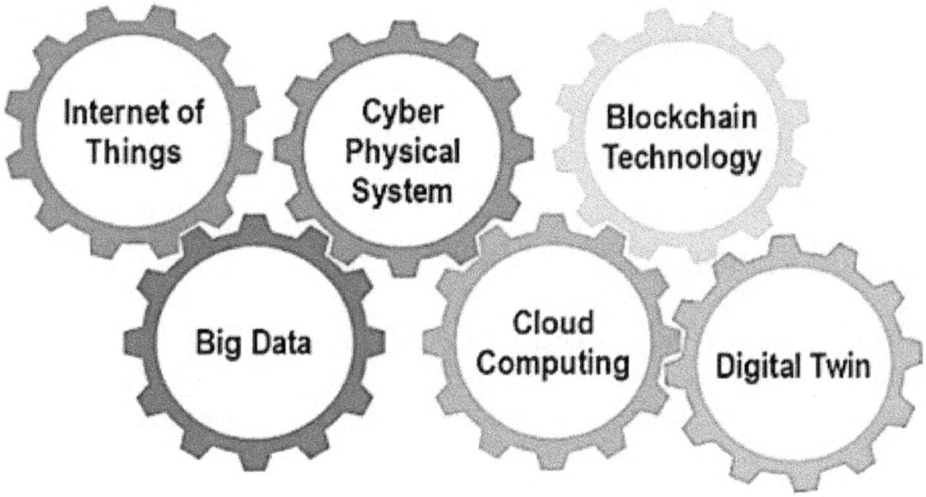

5.2 Cyber Physical System

In order to facilitate the monitoring and control of physical cycles utilizing nearby and remote PC models on network PCs, CSP is defined as the balance of calculation and physical cycles. Examples of real-world applications for CSP include smart city developments, framework control frameworks, and electronic medical devices. The physical organizational framework that is tied to the smart city is Wreck Watch, a tool for tracking automobile accidents. The current speed and acceleration rate of the driver can be determined using the accelerometer and GPS in cell phone-produced gadgets.

5.3 Big Data

Smart cities are now using big data approaches to convey the enormous volumes of data produced by urban technology. The sensor network regularly provides information on the weather in cities, such as temperature, precipitation, and air quality. Mobile devices and social networks are used by residents to share information, while buses and taxis are commonly used to transmit messages. Another crucial aspect of information for smart cities is the idea of open information. By integrating the open access of residents, groups, and NGOs, urban communities can leverage inventive biological systems to offer novel answers to urban challenges.

5.4 Cloud Computing

It offers a foundation for data storage and computing that is incredibly broad, adaptable, and widely available. It is crucial for complex smart city frameworks. A manufacturer can also help with the framework reconfiguration required for widely duplicated processes to be a digital twin of the physical process by developing a virtual representation of its manufacturing process. A digital twin is a profile of the previous and current states of the actual procedure or physical thing.

The dynamics and characteristics of an IoT device's life and operation are shown in his virtual image. The digital twin can offer the location, state, and/or status of physical assets in real time due to continual learning and advancements. Organizations are now able to monitor systems, devise plans of action, and foresee issues before they arise because to the integration of the physical and digital worlds. Digital twin technology is used to produce digital twins. This technology combines the internet of things (IoT), artificial intelligence, software analytics, and network infrastructure graphs (Khang & Misra et al., 2023).

6. SMART CITIES IN INDIA

Following the Prime Minister of India's statement that 100 smart cities will be initiated and developed throughout India, the term "smart city" quickly gained popularity (Khang & Gupta et al., 2023). The goal of the smart city program includes specified features, selection criteria, and evaluation criteria as shown in Figure 5.

6.1 Features of Smart City

Figure 5. Smart city features of India

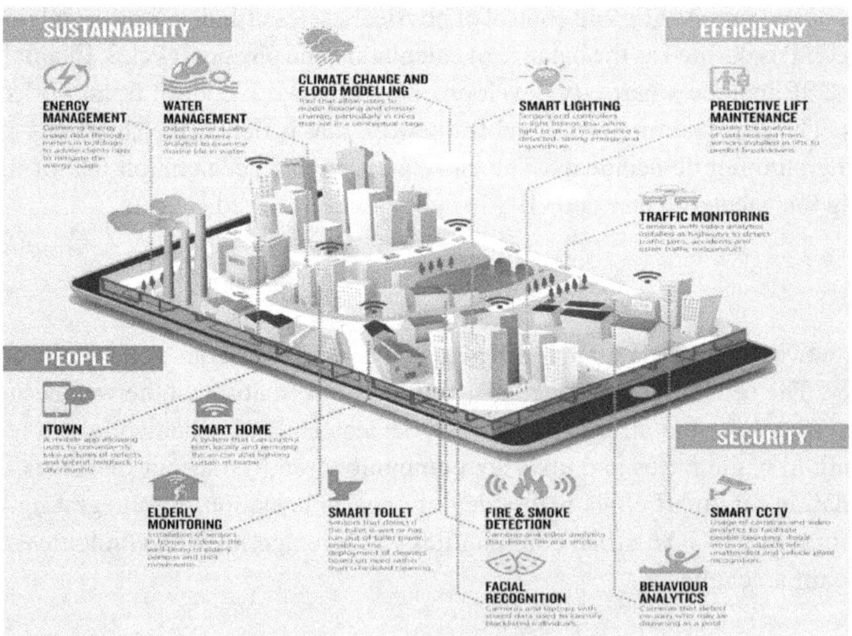

- Encouraging mixed-use development in localized areas
- Inclusivity and housing
- Making places that people can walk to
- Maintaining and creating open places
- Encouraging a range of transportation choices

- Making government more cost- and citizen-friendly
- Giving the city an identity
- Implementing Smart Solutions for the area's infrastructure and services

The features used for the building of smart cities in India were described above. These characteristics consist of environmentally friendly technological, governmental, and economic actions. Both planned and unplanned regions should make efficient use of available land. All city residents could benefit from the expansion of housing and other infrastructure facilities, and such flexibility will contribute to a higher standard of living. The construction of service facilities within cycling or walking distances is another crucial aspect. This decreases the usage of cars, hence indirectly reducing resource use and air pollution. These characteristics will aid in growing the economy (Khang & Rani et al., 2022).

The establishment of public utility facilities like playgrounds, parks, and leisure areas in the targeted local housing communities are a few more noteworthy characteristics, enabling intelligent transportation systems, particularly for high-quality, reasonably priced public transit. Activities that contribute to the development of the economy, such as those in the industrial, tourism, educational, and sporting sectors, will be enhanced through good governance practices, such as public-private partnerships. Technology should be used to make administrative and government services more transparent and efficient. The few qualities mentioned above are those that the smart city mission or program in India has implemented.

6.2 Eligibility Criteria

The smart city's features determine eligibility and selection criteria, and this process is genuinely driven by a competitive challenge. The concepts of competitive and cooperative federalism are used in the selection of smart cities in India. The smart city initiative in India decided on a three-step selection process to pick and distribute cash in order to build

- **Step 1**: To begin, smart city plans are scanned by validating their comprehensiveness, clarity, etc. Following the wetting procedure, each proposal is given a special identifying number. The government's Ministry of Urban Development handles all aspects of the first step from India (MoUD-GoI).
- **Step 2**: The expert committee members from both India and abroad evaluate smart city plans in the second step.
- **Step 3**: The Apex Committee, a subcommittee of the Ministry of Urban Development, Government, considers the expert committee's recommendations before deciding out of India (MoUD-GoI).

The detailed selection process for a city to be selected for the development of smart cities in India is shown in Figure 6.

6.3 Smart City Mission, Policies, and Programs in India

India's Smart Cities Mission was announced and started by the Honorable Prime Minister of India (SCM). The launch date for India's five-year Smart Cities Mission is June 25, 2015. The SCM's main objectives are to develop 100 sustainable and smart communities through urban initiatives for regen-

Figure 6. Selection criteria in India

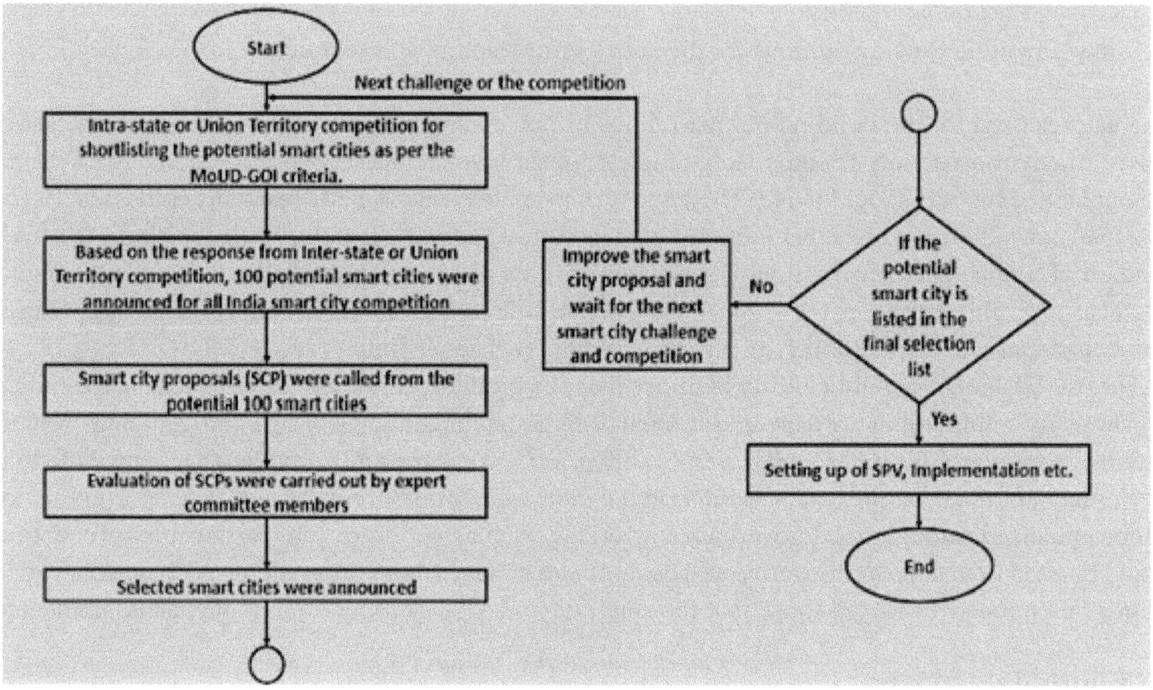

eration and modernization. The SCM is carried out by the Ministry of Urban Development (MoUD), a government body.

The Indian government, in collaboration with state governments and union territories (GoI). The following components of the Smart City Mission Program had few programs or policies that were thought to be helpful.

- National Policy on Transit-Oriented Development (TOD)
- Atal Mission for Urban Renewal and Transformation (AMRUT);
- Public-Private Partnerships
- National Sustainable Housing Mission
- National Urban Transportation Policy
- Policy for National Urban Sanitation (PPP).
- Proposals to create Special Purpose Vehicles [SPV]
- PAN city initiatives;
- Urban local bodies (ULBs).

6.4 Present Situation

The estimated overall cost for building smart cities around the world is INR 189155 Cr. If more money is needed for the development of smart cities, the state governments must raise it and then create their plans. The cabinet approved the estimated investment. This projected investment, which has been divided into two categories and totals INR 152499 Cr, is only meant for the specific area-based projects that have been selected and given the green light under the smart cities mission program. The Pan City Initiative Program will use the remaining INR 36656 Cr to promote smart projects. Based on the eligible and selection criteria, a list of potential smart cities for the state level as well as the All India Level Smart Cities Competition was selected in accordance with the smart city mission program.

The final list of eligible smart cities for "100 new smart cities in India" was made public over the course of three consecutive rounds. Thirteen additional cities were chosen for the fast track phase of the Smart Cities Mission, bringing the total number of shortlisted cities for Round I to 20. A total of 63 further potential smart cities were shortlisted in Round-II, while 30 more were identified and selected in Round-III. It was asked that, in compliance with the 2013 Companies Act, each of these selected and approved towns create a special purpose vehicle (SPV) that functions as a limited company inside the city where the envisaged smart city is situated. This company would be in charge of carrying out the designs for the smart city and constructing the city in accordance with the suggested layout.

The majority of Indian cities are vying for development in crucial economic areas like banking, information technology, energy, power, hotels, media and television, the film industry, cuisine, tourism, cultural heritage, telecom, and textiles, among others, given the current state of Indian cities. Several cities, including Delhi, Hyderabad, Pune, Mumbai, Bangalore, Chennai, and Kolkata, were advancing very quickly in a variety of fields.

6.5 Challenges

There are a few issues that must be resolved in order for smart cities to advance more effectively.

Here is a list of these challenges

- The absence of robust interdependent and related work between central and state cultures.
- A lack of vision and astute leadership.
- Inability to frame and plan for the smart city factors that define success
- A lack of knowledge of the technological components of elements of smart cities.
- The notions of "retrofitting, redevelopment, and greenfield development" are not well understood.
- A lack of visionary policymakers for smart cities; stakeholders, implementers, and evaluators.
- Insufficient investments
- Poor collaboration with the private sector or others connected regulatory organizations.
- A lower level of resource management awareness.
- A dearth of intelligent participants in reforms and government.
- Less assistance from the populace or the citizenry.
- Less improvement in the efforts for strategic development

7. REQUIREMENTS OF SMART SYSTEM

7.1 Four Pillars of Smart City

Institutional infrastructure, physical infrastructure, social infrastructure, and economic infrastructure are the cornerstones of a smart city. No matter their social standing, political views, age, income, gender, or

Figure 7. Four main pillars of smart city

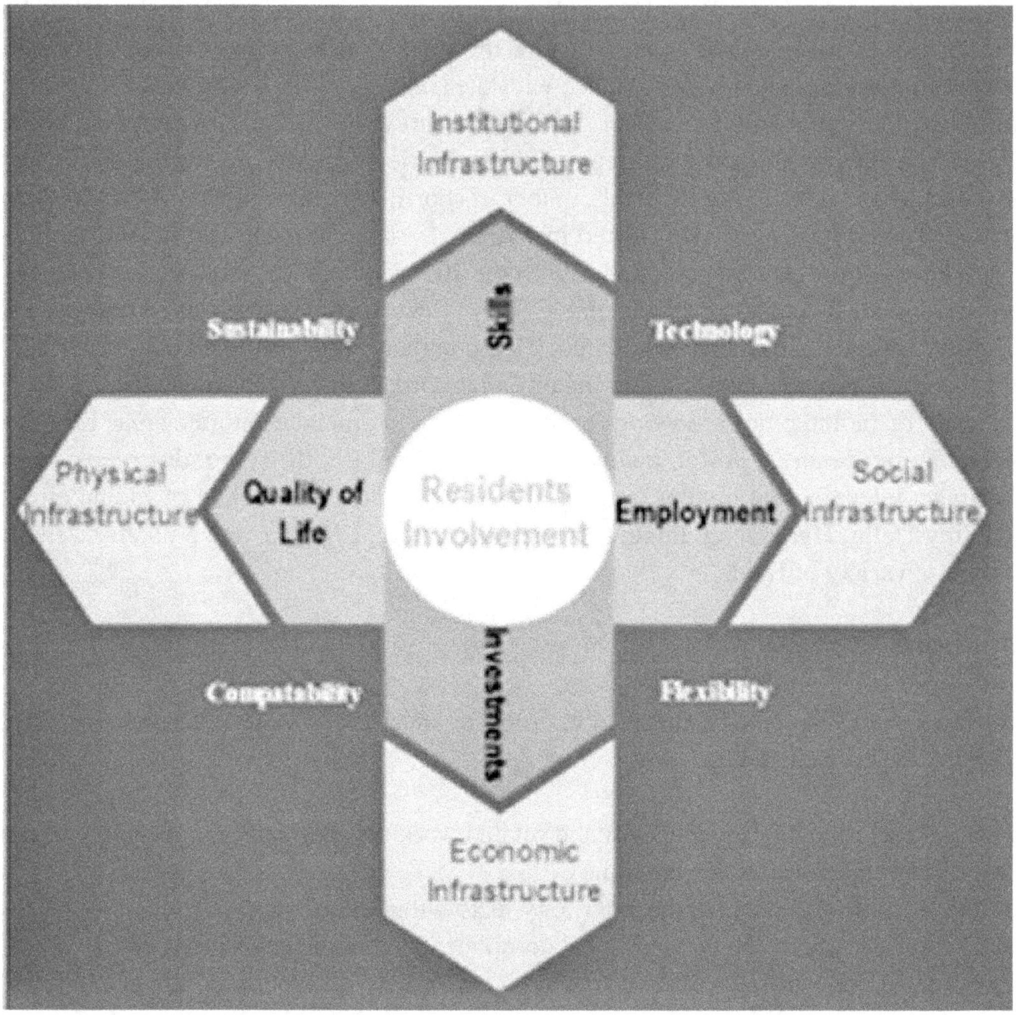

other characteristics, citizens of a smart city are given the greatest possible living conditions as shown in Figure 7.

7.2 Institutional Infrastructure

The word used to describe the procedures used in managing and administering metropolitan areas. Information and communication technology (ICT) have given this system a new dimension, making it more efficient, responsible, transparent, and citizen-focused. They include inclusive governance, participatory governance, and e-government.

7.3 Physical Infrastructure

This phrase refers to intelligent and cost-effective physical infrastructure, which includes urban mobility systems, sewage systems, drainage systems, water supply systems, and solid waste management systems. All of these systems are well integrated using technology.

7.4 Economic Infrastructure

To attract business prospects, investments, and the right kind of employment, a city must first identify its core competencies, comparative advantages, and strengths. It must also assess its capacity to provide the right kinds of economic activities.

7.5 Social Infrastructure

It has to do with the growth of social and human capital. It also covers sports, children's parks, gardens, and open spaces, as well as healthcare and education. The city has to have resources that integrate underprivileged groups, such as SCs, STs, minorities, people with disabilities, and women, into the mainstream of development.

8. SMART CITY INFRASTRUCTURE

8.1 Smart Infrastructure and Building

In order to harness technology improvements to minimize extreme weather brought on by climate change, cities are incorporating green and smart infrastructure into their urban planning strategies. These modifications are being made to both new and existing infrastructure systems.

According to research, smart green infrastructure (SGI) increases climate change resilience while also providing a number of health and wellbeing advantages that improve people's quality of life. Gray infrastructure describes traditional systems of man-made infrastructure. Green infrastructure incorporates both natural and artificial elements, as well as multipurpose green spaces.

Despite the expected lifespan, depreciation, and return on investment of the grey infrastructure assets. The use of hard surfaces like asphalt and concrete reduces surface water infiltration, which is crucial given the current urbanization and unfavorable weather. Grey infrastructure is viewed as stiff investments since it requires the integration of green and grey, which is known as next-generation infrastructure.

Green solutions were traditionally ignored since there were no trustworthy means to compare them to grey solutions. Yet, the development of green-gray analysis methods has demonstrated the long-term benefits of upscaling and investing in green infrastructure. By integrating these new infrastructure options with cutting-edge technological systems that are "smart" and supported by data and communications, future cities can fulfil their long-term sustainability goals. The benefits of using integrated green and smart infrastructure solutions as opposed to grey, green, or smart silo applications and makes recommendations for how to put these fixes into action. Interactions between smart, green, and grey technologies are evolving.

In contrast to grey, green, or smart silo usages, this paper focuses on the benefits of combining green and smart integrated infrastructure solutions and makes recommendations for how to implement these solutions. Interactions between smart, green, and grey technologies are evolving. Smart networks, a collection of interconnected devices that allow data transfer and the sharing of many forms of information, such as connection, features, and performance, are one of these solutions.

As a result of the blending of digital and physical urban infrastructure, the concept of "smart infrastructure" has emerged, where sensing technologies are embedded into the infrastructure and the equipment they interact with to enable real-time data collecting and analysis. To help infrastructure providers make decisions on the structural health and maintenance of their assets, these systems gather, analyze, distribute, and make sense of data that is reliable, trustworthy, and valuable. The benefits of smart and environmentally friendly infrastructure have been widely examined, researched, and documented. By assisting in maintaining the equilibrium of the ecosystem, green infrastructure benefits society, the economy, and the environment.

Smart infrastructure largely uses technology to provide these benefits. Both systems provide benefits and increased production (Khang & Kali et al., 2023). In addition, green infrastructure has the capacity to offer win-win solutions by tackling several problems at once, yielding a variety of benefits while staying within a financially viable framework. Nonetheless, many instances of the green and intelligent infrastructure efforts that are highlighted rely largely or entirely on ICT. These initiatives' main areas of focus are:

- **Mobility and Connectivity**: which aims to enhance how people interact and travel;
- **Healthy-Healthy Aging**: Improving wellbeing, mental health, and quality of life;
- **Information marketplaces**: a program of activity that facilitates the extraction and/or exchange of data by numerous partners and stakeholders with a focus on the aforementioned themes.
- **Skills and Enterprise:** managing supply and demand, upskilling the local population, developing innovative talent, raising enterprise levels, and promoting sustainable start-up and business growth.

Using intelligent technology and IT systems, smart green infrastructure combines environmentally beneficial components to address urban and environmental concerns (SGI). The two ideas now operate in parallel and primarily separately, despite the various long-term benefits that can be realized with their merger. This essay highlights this opportunity and looks at how combining smart and green can provide complete solutions for complex problems facing large cities while boosting resilience.

SGI components can aid in reducing greenhouse gas emissions, enhancing urban heat islands and flooding, enhancing city beauty, and enhancing the health of city dwellers. Given how swiftly urbanization is developing, natural green areas supported by smart services will be essential to the sustainability and well-being of future cities. It might be required to pinpoint places where green designs' efficacy could be increased in order to close the gap between smart and green. These discoveries will improve the use of water and energy, reduce air pollution, and boost biodiversity. Planners, facilities and estate managers, and the building industry will all gain from mapping areas of deficiency and flagging potential improvement areas in order to scale up SGI development.

8.2 Concepts for a Smart, Green, and Grey Infrastructure

Given how swiftly urbanization is developing, natural green areas supported by smart services will be essential to the sustainability and well-being of future cities. It might be required to pinpoint places where green designs' efficacy could be increased in order to close the gap between smart and green. These discoveries will improve the use of water and energy, reduce air pollution, and boost biodiversity.

Planners, facilities and estate managers, and the building industry will all gain by identifying areas of deficiency and flagging potential improvement areas in order to scale up SGI development. Yet, green infrastructure has been incorporated into planning theories and guidelines to encompass wide development ideas considering a range of challenges, including economical, recreational, and ecological repercussions. Urban wetlands, community and tenement plots, private gardens, green walls and rooftops, urban parks, and other interconnected spaces are all managed to offer a variety of environmental, sociological, and economic advantages. This network can be described as green infrastructure. This network offers opportunities for the production of food, the capture of carbon dioxide, the regulation of body temperature, the reduction of flood and wind danger, and other ecosystem services. The "smart city" agenda seeks to make use of technology, complex data, and information platforms in order to facilitate decision-making and promote efficient operational networks in the city.

Energy grids, sewage systems, smart pricing (parking, transportation, energy, and water), transport, and communications are just a few of the businesses that can utilize these networks to boost productivity and save costs. These sectors are systems with intelligent infrastructure. By using green infrastructure to

Figure 8. Concept of grey, green, and smart infrastructure

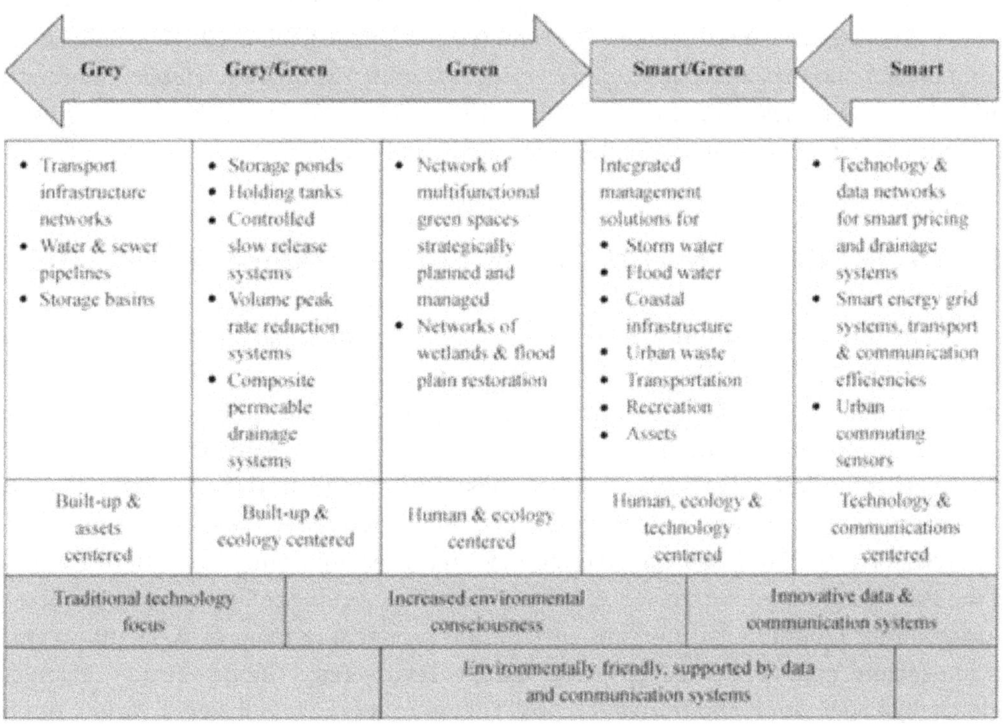

increase city resilience, the resulting integrated systems are capable of handling challenging municipal issues and ensuring long-term sustainability as shown in Figure 8.

8.3 Need to Implement Smart and Green Infrastructure

The "smart city" agenda seeks to make use of technology, complex data, and information platforms in order to facilitate decision-making and promote efficient operational networks in the city. Energy grids, sewage systems, smart pricing (parking, transportation, energy, and water), transport, and communications are just a few of the businesses that can utilize these networks to boost productivity and save costs. These sectors are systems with intelligent infrastructure. By the implementation of green infrastructure to increase city resilience, the resulting integrated systems are capable of managing complex municipal concerns and ensuring long-term sustainability.

Severe weather brought on by climate change poses significant infrastructure challenges for cities worldwide, and this subject has been well studied. Flooding has increased in frequency, resulting in interruptions and expensive repairs that are expected to cost £1 billion year (Department for Environment, Food & Rural Affairs, 2012). A sizable portion of assets are located in areas that are vulnerable to flooding, and the risk is anticipated to increase. By 2080, climate change is predicted to increase the likelihood of flooding in the UK by up to 4°C.

Smart technology solutions can be included into the current infrastructure to enable monitoring, reporting, and even the provision of preventive actions, increasing the capacity and longevity of existing assets. The risk of coastal and riparian flooding, for instance, can be reduced by natural ecosystems and wetland characteristics (UK Green Building Council, 2015). Heavy rain may result in increased sewer flooding and a disruption of transport. The expected rise in sea levels will result in an increase in the proportion of assets susceptible to coastal floods (Committee on Climate Change, 2017a).

The government assesses the risks every five years in order to address the severe effects of climate change on infrastructure. These assessments (Committee on Climate Change, 2017b) indicate that including SGI components into urban construction is a critical step in this process. Hence, in order to manage complicated municipal issues in the future, we urgently need to include smart solutions into networked SGI systems rather than continuing to work in silos and combining.

9. BENEFITS OF INSTALLING SMART AND GREEN INFRASTRUCTURE

Cities of today require infrastructure that is resilient, climate-smart, adaptable, and sturdy while still being cost-efficient and efficient. Performance, efficiency, and return on investment can all be improved by adopting green infrastructure. Green infrastructure upgrades alone can have a favorable economic impact by increasing commercial trading by up to 40% in town centers. Green infrastructure can be used in conjunction with the following smart technologies: The administration of public areas and utility services, citizen involvement in decision-making, and online activities like e-commerce.

Research (NENW, 2008) that considered integrated smart and environmentally friendly infrastructure identified the following unique regions of economic benefits: wealth creation, asset, property, and real estate values, worker productivity, tourism, agricultural products (Khang & Agriculture et al., 2023), recreation, and leisure, quality of place, property, and species diversity, flood control, water management, and climate change adaptation and mitigation. The literature review identified a variety of benefits for combining or independently implementing green infrastructure with smart city applications. The advantages of combining smart and green together demonstrate that doing so results in the most effective and efficient outcomes for the environment, society, economy, and health.

10. INTEGRATION OF SMART AND GREEN INFRASTRUCTURE APPLICATIONS

Cities are in grave danger of losing their socioeconomic and social futures due to a number of environmental problems. Threats to groundwater resources, natural habitats, and climate risks like storms, flooding, cyclones, and coastal erosion are some of these challenges. Also, they have bigger carbon and energy footprints. Because many infrastructure systems depend on one another, their failure could have an effect on other systems. Flooding is more likely than any other climate threat to have an impact on the UK's infrastructure. Storms, floods, and the destruction of coastal infrastructure all have a direct influence on the transportation networks that are necessary to many important metropolis services. Similar to this, storms and flooding rapidly damage infrastructure for transportation, including bridges, highways, ports, harbors, underground buildings, and tunnels.

10.1 Management of Storm Water

Green infrastructure elements, such flora and strata of soil systems, are essential for sustainable water drainage because they may soak up rainwater by simulating natural filtering. Natural green spaces are thought to experience runoff at a rate of 10% to 20%, compared to 65% on artificial surfaces. Real-time sensors that track water levels can help make urban community gardening plots, which offer several advantages such as food production, leisure activities, recycling of garbage, and rainwater, more effective. Integrating and combining these environmentally friendly technologies with intelligent "Continuous Monitoring and Adaptive Control" systems is a practical solution to increase the effectiveness and value of present storm water management infrastructure. To monitor and actively manage storm water storage and flows, this gadget integrates information from field-installed environmental and water-level monitoring with data from real-time weather forecasts.

10.2 Management of Flood Waters

By the 2080s, the number of assets vulnerable to flooding could treble due to expected climatic changes. Dams, railway tracks, bridges, and other constructions are all more susceptible to runoff from rain. Many hard, built-up surfaces are non-porous, which raises the risk of flash floods. Sensors, valves, and soft-closing vent systems are examples of smart infrastructure components composed of environmentally friendly materials that can be used in flood control systems.

Building information modelling (BIM) has been utilized in flood prevention in Ipswich, UK, to oversee the development of a new coastal shield along the River Orwell. For health and safety briefs, the 3D models have benefited in decision-making, building planning, and monitoring the geographical synchronization of the workforce. The best results for minimizing the consequences of harsh weather are achieved by combining parts of green infrastructure inspired by nature with cutting-edge technology.

10.3 Management of Coastal Infrastructure

Infrastructure near the ocean has remained vulnerable to changes in the coast caused by erosive pressures, storm surges, or persistent flooding. The rate of coastal erosion is significantly impacted by sea level rise brought on by climate change. Historically, grey infrastructure assets have served as coastal defenses, adopting an engineering method to build floodwalls and construct embankments to fend off storms.

Integrated green solutions utilize native wetlands and bushes between the beach and land as buffers instead of erecting brittle, artificial barriers that won't be able to change with the environment over time. By combining these naturally occurring solutions with cutting-edge mapping technologies like Geospatial Information Systems (GIS) and modelling techniques like Building Information Modelling, a thorough understanding of the response to coastal erosion may be attained (BIM). In order to plan for future severe weather occurrences, intelligent devices and drones that undertake surveys give data for modelling how coastal areas and the seabed behave during storm surges.

10.4 Waste Management in Cities

The recycling and repurposing of waste, is a huge challenge in modern cities and could be helped by clever waste management techniques. Water can be filtered and used to service green, blue, or smart infrastructure components. In Southern California, USA, wastewater can be successfully converted into drinking water. The Internet of Things has made "Smart Urban Waste Management" approaches possible by facilitating efficient waste collection and performance optimisation (IoT). The phrase "Internet of Things" (IoT) is used to refer to the wireless communication networks and legacy system hardware that communicate sensor data to computing devices for real-time analysis. Route selection is improved in accordance with trash fill levels using these clever techniques.

10.5 Transportation Administration

Transportation is one of the biggest contributors of greenhouse gas emissions. In the UK, road travel increased by 6% between 1990 and 2017. The chance to reduce automobile traffic is tremendous thanks to the quick development of smart, sustainable mobility. To enhance the quality of life for today's urban workers, the daily travel time and conditions must be improved. Long-term sustainable green city goals promote walking, biking, carpooling, public transportation, and other low-pollution modes of mobility.

Depending on the city's current infrastructure, population, capacity, and traffic volumes, smart applications may reduce traffic and commute times. E-commerce has changed how people move as more people share rides, integrated timetables and ticketing mechanisms, electric cars, and intelligent parking systems become more widespread. Examples of integrated green and smart mobility include Zipcars (US-based shared cars with ICT), Uber (data-driven transit services), and neighborhood bike-sharing. These innovative transportation systems have given locals several options for moving about without a private automobile.

10.6 Management of Recreation and Outdoor Areas

A town needs thoughtful parks, wetlands, and recreational amenities to improve its quality of life. Increased leisure and physical activity among a city's residents is linked to improvements in health and wellbeing and cheaper health care costs because of a more active population. Gardens and green roofs used for storm water drainage can be used as urban recreation areas because of their aesthetic value. By adding sophisticated security lighting systems, these places can be made appealing for use by the general public. Some examples of intelligent and environmentally friendly architectural components include energy-generating exercise equipment, interactive play areas for children, the elderly, and persons with disabilities, intelligent controls for water features, and automated gardening maintenance techniques.

10.7 Asset Administration

By providing comprehensive, real-time data on asset usage, smart technology can improve capacity and efficiency by enabling infrastructure providers to effectively plan and manage capacity. Infrastructure assets comprise both above- and below-ground buildings, as well as structures associated to transportation, such as auxiliary structures and tunnels, which have substantial development, maintenance, and rehabilitation costs. The risk of storm surge, floods, and sea-level rise increases the vulnerability of the transportation infrastructure, which includes bridges, roads, tunnels, ports, and harbors. Hence, coordination amongst stakeholders with the aid of digital intelligent technology is essential for these interconnected and reliant systems to function effectively. Sensors incorporated into built assets can provide monitoring input in real time. New maintenance, inspection, and renovation routines with a preventative maintenance focus are made possible by proactive systems.

The UK is a strong proponent of "Green Asset Management" concepts, advocating the use of weather accounting data to help fund managers identify green and low-carbon investment opportunities to boost investment into cutting-edge clean technologies through government funding and grant finance. London,

Figure 9. Drivers and integrated infrastructure systems that help create future cities

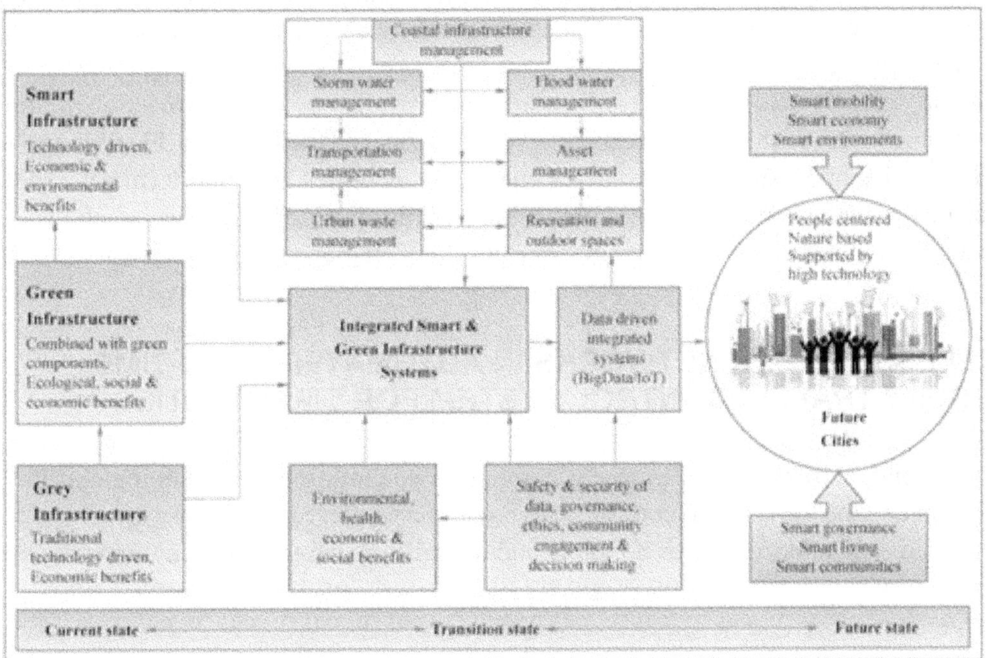

UK, is acknowledged as the world's top hub for green consultancy services, green project financing, and financing climate adaptation due to its vast range of possibilities for green asset managers as shown in Figure 9.

11. SMART TRANSPORTATION

Since the beginning of civilization, people have relied on transportation. This requirement has grown as a result of technological improvement to encompass land, maritime, rail, and air transportation. The world's conventional modes of transportation were neither linked nor connected. The idea of everyday objects interacting with one another has changed the conventional transportation framework into a modern interconnected transportation framework. Because of this, several correspondence and route frameworks incorporate contemporary vehicles. This is the link between every molecule of a certain transmission type. By enlarging the network inside a single transporter, several transporters can be connected to create a global transportation system.

The Ad Hoc Vehicle Network (VANET) has received much discussion from Intelligent Transport Systems (ITS). VANET has typically been used to control rural traffic through the use of vehicle-to-vehicle (VV) and vehicle-to-infrastructure (VI) communication capabilities. The transportation system may successfully use the real-time database for the actual communication function. Users are also informed of traffic congestion, detours, and alternative routes by intelligent transportation systems.

The smart transportation system incorporates safety measures for passengers and pedestrians in order to improve performance. The newest framework offers global hubs, smart street organizations, intercity train organizations, tram and metro organizations, coordinated public security transport, guaranteed bike pathways, and guaranteed passer-by courses as a result. In conclusion, the incorporation of smart transportation frameworks into cities improves the operational effectiveness of metropolitan regions as well as the effectiveness, affordability, and safety of urban transportation provided a calculation for crash identification for contemporary air travel. This is avoided by anticipating aviation crashes and sharing information among them.

Several scientists also deal with managerial issues and organizational impediments in the train industry. There have been numerous studies on executives, street traffic, and traffic safety. Flexible applications for hiring and following taxis, stopping executives, managing identities at airport terminals, and RFID cost assortment are further new transportation applications. While abusing resident satisfaction, metropolitan supportability is ensured by using these new transportation framework measurements.

12. WIRELESS SENSOR NETWORKS IN AUTOMATED REAL-TIME INTELLIGENT TRAFFIC CONTROL SYSTEM FOR SMART CITIES

This system is unique from others because it combines two cutting-edge technologies—the intelligent parking system and the traffic light control system—into a single, revolutionary system to connect city residents to roads and parking spaces remotely and in real-time using just a single mobile application.

Figure 10. Intelligent traffic control system

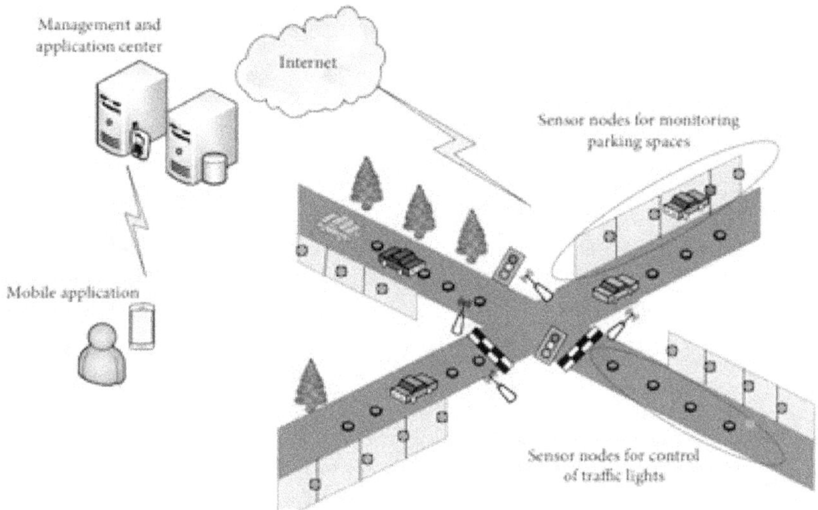

12.1 The Intelligent Traffic Control System's Architecture

The three primary elements of the proposed system are a parking space management center, a traffic signal control center, and a global information and management center as shown in Figure 10.

Alt-text: Figure 10 depicts he parking space management center is built around the deployment of WSNs in each parking space to aggregate all the availability states of areas in each area of the town for those transmitted to the corresponding entry point (sink). These will then be converted to the global data and oversight center to be used by and given access to drivers and citizens.

Figure 11. Car presence detection using hybrid sensor

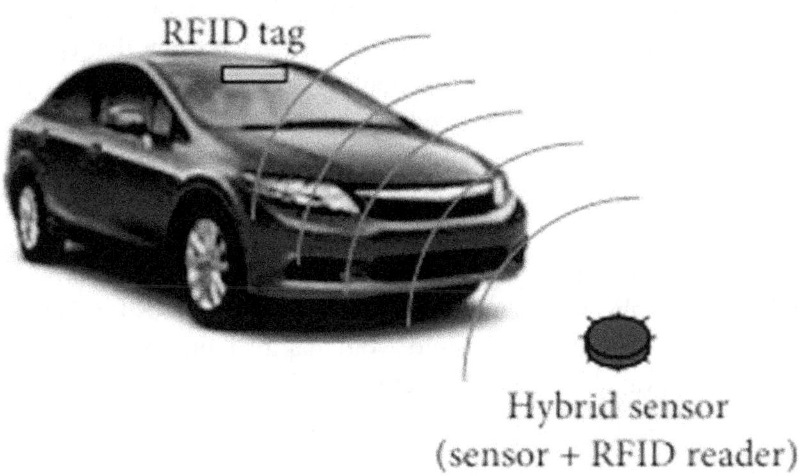

Figure 12. The angular value of each node of the first level

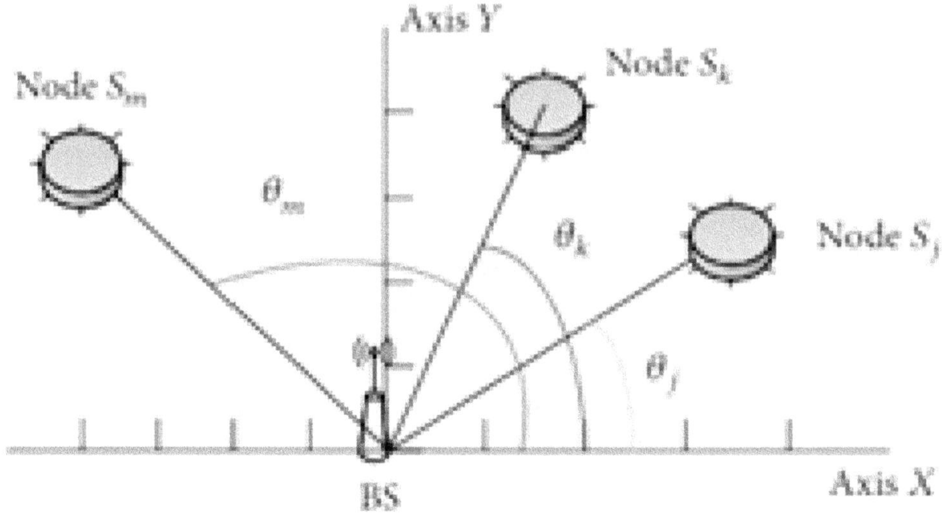

$$\theta_i = \begin{cases} \arctan \dfrac{y_i - y_{bs}}{x_i - x_{bs}}, & x_i > x_{bs}, y_i > y_{bs}, \\ 90 + \arctan \left| \dfrac{y_i - y_{bs}}{x_i - x_{bs}} \right|, & x_i < x_{bs}, y_i > y_{bs} \end{cases}$$

The used sensor nodes are hybrid sensors (presence sensor + RFID readers), making it possible to both identify cars using their registration information stored in their RFID tags and sense their presence as shown in Figure 11.

If hybrid sensors detect the presence of a car without an RFID tag, the system tells the parking attendant to enter the registration number of the vehicle parked in the system. The traffic light management center is in charge of managing and regulating the traffic lights at road intersections in order to lessen traffic congestion and preserve the city's traffic flow. This center uses sensor networks to collect information on the density and volume of cars travelling via each route that constitutes a crossroads in order to establish the maximum amount of time that a traffic light can remain green. This center is based on original and trustworthy calculations that raise the effectiveness and efficiency of our system for managing traffic on the roads.

The global information and management center, which also regulates the traffic lights to enhance traffic flow in the city, is made up of a database of all the information obtained from all the sensors placed around the city. It provides a summary of the available parking places. This center provides drivers with an Android mobile application that identifies free parking spaces for their destinations as well as the traffic rate in real time. This allows drivers to avoid making unnecessary trips, search for alternate routes, and find other places available in locations closer to their destinations. This enables motorists to avoid being snarled in traffic and adding to the city's increasing traffic congestion.

Figure 13. Messages exchanged in the first phase

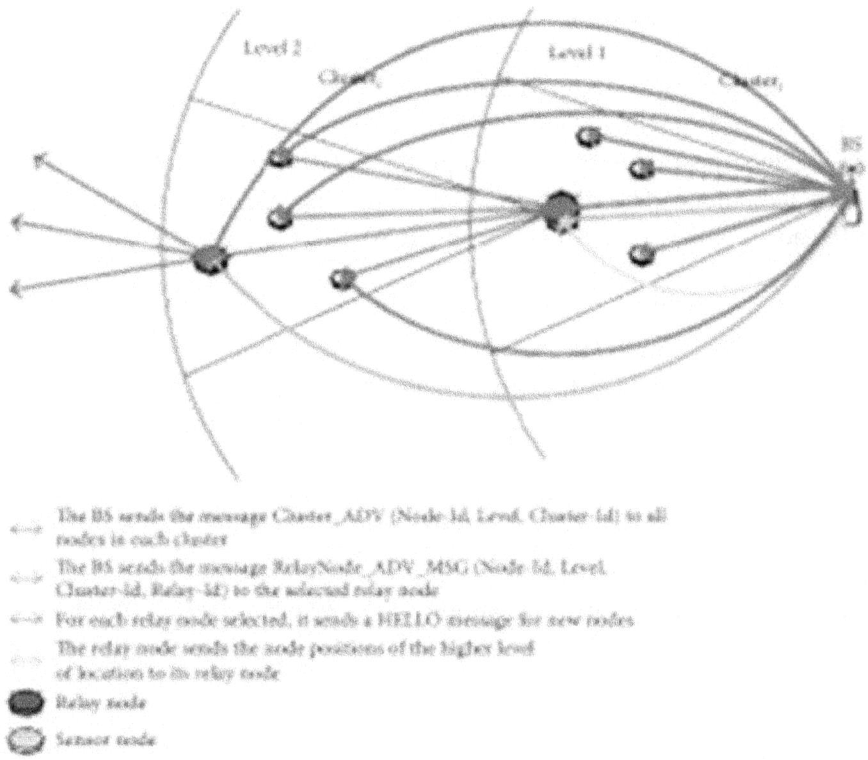

12.2 Implementation of Sensor-Based Smart Parking

Figure 12 illustrates how Micro I/O developed a smart parking system that functions both indoors and outdoors and Figure 13 depicts the message exchanged in the first phase.

To detect the proper occupancy, the suggested system uses wireless magnetic sensors that are positioned on the ground in each parking space. A variety of auxiliary technologies, such as LED panels, tags, indicators, alarms, and obstacles, are compatible with the system. A range of user interfaces for management and driving guidance to open areas are also compatible with it. The three main levels of the system's architecture are as follows:

- The endpoints layer contains the parking sensors and actuators.
- The communication layer, which consists of repeaters and gateways, permits communication between endpoints and the rest of the system.
- The software layer consists of all cloud-based components required for data collection, processing, storing, and providing, from user platforms and databases to web services.

The system has the following features and information:

12.2.1 Encapsulation

Since the sensor includes wireless communications and internal batteries for power, it does not need wiring. A plastic capsule containing the gadget is then filled with epoxy, cured, and finally sealed shut. The encapsulation enables complete impermeability and excellent pressure endurance (as the weight of the vehicle). The sensor is quick and simple to deploy after being encased. Asphaltic glue was only needed to be applied underneath the sensor, mild pressure to the ground, and a brief waiting period for the adhesive to dry and harden. The firmware of the sensor can be remotely updated while it is being programmed. The gadget is FOTA (Firmware over the Air).

12.2.2 Initialization

The sensor must be roughly aligned with a given magnet in order to be initialized. A reed sensor built into the gadget will sense the magnet and trigger the gadget's activation.

12.2.3 Detection

The sensor detects changes in the Earth's magnetic field brought on by the passage or presence of a vehicle above it using a 3-axis electronic magnetometer (Freescale MAG3110). The sensor has a predetermined sample rate, and the system is alerted to a status change if the amplitude of the data exceeds the defined thresholds of the algorithm.

12.2.4 Temperature Adjustment

The ambient temperature and solar exposure both have an effect on the magnetometer's sensitivity, which changes with temperature. To work correctly, the sensor must be calibrated and temperature changes must be considered. For this reason, the MAG3110 incorporates a temperature sensor, and samples are also collected in addition to the magnetic readings in case correction or recalibration is necessary.

12.2.5 Communication

The application's criteria for communications range and device autonomy led to the selection of LoRa over 868MHz as the preferred communication technology. With this technique, the communication range can be extended to 200 meters while using less electricity. Typically, there are two ways to communicate: Status change notification: Following a sampling event, the data is evaluated, and if a status change is discovered, the sensor provides the proper message or notice to the system. I'm alive alert: If there hasn't been a status change for a predetermined period of time, the sensor alerts the system that it is operating normally.

12.2.6 Power Management and Autonomy

The autonomy of a gadget is a measure of its lifetime. The sensors can run on up to four AA low self-discharge Li-SOCl2 batteries (SAFT LS14500, 3.6V, 2.6Ah), however they cannot be replaced or recharged because they are permanently encased in epoxy. Power management is essential as a result. To increase the device's lifespan, the procedures for sampling, data processing, and communications were improved:

- **Sampling**: To reduce the amount of time the CPU is awake without sacrificing the calibre of service levels, the prescribed sampling periodicity is defined as 4 seconds. The microprocessor will wake up from sleep after device activation every 4 seconds, take a sample, and if no change is discovered, go back to sleep.
- **Processing**: To reduce power consumption, the (Low Power) Microchip PIC18LF26K22-I/SS microcontroller was selected. It operates most of the time in sleep mode, which consumes roughly 1muA of power.
- **Communication**: The activity that consumes the most energy is communication. Because it supports low power communications, LoRa on the 868 MHz band was chosen as the communication technology. However, the frequency and runtime (size) of the communications were also optimized for greater power savings. Based on the specified periodicity of the "I'm alive" messages, for example, with intervals not less than 30 minutes, the sensors can achieve an autonomy of more than 8 years.

13. SMART ENERGY

For every work to be completed, energy is a necessity. There are a variety of sustainable and non-sustainable fuel sources. Inexhaustible resources (such as solar, wind, and geothermal energy) do not run out like sustainable resources (such as petroleum goods) do because of the concept of recovery. In order to draw attention to issues and enhance the most effective energy-use practices, specialists have recently expanded the ideas of smart energy, green energy, and sustainable energy. Using energy with little negative environmental impact is the goal of eco-friendly energy.

A clear objective of maintainable energy is to assist people and reduce current reliance on non-sustainable power sources. The idea of smart energy is frequently more appealing since it emphasizes a holistic strategy to dealing with the synchronization of environmentally friendly power energy, maintainable energy, and a sustainable power source. Simply said, smart energy should be used to meet energy needs while minimizing negative natural repercussions, such as lowering carbon dioxide outflows, and keeping the manageability of non-sustainable power sources. Since it is a sustainable power source, as was said above, it can meet all of the world's energy needs. The growing interest in energy has increased the need for sustainable power sources recently. In order to incorporate sustainable power sources into intelligent structures, a lot of study has been done in this area. Sustainable power plants may be integrated into smart frameworks or smart structures may be connected to sustainable power sources.

A suggested microgrid energy board structure coordinates a photovoltaic (PV) generator and a substantial storage unit. Here, solar and wind energy are combined in a similar manner to power the residence. In order to provide CEOs with efficient energy in beautiful settings while integrating sustainable power sources, many investigations have been directed. This has a significant impact on overall energy consumption, thus different energy board frameworks are adjusted to the state of the home. According to the anticipated energy assessment, domestic energy demand would rise by 24% over the next few years.

As a result, the fundamental management of energy utilization is reasonable at the city level and looks prospective for increasing energy for the executives. Analysts have expanded their responsibilities to oversee intelligent frameworks for bettering energy consumption. The energy usage arranging method based on the game theory is suggested in. This methodology reduced total energy demands, overall energy costs, and daily power rates per capita. Similar issues were raised when discussing how to successfully synchronize flexibility and demand, cut energy costs, and reduce carbon emissions. By distributing HEM for network energy, the board increases energy consumption at the local level.

14. SMART HEALTH

There are currently a lot of healthcare contacts due to rapid population expansion. The world's population has medical needs, and traditional medical procedures are inadequate, out-of-date, and ineffectual in providing for those requirements. The problem has gotten worse as a result of the fact that the proportion of doctors in the health sector to the population has not increased. The risks of prescribing improper medication, receiving a misdiagnosis, and misinterpreting an infectious or communicable sickness are next. The disparity between aspirations and actual treatment is widened by a lack of resources and high demand. The solutions provide a smart framework for medical services that balance requests for medical care and difficulties with grace while maintaining productivity, accuracy, and maintainability.

The use of advanced clinical mediation techniques in conjunction with customary clinical processes is referred to as "Smart Medical Care" (clinical gadgets, sensors, cell phones, crisis administrations, ICT, etc.). To answer the demand and enhance the character of administration, modern smart wellness administrations blend sensor organizations, ICT, distributed computing, computer fog, mobile applications, and astonishing information management systems. Smart Healthcare Services promoted ongoing status choice tolerance by disclosing sensitive patient information to authorized clients' specialists, medical assistants, and lab experts through the secure arrangement of clinic frameworks. Integrated Electronic Clinical Records (ECRs) also provide consistent decision-making based on the newest data.

The proposed CoAP-connected RFID, WSN, and clever, adaptable technologies are used in the smart medical system. Another sophisticated healthcare architecture that envisions cloud- and mobile-based frameworks for amazing medical services has been put out. Handled ambiguity in a variety of circumstances. Information fusion is a complex healthcare system that has been proposed. Proposed an alternative approach to establishing sensitive portable wellbeing in remarkable urban areas. Increasing healthcare administrations also involves increasing the personal enjoyment of city dwellers. Along these lines, coordinating excellent medical care in smart urban communities is necessary to demonstrate considerable progress in getting the smart city concept recognized internationally.

IoT and the medical industry IoT applications in the healthcare sector have the potential to benefit the majority of people. The main effect is on health monitoring, particularly when there are silent symptoms present. An early diagnosis may trigger disease and even death preventative measures. Advances in IoT and digital technologies were essential for improvements in data processing, storage, and analysis. Benefits span a broad range of industries, including those in business, communications, energy, education, health, and human services as well as sustainability, transport, and water. The majority of e-health apps now available focus on managing and monitoring medical care and disease treatment. In conclusion, more IoT integration seems desired, especially given the obvious shift toward health care at home in the global health system as shown in Figure 14.

Several communicated areas are covered by a dispersed node set. Each node may cooperate with additional elements, including telemedicine, researchers, nurses, doctors, and ambulances, among others.

Figure 14. Applications for IoT in healthcare

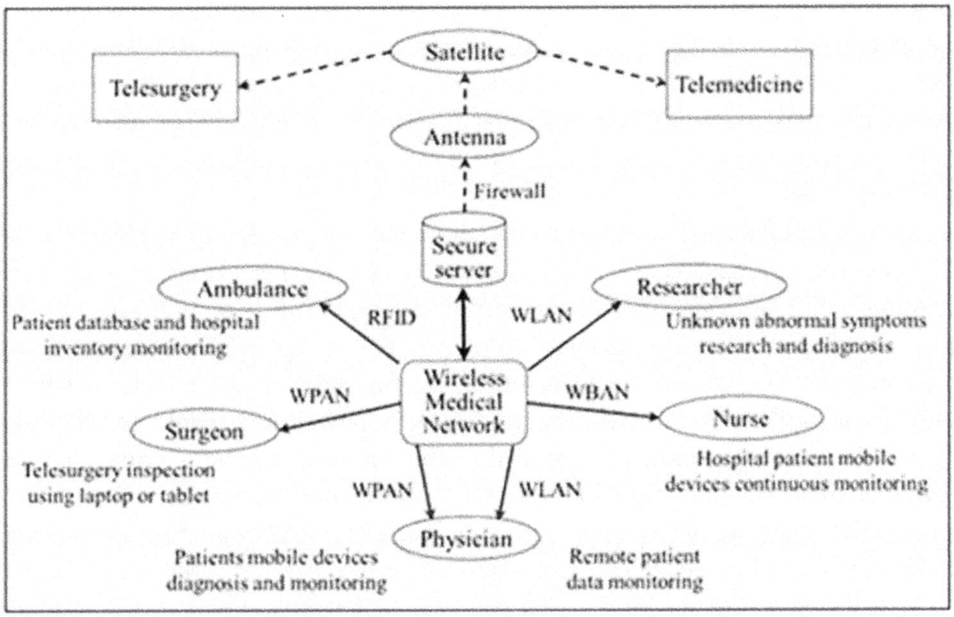

Hospital patient control through monitoring and control systems may have a wide range of uses. Numerous networks, including radio-frequency identification (RFID), wireless local area network (WLAN), wireless personal area network (WPAN), wireless body area network (WBAN), and others help with automated identification and data collection from electronic devices.

S-health and the Internet of Things, the ease of joining nodes is one of the fundamental features of the Internet of Things, which makes changing topologies and creating new communication protocols challenging. WSNs satisfy several criteria outlined by Internet technology solutions. A distributed network of nodes connected by impromptu communication that promotes node cooperation makes up a WSN.

The wireless technologies for biomedical sensor systems include WPAN, WBAN, also known as body area network (BAN), medical body area network (MBAN),16, WLAN34,35, and WiFi36, to name a few. Applications for IoT in healthcare. Global Microwave Access Interoperability and Bedo'n-Molina et al (WiMAX). Low-power digital radio signals based on the IEEE 802.15.4 standard include ZigBee, Bluetooth, Adaptive Network Topology (ANT), ultrawideband (UWB), endoscopic capsules, implantable and ingestible sensors, epidermal electronics, smart bands, smartphone apps, RFID used for object automatic identification and location, real-time location system (RTLS), and indoor positioning system. RFID and ZigBee technologies, which connect smart devices to the Internet, save working hours and costs in addition to utilizing cloud computing.

15. ADVANCED MEDICAL SERVICES

The effectiveness of any disease therapy depends on the medicine following the prescription exactly. According to Zhibo et al. 56, patient-related issues may cause patients to forget to take their prescriptions. These elements could be psychological, linked to motivation to recover from disease, worry, and, in the case of the elderly, forgetting the timetable. It can be challenging for a doctor to determine whether a patient is taking their prescription appropriately during a routine appointment unless the patient specifically mentions it. The patient's recuperation could be negatively impacted by this. As a result, technologies to track patients' medicine at home evolved. As a smart prescription box and smart pharmaceutical package with RFID communication capabilities, Zhibo et al created iHomeHealth-IoT. The patient receives medication timing advice from the smart box. The box sends a text message to the doctor to alert him as well.

Telemedicine is the practise of medicine using computers, communication networks, medical technology, and health professionals to provide remote care to patients. The provision of consultation, diagnosis, prescription, health information, and prescribing services. The primary means by which patients, particularly the elderly, can be treated using e-health apps is ICT that is 3G, 4G, and 5G-based, including IoT and communication protocols. The sick room performed well when using IoT based on RFID technologies. One of India's biggest public health issues is cancer, and thanks to wireless technology and oncological telemedicine, remote professional medical advice may be provided. This aids in early detection, palliative treatment, and rehabilitation in difficult-to-reach locations.

The ECG, a mobile cardiology device that transmits the electrocardiography signal through a smartphone in real-time, is a widely used telemedicine application. a research project in Bangladesh, where the majority of people reside in underserved rural areas. Telemedicine technologies would make it possible for hospitals and doctors to work together to offer healthcare to rural residents. The article describes a distributed information system for several hospitals' telemedicine that links patients in rural areas with medical specialists as shown in Figure 15.

Figure 15. Home based care in smart health sector

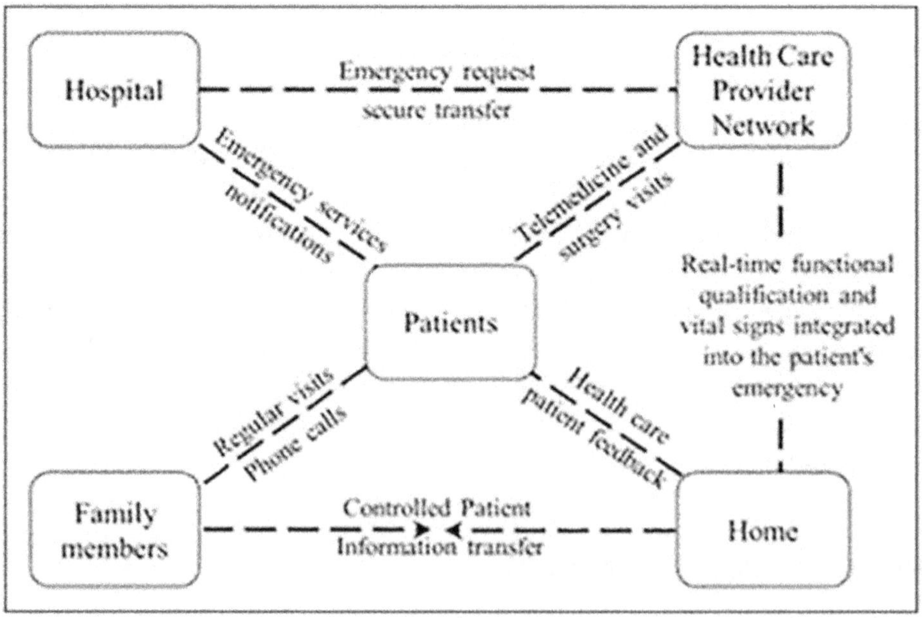

16. IOT IN SMART CITY

16.1 A Smart City IoT Architecture for Measuring Road Safety

The World Health Organization (WHO) noted in its Global Status Report on Road Safety for 2015 that the total number of road traffic fatalities globally has stabilized at 1.25 million per year, with tens of millions more either injured or disabled. Various programs, notably the United Nations' 2011-2020 Decade of Action for Road Safety effort, have enhanced road safety legislation and regulations. Despite acknowledging that the progress has been minimal, the WHO has persisted in calling for immediate action to cut these rates. There are significant financial losses as a result of medical expenses, infrastructural repairs, and delayed productivity in addition to the losses in human lives and wellbeing.

Despite the fact that global data have plateaued, the Global Status Report does show an increase in the number of road fatalities and injuries in low-income countries. These variations emphasize how challenging it is for low-income countries to follow high-income countries' tactics for improving road safety. The WHO highlights numerous activities that can be implemented to increase road safety with little to no cost impact in its "Save LIVES: Road Safety Technical Package." A key component of these efforts is the development of economic mechanisms for "monitoring road safety by enhancing data systems." A significant focus of the package is encouraging the adoption of the Safe System approach, a comprehensive plan for enhancing road safety that varies from typical management approaches by placing a premium on security.

16.2 Approach Using a Safe System

The Safe System (SS) approach to transportation networks originated with the "Safe Road Transport System" concept developed by the Swedish Transport Agency. The approach basically abandons the viewpoint that accidents are primarily and automatically the responsibility of the driver in favor of a viewpoint that analyses and evaluates the true causes for accidents. By dividing safety into the three categories of vehicle, road, and road user safety, SS reduces fatalities and injuries by regulating speeds and permitting quick emergency response. The WHO is now using the model as a foundation for planning, creating policies, and enforcing laws related to road safety. The model has been widely adopted since its introduction as shown in Figure 16.

Alt-text: Figure 16 shows the components of the safety system. Speed is emphasized heavily in the SS approach since it is the most important and powerful factor that determines whether a fatality occurs.

Given the vulnerability of the human body, an unprotected impact at a speed of more than 30 km/h is unlikely to result in survival; at lower speeds, it is more likely that either death or serious injury will happen. The three model components are designed and monitored by the SS technique to proactively avoid lethal speeds, which reduces the amount of time emergency responders need to get at the scene of an accident. Some SS approach elements are as follows.

16.2.1 A Secure Vehicle

The importance of vehicle safety is confirmed by regulations that require testing and grading, as well as by innovations like electronic stability control. In addition, mandatory inspections (such those performed at license renewals) and on-the-road reports are used to assess the state of vehicle safety.

Figure 16. The components of the safe-system-based safe road transportation systems are a safe vehicle, a safe route, and a safe road user

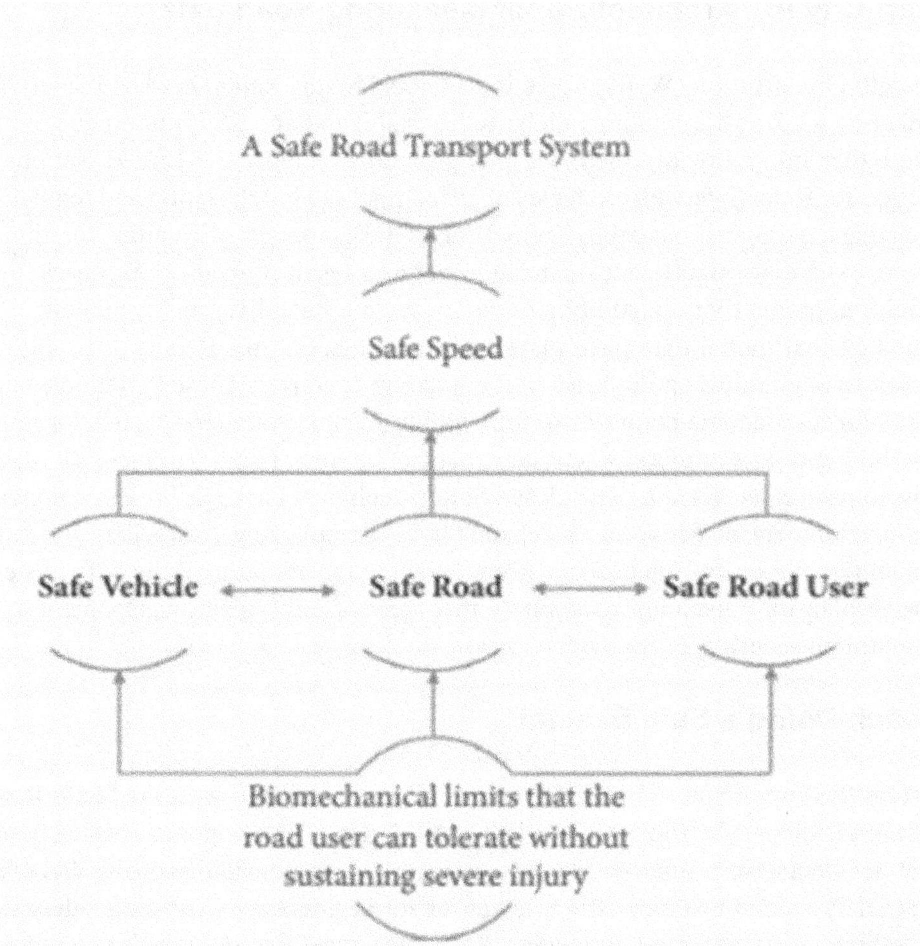

16.2.2 Secure Road

Road (or network) safety is evaluated in a variety of ways. Road inspection offers a direct and unambiguous assessment of the condition of the road and determines whether repairs or adjustments are necessary. By dividing the road network into "Traffic Analysis Zones (TAZs)," the structure of the network can be evaluated for safety. The European Road Assessment Program created the European Road Safety Atlas for EU countries in 2011. (EuroRAP). The atlas appraised the safety of the roads and assigned them a star rating based on vehicles with specific equipment for aggregating multimedia-based data. Along with a number of additional national projects under the International RAP, or iRAP, project, the EuroRAP activities continue to implement an SS strategy throughout the EU.

16.2.3 Safe Road User

Road user safety has several different facets, including initiatives for sober driving, travel distance, exposure, licensing, and enforcement. The necessity for such characterization becomes increasingly urgent since city-based crash record analyses often reveal a crucial reliance on either driving conduct or driver awareness. These studies also highlight the critical need for creative approaches to promote safe driving during the application and post-application processes.

16.2.4 Contributions

The figure 16 illustrates how the medium-to-long term perspective of the SS method facilitates systemic actions that are taken to guarantee the safety of the road network. Although using "data monitoring systems" in shorter-term contexts is recommended and suitable, the main emphasis is still on medium- to long-term reaction cycles. Our intention is to extend SS to the short-to-medium term by utilizing recent advancements in the Internet of Things (IoT) and Intelligent Transport Systems frameworks (ITS). This aligns with the view of "smart cities," which places a strong focus on automation to deal with the increasingly dynamic nature of urban elements. To this purpose, the contributions made to this effort are as follows:

- To assess the safety of the transportation road network, a study offers a comprehensive, IoT-based design.
- It also completes the SS method by filling in the gaps in its short-to-medium term scope of concerns, further adapting the approach to the dynamic nature of smart cities. As a result, the suggested architecture is entirely consistent with the SS approach (note the scopes illustrated in Figure 16).
- The potential of a cost-effective road safety monitoring system is demonstrated by the proposed architecture, particularly those targeted at building smart cities.

The core of the proposed architecture uses machine learning in a novel way to measure road safety. This tool makes it easier to conduct rigorous and dynamic assessments. Here is the demonstration of use of the created core for smart city route planning using a safety-based approach as shown in Figure 17.

17. A DYNAMIC ROAD TRANSPORT SYSTEM SAFETY ASSESSMENT IOT ARCHITECTURE

In our evaluation of pertinent works in the preceding part, we demonstrated how several innovations are making it possible to analyze the safety of cars, roads, and drivers. The purpose of this part is to demonstrate a state-of-the-art, flexible IoT architecture that enables you to assess the level of safety on a city's road system. We go over the evaluation elements in more detail, as well as how they may be merged to provide a single, useful safety indicator. We also discuss the components of the design and how they interact, including a robust computational engine for safety assessment.

Figure 17. Aspects of the safe system (SS) for short-, medium-, and long-term considerations

17.1 Elements of Assessment

Figure 17 illustrates how the three SS components—a safe vehicle, a safe road, and a safe driver—allow for a hierarchical safety assessment approach in which the security of the individual components can act as a general signal of security for the road network as shown in Figure 18.

This signal can then be combined with the evaluation of particular road segments, routes, and the entire road network. The OBD-II device would provide up-to-date data, the vehicle's Vehicle Identification Number (VIN), which identifies the make, model, and basic safety rating, and the regulator's information (such as the findings of the most recent regulatory check) (on board diagnostics). The severity of

Figure 18. Aspects of the safe systems method to evaluate the safety of the road network

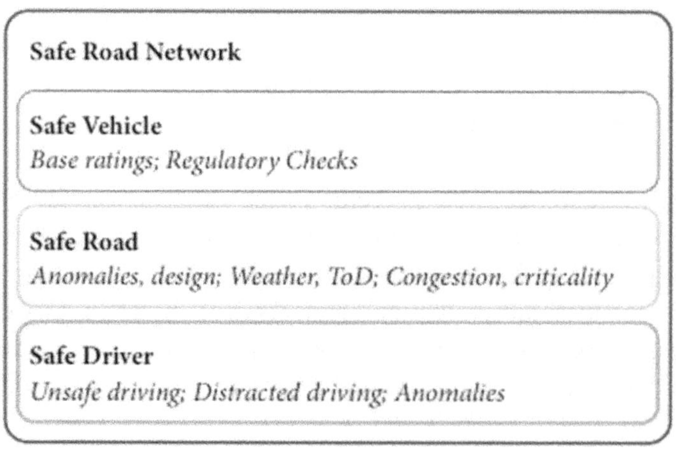

curves, the presence of shoulders, and whether the road is on a hillside may all be determined for roads by using both historical and recently detected data. It can aid in understanding the context of the route and the network map of the roads.

Hence, locating a vehicle's location using an on-board GPS would enable the first level of this comprehension (whether dedicated or via a smartphone). Road anomalies can be notified while this is happening using in-vehicle sensing (e.g., potholes). OBD-II interface-based sensing can detect skidding incidents, which can then be connected to news sources or the weather. Weather, time of day (ToD), and data on the presence and/or condition of road lights can all be used to determine visibility. Basic traffic information can also show the severity of traffic bottlenecks, the importance of a road or segment, or how frequently hastily made route plans choose a certain road or section. Due to DBM's developing technology, drivers will have an easier time identifying distracted or unsafe driving behavior.

Also, it will make it simpler to spot new problems, including when a driver frequently switches lanes to avoid a fresh block on the road brought on by a falling tree trunk or other road debris. Additionally, it will enable the detection of certain driving behavior irregularities. Depending on the actual (or real-time) state of the road, a meaningful safety statistic may be taken into account. The level of safety of a particular section or road, for example, depends on the general level of safety of the vehicles using it, the quantity of potholes and/or how wet or slick the surface is, as well as the safety and attention of the drivers on the road.

The design of our architecture makes use of three significant dependencies. The first is between SS components, such as how well a vehicle can handle a certain route or how some drivers act more cautiously when visibility is better. The second reliance is between succeeding segments/roads, especially in terms of moving cars and drivers. The third reliance grows gradually but is identical to the second. Therefore, abrupt changes in safety levels can be seen as an anomaly (outlier) or extrapolated to mean that the road environment has undergone a considerable change.

17.2 Considerations for Architecture

The key component of the proposed design is a computational core that operates in two layers. Dynamically assessing each SS component's safety is the top goal. The second uses the outcomes of these three evaluations to produce a single rating for the route as shown in Figure 19.

Alt-text: Figure 19 shown an illustration of an architectural consideration.

There are two categories that can be applied to the many sources considered in each unique assessment. The first is obtained by in-vehicle sensing, whereas the second is derived from sources that outline the route's general environment. Examples of this include traffic problems, regulatory reviews, crash reports, processing of news and social media, weather updates, and time of day. As was previously said, all three SS components' inferences and assessments can be made using the same sources. This determines where in the architecture the data fusion and correlation element is located as shown in Figure 20.

Alt-text: Figure. 20 illustrate how to combine, correlate, and evaluate a Safe System component.

While data fusion joins sources that provide the same knowledge synthesis to strengthen the dependability of that knowledge synthesis, correlation focuses on cotiming and colocation of sensed data as well as cross-validation of incoming information streams. The results from the three assessment modules can be used to create a live representation of the safety of a sector. We discuss a model for producing this representation in more depth below. The values can then be provided to a variety of services, including driver assistance programs and those that result in calls for emergency assistance or road maintenance,

Figure 19. Considerations for architecture in dynamic road network safety evaluation

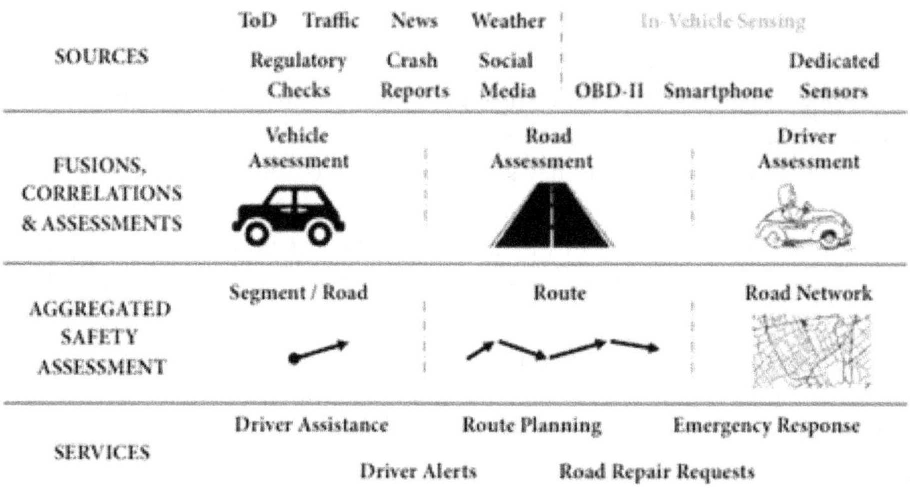

once the aggregated safety evaluations have been established. In the section that follows, we demonstrate how this evaluation core can be used in a safety-based route planning application.

17.3 Vehicle Sensing

A hypothetical in-vehicle sensing system consists of three parts: an OBD-II dongle, a specialized sensor, and a smartphone. The specialized sensor was an Aaronia GPS logger that included a pressure sensor, compass, accelerometer, gyroscope, and sensors for GPS localizations. The logger lacks wireless connectivity features but can be connected to a PC through USB. The networking and sensor capabilities of contemporary cell phones running iOS and Android were utilized. Although one of the objectives of the work is to find a cost-effective solution for dynamic evaluation, the arrangement described above was designed with both exploratory and validation purposes in mind.

The connection was allowed in cases where the sensing components could connect directly to the cloud repository, such as when utilizing a smartphone or the music box dongle. If not, a connection was formed through another device in an indirect manner, such as the logger via a laptop running a PC or the Scan Tool dongle via a smartphone. A PostgreSQL 9.6 database built on a Red Hat Enterprise Linux Developer OS for the backend will receive and store sensing data. Access to an Apache web portal would

Figure 20. Combination of correlate, and evaluate a safe system component

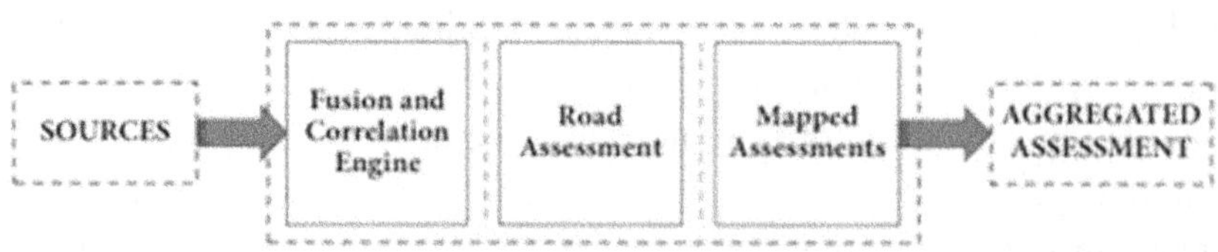

be made possible by this arrangement, encouraging quick prototyping and the simplicity of concurrent data reports from numerous sources.

17.4 Core Assessment

Road networks are dynamic, intricate structures, thus modelling any aspect of them requires care. This is especially true in densely populated urban areas where other aspects of city life are linked and entwined with traffic dynamics. A realistic model to assess the safety of a road network should consider interelement connections in SS as well as relationships in both space and time, as we noted in our prior discussion of dependencies. Furthermore, it ought to make it simpler and less expensive to quantify these interdependencies.

18. BLOCKCHAIN TECHNOLOGY IN SMART CITY

18.1 Blockchain and IoT Integration Underpinning 6g Communication for Smart Cities

The exponential urbanization that emerges from the increase in global population over several decades has a huge detrimental effect on people's day-to-day lives in terms of social, economic, and environmental difficulties. According to a United Nations estimate, by 2050, more than 70% of the world's population would live in urban areas. Transportation, education, health, economy, and working and living conditions are only a few of the areas where the expansion of urbanization has considerably improved peoples' living standards (Khanh & Khang et al., 2021).

Nonetheless, the higher population density creates substantial challenges for the modern technology-based effective distribution of available resources. Environmental resource limitations including traffic congestion, trash disposal, air pollution, and greenhouse gas emissions are also influenced by the rapid urban population expansion. To address all of these problems and challenges and allow cities to expand sustainably, the concept of the "Smart City" was created. It provides opportunities to address the aforementioned difficulties and provide a high-quality life by utilizing ICT and IoT intelligent services as shown in Figure 21.

Alt-text: Figure 21 depicts five different scenarios for smart cities, each of which shows a specific IoT-envisioned application that will help to improve the intelligent services offered by smart cities. They are *(i) Emergency Communication, (ii) Healthcare 5.0, (iii) Smart Agriculture, (iv) Smart Grid, and (v) ITS.*

Recent advancements in IoT and ICT devices offer sustainable smart city features like better health, energy, and basic amenities, as well as enhanced public transparency, increased community participation, resource efficiency, effective traffic and public transportation management, improved educational services, and many others. IoT has expanded Internet architecture even though it connects computing devices (i.e., things) to form connections for transmitting and receiving data with little to no human engagement. When IoT data is processed, stored, and retrieved at a different server over the Internet (using cloud/fog computing), security vulnerabilities are present.

Current cloud-based solutions have an issue with single-node failure, and IoT data are susceptible to hacks such as data tampering and fake data injection. Cloud-based solutions often cannot fully ensure data availability, integrity, and security for IoT-based smart cities. The state-of-the-art and decentral-

Figure 21. Smart city perspective that is application-specific

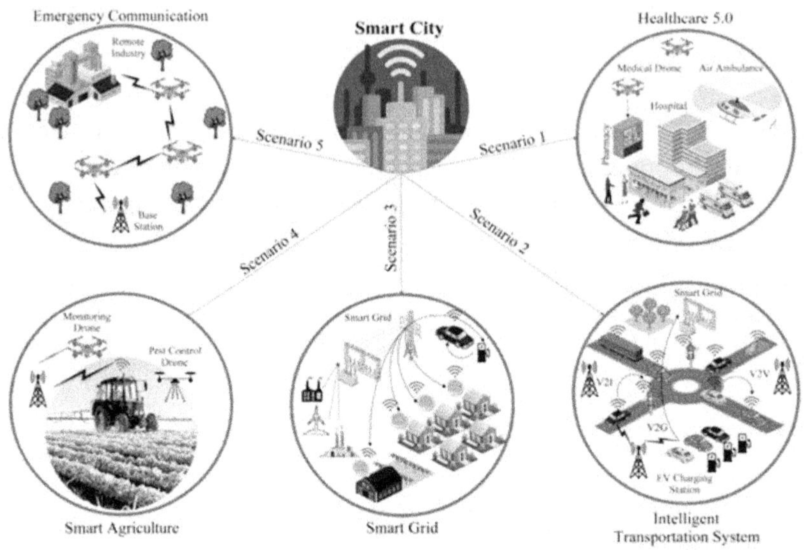

ized blockchain technology can be utilized to solve the aforementioned issues for the establishment and improvement of smart city services.

The first time a Distributed Ledger Technology (DLT) was applied was in 2008. Here, each transaction is stored in a block and made visible to the public (who is part of the blockchain network). Decentralization, trust lessness, democratization, automation, data immutability, data integrity, transparency, and pseudonymity are just a few of the aspects that can be implemented with blockchain technology in the IoT-envisioned smart cities. Without a Trusted Third Party, the system normally operates peer-to-peer (P2P) (TTP).

Pseudonymise makes it simpler to use public pseudonymous addresses while hiding real identities. Blockchain-based smart contracts (SCs) can also perform transactions between two parties on their own and store data. SCs are groups of code that can be written in a number of different programming languages, such as Solidity and others. Owing to the decentralized nature of the DLT, it is essential to guarantee consistency by using several consensus techniques, including Proof-of-Work (PoW) and Proof-of-Stake (PoS), among decentralized nodes.

In addition to security issues, smart city apps encounter network latency, reliability, and connectivity issues. Traditional communication networks like 4G, LTE-A, and 5G are not entirely appropriate for mission-critical applications like border monitoring, healthcare drone delivery, telesurgery, etc. as a result of issues with connection density (the number of devices per square kilometre) and high latency (round-trip time). These issues can be solved by the integration of the sixth generation (6G) communications network with the IoT-enabled smart city development.

Figure 21 displays the features of the 6G network. As a result, this study emphasizes both data security and network performance concepts. In the context of the sustainable development of IoT-enabled smart cities, this paper presents the architecture for smart grid, ITS, and healthcare 5.0 from a blockchain-envisioned perspective. Moreover, as previously said, carefully studied cutting-edge techniques for each application

19. IOT, BLOCKCHAIN, AND 6G HISTORY FOR SMART CITIES

The history of blockchain, IoT, and smart cities is highlighted in this subsection (Khang & Chowdhury et al., 2022). The following is a more thorough explanation of these technologies:

19.1 Smart City

A smart city is one that has integrated ICT and IoT into its design to improve livability, increase operational efficiency, and promote economic growth. It coordinates and automates a range of services (including dependable electricity and water delivery, efficient public transit, and many others) to make the best use of available resources, minimize environmental impact, and reduce living expenses for a more sustainable growth of the city.

The following are the essential elements of an infrastructure for a smart city: Good healthcare and education services, affordable housing for the poor, sufficient electricity, intelligent public transportation and urban mobility, a reliable water supply, good e-government with citizen participation, sanitation and waste management, citizen safety and security, a sustainable environment, and vigorous economic growth are all essential. By addressing issues and providing solutions utilizing the data generated by IoT-envisioned devices like smartphones and sensors, any urban area may become a smart city. IoT and blockchain are two disruptive technologies that are key to this shift.

19.2 Internet of Things

It is a network of connected devices with Internet Protocol (IP) addresses that gathers and sends data. Every aspect of human life has the potential to change as a result, with healthcare, smart environments (including smart cities), transportation and logistics, and personal and social applications being the main areas of change. The requirements of a variety of applications, including ITS and the smart grid, are satisfied by the networking architecture, different endpoints, numerous protocols, and communication technologies.

The two key components that differ between various applications for smart cities are as follows: The three IoT-envisioned applications with mobile and stable topologies are smart buildings, vehicle ad hoc networks (VANETs), and intelligent transportation systems (ITS). IoT-enabled devices used in the smart building are steady because of a stable network, in contrast to VANETs, which require a time-varying topology for fast moving vehicles. Managing entities is very challenging because of the mobility of the IoT endpoint devices, which makes network connectivity unpredictable. Because they have a limited amount of storage and incredibly little processing and communication power, IoT devices are inexpensive. High-capacity performance is (ii). It is consequently easy to use and can detect humidity, pressure, temperature, and other variables on a large scale. They are heterogeneous by nature since they use a variety of hardware and software platforms.

19.3 Blockchain Technology

Although the inner workings of blockchain are rather complicated, the basic idea of distributed ledger technology (DLT), which enables the decentralization of data storage so that data cannot be modified or controlled by a central organization, is simple. Satoshi Nakamoto first presented the Bitcoin framework in the year 2008. It is a series of blocks put together within the blockchain network using the hash value of the previous block. The block header, transactions, hash values, and other kinds of data are all included in a block. Each blockchain participant must use their digital signature to execute a transaction, which ensures data security, data immutability, and safe participant identity.

A SC is a self-executable, self-enforcing digital contract between two parties that is made possible by distributed ledger technology. A digital agreement created with SC does not require the involvement of a third party, and it is irrevocable once signed. It is a piece of programming that can be done in a particular language, such as C++ or simplicity. Data immutability security is provided by DLT using a variety of consensus procedures, including Proof-of-Work (PoW), Proof-of-Authority (PoA), and others. There are three major types of DLT: private blockchains, owned by a single company and permissioned, and consortium blockchains. Public blockchains are entirely distributed, decentralized, and permission less, similar to Bitcoin and Ethereum (permissioned blockchain involves multiple organizations and similar to the private blockchain).

19.4 6G Communication Network

The 5G technology is almost ready for deployment, despite the research community's interest in the 6G technology. The potential of the 6G communication network will undoubtedly catch the eye of app developers in the near future. The massive connectivity required for data-explosive applications like telesurgery, holographic communication, and gaming won't be available on the 5G network.

Figure 21 depicts a comparison of the 5G and 6G communication networks. A few of the real-time, data-intensive applications that 6G enables include haptic communication, IoT-enabled smart cities, smart homes, autonomous vehicles, smart healthcare, and ITS (Khang, Vugar and Abuzarova et al., 2023). The two distinct IoT-envisioned applications with mobile and stable topologies are vehicle ad hoc networks (VANETs), intelligent transportation systems (ITS), and smart buildings.

Since the network is stable, the smart building uses IoT-enabled devices that are steady, whereas VANETs call for a time-varying topology for quickly moving vehicles. The mobility of IoT endpoints makes network connectivity unpredictable, which makes controlling entities very challenging. (ii) The low cost of IoT devices is a result of their tiny computing, communication, and storage capabilities. As a result, it may be used to measure temperature, pressure, humidity, and other factors on very large scales. They are heterogeneous by nature since they use a variety of hardware and software platforms.

20. IOT-ENVISIONED SMART CITY SECURITY CHALLENGES

IoT's unique and specific characteristics cause major data security risks. When IoT devices are used in unfavorable human situations or are left unattended, they are vulnerable to a variety of risks, making management of the huge number of these devices difficult. To attack IoT networks, for example, enemies may physically capture and take control of these IoT devices. Furthermore, data from sensors may be captured, shared, and processed at several intermediate systems, increasing the possibility that it may be tampered with or fabricated. The risks to safety and security are exacerbated by poor wireless communication techniques. To overcome the aforementioned problems, IoT system security must be improved.

20.1 Security Issues

The following information clarifies the potential security threats to the environment of smart cities.

- Jamming attack: The connection of smart cities is being attacked using data collision, data packet flooding, excessive outside noise, or channel interference. It is one of the simplest assaults to carry out because it doesn't need for a thorough understanding of the victim system.
- Hijacking attack: The intruder wants to seize entire control of smart city IoT apps and update the data that has been stored. For instance, a hacker may alter energy consumption data from a smart meter.
- Eavesdropping: In this case, an attacker attacks the insecure communication channel used in a smart city environment to send and receive data. The transmission over the network won't be harmed.
- The Man in the Middle Attacker manipulates the communication system of smart cities by listening in and sending fake commands.
- Replay attack: In this kind of attack, the real data that was previously taken from a smart city communication network and replayed when there was suspicious activity.
- Data traffic analysis: In this technique, hackers passively monitor open communication channels, much like eavesdroppers.
- DoS/DDoS: In this type of assault, fictitious requests totally clog the communication channel, making the environment for smart cities useless.

The atmosphere of smart cities is negatively impacted by different attacks, such as spoofing attacks and many others. The following summarizes typical threats to fully IoT-envisioned smart cities:

20.1.1 IoT Device Attacks

By using node-capture techniques, hackers can physically take and control IoT devices. The attackers will have access to the private information stored on the seized nodes, such as certificates, keys, etc. Additionally, they might utilize the information gathered to pose as trustworthy nodes and launch other attacks (such a malicious data injection attack).

20.1.2 Harmful Software Attacks

Software assaults are a series of attacks that alter a piece of software's control and functionality by accessing the software's back end; well-known examples of common software assaults include several scripts, viruses, worms, and many more. Traditional online security measures and intrusion detection systems (IDS) are employed to thwart such attacks. Security is a top concern for IoT applications, especially the dependability of IoT-enabled devices and the integrity of sensory data. For instance, the primary assurance for IoT operational security is provided by actuator commands and sensor data.

To protect IoT-based communications, effective methods for assuring non-repudiation, authenticity, integrity, and secrecy of data transfers should be used. Since data integrity often rely on TTP, such as identity providers, identifying IoT devices ensures data reliability from the source. As a result, methods for authentication and encryption were developed and are now being utilised to protect IoT data. As a result, the data storage service is in charge of ensuring the security of sensory data when it is supplied to the data storage system.

20.1.3 Combining IoT and Blockchain Assets

In the context of smart cities, DLT may hold the key to managing and regulating the scalability, privacy, dependability, and—most importantly—security challenges of IoT. IoT device security in smart cities is made possible by a few key blockchain characteristics. Huge amounts of data are stored in the IT systems of smart cities, making them a gold mine for hackers and raising worries about data security and privacy.

20.2 Using Blockchain and IoT Assets Together

This section focuses on how the blockchain could improve the management of energy demand and response in a smart grid context by lowering security concerns. Fossil fuels including oil, coal, and natural gas are used to produce the bulk of energy, which results in pollution and greenhouse gas emissions. In order to protect the environment, renewable energy sources like solar and wind power are being used more frequently all over the world.

More people are becoming prosumers who generate and store their own energy as energy storage technology advances. In order to create power grids that are safe, efficient, sustainable, and economical, the smart grid concept was developed because the transitional grid must manage such a vast number of consumers and prosumers. By transforming energy users into prosumers, the widespread installation of rooftop PV solar panels is also fostering the development of decentralized energy systems (consumers who can also produce energy).

Peer-to-peer (P2P) energy trading is a new type of business model developed by prosumers who are keen to reduce their energy expenses and sell surplus solar power to consumers, other prosumers, and the grid (ET). These ETs improve grid resilience while improving the scalability, speed, and security of energy distribution. Yet, purchasing and selling energy is a difficult process because of centralized design and problems with security and privacy. Cheng et al. introduced a distributed ET model based on blockchain with the aim of facilitating the balance of energy production and consumption and maintaining the stability of the energy market. To provide a secure and decentralized control over the energy exchange, it is suggested to employ the Digital grid router in a blockchain-based peer-to-peer ET system.

A decentralized transactive energy system concept was developed by Laszka et al. that involved people actively taking part in both energy production and consumption. Data security is also a crucial component of the smart grid system. A consumer data protection system that tracks energy use and feeds it to the grid is provided via blockchain. Every activity in the smart grid is automatically monitored by SC, which can spot unauthorized energy use. A data protection solution based on blockchain is created to increase the security of energy data.

A proof-of-concept for things-to-things energy payment is also being considered because of the rise in IoT energy system devices, which makes blockchain a viable option for machine-to-machine (M2M) energy trading. Electric Vehicle (EV) charging, Vehicle-to-Grid (V2G), Grid-to-Vehicle (G2V), Vehicle-to-Vehicle (V2V), and many more services offered by the smart grid system can all be improved by the usage of blockchain technology in the context of smart cities.

Figure 22. Smart grid architecture based on blockchain for peer-to-peer energy trade

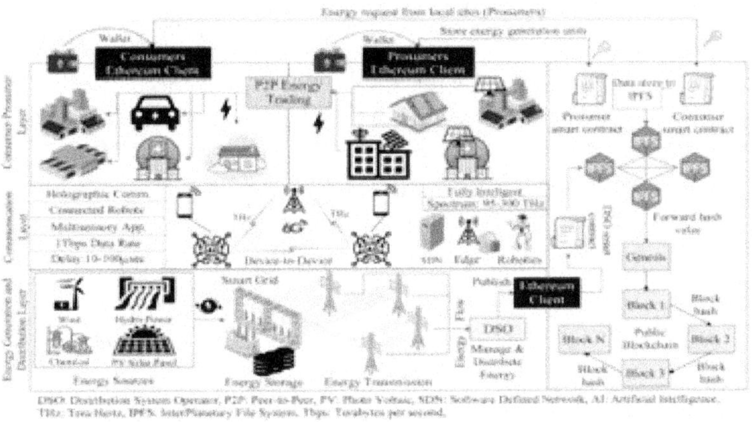

20.3 A Structure Based on Blockchain

For the smart grid system envisioned by IoT as shown in Figure 22.

Three levels make up the suggested architecture:

I. The layer that generates and distributes energy.
II. The layer of communication
III. And the stratum of consumers and producers.

20.4 Layer for Energy Production and Delivery

This tier includes energy production and delivery through the distribution system operator (DSO). The management of energy distribution at medium, low, and voltage levels is the responsibility of DSOs (specific regions). Initially, energy is created from many different sources, such as hydropower plants, wind farms, PV solar power plants, and more. The DSO then delivers and disperses the energy to the ultimate customers (consumers and prosumers). The proposed architecture enables the seamless exchange of energy between two parties—the energy customer and the energy vendor. An Ethereum client is used to connect each participant, including the Distribution System Operator (DSO), prosumer, and customer, to the blockchain network. The proposed solution makes use of an asymmetric cryptography approach based on the Elliptic Curve Digital Signature Algorithm (ECDSA). It begins by creating a random private key using SHA256, creates public keys using secp256k1, and creates Ethereum addresses using the keccak256 algorithm.

20.5 Transmission Layer

For trustworthy and efficient communication, each component in this tier uses the 6G communication medium to connect with the others. Because of its 107 connections per km2 connection density, incredibly low latency, peak data throughput of 1 Tbps, user experience data rate of 1 Gbps, and mobility of 1000 km/h, the 6G communication channel was selected. Using intelligent decision-making, this layer

Figure 23. Blockchain's potential in a smart grid system

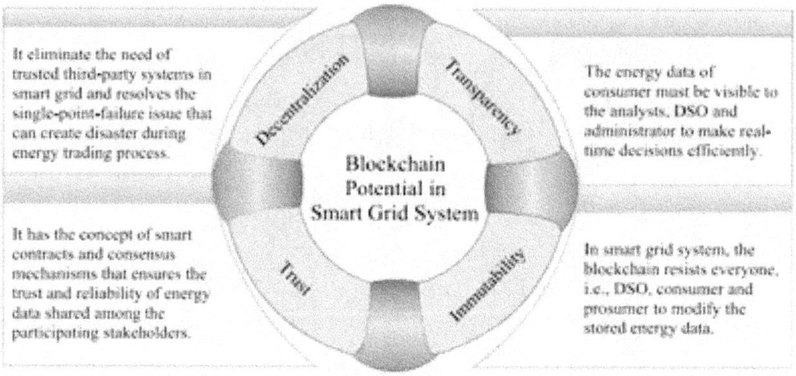

manages the entire smart grid network infrastructure, including load balancing, switching, access control, and data packet forwarding. Applications utilizing multiple senses and holographic communication can both profit from 6G connectivity. For efficient network administration, 6G technologies similar to those of 5G include edge AI, slicing, and network virtualization (SDN/NFV).

20.6 Layer of Consumer-Prosumer

This layer includes of entities that engage in peer-to-peer (P2P) energy trading within the blockchain network. These entities are consumers and prosumers connected through an Ethereum client. It consists of sensors and smart meters, among other IoT components. Each entity has a wallet to make energy transfers. The energy generated by consumers is stored at their end in a local battery. The blockchain records the energy exchange and transaction settlement when a buyer and a seller transact in energy. The energy exchange is then automatically monitored using SC as shown in Figure 23.

Transaction data is stored off-chain via the IPFS protocol to provide quick access and reduce communication lag. Following each energy transaction, the Ethereum wallets of the buyer and the seller, i.e., the buyer and seller wallets, are updated. It is simpler to establish an accurate, real-time energy consumption bill with a real-time settlement when smart meters that can be connected to the internet of things (IoT) are used. This eliminates the need for TTPs like banks or financial institutions.

20.7 Proposed Architecture Goes Through a Lengthy Process and Includes

Prosumers sign up with the Ethereum Blockchain and use smart contracts that specify terms for selling energy. Prosumers make the amount of energy (in Watt) and reservation prices (in Ether) they have purchased visible to all stakeholders. Consumers can submit their purchase intentions if they are interested in making a purchase. P2P trading takes place when the Buy Price is higher than the Reserve Price. The consumer receives energy, the information is entered into IPFS, the transaction is complete, and the prosumer receives ether (the cost of the energy) in his wallet. The suggested architecture supports a wide range of smart grid features, including things-to-things ET, EV charging, data security and stability, and many more. They are described as follows:

20.7.1 Trading in Secure Energies

Blockchain technology is being used to create a decentralized P2P energy market. Consumers and prosumers can trade on a trustworthy, decentralized, and transparent ET market in the smart grid area thanks to the DLT. Every time energy becomes available, prosumers start auctions and post the available energy to the Ethereum blockchain. After that, the essential buyers (consumer/prosumer, grid, or EV owner) monitor the energy auctions and put bids as required. IoT-enabled smart meters are necessary for tracking how much energy is consumed or sold over a specific period of time.

The IPFS off-chain storage solution maintains all energy transaction data from buyers, sellers, and smart meters on the Ethereum blockchain. Because there are no local charging facilities, the safe P2P ET system enables EV owners to plan long-distance trips without worrying about their vehicles running out of charge during commutes. EV drivers utilize the P2P ET system to charge their vehicles and search the surrounding areas for accessible chargers.

20.7.2 Data Security and Smart Grid Stability

Smart grid stability and data security are crucial for all parties involved, including energy suppliers and consumers/prosumers. One party, the consumer, wants to prevent overspending on energy and needs accurate information about how much energy their appliances use. On the other hand, inaccurate data could lead to bad decisions, grid disruption, and financial losses. Data security, integrity, and accuracy must be guaranteed for the benefit of energy providers, consumers, and prosumers. Blockchain's compelling features, such as non-repudiation, immutability, and decentralization, can be used by smart grid systems to ensure data accuracy, security, and integrity.

20.7.3 Energy Exchange Between Items

The exponential proliferation of IoT devices makes it difficult to manage them using the current method, so a decentralized IoT system is the way of the future (i.e., centralized). Blockchain is helpful in this situation and can establish an M2M energy exchange market. Energy is transferred from thing to thing without the assistance of a human using smart wires and a connected smart socket.

21. AN IOT-ENVISIONED INTELLIGENT TRANSPORTATION SYSTEM BUILT ON A BLOCKCHAIN

The suggested blockchain-based safe ITS for the 6G context serves as the foundation for many applications, such as traffic monitoring, EV charging, emergency vehicle services, moving medical supplies, and more. It adds intelligence, adaptability, trust, security, reliability, and efficiency to the ITS communication system. The suggested design is connected between three logical layers: the application layer, the communication layer.

Figure 24. ITS architecture using 6G with blockchain-based UAV support

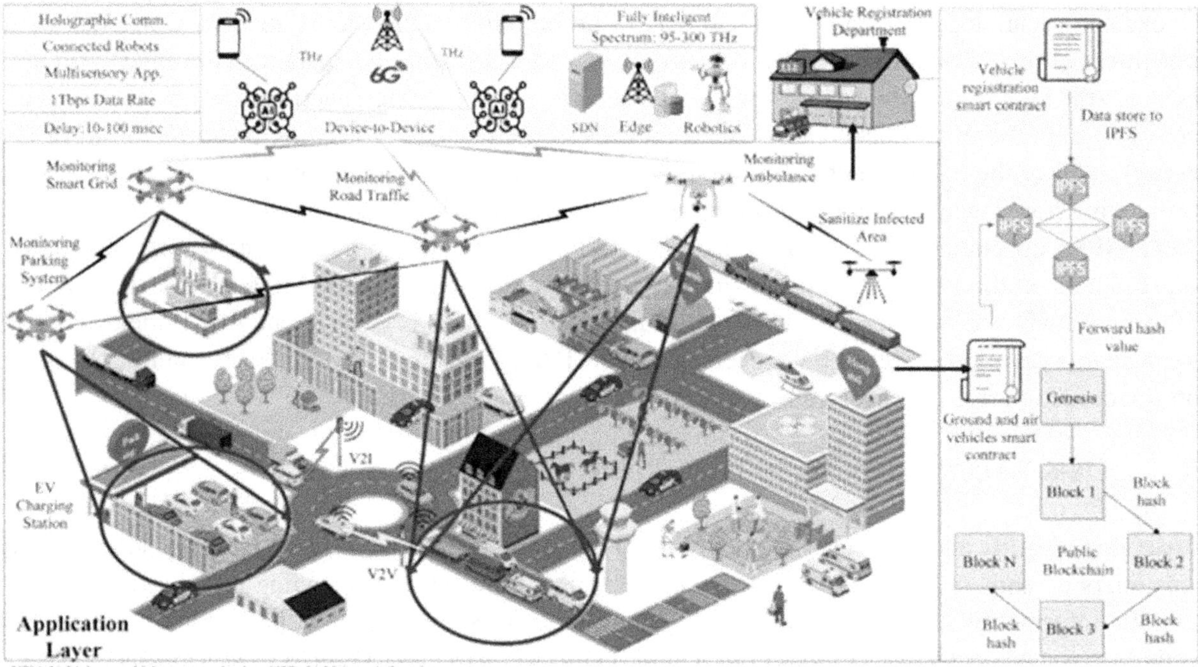

V2V: Vehicle-to-vehicle communication, V2R: Vehicle-to-infrastructure

21.1 Layer of Application

The Internet of Things (IoT) environment for the ecosystem of intelligent mobility is built at this layer, which houses the real physical sensors and actuators. It is made up of various surfaces for ITS applications that can be utilized to solve parking and traffic issues. Autonomous vehicles (AVs) and connected autonomous vehicles (CAVs), which are independent cars that can make decisions based on their surroundings and information from other autonomous vehicles, are the main advancements in ITS. Many countries have traditionally erected huge displays or announcement speakers on the road to alert the public of the current traffic conditions so they can choose alternate routes to their destination. Going through those congested routes also saves the general public a lot of time, but its installation is very expensive. The suggested system is meant to assist applications including traffic monitoring, EV charging, smart parking, monitoring emergency vehicles, and healthcare delivery (Khang & Medicine, 2023).

Making the transportation system intelligent and effective requires installing roadside infrastructure (RSI) with sensors for diverse functions, such as a DHT sensor (to monitor temperature and humidity), a gas sensor, and an ultrasonic sensor (to calculate the distance between objects). The traditional method requires the installation of numerous RSI units at various key locations, which raises the cost of the system. However, the suggested system makes use of UAVs for each application area to reduce the cost of system implementation.

UAVs can be employed in a variety of applications depending on their various properties, which include flying altitude, geographic range, height, width, and weight. Here, the proposed approach takes use of an asymmetric cryptography method based on ECDSA. It begins by creating a random private key

Figure 25. Shows the possibilities of blockchain-based ITS functionalities

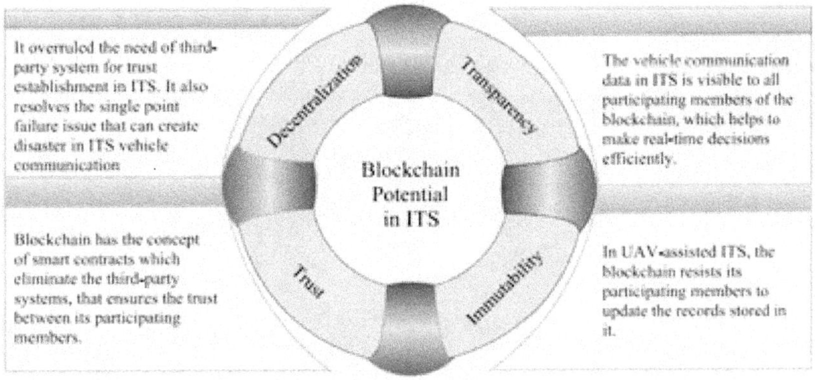

using SHA256, creates public keys using secp256k1, and creates Ethereum addresses using the keccak256 algorithm. This layer generates a lot of data, which it subsequently transfers through the communication layer and stores in the blockchain layer. Communication with the blockchain is only permitted by UAVs and CAVs that have been registered with the vehicle registration authority.

21.2 Layer of Communication

The information interchange between ITS participants, UAVs, and IoT devices depends on it. The application layer below will generate enormous volumes of data, which will be stored there utilizing a 6G communication channel. The features of 6G include massively low latency (10-100 s), extremely high dependability (109), a remarkably wide spectrum to accommodate more users (10 times more than 5G), a peak data rate (1 TBps), and full intelligence. It works perfectly for time-sensitive applications like AVs and CAVs in ITS, where a millisecond of delay might lead to a serious situation. Networked robotics, multisensory applications, and holographic communication can all profit from 6G connectivity. Similar techniques are applied in 5G and 6G for efficient network management, including SDN, NFV, Edge AI, and slicing.

21.3 Layer of Blockchain

This layer allows for the secure cryptographic distribution and storage of data. The data that was collected from the application layer in the form of a transaction is only retained if it complies with the SC's requirements. A SC is a piece of software code created specifically in the solidity, java, or kotlin languages. Since the blockchain is constantly self-executable and self-enforcing, there is no longer a need for centralised third-party mechanisms to foster confidence among its participants. PoW and PoA consensus protocols are employed to uphold the accuracy of the data stored throughout the blockchain. Due to its many benefits, including immutability, transparency, and distributed computation, and trust, blockchain is safe and efficient as shown in Figure 25.

We have included the Ethereum public blockchain, which enables anybody to join the network and store data as a transaction by paying a transaction fee, in the suggested architecture. The ITS is distributed

Figure 26. Application of quantum computing in smart city planning and logistic sector

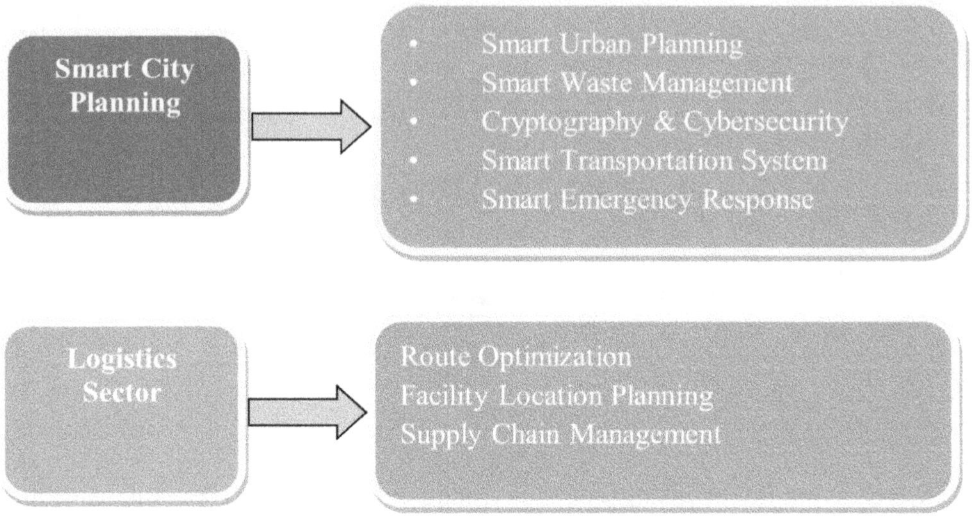

and decentralized in the correct way by Ethereum, ensuring that assets are always completely safe and transferable. Blockchain has improved the security and privacy of the data utilized in ITS communications. A blockchain cannot ever change data that has already been entered. The information gathered from the vehicle must include the token proving that the car is registered with the transport authority.

Otherwise, the entry will be rejected by the SC. The majority of the data on the blockchain network cannot be stored by an individual or an organization since it costs roughly USD 550 to keep data directly in Ethereum. The aforementioned issue will be resolved by maintaining the data in the decentralized, unchangeable, and secure off-chain storage system known as IPFS. The adoption of IPFS is also influenced by ITS data access with low latency and high throughput. The proposed architecture's workflow involves storing UAV-assisted ITS data on the blockchain and distributing blockchain data to all stakeholders via smart contracts. For instance, the vehicle registration bureau must use smart contracts to register new vehicles. Data must then be transferred to an off-chain storage solution like IPFS, which provides quick and secure access to data whenever it is needed (Ushaa, Khang et al., 2023).

22. CONNECTION BETWEEN SMART CITIES AND QUANTUM COMPUTING

Quantum computer technology is a branch of science that employs quantum theory fundamentals. It describes the energy, material, and their interactions at the level of atoms and subatomic particles. It can have an impact on many aspects of smart city planning by enabling more efficient, enhanced, and ecologically sound solutions to multifaceted issues. Quantum-inspired algorithms may result in new approaches to optimizing processes, accelerating transfers, innovating product development, and shortening overall time durations (Shah & Khang et al., 2023).

The science of quantum computing excels at solving the problems that are complex. It specifically addresses optimization-related issues ranging from privacy to financial transactions. It can run algorithms and computations at unprecedented speeds, permitting it to solve problems that traditional computers cannot. It improves machine learning capabilities, recreates quantum mechanical structures, optimizes obstacles, and helps to advance the fields of science. It also offers developed encryption of data as well as safe communication. Some of the primary applications of incorporating quantum computing into smart cities as shown in Figure 26.

22.1 Quantum Computing in Smart Cities

a) **Smart Urban Planning:** It has the ability to analyze urban systems and make decisions that are effective by processing an enormous amount of data at the same time. All of these cutting-edge solutions improve energy efficiency, mobility, and sustainability.

b) **Smart Waste Management:** In waste management, quantum computing promotes real-time information analysis and decision-making. Quantum computers, by handling massive amounts of data, are able to interpret sensor data collected in real time from disposal containers such as waste composition and capacity in order to optimize the garbage collection and processing for recycling.

c) **Cryptography & Cyber security:** Normal computers are struggling to cope with the rising threat of cyber-attacks. Machine learning, Quantum computing and artificial intelligence (AI) is examples of emerging technologies that can aid in the development of advanced techniques for mitigating cyber threats. Furthermore, it creates new methods for protecting data such as quantum cryptography. Asymmetric techniques based on complex problems in mathematics such as converting large numbers are used in many types of encryption. From a theoretical standpoint, quantum computers can solve these challenges in hours, whereas traditional computers take many years.

d) **Smart Transportation System:** Quantum algorithms can analyze highway traffic patterns and indicate congestion, permitting smart routing that minimizes the waiting time for citizens.

e) **Smart Emergency Response:** Quantum computing has the potential to significantly aid disaster preparedness and emergency management. Analyzing various data sources, forecasting the weather, population density, and real-time information provided by sensors can effectively aid in crisis management, evacuation route selection, and allocation of resources.

22.2 Quantum Computing in the Logistics Sector:

a) **Route Optimization:** Quantum computing is able to quickly determine the fastest and most effective transportation routes, centers considering factors like traffic, geographical separation, and duration, reducing energy consumption as well as delivery times.

b) **Facility location Planning:** Quantum algorithms can determine the best locations for warehouses, fulfilment centers, and a variety of logistics centers by analyzing many different factors such as population density, convenience, and the built environment.

b) **Supply Chain Management:** Quantum computing has the potential to improve the handling of inventory, prediction of demand, and the use of resources, resulting in a streamlined supply chain that reacts to market variations in real time.

23. CONCLUSION

Smart towns and cities have enhanced infrastructure and established technological integration in administration, security mechanisms, management of waste, producing goods, cost-effective transportation, and a greater degree of connectivity. Urban areas are likely to benefit from smart technologies by saving capital, reducing greenhouse gas emissions and managing traffic flows (Khang & Quantum, 2023).

However, the agenda's complexity is impeding its development. Quantum information technology has enormous possibilities of transforming smart city planning. It has the potential to play a significant role in addressing difficult issues in a variety of fields. It provides sustainable, optimized, and practical solutions that surpass many technical obstacles in the design of smart cities by utilizing algorithms and advanced technology, simulations of real-life phenomena, and massive amounts of complex information processing (Khang & Kali, 2023).

The worth of technology is determined by how it is applied instead of the extent to which technology is available. Advancements in computational and analytical capabilities, it opens up new possibilities for cryptography and cybersecurity, smart waste management, smart emergency services, decreased energy usage, and other applications (Khang & Vugar et al., 2023).

REFERENCES

Abd El-Latif, A. A., Abd-El-Atty, B., Mehmood, I., Muhammad, K., Venegas-Andraca, S. E., & Peng, J. (2021). Quantum-Inspired Blockchain-Based Cybersecurity: Securing Smart Edge Utilities in IoT-Based Smart Cities. Information Processing &amp. *Information Processing & Management*, *58*(4), 102549. doi:10.1016/j.ipm.2021.102549

Alam, M., Moroni, D., Pieri, G., Tampucci, M., Gomes, M., Fonseca, J., & Ferreira, J. (2018). Real-Time Smart Parking Systems Integration in Distributed ITS for Smart Cities. *Journal of Advanced Transportation,* 1–13. doi:10.1155/2018/1485652

BedÃn-Molina, J., Lopez, M., & Derpich, I. (2020). A home-based smart health model. *Advances in Mechanical Engineering, 12*(6), 168781402093528–. . doi:10.1177/1687814020935282

Eswaran, U., Khang, A., Eswaran, V. (2023). Role of Quantum Computing in The Era of Artificial Intelligence. Applications and Principles of Quantum Computing. IGI Global Press. https://doi.org/ doi:10.4018/979-8-3693-1168-4-ch003

Hilmani, A., Maizate, A., & Hassouni, L. (2020). Automated Real-Time Intelligent Traffic Control System for Smart Cities Using Wireless Sensor Networks. *Wireless Communications and Mobile Computing*. . doi:10.1155/2020/8841893

Khang, A. (2022). Cyber-Physical-Social System and İncident Management. AI-Centric Smart City Ecosystems: Technologies, Design and Implementation, 7(12). CRC Press. doi:10.1201/9781003252542-2

Khang, A. (2023). AI and IoT-Based Technologies for Precision Medicine. IGI Global Press. doi:10.4018/979-8-3693-0876-9

Khang, A. (2023). AI-Aided Data Analytics Tools and Applications for the Healthcare Sector. AI and IoT-Based Technologies for Precision Medicine. IGI Global Press. doi:10.4018/979-8-3693-0876-9.ch018

Khang, A. (2023). *Advanced Technologies and AI-Equipped IoT Applications in High-Tech Agriculture* (1st ed.). IGI Global Press. doi:10.4018/978-1-6684-9231-4

Khang, A. (2023). Applications and Principles of Quantum Computing. IGI Global Press. https://doi.org/ doi:10.4018/979-8-3693-1168-4

Khang, A. (2023). Quantum Mechanics Primer - Fundamentals and Quantum Computing. Applications and Principles of Quantum Computing. IGI Global Press. https://doi.org/ doi:10.4018/979-8-3693-1168-4-ch001

Khang, A. (2023). Introduction to Quantum Computing and Its Integration Applications. Applications and Principles of Quantum Computing. IGI Global Press. https://doi.org/ doi:10.4018/979-8-3693-1168-4-ch002

Khang, A., Chowdhury, S., & Sharma, S. (2022). *The Data-Driven Blockchain Ecosystem: Fundamentals, Applications, and Emerging Technologies* (1st ed.). CRC Press. doi:10.1201/9781003269281

Khang, A., Gupta, S. K., Rani, S., & Karras, D. A. (2023). Smart Cities: IoT Technologies, Big Data Solutions, Cloud Platforms, and Cybersecurity Techniques. CRC Press. doi:10.1201/9781003376064

Khang, A., & Kali, C. R. (2023). Enabling the Future of Manufacturing: Integration of Robotics and IoT to Smart Factory Infrastructure in Industry 4.0. AI-Based Technologies and Applications in the Era of the Metaverse. IGI Global Press. doi:10.4018/978-1-6684-8851-5.ch002

Khang, A., & Muthmainnah, M. (2023). AI-Aided Teaching Model for the Education 5.0 Ecosystem" AI-Based Technologies and Applications in the Era of the Metaverse. IGI Global Press. doi:10.4018/978-1-6684-8851-5.ch004

Khang, A., Rani, S., & Sivaraman, A. K. (2022). *AI-Centric Smart City Ecosystems: Technologies, Design and Implementation* (1st ed.). CRC Press. doi:10.1201/9781003252542

Khang, A., Shah, V., & Rani, S. (2023). AI-Based Technologies and Applications in the Era of the Metaverse. IGI Global Press. doi:10.4018/978-1-6684-8851-5

Khanh, H. & Khang, A. (2021). The Role of Artificial Intelligence in Blockchain Applications. *Reinventing Manufacturing and Business Processes through Artificial Intelligence*, *2*(20-40). CRC Press. doi:10.1201/9781003145011-2

Kumari, A., Gupta, R., & Tanwar, S. (2021). Amalgamation of blockchain and IoT for smart cities underlying 6G communication: A comprehensive review. *Computer Communications*, *172*, 102–118. Advance online publication. doi:10.1016/j.comcom.2021.03.005

Shah, V., & Khang, A. (2023). Metaverse-Enabling IoT Technology for a Futuristic Healthcare System. AI-Based Technologies and Applications in the Era of the Metaverse. IGI Global Press. doi:10.4018/978-1-6684-8851-5.ch008

Taha, A. M. (2018). An IoT Architecture for Assessing Road Safety in Smart Cities. *Wireless Communications and Mobile Computing*. . doi:10.1155/2018/8214989

Chapter 11
Quantum Leap in Healthcare: Unleashing AI's Epoch of Unprecedented Medical Metamorphosis

Sudipta Adhikary
Department of Law, Brainware University, India

Kaushik Chanda
Department of Computational Science, Brainware University, India

Kaushik Banerjee
Brainware University, India

Gunjan Mukherjee
https://orcid.org/0000-0002-3959-3718
Department of Computational Science, Brainware University, India

Avijit Kumar Chaudhuri
https://orcid.org/0000-0002-5310-3180
Department of Computer Science and Engineering, Brainware University, India

ABSTRACT

Artificial intelligence (AI) stands out as a potent asset across diverse industries, with its potential within the realm of medicine proving to be exceptionally profound. This chapter takes a comprehensive journey through the myriad applications of AI in healthcare, illuminating its profound influence on patient care, disease diagnosis, therapeutic interventions, and advanced medical exploration. The fusion of AI algorithms, machine learning paradigms, and deep learning strategies empowers healthcare providers to amplify operational efficiency, precision, and the very bedrock of decision-making processes. Within this document, an intricate examination of the merits and hurdles entailed in the integration of AI into the medical domain ensues, concurrently dissecting ethical quandaries and prospective trajectories for evolution.

DOI: 10.4018/979-8-3693-1168-4.ch011

1. INTRODUCTION

The landscape of healthcare is in a perpetual state of metamorphosis, impelled by the strides taken in technology and an escalating demand for streamlined, result-oriented patient care. In thc midst of these revolutionary tech advancements, one entity reigns supreme: Artificial Intelligence (AI). This juggernaut, encompassing the realms of machine learning, deep learning, and data-driven algorithms, possesses the latent ability to transmute multifarious facets of healthcare administration. From the intricate realms of diagnosis and intervention to the expanse of medical exploration and the intricate fabric of decision-making, AI holds the potential to orchestrate a complete overhaul (Khang & Quantum, 2023).

The advent of AI and its continuous progress unfurled so many directions in medical domain which could easily overcome so many hitherto unsolved problems. The criticality of diseases increases day by day with the adverse trend of lifestyles. So many drugs have been invented so far. The intervention of AI in the current realms of health sector research domain helped in accurate diagnosis and analysis of disease specific report and provide the probable solution with certainty. In every spheres of related domain in medical sector, the vicinity of AI and its application aspect has been evidently observed which can provide the much-needed facility toward the recovery of patient with data-oriented approach.

2. BACKGROUND

The realm of healthcare has borne witness to remarkable progressions across time, with technology emerging as a pivotal force in reshaping the landscape of patient care delivery. Amidst these transformative tides, the integration of Artificial Intelligence (AI) into the domain of medicine surfaces as an exceptionally ground-breaking stride. AI, a concept encapsulating the emulation of human cognition within machines, endows them with the prowess to undertake tasks traditionally requiring human intellect, encompassing facets like learning, reasoning, and intricate problem-solving.

The surge in AI's prominence and renown is attributed to its prowess in processing massive troves of data, identifying intricate patterns, and unearthing profound insights. Within the precincts of healthcare, AI harbors a profound potential to overhaul sundry dimensions, spanning the gamut from the expanse of medical imaging and diagnostics to the intricacies of devising treatment strategies and uncovering novel pharmaceutical compounds (Khang & Smart Healthcare, 2024).

Nurtured by the capabilities of machine learning algorithms and the intricate latticework of deep neural networks, AI holds within its grasp the means to augment the capacities of healthcare practitioners, resulting in elevated patient outcomes and more streamlined healthcare frameworks. The propulsion of AI into the medical sphere is propelled by the astronomical proliferation of medical data, encompassing electronic health archives, scans delineating medical imaging, and the strata of genomic information. The conventional manual methodologies of scrutinizing and construing these troves of data often transpire as laborious, susceptible to inaccuracies, and tinged with the biases inherent to human cognition. Herein, AI-driven systems manifest their potential, facilitating healthcare providers in expeditiously dissecting and parsing this colossal corpus of data with precision, celerity, and impartiality (Khang, Vugar and Abuzarova et al., 2023).

In addition, the tapestry of AI technologies unfurls the realm of predictive modelling and data-induced decision-making, furnishing healthcare experts with the tools to fashion pinpoint diagnoses, custom-tailor treatment regimens, and prognosticate the trajectory of maladies. AI's intricate algorithms unearth veiled intricacies and correlations concealed within medical datasets, thereby abetting in the unearthing of incipient markers of ailments and contributing to the prompt identification and pre-emption of diverse medical conditions. Notwithstanding the burgeoning potential of AI in the medical arena, several hurdles and ethical dimensions warrant attention. Safeguarding patient privacy, shoring up data fortifications, mitigating algorithmic prejudices, and ensuring the presence of human oversight linger as cardinal quandaries tethered to the infusion of AI into healthcare. Striking a harmonious equilibrium between harnessing the virtues of AI and preserving the trust, confidentiality, and welfare of patients stands forth as an imperative.

Against this backdrop of advances and considerations, the nucleus of this thesis orbits around a voyage traversing the manifold applications of AI within the medical frontier. It delves deep into the ripples wrought by AI across patient well-being, the terrain of diagnostics, the contours of curative interventions, and the echelons of medical exploration. By subjecting the benefits, obstacles, and ethical nuances to a discerning analysis, this scholarly pursuit aspires to cast light upon AI's latent potential to transfigure healthcare paradigms, thereby casting its die in the shaping of medicine's imminent future (Khang & Vugar et al., 2024).

The AI based expert system has become an additional feather to the crown of medical science. The highly efficient tool and its real-life activities with full human like cognition and also provides the facts and heuristics like the human expert. Some expert systems like Dendral, Mycin, Pxdes, and Cadet are much popular and shows completeness in the performance, understandability, reliability and prompt responsiveness towards disease tracing and other allied diagnosis of patients.

2.1 Purpose and Objectives

The prime directive of this thesis is to embark on a comprehensive odyssey through the manifold applications of AI within the medical realm, unraveling the threads that hold the potential to transmogrify the very essence of healthcare dispensation. By plunging into the intricate tapestry woven by the fusion of AI and medical practices, this scholarly endeavor aspires to weave a holistic comprehension of the dividends, obstacles, and moral intricacies that accompany the assimilation of AI into healthcare's embrace. The compass of this research charts its course according to the subsequent waypoints:

To traverse the realm wherein AI is enshrined in medical imaging, casting a discerning gaze upon the echoes it leaves within radiological practices. This compass point seeks to dissect how the symphony of AI algorithms and the choreography of machine learning choreography choreography can amplify the veracity and efficiency governing the scrutiny of medical imagery. The objective unfurls as an exploration into how this tandem of AI and machine learning orchestrates precise medical image scrutiny, arming the medical fraternity with sharper tools for disease detection and diagnosis.

To penetrate the sanctuary wherein AI fuses with diagnostic systems, embarking upon a journey to unveil its prospective contribution to the annals of clinical decision-making. This epiphany strives to decode the harmonious interplay between AI and healthcare professionals, especially within the sphere of disease prognosis, early detection, and laboratory experimentation. Ultimately, it seeks to illuminate a trajectory wherein these synergies culminate in more punctual and precise diagnoses.

To venture into the precincts where AI intertwines with robotic surgery, casting a spotlight on the seismic perturbations it might engender in the domain of surgical practices. This expedition is an odyssey to decipher how AI-infused robotic envoys have the potential to infuse surgical precision with an elixir of unparalleled efficacy. Further, it envisions the prospect of surgical procedures orchestrated remotely, alluding to a radical metamorphosis on the surgical landscape (Khang & Abdullayev et al., 2024).

To dissect the chronicles wherein AI harmonizes with drug discovery and precision medicine, with the central theme being the expeditious journey from drug conceptualization to personalized therapeutic regimens. This chapter aspires to fathom how AI's algorithms, akin to alchemists of old, transmute data into potential drug leads and tailor-fit therapies for individual patients through the prism of large-scale data analysis and prophetic modelling.

To convene a symposium where ethical considerations and conundrums precipitated by AI's entry into the medical panorama take center stage. This focal point seeks to erect a pedestal for discourse on patient confidentiality, data bastions, algorithmic predilections, and the unyielding necessity for the human watchtower. The narrative strives to accentuate the conscience that underpins the judicious and ethical incorporation of AI into healthcare.

In welding these objectives, this thesis augments the corpus of knowledge underpinning AI's dominion over the medical landscape, expounding upon its metamorphic potential while diligently navigating the chasms of ethics and complications that warrant resolution for a harmonious assimilation.

The dire need of holistic and realistic approach to the diseases of varying degree of criticality of human patient has been the bed rock of necessity for exposition of AI. The concept of expert system has become the solid tune in delving information of patient with typically crafted interface. The need of the conducive user interface for capturing patient related data and integration of inference engine and knowledgebase for the efficient approach to the voluminous patient data with utmost rationality has become the prominent objective of AI based approach (Anh & Vladimir et al., 2024).

2. Application of AI in Medical Imaging

2.1 AI and Radiology

In the realm of specialized medical fields, radiology emerges as a tapestry woven from the threads of medical imaging, meticulously crafted to unveil the secrets harbored within the human body. Yet, the advent of Artificial Intelligence (AI) stirs the embers of revolution, poised to paint the canvas of radiology with hues of innovation, fortifying the prowess of radiologists and imbuing the panorama of image interpretation with an unparalleled precision and alacrity.

In the theater of diagnostic medical imaging, AI takes center stage, the protagonist of a saga being woven through rigorous evaluation. AI unfurls its mantle, adorned with auras of impressive accuracy and sensitivities, seamlessly distinguishing the threads of normality from the aberrations woven by diseases. This techno-mage pledges to metamorphose the very essence of tissue-based detection and characterization, a conjurer of insights mining through the tapestry of body scans, unlocking the troves of invaluable knowledge they hold.

AI stands as a sentinel poised to carve transformative marks on the edifice of medicine, a harmonious duet echoing between practitioners and patients. The chapters of prospective studies and leaps in medical image analysis confluence, bridging the chasm between experimentation and implementation. AI, once elusive, inches closer to its crowning moment of practical application within the radiological domain.

Chief among AI's envoys in the realm of radiology stands its role in the crusade against cancer. Deep neural sentinels, honed through the alchemy of training, undertake the sacred task of deciphering radiological hieroglyphs and digitized pathology scrolls, delving into the shadows to unearth the specter of cancer. In the dance of detection, AI's artistry rivals even that of certified screening radiologists. The pages of this chronicle bear the promise of heightened early detection and a symphony of efficiency in the grand overture of screening endeavors.

Yet, the saga goes beyond detection, delving into the folds of treatment. AI, a modern oracle, peers into the intricacies of clinical and molecular realms, illuminating paths that guide responders towards the light of personalized therapies. Immune therapies and chemotherapies become an Amphitheatre where AI's spotlight shines, revealing the individuals destined to embrace the fruits of these treatments, an ensemble orchestrated with a finesse only AI can command.

Casting a broader net, AI unveils new acts in the grand drama of radiology. The algorithms, like scholars of the microscopic, dissect tumor genomics, unravel the nuances of the tumor's habitat, and unveil the biomarkers that foretell fate. The searchlights of AI extend even to the depths of drug discovery, assisting scientists in their quests to unearth the elixirs that may yet vanquish cancer's cloak.

Bound by the embrace of AI, radiology flourishes as a garden of augmented accuracy and streamlined efficiency, where the blossoms of patient outcomes unfurl in unparalleled splendor. This technological sentinel, this guardian of insight, has the power to enrich the tapestry of healthcare, infusing each thread with renewed precision and each chapter with the promise of a brighter tomorrow, particularly within the realm of cancer detection and its tenacious battle against time.

The core principle of radiology is based on the histopathology or high-resolution image sensing operation. The identification of the correct object or the correct class of belongingness can be ascertained with the help of the machine learning or deep learning concept. The deep learning model can be used with sufficient training data to have the good and distinct accuracy for classification or identification process. The complexity and intricacies of images to be classified can be the good subject for deep learning model. The model optimisation can be taken into consideration to have more prospect for the classifier.

2.2 Progress in Computer-Aided Diagnosis

Advancements in the realm of Computer-Aided Diagnosis (CAD) depict a swiftly unfolding saga, where the tapestry woven by artificial intelligence (AI), computer vision, and the art of medical image processing converges to furnish a symphony of aid to healthcare practitioners in their pursuit of precision and swiftness. The evolution, particularly with AI's infusion, is akin to a comet's trail, promising heightened diagnostic accuracy and a choreography of streamlined workflow.

Within this realm, AI's infusion into CAD stands as a herald of transformation, hinged upon the intricate algorithms of deep learning. The resonance of AI reverberates through the corridors of medical images, where the alchemy of deep learning unravels the secrets concealed within the pixels. These algorithms, as students of vast annotated datasets, learn to decode the enigmatic patterns hinting at specific conditions. These insights, bestowed by AI onto medical imaging, empower CAD systems to cast a spotlight on potential concerns, a virtuoso performance enhancing the diagnostic narratives painted by radiologists and the medical fraternity (Khang & Vugar et al., 2024).

The distinguishing feature of AI-enhanced CAD is its dexterity in taming complex datasets with speed and precision. The antiquated CAD systems depended upon rule-based edicts and predefined attributes, their adaptability tethered. Yet, the AI algorithms emerge as adaptable scholars, capable of mastering new data, thus refining their accuracy and universality. This limberness allows CAD systems to converse fluently in the languages of diverse imaging modalities: X-rays, CT scans, MRIs, and ultrasounds. This versatility renders them allies in diagnosing a spectrum of diseases across varied medical domains (Khang & Ragimova et al., 2024).

In the realm of cancer detection, the AI-driven CAD orchestrates a melodic crescendo. The technology reveals its virtuosity in detecting anomalies within mammograms, lung CT scans, and digital pathology slides. The algorithms weave tales of subtleties and irregularities hidden within these visual narratives, guiding radiologists to the hidden whispers of cancer's presence. This orchestration of early detection could alter the trajectories of screening programs, vanquishing the specter of false negatives and sculpting improved patient narratives.

Furthermore, the AI-led CAD symphony extends its resonance to diagnoses beyond cancer. The realms of cardiovascular disorders, neurodegenerative conditions, and musculoskeletal aberrations experience a transformative overture. Through the harmonious matrimony of AI algorithms and sophisticated imaging techniques, CAD's voice guides the detection of coronary artery afflictions, the dissection of cerebral structures and functions, and the revelation of fractures or joint anomalies. These movements harmonize to equip medical maestros with more nuanced decisions and blueprints of personalized treatments.

However, amidst these crescendos, the labyrinth of AI-led CAD presents its own conundrums. The veils shrouding the interpretability of AI's orchestration emerge as a subplot. The arcane rituals of deep learning often operate as enigmatic riddles, curtailing the understanding of the basis for their diagnoses. The quest for elucidation and reliability in AI's outputs becomes a cornerstone for CAD's assimilation into the clinical tapestry.

Furthermore, the integration of CAD within the medical choreography necessitates a choreographer's skill. Challenges encompass data synchronicity, the interlacing of electronic health records, and a user interface design that bridges the chasm between AI and medical practitioners. Overcoming these technical and logistical barriers forms the prologue to CAD's successful integration into the healthcare narrative. The integration of AI into the CAD structure is very much vital in order to bring in the concept of automation. The CAD module has been transformed into 3-dimensional perception using AI concept and shows the much-improved kind of engineering performance towards solutions of myriad problems. The generative design has become the successor of CAD based approach (Khang & Vladimir et al., 2024).

3. DIAGNOSTIC SYSTEMS EMPOWERED BY ARTIFICIAL INTELLIGENCE

3.1 Artificial Intelligence Enhancing Clinical Decision Support

AI incorporated into clinical decision support (CDS) systems holds immense potential for elevating the diagnosis, treatment, and prognosis of diverse medical conditions. These systems employ AI algorithms to scrutinize biomedical imaging data and anticipate medical outcomes or disease risks. The AI-powered CDS empowers clinicians to grapple with vast patient data, fuelling informed inferences and refined clinical decisions.

The fusion of AI with clinical decision support promises an upswing in patient care and outcomes. AI algorithms dissect intricate datasets including genomics, biomarkers, phenotypes, electronic health records, and care histories, proffering clinicians invaluable insights and forecasts. Capitalizing on these AI-fueled systems, clinicians personalize treatments, fine-tune management, and make enlightened decisions.

Yet, embedding AI within clinical decision support orbits around hurdles. Prudent navigation involves architecting, cultivating, selecting, deploying, and perpetually monitoring AI systems. Certifying the safety and efficacy of AI-fueled CDS systems is paramount, considering their dynamic nature and the diverse reservoirs of data they gulp. Sturdy evaluation frameworks are imperative to gauge their real-world performance, reliability, and versatility.

Additionally, AI assimilation in CDS drapes itself in ethical and legal robes. Safeguarding privacy, upholding security, and wielding data governance emerge as cornerstones, as sensitive patient information grapples with the labyrinth of regulations. The luminosity of AI algorithms should extend to their transparency, enabling clinicians to unravel the rationale behind system counsel and cultivate reliance on its capabilities.

Moreover, grafting AI into clinical decision support commands a harmonious orchestra involving healthcare professionals, data virtuosos, and architects of technology. Synthesis with prevailing clinical workflows, electronic health record systems, and interoperability with a symphony of healthcare technologies is pivotal for the seamless symphony and triumphant utilization of AI-laden CDS systems. The CDSS based system has been integrated with the electronic health records and computerized provider entry system which can be administered through the different gazettes like desktop, smart phones, tablets etc. The CDSS works also on the principles of non-knowledge-based approach requiring the datasets. The concept of Artificial intelligence and machine learning or the statistical pattern recognition can follow the expert medical knowledges. The CDSS based on the non-knowledge based approach has curved a niche in the medical domain and has been improved with AI based approach to meeting the criterion for real life-oriented development.

3.2 Anticipating Illness and Timely Identification

The realm of AI presents an extraordinary potential for reshaping the landscape of disease anticipation and the prompt unveiling of maladies. By harnessing the intricacies of machine learning, the tapestries woven from diagnostic patient datasets, such as ECGs, EEGs, or X-ray images, unfold before us, revealing the subtle nuances betraying the inception of ailments. This metamorphic attribute arms healthcare custodians with the power to intervene swiftly, orchestrating timely interventions and engineering therapeutic plans that steer patient outcomes toward favorable horizons.

A notable arena where AI unfurls its banner of disease prediction is the terrain of machine learning models birthed from diagnostic laboratory assays. These models, akin to watchful sentinels, have achieved extraordinary feats in prophesying the advent of illnesses. Witness an optimized ensemble model, resplendent with an F1-score of 81% and a predictive acumen of 92% for the quintessential quintet of common diseases. This realm of precision bestows upon medicine the gift of proactive interventions and bespoke treatment roadmaps. However, the integration of AI into the domains of disease prediction and early detection does not transpire devoid of challenges. A formidable technical hurdle lurks in the task of embedding AI within the corporeal abode of the human body. The endeavor of seamlessly knitting AI with internal sensors or real-time monitoring devices, while progressing, remains a formidable undertaking.

Moreover, AI algorithms thrive on a plenitude of data to sculpt robust and precise prognostic frameworks. Yet, the quest for such datasets navigates the labyrinth of privacy concerns and the requisite patient acquiescence. Striking the harmonious chords between data confidentiality and the accrual of ample data for AI-driven disease soothsaying becomes the fulcrum of this endeavor. Concomitant with these pursuits, the evolution of effective AI algorithms for disease prediction and early identification calls for a convergence of interdisciplinary minds. The tapestry weavers span healthcare maestros, data visionaries, and AI virtuosos, harmonizing their efforts to ensure the integrity, accuracy, and elucidation of the AI compositions.

Transparent and decipherable AI codes emerge as the cornerstone, nurturing a trust-filled kinship between the healthcare custodians and their patients. The accuracy and timeliness can be enhanced using the deep learning model concept. The increased size of the data set easily support the training accuracy enabling the model to meet much improved accuracy. The realization of predictive nature of model can easily confer the anticipatory behavior of impending illness in human patients.

3.3 Artificial Intelligence in Laboratory Examination and Interpretation

The potential of AI to reshape the landscape of laboratory testing and analysis rests on its adept utilization of intricate algorithms, fostering amplified diagnostics, profiling patient risks, dissecting laboratory outcomes, and orchestrating financial evaluations. The union of AI with laboratory environments harbors a wealth of promises, envisioning elevated accuracy, streamlined efficacy, and an elevated echelon of quality within the realm of laboratory testing.

Diagnostics emerge as a star in AI's constellation within the laboratory domain. AI's intellect, nurtured on colossal troves of patient narratives spanning medical records, laboratory findings, and images, dances to a tune of accuracy and promptness. Leveraging the tenets of machine learning, AI becomes a guide, ushering healthcare custodians toward enlightened decisions that unfurl the tapestries of improved patient destinies.

Yet, the journey to embedding AI within laboratory testing and analysis carves its own trajectory lined with challenges. Education takes center stage – a symposium where the laboratory community learns the lexicon and essence of AI (Khang & Muthmainnah et al., 2023). The stakeholders, healthcare custodians, and laboratory artisans alike, must embrace AI's symphony to seamlessly assimilate and weave it into the laboratory's tapestry. Furthermore, the canvas of research beckons, urging the generation of clinical evidence and strategies to surmount the challenges intrinsic to AI's integration within laboratories.

Beyond academia, the theater of technology unfurls. The interplay of AI within the pre-existing cadence of laboratory workflows demands meticulous choreography. Harmonizing AI with the laboratory's essence mandates reconfigurations in data assimilation, analysis processes, and information architecture. The interplay with the bedrock infrastructure, exemplified by the laboratory information management systems (LIMS), necessitates delicacy.

Yet, the siren's call of potential triumph overcomes these trials. AI stands as a catalyst, refining the tempo and accuracy of laboratory testing. Swift diagnoses, painted by AI's brush, offer rays of hope. This symphony of data parsing and interpretation acts as a guiding beacon, steering healthcare practitioners towards better decisions and the realm of personalized treatments. And the echo extends beyond diagnostics – AI's financial acumen traverses the laboratory tapestry, sculpting optimal resource allocation and financial equilibrium. The AI in interpretation of histopathological image can divulge out many relevant information with the automated technique of non-invasive approach. The machine learning concept extension obviously creates the pathways towards the better accuracy and achievement.

4. SURGERY AIDED BY ARTIFICIAL INTELLIGENCE AND ROBOTICS

4.1 Elevated Exactness and Precision

AI-driven robotic surgery holds the promise of transforming surgical landscapes through the augmentation of precision and accuracy. Through the harnessing of artificial intelligence, surgical automatons emerge as conduits of data-guided decision-making, furnished with the prowess of cognitive robotic collaboration. These orchestrated machines assume the mantle of surgical support, liberating surgeons to navigate the intricate domains of surgeries, enriched by the symbiotic alliance with AI.

Foremost among AI's endowments to robotic surgery is the finesse it lends to precision. AI algorithms dissect real-time data plucked from imaging systems and sensors, enabling the robotic counterpart to execute deft maneuvers and nuanced adjustments, intricately tailored to the patient's anatomy and the surgery's dictates. This elevation in precision bequeaths a bounty of improved surgical outcomes, minimized trauma, and expedited patient convalescence. Furthermore, AI extends the hand of decision support to surgeons, providing sagacious guidance during surgery. Sifting through vast reservoirs of patient records and images, AI extracts pearls of insight, directing surgical narratives. These data-imbued suggestions unfurl surgical efficiencies and amplify surgical achievements.

However, the voyage of AI into robotic surgery is not devoid of navigational complexities. Surgeons, computer savants, and engineers must harmonize their expertise to blueprint the contours of AI's surgical presence. Additionally, the confluence of AI systems within pre-existing surgical choreographies harbors intricacies; a labyrinth where data collection and analysis metamorphose, demanding substantial adaptations.

Yet, the alchemical success stories of AI within robotic surgery echo with promise. The real-time responsiveness AI imparts to surgical automata empowers them to swivel, adapt, and even switch surgical stratagems mid-course, all guided by immediate feedback loops. The AI's integration renders these robotic assistants not just appendages within the hospital ecosystem but interactive entities, autonomously drawing insights from patient records to inform their surgical interventions. The uniqueness in robotic movement can better be divulged out using the concept of AI. The concept of mechatronics and its effects on the robot movement has shed enough light to ease out the criticality of any manual operation which can inculcate the huge risk factor on accessing most delicate organ parts of human physiology.

4.2 Surgery at a Distance and Virtual Medical Care

The domains of remote surgery and telemedicine are poised to reshape the healthcare landscape, unfurling a panorama where medical practitioners traverse virtual distances to deliver care to far-flung patients. Telemedicine, in particular, has cast its influence even upon surgical tutelage, weaving a narrative where remote surgeries burgeon. Its script encompasses mentoring, proctoring, and virtual conferences, creating a nexus for healthcare professionals to coalesce and exchange wisdom.

AI's partnership with remote surgery and telemedicine paints a canvas rich with advantages. It serves as the bridge for physicians to reach patients tucked in remote corners, slashing healthcare disparities and enriching medical accessibility for marginalized populations. Through the symphony of AI algorithms, remote surgeries acquire an aura of precision, their strokes guided by decision support systems and the finesse of cognitive robotic collaboration.

However, challenges stand in this union of AI with remote surgery and telemedicine. The technology underpinning remote surgery is still in its nascent stages, demanding further refinement to orchestrate seamless communication, minimal latency, and unwavering network connections. Furthermore, the scaffolds of regulations and guidelines require erection, sculpted to address labyrinthine concerns like licensing, responsibility, and reimbursement within the sphere of remote healthcare services.

Yet, even in the face of these trials, triumphs emerge. The narrative of AI within remote surgery and telemedicine chronicles optimized surgical choreographies, elevated precision within remote surgical endeavors, and the dawn of data-guided decision-making. This partnership, where AI's acumen intertwines with the fabrics of remote surgery and telemedicine, has the potential to metamorphose patient destinies and unravel access to specialized medical sanctuaries.

4.3 The Outlook for AI's Role in Surgical Advancements

The trajectory of AI in the realm of surgery paints a promising picture. With technological strides in imaging, navigation, and robotic interventions, AI's touch is subtly reshaping surgical paradigms. This evolution holds the promise of ushering modern surgery into an era of precision and autonomous interventions, catering to both acute and lingering maladies.

Already, AI has etched its mark on surgery, unfurling its wings to underpin clinical decisions and advance multifarious domains like risk assessment, genomics, imaging, diagnosis, tailored medicine, and drug discovery. The orchestration of AI algorithms presents surgeons with the treasure trove of insights and models, beacons that illuminate surgical pathways, elevate outcomes, and enrich patient care.

Nonetheless, challenges persist in the seamless amalgamation of AI into surgical practice. The symphony of collaboration, fusing surgeons, computer luminaries, and engineers, forms the bedrock, carving out requisites and ensuring AI harmonizes with the very heartbeat of surgical procedures. Ethical nuances, data sanctity, and regulatory frameworks stand as sentinels, their gaze unwavering as AI's foray into surgery progresses.

Amid these trials, the tapestry of AI's future in surgery glimmers with promise. As AI's facets continue to unfurl, surgical landscapes are poised to transmute into realms of precision, tailor-made care, and augmented efficiency. The paeans of AI-driven ingenuity, manifesting as real-time decision aids, intelligent surgical automata, and automated data interpreters, could orchestrate a renaissance in surgical landscapes. The future they paint is one where patient outcomes gleam brighter, surgical techniques evolve, and the very essence of surgery undergoes a profound transformation. The risk abetment has become the key aspect of medical domain in order to save life of human beings. The robot-based mechanism for operation of any internal delicate organs of human being can easily be executed with the help of robots.

5. AI-POWERED EXPLORATION OF PHARMACEUTICALS AND TAILORED MEDICAL CARE

5.1 Expediting the Progression of Drug Development

AI-powered exploration of pharmaceuticals and precision medicine stand poised to revolutionize and expedite the journey of drug development. AI's integration into the realm of early drug discovery not only catalyzes the search for drug candidates but also takes the helm in driving biomarker exploration and drug design, ushering in an era of personalized medical interventions. The multifaceted prowess of AI already extends to supporting clinical decision-making and spanning domains like risk stratification, genomics, imaging, diagnosis, precision medicine, and drug discovery (Khang & Kali et al., 2023).

Within the nexus of drug discovery and precision medicine, the amalgamation of AI yields a bouquet of advantages, an orchestration that hastens the emergence of novel therapies. AI algorithms, endowed with the acumen to analyze extensive datasets encompassing genomic profiles, molecular structures, biological pathways, and clinical narratives, set forth a more efficient course for identifying potential drug candidates. This computational artistry aids researchers in refining their focus on the most promising contenders, thereby optimizing the nascent phases of drug discovery.

Furthermore, the AI landscape broadens to embrace biomarker discovery – those quintessential fragments that inform patient stratification, foretell treatment responses, and pinpoint therapeutic targets. Delving into intricate datasets, AI uncovers nuances and interrelations often eluding traditional approaches, birthing novel biomarkers that illuminate the path to bespoke treatments, a leap towards precision medicine.

In the realm of drug design, AI wields its transformative wand. By leveraging the finesse of machine learning and computational modelling, AI predicts the characteristics and propensities of drug candidates, thereby casting a prescient light on their efficacy, safety, and potential side effects. This predictive canvas guides researchers in deftly calibrating their experimental pursuits, curtailing the temporal and financial outlay that might accompany unsuccessful candidates.

Yet, traversing the realm of AI in drug discovery and precision medicine is not devoid of obstacles. The onus rests on seamless collaboration among medical luminaries, computer wizards, and engineering visionaries to meticulously delineate the prerequisites for effective AI partnerships. This interdisciplinary dance underscores the imperative of harnessing AI's full potential via a harmonized team approach.

Nonetheless, the tapestry of AI's triumphs continues to unfurl. AI has, and continues to, cast its die in drug discovery and precision medicine, underscoring its role in clinical decision-making, risk assessment, genomics, imaging, diagnosis, and beyond. In these milestones, one glimpses the burgeoning future where AI-wrought paradigms expedite drug development, imbuing it with efficiency, and sculpting bespoke treatment pathways. The significant contribution of AI in drug discovery is the classification and mixing of drug components and other ingredients. The deep learning approach backed up by the machine learning helps in fine grained selection of proper components and their judicious mixing to get the medicinal product ready (Khang & Shah et al., 2023).

5.2 Tailored Therapeutic Strategies

AI stands at the precipice of transforming personalized treatment approaches within the healthcare realm. Precision medicine's pursuit involves identifying patient phenotypes with idiosyncratic treatment responses or distinctive healthcare requisites. Here, AI, armed with its intricate computational tapestry and inferential prowess, emerges as the harbinger of profound insights, the wellspring of rationality and learning, bestowing clinicians with augmented intelligence to navigate decision-making waters.

Within the terrain of healthcare, AI-embellished personalized treatment strategies extend a bounty of advantages. The fusion of vast datasets and intricate algorithms empowers AI to unravel the labyrinth of intricate patient data – genomic signatures, medical imagery, electronic health records, and lifestyle narratives. In this tapestry, AI perceives patterns and correlations that kindle the beacon of personalized treatment trajectories. This bespoke approach holds the potential to elevate patient care to unprecedented heights, potentially illuminating pathways to enhanced treatment outcomes and patient contentment.

Nonetheless, traversing the realm of AI-driven personalized treatment isn't without hurdles. The convergence of multimodal data from disparate sources – genetic tests, medical imagery, wearables, and patient testimonials – begets a complex puzzle of integration and analysis, entwining technical intricacies with logistical jigsaws. Meanwhile, ensconcing sensitive patient data in the safety of AI's arms poses paramount concerns of privacy and security.

Another precipice to navigate is AI's potential to introduce bias into personalized treatment methodologies. The algorithms that form the bedrock of AI systems draw wisdom from historical data, a repository that might bear the blemishes of biases and inequalities. Left unattended, these biases might taint the counsel AI offers, propagating treatment disparities.

Yet, within these conundrums, AI's mettle shines. AI seamlessly absorbs inputs from myriad structured and unstructured sources, unravelling semantics, engaging in computer vision, embracing reading comprehension, conversational systems, and delving into multimodal vistas to infuse healthcare practitioners' decisions with enlightenment. This marriage of AI with clinical cadence kindles hope – hope for optimized treatment blueprints, hope for a reduction in the trials and tribulations of trial-and-error, and hope for the refinement of patient outcomes.

In the microcosm of personalized treatment, AI's sway is palpable, most notably in the realm of oncology. Here, AI algorithms unravel genetic signatures, clinical annals, and treatment repercussions, conjuring predictions of individual patient retorts to cancer therapies. This alchemy permits oncologists to craft treatment regimens on the loom of genetic footprints, charting a course toward heightened efficacy and minimized side effects.

The chronicle of personalized treatment's future is interlaced with AI's ascent, a symbiotic dance between technology and clinical orchestration. The crucible of success resides in addressing knotty conundrums like data fusion, security, bias, and ethical compass – the four cardinal points guiding AI's voyage toward delivering care as unique as each patient's narrative.

5.3 AI's Role in Genomic Healthcare

AI's emergence has sparked a potential revolution in the realm of genomic medicine, orchestrating intricate algorithms and computational might to scrutinize and decipher sprawling genomic datasets. In the crucible of clinical genomics, the spotlight falls on a particular AI breed - deep learning algorithms - adept at unravelling complex genomic intricacies. Their prowess lies in unveiling cryptic patterns and correlations that unveil insights for diagnoses, treatment strategies, and the realm of personalized medicine.

AI's arsenal, encompassing machine learning and deep learning, showcases its mettle in the corridors of genomic medicine. These techniques offer respite, streamlining the labyrinthine analytics that genomic data presents. They unearth genetic variations, categorize genomic profiles, and offer crystal ball glimpses into disease predispositions, championing a more tailored and precise avenue for healthcare interventions.

Yet, AI's journey in genomic medicine is not devoid of obstacles, demanding careful navigation. A beacon of responsible and ethical AI adoption must illuminate this path. Data privacy, informed consent, algorithmic transparency, and the spectre of bias lurking in training data all beckon rigorous consideration. These tenets are cardinal for nurturing patient trust and steering healthcare outcomes toward equitable shores.

Furthermore, the embrace of AI within clinical workflows and the assimilation of AI-driven genomic medicine aren't devoid of complexities. The convergence demands harmonious symphonies among diverse players - researchers, clinicians, bioinformaticians, and the guardians of regulatory landscapes. A parallel need for standardized, interoperable genomic data formats and robust infrastructures underscores the essence of seamless data sharing and analysis.

Notwithstanding the impediments, AI has already donned laurels in genomic medicine. Its prowess lies in deciphering genomic jigsaw puzzles, identifying disease-linked genetic nuances, forecasting patient retorts to bespoke treatments, and illuminating novel gene-disease crossovers. This pantheon of accomplishments bespeaks of amplified genomic analysis, resulting in refined diagnoses, prudent treatment selections, and the elevation of patient outcomes.

AI's sonic waves have also reverberated in the realm of precision oncology. Parsing genomic data from cancer patients, AI unearths potential therapeutic crosshairs, augurs drug retorts, and wields the sceptre in steering treatment expeditions. The result: a poignant tapestry of personalized, honed, and precise cancer care strategies.

The future tableau of AI in genomic medicine unfurls a vista of advancing algorithms, knitted with the warp and weft of amplified data integration and sharing. The linchpin is the harmonious ballet between computational maestros, bioinformaticians, and clinicians. A concerted effort, bolstered by fortitude against data privacy challenges, algorithmic opaqueness, and the orchestration of seamless clinical assimilation, holds the promise to transform the contours of genomic medicine, elevating it to a realm of precision and personalization. The supremacy of AI as the technological pathways, provides the accuracy value to the ulterior scale. The concept helps in minimizing the risk factors of causality of patients by means of detection and identification of the complex genomic structure and its mapping to the drugs with the ultimate accuracy.

6. ETHICAL REFLECTIONS AND HURDLES

6.1 Restating: Ensuring the Confidentiality of Patient Information and Safeguarding Data Integrity

Preserving the confidentiality of patient records and the integrity of data stands as paramount imperatives as artificial intelligence finds its place in the realm of healthcare. Employing AI technologies to sift through and dissect sensitive patient information mandates unwavering adherence to privacy statutes and the deployment of robust security measures to shield patient data from harm's way.

Incorporating AI into healthcare doesn't come without its challenges when it comes to upholding patient privacy and data security. One hurdle involves ensuring conformity with privacy legislations like the United States' Health Insurance Portability and Accountability Act (HIPAA) or the European Union's General Data Protection Regulation (GDPR). AI systems must be purpose-built and executed to envelop patient data in a protective cocoon, ensuring that only those with authorization can lay hands on this delicate information.

The engagement of external AI solution providers introduces its own set of risks to patient confidentiality and data safety. Organizations must meticulously scrutinize the privacy protocols and security methodologies of these vendors to mitigate latent vulnerabilities and to shield patient data from any possible compromise.

However, despite these roadblocks, instances of triumphant AI deployments have come to light, underlining improvements in patient privacy and data security. AI algorithms can play an instrumental role in identifying and deterring potential data breaches, pinpointing unusual behaviors, and amplifying the capacity to detect threats, thereby mounting a staunch defense against unsanctioned access or malicious cyber onslaughts.

Moreover, AI can take the lead in rendering data anonymous and stripped of identifiers, thus facilitating the use of expansive datasets for research and analysis while serving as a guardian for patient privacy. Techniques such as differential privacy and federated learning offer mechanisms to preserve privacy, enabling cooperative analyses devoid of exposure of personal patient information.

Adding another layer of security, AI-powered systems can lend healthcare providers a hand in enforcing rigorous access control protocols, encryptions of various stripes, and real-time monitoring tools, thereby ensuring data security across its lifecycle. Such systems can also bring to light any frailties in the security posture, enforce data usage principles, and instantaneously notify stakeholders about impending breaches, thereby giving an all-encompassing boost to the domain of data security.

In this AI-driven age, cementing patient privacy and data security necessitates the erection of comprehensive governance frameworks and regulatory edicts. Organizations must accord primacy to data safeguarding, conduct meticulous assessments of potential risks, and remain in a state of perpetual vigilance, tweaking security blueprints to keep pace with the ever-evolving landscape of threats and susceptibilities.

6.2 Addressing Bias and Ensuring Clarity in Algorithms

Navigating the realms of bias and algorithmic transparency stands as a pivotal ethical concern when it comes to harnessing the potential of AI. As AI systems progressively extend their influence across diverse sectors, such as healthcare, the imperative to uphold principles of equity, impartiality, and lucidity becomes paramount in order to foster unwavering confidence in the capabilities of AI. The landscape becomes intricate when grappling with the intricacies of bias and algorithmic transparency within the AI landscape.

Embedded bias within training data surfaces as a formidable challenge, often culminating in skewed predictions and decision-making processes. Identifying and alleviating these biases during the stages of data accumulation and model construction emerges as a critical task, aiming to forestall the perpetuation of discriminatory outcomes. Equally formidable is the pursuit of algorithmic transparency. The complexity and opaqueness characterizing AI models impede a clear comprehension of the decision-making mechanisms at play. This opacity acts as a stumbling block, impeding the ability to pinpoint and rectify latent biases, while also constraining accountability and the ability to offer comprehensible explanations.

Moreover, the prospect of unveiling the inner workings of AI models bears certain risks. Elaborate elucidations could inadvertently unveil proprietary insights or furnish adversaries with tools to exploit vulnerabilities. Striking a harmonious equilibrium between transparencies and safeguarding sensitive intelligence emerges as a delicate tightrope walk, demanding meticulous assessment.

Nonetheless, glimmers of success in addressing bias and amplifying algorithmic transparency have come to the fore. AI algorithms can be meticulously fashioned to discern and alleviate biases, leveraging methodologies such as bias detection, equity metrics, and algorithmic recalibrations. Such endeavors aim to usher in a landscape of impartial and unbiased decision-making, applicable across diverse demographic strata.

Simultaneously, strides are being taken to augment the translucency of algorithms. Research in the realm of explainable AI (XAI) endeavors to design intelligible models that cast light on the rationale underpinning decisions. Techniques spanning rule-based explanations, attention mechanisms, and visualization aids serve to empower users with a grasp of the logic steering AI predictions.

Organizations and scholars are also delving into mechanisms for evaluating and certifying the equity and transparency of AI systems. This encompasses the evolution of standardized evaluation frameworks, audit mechanisms, and guiding principles, all of which foster the responsible development and integration of AI technologies.

In the quest to effectively combat bias and cultivate algorithmic transparency, collaboration emerges as a linchpin. Close-knit cooperation between AI developers, domain specialists, ethicists, and policymakers can pave the way for the formulation of guidelines, regulatory frameworks, and exemplar practices that ardently champion fairness, transparency, and accountability within the realm of AI systems.

6.3 Securing Supervision by Humans and Establishing Responsibility

Upholding vigilant human oversight and establishing a robust sense of accountability assumes paramount importance when delving into the realm of AI deployment. The notion of accountability stretches to encompass the duty of assigning responsibility to individuals or entities for the outcomes and choices borne out of AI systems. This facet emerges as a linchpin, nurturing an ecosystem characterized by transparency, impartiality, and dependability.

The landscape brims with challenges when the aspiration is to enshrine human vigilance and accountability within the AI domain. One hurdle lies in the necessity of outlining and cementing definitive directives and legal regulations for holding parties accountable. Given the intricate tapestry woven by AI systems, interweaving algorithms, datasets, and human interactions, it becomes imperative for stakeholders such as policymakers, ethical theorists, and technologists to engage in harmonious collaboration, birthing potent frameworks that can withstand the test of scrutiny.

Another stumbling block involves the delicate equilibrium between human agency and the autonomy granted to AI. While AI systems are endowed with the potential to automatize the decision-making apparatus, it remains indispensable to ensure that humans stand poised to exercise oversight, wielding the capacity to intervene or supersede AI-derived outcomes whenever the need arises.

Yet, defying these challenges, instances of triumphant integration of human oversight and accountability within the realm of AI have come to the fore. An illustrative instance can be discerned in the domain of decision support systems, where AI algorithms tender recommendations or insights to human decision-makers. This dynamic interplay embodies human surveillance, thus effectuating decisions harmonious with moral and legal tenets. Here, AI assumes the role of an augmenter, amplifying human prowess, thus enriching the accuracy and efficiency that underscore decision-making.

Delving deeper, the infusion of accountability mechanisms such as audit trails and transparent logs emerges as a potent instrument for strengthening human vigilance. These mechanisms furnish the means to monitor and track the comportment of AI systems, thereby streamlining the identification of glitches, prejudices, or even malevolent intent, subsequently affording the opportunity for requisite corrective actions.

Furthermore, ethical evaluation panels and regulatory enclaves wield substantial influence in ensuring the perpetuity of human vigilance and accountability. These entities take upon themselves the mantle of evaluating the repercussions and latent perils aligned with AI integration. The purview extends to scrutinizing ethical dimensions, and thus, charting out guidelines that foster responsible and accountable application of AI across diverse sectors.

In confronting these challenges, the axiom of collaboration assumes a cardinal role. A multidisciplinary approach, heralding the union of divergent disciplines spanning ethics, law, technology, and the social sciences, ushers in a holistic paradigm for AI governance. This union effectively culminates in the erection of comprehensive accountability structures that weather the tides of complexity and usher in a new era of accountability and transparency.

7. EMERGING AVENUES AND FINAL THOUGHTS

7.1 Advancements in Artificial Intelligence Technologies

The swift strides taken in the realm of AI are further amplified through the infusion of an array of burgeoning technologies, thereby unlocking fresh avenues and prospects. Below are significant insights and details pertaining to the convergence of AI with emerging technologies?

- **Cloud Computing**: Cloud computing extends on-demand access to computational resources, granting scalability and adaptability to AI applications. This spares organizations the need for hefty investments in infrastructure while still capitalizing on AI capabilities.
- **Fog Computing**: Fog computing pushes the frontiers of cloud capabilities to the network's periphery, in proximity to data sources. This catalyzes real-time data processing, curbing latency, and enriching AI applications in contexts where immediacy matters, like autonomous vehicles and industrial automation.
- **Edge Computing**: Edge computing ushers in data processing at the source, close to the network's edge, obviating the reliance on centralized cloud structures. This accelerates data analysis, reduces dependence on ceaseless connectivity, and elevates privacy and security levels for AI applications in IoT devices.
- **Serverless Computing**: Serverless computing, synonymous with Function-as-a-Service (FaaS), empowers developers to focus on coding, extricating them from managing the underpinning infrastructure. It paves the way for deploying AI algorithms as microservices, fostering scalability and cost-effectiveness in AI applications.
- **Quantum Computing**: Quantum computing has the potency to redefine AI's horizons by leveraging quantum principles to execute intricate computations. Quantum algorithms stand to significantly heighten the potency of machine learning, optimization, and pattern recognition tasks, potentially heralding breakthroughs in AI capabilities.

7.2 The Landscape of Challenges in Utilizing AI With Emerging Technologies

- **Integration Complexity**: The amalgamation of AI with emerging technologies necessitates prowess across multiple domains encompassing AI, cloud computing, networking, and security. Collaboration and harmonious coordination among diverse stakeholders are quintessential to fashion effective integration strategies.

- **Ethical Reflections**: The infusion of AI within emerging technologies brings forth ethical quandaries pertaining to privacy, transparency, bias, and accountability. Erecting robust safeguards, regulations, and ethical frameworks serves as the crux to surmount these challenges.
- Triumphant Applications of AI within Emerging Technologies:
- AI-driven IoT: AI augments IoT devices by facilitating real-time data processing, predictive analytics, anomaly detection, and astute automation. The outcome is elevated operational efficiency, fine-tuned resource utilization, and enriched user experiences.
- **Intelligent Automation**: The confluence of AI with robotic process automation (RPA) yields intelligent automation of repetitive, rule-bound tasks. This streamlines business processes, truncates errors, and amplifies productivity across diverse industries.
- **Natural Language Processing (NLP) and Voice Recognition**: AI-infused NLP and voice recognition technologies disrupt human-computer interaction. They energize virtual assistants, chatbots, voice-command systems, and language translation services, thereby enriching user encounters and engendering more intuitive interfaces. The knowledge of natural language and its application with the tint of unambiguity has been performed with the advent of AI which also helps in overcoming the difficulties faced due to language barriers in the proper interpretation of medical reports and the consequential approaches to the remedies of patients of varying complicacies.
- **Predictive Analytics**: AI algorithms seamlessly melded with emerging technologies bring forth predictive analytics, empowering organizations with data-driven decision-making and forecasts. The applications span varied sectors, encompassing healthcare, finance, supply chain management, and the analysis of customer behaviors.

As the symphony of AI harmonizes with emerging technologies, the vista of possibilities continues to expand across industries. In this evolution, the imperative lies in addressing challenges while ensuring the judicious and ethical deployment of AI, thereby harnessing their latent potential for the greater societal good.

7.3 Healthcare Systems Enhanced by Artificial Intelligence Integration

The integration of artificial intelligence into healthcare systems holds the potential to revolutionize the landscape of the healthcare industry, ushering in a wave of benefits and unprecedented advancements (Khang & Medicine, 2023). Here's a comprehensive exploration of the pivotal aspects to contemplate:

- **Metamorphosing Healthcare Practice**: The infusion of AI technologies has the transformative prowess to elevate healthcare practice by supplementing human capabilities. This encompasses heightened diagnostic precision, refined treatment strategizing, optimized resource distribution, and the facilitation of tailor-made care. By dissecting vast troves of patient data, AI discerns patterns, furnishing invaluable insights that bolster the fabric of clinical decision-making.
- **Elevated Efficiency and Amplified Productivity**: AI is a catalyst for streamlining healthcare workflows, alleviating administrative burdens, and fueling efficiency and productivity. Automated frameworks adeptly manage mundane tasks, emancipating healthcare professionals to concentrate on intricate facets of patient care. AI's reach extends to tasks like automating medical record documentation, data interpretation, and workflow fine-tuning.

- **Hurdles on the AI-Healthcare Horizon**: Amid the optimism lie critical hurdles intrinsic to assimilating AI into healthcare systems. Foremost among these is the imperative of ethically and judiciously deploying AI. This realm encapsulates a gamut of concerns encompassing data privacy, security, algorithmic prejudice, and the potential to supplant human decision-making.
- **Upholding Data Confidentiality and Privacy**: The synergy of AI with healthcare systems mandates the collation and scrutiny of extensive patient data. Ensuring the sanctity of patient privacy and data security surfaces as a non-negotiable. Adhering to proper data anonymization, encryption protocols, and compliance with privacy regulations emerges as the bedrock for nurturing patient confidence and shielding sensitive information.
- **Tales of Triumph**: The impact of AI in healthcare is already palpable, manifesting across a spectrum of domains. For instance, the prowess of AI algorithms shines in radiology, elevating image interpretation accuracy and aiding early identification of afflictions like cancer. Predictive analytics buoyed by AI guide patient monitoring and disease trajectory forecasts, priming the ground for timely interventions.
- **Remote Monitoring and Telehealth at the Fore**: AI-marinated systems catapult remote monitoring and telehealth services into the spotlight, a boon for individuals residing in remote or underserved pockets. AI algorithms orchestrate real-time scrutiny of patient data, promptly notifying healthcare providers about alarming deviations. This underpins timely interventions and slashes the need for in-person consultations.
- **Precision Medicine Forging Ahead**: Precision medicine carves a trajectory further propelled by AI's clout. Through dissecting voluminous genomic datasets, interwoven with clinical records, AI unravels biomarkers, forecasts treatment responses, and steers personalized therapeutic modalities. This enables healthcare practitioners to administer targeted interventions, thereby enhancing patient outcomes. The use of machine learning algorithms has been much sharpened with the advent of AI approach. The optimality enhancement has added the leverages to the procedure for accurate classification with the precision knowledge boundaries. The detection of outlier has become the stern task but can fairly be achieved by means of AI and machine learning approach. The expression and interpretation of the same can be recognized by means of many different metrics leading to the mitigation of risks involved (Khang & Rana et al., 2023).

8. CONCLUSION

The healthcare landscape has been indelibly altered by the disruptive influence of AI, ushering in a panorama of prospects to heighten patient well-being, streamline operational efficiency, and institute the era of personalized medicine. The expanse of this discourse has delved into myriad facets of AI's imprint on healthcare, encompassing clinical verdict formulation, diagnostics, remote surgical interventions, pharmaceutical discovery, precision medicine, patient confidentiality, algorithmic lucidity, and the tenet of human supervision (Khang & Kali, 2023).

Yet, while the horizons brim with the potential benefits of infusing AI within healthcare, a judicious acknowledgment of the challenges intertwined with its assimilation is indispensable. Within this labyrinth, the contours of challenges extend to safeguarding patient privacy and data sanctity, navigating the tides of prejudice, and fostering the transparency of algorithms. Equally, they encompass sustaining the vigor of human surveillance and upholding the mantle of accountability. The pivotal touchstone in the journey to fully harness the might of AI within healthcare rests upon the bedrock of cooperation—conjuring a symphony that harmonizes the chords of medical practitioners, computer visionaries, engineers, policymakers, and an array of stakeholders. This orchestration necessitates the erection of ethical paradigms, regulatory benchmarks, and guiding principles that stand sentinel over patient prerogatives, nurturing impartiality and clarity, all while nipping potential hazards in the bud (Khang & Vugar et al., 2023).

Amidst these crucibles, however, a tapestry of triumphs spotlighting AI within healthcare has already begun to unfurl. AI finds its vantage as an instrument for sharpening diagnostic precision, optimizing therapeutic blueprints, underpinning clinical verdicts, steering telehealth provisions, and propelling the envelope of precision medicine. These veritable feats unfurl the promise AI bears to reconfigure the scaffolding of healthcare dispensation, all the while elevating patient outcomes (Prabu, Khang et al., 2023).

As AI's evolutionary march deeper into healthcare echelons continues unabated, the compass points towards an equilibrium—where technological leaps are calibrated by ethical mindfulness. Through surmounting challenges, nurturing synergies, and harnessing AI's potential with circumspection, we are poised to unlock the doorways to AI-integrated healthcare systems. The vista that lies beyond is one marked by heightened efficiency, unwavering precision, and an aura of care personalized for patients spanning the globe (Ushaa, Khang et al., 2023).

REFERENCES

Aggarwal, C. C. (2011). An introduction to social network data analytics. In C. C. Aggarwal (Ed.), *Social network data analytics* (pp. 1–15). Springer. https://link.springer.com/chapter/10.1007/978-1-4419-8462-3_1 doi:10.1007/978-1-4419-8462-3_1

Aggarwal, C. C. (2018). Big Data in Healthcare: Challenges, Opportunities, and Future Directions. *ACM Transactions on Management Information Systems*, *9*(4), 1–7. https://www.sciencedirect.com/science/article/pii/S1474034616301938

Anh, P. T. N. (2024). AI Models for Disease Diagnosis and Prediction of Heart Disease with Artificial Neural Networks. Computer Vision and AI-integrated IoT Technologies in Medical Ecosystem. CRC Press. https://doi.org/ doi:10.1201/9781003429609-9

Barbier, G., & Liu, H. (2011). Data mining in social media. In C. C. Aggarwal (Ed.), *Social network data analytics* (pp. 327–352). Springer. https://link.springer.com/chapter/10.1007/978-1-4419-8462-3_12 doi:10.1007/978-1-4419-8462-3_12

Beaver, D., Kumar, S., Li, H. C., Sobel, J., & Vajgel, P. (2010). Finding a needle in haystack: Facebook's photo storage. In *Proceedings of the nineth USENIX conference on operating systems design and implementation* (pp. 1–8). USENIX Association. https://www.usenix.org/event/osdi10/tech/full_papers/Beaver.pdf

Chen, H., Chiang, R. H. L., & Storey, V. C. (2012). Business intelligence and analytics: From big data to big impact. *Management Information Systems Quarterly*, *36*(4), 1165–1188. https://www.jstor.org/stable/41703503. doi:10.2307/41703503

Cukier, K. (2010). Data, data everywhere: A special report on managing information. *The Economist*. https://www.economist.com/node/15557443

Diebold, F. X. (2012). *A personal perspective on the origin(s) and development of "big data": The phenomenon, the term, and the discipline* (Scholarly Paper No. ID 2202843). Social Science Research Network. http://papers.ssrn.com/sol3/papers.cfm?abstract id=2202843

Eswaran, U., Khang, A., & Eswaran, V. (2023). Role of Quantum Computing in The Era of Artificial Intelligence. Applications and Principles of Quantum Computing. IGI Global Press. https://doi.org/ doi:10.4018/979-8-3693-1168-4-ch003

Fan, J., Han, F., & Liu, H. (2014). Challenges of big data analysis. *National Science Review*, *1*(2), 293–314. https://academic.oup.com/nsr/article-abstract/1/2/293/1397586. doi:10.1093/nsr/nwt032 PMID:25419469

Gandomi, A., & Haider, M. (2015). Beyond the Hype: Big Data Concepts, Methods, and Analytics. *International Journal of Information Management*, *35*(2), 137–144. https://www.sciencedirect.com/science/article/pii/S0268401214001066. doi:10.1016/j.ijinfomgt.2014.10.007

Gil, Y., & Zhou, M. (2019). Cloud AI: Making Artificial Intelligence Accessible to Everyone. *IEEE Intelligent Systems*, *34*(5), 2–5. https://ieeexplore.ieee.org/abstract/document/6547979/

He, W., Zha, S., & Li, L. (2013). Social media competitive analysis and text mining: A case study in the pizza industry. *International Journal of Information Management*, *33*(3), 464–472. https://www.sciencedirect.com/science/article/pii/S0268401213000030. doi:10.1016/j.ijinfomgt.2013.01.001

Khang, A. (2023). Enabling the Future of Manufacturing: Integration of Robotics and IoT to Smart Factory Infrastructure in Industry 4.0. AI-Based Technologies and Applications in the Era of the Metaverse. IGI Global Press. doi:10.4018/978-1-6684-8851-5.ch002

Khang, A. (2023). Applications and Principles of Quantum Computing. IGI Global Press. https://doi.org/ doi:10.4018/979-8-3693-1168-4

Khang, A. (2023). Quantum Mechanics Primer - Fundamentals and Quantum Computing. Applications and Principles of Quantum Computing. IGI Global Press. https://doi.org/ doi:10.4018/979-8-3693-1168-4-ch001

Khang, A. (2023). Introduction to Quantum Computing and Its Integration Applications. Applications and Principles of Quantum Computing. IGI Global Press. https://doi.org/ doi:10.4018/979-8-3693-1168-4-ch002

Khang, A. (2024). Application of Computer Vision in the Healthcare Ecosystem. Computer Vision and AI-integrated IoT Technologies in Medical Ecosystem. (1 Ed.) (2024). CRC Press. https://doi.org/ doi:10.1201/9781003429609-1

Khang, A. (2024). Using Big Data to Solve Problems in the Field of Medicine. Computer Vision and AI-integrated IoT Technologies in Medical Ecosystem. CRC Press. https://doi.org/ doi:10.1201/9781003429609-21

Khang, A. (2024). Medical and BioMedical Signal Processing and Prediction. Computer Vision and AI-integrated IoT Technologies in Medical Ecosystem. CRC Press. https://doi.org/ doi:10.1201/9781003429609-7

Khang, A., & Abdullayev, V. A. (2024). *Computer Vision and AI-integrated IoT Technologies in Medical Ecosystem* (1st ed.). CRC Press. doi:10.1201/9781003429609

Khang, A., & Muthmainnah, M. (2023). AI-Aided Teaching Model for the Education 5.0 Ecosystem. AI-Based Technologies and Applications in the Era of the Metaverse. IGI Global Press. doi:10.4018/978-1-6684-8851-5.ch004

Khang, A., Rana, G., Tailor, R. K., & Hajimahmud, V. A. (2023). Data-Centric AI Solutions and Emerging Technologies in the Healthcare Ecosystem. CRC Press. doi:10.1201/9781003356189

Khang, A., Shah, V., & Rani, S. (2023). AI-Based Technologies and Applications in the Era of the Metaverse. IGI Global Press. doi:10.4018/978-1-6684-8851-5

Li, Z., Chen, C., & Li, M. (2021). A Comprehensive Review of Artificial Intelligence in Medicine. *Artificial Intelligence in Medicine*, *113*, 102043. https://www.sciencedirect.com/science/article/pii/S002604951730015X

Nasir, M., Shao, Z., Zhang, Y., Zhang, J., Yang, Z., & Vasilakos, A. V. (2021). The Role of Edge Computing in Internet of Things. *IEEE Network*, *35*(1), 68–75. https://ieeexplore.ieee.org/abstract/document/8450541/

Prabu, B., & Khang, A. (2023). Quantum Computing Smart City: Fostering Sustainability and Green Infrastructure with Smart City. Applications and Principles of Quantum Computing. IGI Global Press. https://doi.org/ doi:10.4018/979-8-3693-1168-4-ch009

Sheth, A., Anantharam, P., Henson, C., & Sahoo, S. (2020). Next Generation Computing Paradigms: Cloud, Edge, and Fog/Edge Computing for Internet of Things and Artificial Intelligence. *IEEE Internet of Things Journal*, *7*(12), 10977–10980. https://link.springer.com/chapter/10.1007/978-3-030-74402-1_13

Van der Schaar, M. (2019). Machine Learning and AI for Healthcare. *IEEE Journal of Selected Topics in Signal Processing*, *13*(2), 239–241. https://link.springer.com/content/pdf/10.1007/978-1-4842-6537-6.pdf

Zhang, C., & Zhang, X. (2019). An Overview of Quantum Computing: Technologies, Applications, and Challenges. *Future Generation Computer Systems*, *97*, 271–281.

APPENDICES

Appendix A: Glossary of Terms

- AI: Artificial Intelligence
- ML: Machine Learning
- IoT: Internet of Things
- EHR: Electronic Health Record
- NLP: Natural Language Processing
- CT: Computed Tomography
- MRI: Magnetic Resonance Imaging
- FDA: Food and Drug Administration

Appendix B

- AI-Driven Healthcare Use Cases
- AI-assisted diagnosis in radiology
- Predictive analytics for disease outbreak detection
- Personalized treatment recommendation based on genomics data
- Virtual assistants for patient engagement and support
- AI-powered robotic surgery

Appendix C: Regulatory Guidelines for AI in Healthcare

- HIPAA: Health Insurance Portability and Accountability Act
- FDA guidelines for AI-based medical devices
- Ethical guidelines for AI research and deployment

Appendix D: AI Ethics Principles

- Transparency: Ensuring the explain ability of AI algorithms and systems.
- Fairness: Mitigating bias and ensuring equitable outcomes for all populations.
- Privacy: Safeguarding patient data and respecting privacy rights.
- Accountability: Establishing mechanisms for human oversight and responsibility.
- Robustness: Ensuring AI systems are reliable and resilient to errors or attacks

Chapter 12
Quantum Computing Research:
Ontological Study of the Quantum Computing Research Ecosystem

Rakesh Kumar Sheoran
Banasthali Vidyapith, India

Seema Verma
Banasthali Vidyapith, India

Savita Kumari Sheoran
https://orcid.org/0000-0002-0359-4700
Indira Gandhi University, India

Vikesh Yadav
Indira Gandhi University, India

ABSTRACT

Quantum computing is one of the remarkable achievements jousting on the coalesced power of quantum mechanics and information theory to bring innovative paradigm shift in computational regime. It possesses potential to solve computationally complex problems in real-time which were otherwise considered intractable in traditional computer science. The nascent field, originated during the early 1980's, has now revolutionarised the global technological reign by fetching the attention of every government and corporate across the world. Despite huge fund infusions, research advancements, and policy supports, the ideal potential has yet not been realised. This chapter intends to study the ontology of quantum computers to understand its astonishing journey—based on underlying mathematical, scientific principles, technical issues, challenges, and variants—for paving a way for future direction of research. The study is anticipated to come out with the essential research elements required to formulate the hypothesis and address research issues in this field.

DOI: 10.4018/979-8-3693-1168-4.ch012

1. INTRODUCTION

The end of 20th century earmarked a wacky technological leap to drastically overhaul the societal landscape across the globe as the century-long technological advantage has resulted in the culmination of a promising discipline of 'quantum computing' leveraging the principles of quantum mechanics to enhance the capabilities of information science. The existing practices in traditional information sciences were restricted by the narrow computational space available with underplaying cmos technology.

Quantum computing comes up with innovative solutions to harness the principles of quantum mechanics *viz.* interference, parallelism, entanglement, and superposition to overcome the existing limitations and offer enormous computational space with unprecedented speed. Traditional computer machines use the bit as a basic computational unit with only two available states *i.e.* either '0' or '1' offering '2n' computational spaces while quantum computers use a quantum bit (qubit) as a fundamental computing unit comprising of |0, |1 or a superposition of both and offer 2^n computational spaces. This enlarged computational space, exponentially faster the capabilities to solve complex problems in real time, which otherwise were considered NP-Hard or knapsack problems in the earlier regimes. This way quantum computing is poised to bring a paradigm shift in information sciences by relying on an entirely new concept of computation rather than merely a modified version of existing traditional computing (Ying et al., 2010).

The roots of computing based on principles of quantum mechanics can be traced back in Moore's law, a talisman to forecast the future of technology, stating that with the increase in computational power the size of processor will reduce exponentially (Khang and Quantum et al., 2023). The present-day classical processor has reached to nanoscale and the further reduction will enter an atomic scale computing *i.e.* quantum computing. The foundation of quantum computing was initiated with a proposition of quantum automata by Russian Mathematician Yuri Manin in 1980 and a quantum Hamiltonian to simulate Turing machine by Paul Benioff in the same year. This eureka moment was venerated with the idea of great physicist Richard Feynman, in 1982, conceiving that some quantum phenomenon could not be simulated by classical computers without any exponential slowdown.

Feynman's quantum curiosity was further explored by Deutsch for propounding the Quantum Turing Machine as a universally empowered quantum computer based on the technique of quantum parallelism. In 1994, Peter Shor was the first to develop quantum algorithm to address the prime factorization issue which otherwise was infeasible to be solved in real-time (Ying et al., 2010). In 1996, Lov Grove implemented an algorithm that sped up the large database search to real by modifying the classical algorithms. Shor's and Grover's algorithms poised innovative ideas for practical implementation of quantum computing to solve real-life puzzles. This breakthrough in harnessing the weird potential of quantum computing has ignited various scientific and industrial giants across the globe to rope into the quantum race making it a hot issue of strategic diplomacy for governments across the world (Shah & Khang et al., 2023).

D-wave Systems, a Canadian company, claimed creation the first quantum computer in 1999. IBM has already released 400 plus qubit quantum process 'Sprey' in 2022 and paved a road map for another 1121-qubit processor by 2026. Google proved its quantum supremacy in 2019 through its 3-qubit Sycamore processor by carrying out boson sampling in 200 seconds which is not expected from present-day classical computers in a time less than 2.5 billion years while two Chinese teams have claimed such achievements in 2021 (Khang and Quantum et al., 2023) (Abd El-Latif et al., 2018).

However, despite huge fund infusions and swift technological developments, the quantum computing is still at infancy, hindering its application as an autonomous computing machine. The quantum computer prepares qubit states to store and process the information while quantum gates based linear algebra are used for implementing the programs. The hardware used in quantum computer is sensitive to the environment, prone to error, fragile, and ephemeral to limit that even a small change in physical condition may shutter the quantum state completely (Khang & Quantum, 2023).

The measurement of quantum state is also a herculean task as measurement exposes it to the environment resulting in the introduction of error by flipping the quantum state, phase, or both of a qubit. To make quantum computing feasible such evitable errors need to be mitigated below a rate of one in billion against the prevailing rate of one in thousand. Even the traditional computers are also exposed to error while operating but it can be removed by copying the information multiple times. Similar strategy will not work for quantum computers because 'no cloning theorem' restricts the copying of quantum information. Therefore, the quantum computer is hindered by various issues and challenges that need to be addressed to make quantum computation practically feasible.

This chapter intends to carry out an ontological review of the astonishing journey of quantum computing for the escalating future of this technology (Khang & Shah et al., 2023). The subsequent sub-sections will describe the relevant technological and market variants of this quantum computing in detail. Section 2 will review the literature available in reliable databases to identify the prominent research areas in this regime. Section 3 intends to present the essentials of quantum computing with its theoretical and mathematical foundations. Section 4 explored the key challenges in this regime to formulate the research hypothesis for further advancement of this chapter. Section 5 peeps into the astonishing journey of this novel technology while section 6 studies its impact. Section 7 and section 8, respectively, present the important algorithms and tools in quantum computing. Further, the current global and national scenario in this regime is presented in section 9. Section 10, elucidates the important applications of quantum computing. Finally, the chapter is concluded for meaningful inference in section 11.

2. CONTEMPORARY RESEARCH IN QUANTUM COMPUTING

Quantum computing is a nascent field governed by the convergence of independent disciplines of quantum mechanics and information theory. However, a lot of literature is available on the existing base but still there is a serious dearth of the contiguous literature on quantum computing. The authors has identified three areas of advanced research in this domain, reviewed the literature and presented in the subsequent sub-sections.

2.1 Quantum Computing in Routing Problems

Routing problems are very well-known optimization problems that are commonly found in transportation, supply chain management, and logistics. The routing problems address the issues of optimization for finding the shortest and most efficient path only by utilizing the available resources. Generally Bellman-Ford dynamic programming algorithm is deployed to solve such problems with negative cost edges but without negative cost cycles. Such implementations offer a time complexity of (o|V||E|) and require O|v| space.

The application of quantum computing can enhance performance because quantum networks are capable of transmitting data from one node to another using the teleportation and entanglement properties of quantum physics. The efficiency of quantum methods in this regime is envisaged from various instances such as Dijkstra method capable of solving the issue for a graph with non-negative edges in O|v|^2 steps. Similarly, Grover search algorithm in Quantum computing efficiently addresses the shortest path problem with a time complexity of O|v|^(7/4) as compared to the classical Dijkstra algorithm (Krauss et al., 2021). This enhancement in performance is due to the reason that the majority of entanglement distribution studies use the network's present state and do not take historical data into account resulting in the success of connection setups with an increase in the size of the quantum network (Solmeyer et al., 2017).

Intelligent transport system (ITS) is one of the prominent applications of logistic management and is considered a crucial factor for the success of Industry 4.0 and 5.0.A. It delivers better optimization results by applying Ising Hamiltonian hybrid quantum-classical heuristic known as the Quantum Approximate Optimization Algorithm (QAOA). Azad et al., have compared the performance of QAOA with that of traditional solvers like CPLEX and recorded QAOA as a far better performer (Azad et al., 2022).

Wang et al. have studied entanglement and teleportation as essential components to establish communication between two remote communication nodes. They have explored the strategies such as real-time entanglement distribution (R-TED) and pre-established entanglement distribution (PEED) required to establish the entanglement Their study concluded that the results of PEED are better than RTED in terms of efficiency, memory, and stability (Wang et al., 2022).

Le and Nguyen have studied quantum routing and found its significant contribution to the advancement of future network systems. Their work relied upon Deep Quantum Routing Agent (DQRA) utilizing the capabilities of both machine learning and quantum computing to satisfy the maximum number of demands generated in a time frame. The system utilizes a neural network to monitor the current state of the network and adjust to the generated requirements. These demands are subsequently connected using a Qubit-preserved shortest path algorithm (Le et al., 2022).

2.2 Quantum Computing in Blockchain

Blockchain technology is the most promising innovation from the early 20th century, involves digital data storage (block) in a public ledger (chain), it is one of the (Shahriar et al., 2020). The recent years have witnessed a notable interference of quantum computing in the field of blockchain and other Distributed Ledger Technologies (DLTs) because of its potential to address the accountability, redundancy, and transparency issues existing in the former. These advancements suggest the possible application of these technologies in different domains (Khanh & Khang et al., 2021).

The utilization of public-key cryptography and hash functions is implemented inside the framework of the blockchain to confer these characteristics (Khang & Chowdhury et al., 2022). The rapid progression of quantum computing has, however, introduced the possibility of future implementation of attacks employing Grover's and Shor's algorithms. The utilization of these algorithms presents a significant risk to the integrity and confidentiality of public-key cryptography and hash functions, hence necessitating the revision of blockchain architectures to integrate cryptographic systems that can effectively counteract quantum attacks. The cryptographic systems outlined above are occasionally referred to as post-quantum, quantum-proof, quantum-safe, or quantum-resistant cryptosystems (Fernandez et al., 2020).

2.3 Quantum Computing in Cryptography

Cryptography is an essential element for secure transmission of information over computer networks. The existing cryptographic techniques are at serious risk of breach with the invention of quantum computing. Therefore is essentially required to develop cryptography that work in the post quantum era. Gill et al. have develop a cryptographic scheme based on the science of quantum mechanics to encrypt data at the physical network layer (Gill et al., 2022). Cheng et al. have developed a common secret key for two parties in a symmetric cryptosystem used to encrypt and decrypt messages with the capability to decipher the encrypted messages (Cheng et al., 2017).

3. FOUNDATION OF QUANTUM COMPUTING

Quantum computers joust on the principles of quantum mechanics to empower information science for leap and bound computational capabilities deemed computationally unattainable with traditional computers. The underlying quantum circuit can be designed and implemented through linear algebra. The subsequent sub-sections will describe the basic quantum principle, mathematical foundation, hardware, and other conventional representations required for practical quantum computer (Khang & Kali, 2023).

3.1 Principles of Quantum Computing

Quantum mechanics is one of the most astonishing developments of the 20th century which has phenomenally challenged the established scientific laws to solve the most challenging questions (Khang & Vugar et al., 2023). Some of the principles of quantum mechanics which are essential to culminate the quantum computer are depicted below.

3.1.1 Superposition

This principle holds the multiple quantum states together to enhance the exponentially computational space in comparison to traditional computers. Owing to superposition the qubit can exist in both |0 and |1 states together which otherwise has to be in either of '0' or '1' at a time offering 2^n computation space in comparison to 2n spaces (Bhat et al., 2022) as shown in figure 1. For instance, a 256 bit system provides 512 computational possibilities while a 256 qubit system offers 2^{256} possibilities for computation which reduces both the computational time and complexities. Therefore, the superposition of qubits offers computational edge in terms of quantum advantage.

3.1.2 Entanglement

It is the phenomenon of sharing proximity between quantum particles in a group such that their state cannot be expressed independently. It is a bizarre phenomenon where the change happened in one subatomic particle automatically induced in the entangled particles located remotely. Einstein had described it as a spooky action at a distance as envisaged in figure 2.

Figure 1. Bit versus qubit

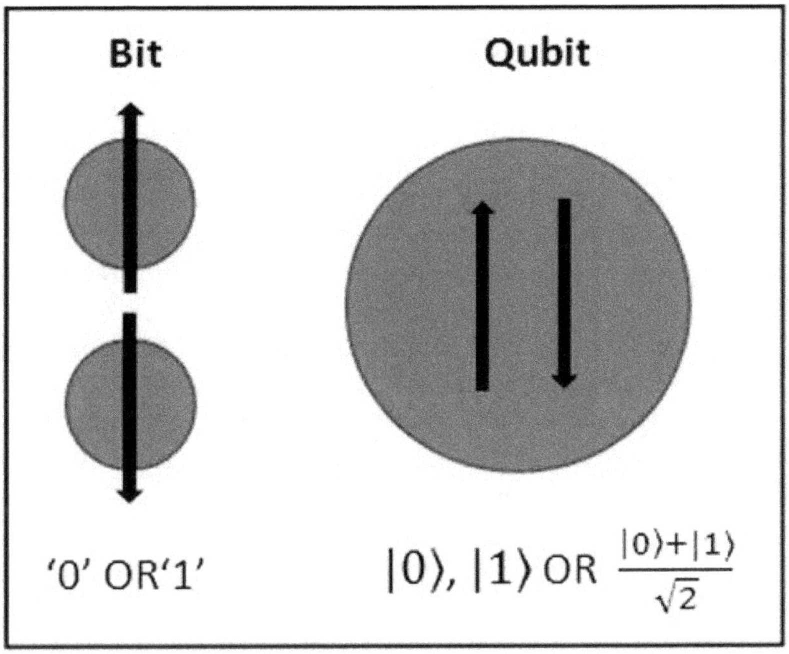

Figure 2. Spooky action at a distance

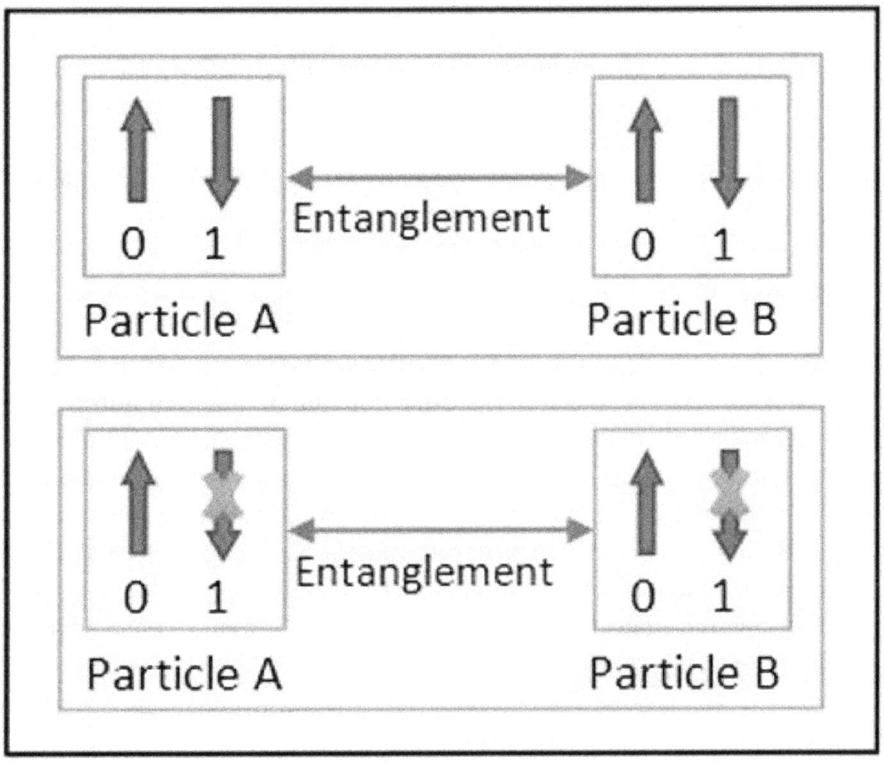

The phenomenon of entanglement was described by Einstein, Podolsky, and Rosen through EPR Paradox in 1935 as the incompleteness of quantum theory while the same was responded by Bell's inequality theorem in 1964 for a computational term as depicted below:-

***EPR Paradox*:** The EPR paradox demonstrates how a "measurement" may be made on a particle without actually disturbing its states. In 1935 Einstein-Podolsky-Rosen (EPR) examined the two entangled photons assuming validity of the locality conditions which states that no significant change can occur in the second system as a result of anything that may be done to the first system because the two systems are no longer interacting at the time of measurement (Zeilinger et al., 1999). The EPR paradox has demonstrated the limitations of quantum mechanics and put a bar on any kind of computations using quantum mechanics.

Bell's Inequality Theorem: In 1964 Bell demonstrated that entanglement correlations cannot be replicated by "local hidden-variable theories and exhibited that our world is non-local if certain predictions made by quantum theory are true. The statement is known as Bell's Inequality Theorem (Wang et al., 2021). Mathematically, $R_{i\alpha,}$ where i=1,2,... and α=a,b,c,... are random variables limiting itself to the values ±1. If R1*a*= *-R2a* ∀ entirely anti-correlated random variables are described by function

$$P\left(R_{1a} \neq R_{2b}\right) + P\left(R_{1b} \neq R_{2c}\right) + P\left(R_{1c} \neq R_{2a}\right) \geq 1.$$

3.1.3 Reversibility

Another important property of quantum computing is the reversibility of logic gates used for its implementation which is the ability to preserve information without loss. The reversible circuits exhibit a one-to-one relationship between the input and output vectors, implying that each input vector can generate a unique output vector and vice versa. The qubits are a two-state quantum system corresponding to the traditional bit values of 0 and 1, realized through elementary unitary operations carried out on the quantum logic gates. All unitary operations are inherently reversible, hence, quantum computers must be constructed using reversible logical elements (Thapliyal et al., 2009).

3.1.4 No Cloning Theorem

In classical computing, error can be mitigated by saving multiple copies of information to retrieve the same in case of information loss due to error. The exact copying of quantum information is impossible due to the inherent impossibility of replicating an unidentified quantum state known as 'No Cloning Theorem' which simply means that quantum error correction is impossible by copying the information and in the fictitious scenario of copying, a quantum state 'E' would automatically turn to replicated state to 'B'. On one side "No Cloning Theorem" restricts the quantum error correction while on the other side it provides full proof security to quantum information as eavesdropping will eventually be detected in term of flipping in the quantum state. This ability to detect eavesdropping offers advantage to quantum cryptography over traditional cryptography (Teja et al., 2007).

3.2 Mathematical Formulation of Quantum Computing

Quantum computers work on the principles of quantum mechanics and require manipulation and analysis of the quantum states and operations. Linear algebra is a mathematical framework that can express the quantum mechanical properties in computational format. The essential mathematics concepts required for the quantum computers are described in below sub-sections.

3.2.1 Inner Product

The inner product of two vectors is a binary operation that accepts two vectors as input for producing a complex output for measurement of similarity or angle between two vectors. An Inner Product is a generalization of dot product operation on complex vector space d is a function: $d*d \rightarrow \mathbb{R}$. The inner product on vector space ∀ (x, y) ∈ ¡ⁿ can be described as $x,y=x^T*y$.

3.2.2 Hilbert Space

Hilbert space H is a complete inner product space such that (Scherer et al., 2019)

$$\Psi + \phi \in \mathbb{H}, \alpha, \beta \in \mathbb{C} \Rightarrow \alpha\Psi + \beta\phi \in \mathbb{H},$$

where dot (.) product satisfies the reproductive property:

$\langle . | . \rangle : \mathbb{H},\ \ \mathbb{H} \rightarrow \mathbb{C}$, and $(\Psi, \phi) \mapsto \langle \Psi \mid \phi \rangle\ \forall \Psi_1, \Psi_2,\ \phi_1 and \phi_2 \in \mathbb{H}$ and $\alpha, \beta \in \mathbb{C}$

3.2.3 Eigenvalue and Eigenvector

These are the values and vectors associated with square matrix, together to provide the structure of that matrix (Abdi et al., 2007). For any matrix A in $C^{m \times n}$, there is c in £, and there exists a vector $V \neq 0$ with $\mathbb{C}^n$ such that $AV=cV$, where 'c' is called the eigenvalue and 'V' is eigenvector of matrix 'A'.

3.2.4 Hermitian Matrix

Any matrix 'A' is called symmetric if $A^T = A$, or $A[i,j]=A[j,i]$. Any n×n matrix is Hermitian if it satisfy condition $A^{\dagger}=A$. The operator represented by a Hermitian matrix is commonly referred to as a self-adjoint operator.

3.2.5 Unitary Matrix

It is a kind of invertible matrix. Any matrix 'A' is invertible if there exist A^{-1} matrix such as: $A \times A^{-1}=A^{-1} \times A$. All the invertible matrices are unitary (Yanofsky et al., 2008).

Table 1. Symbol used in quantum computing (Njorbuenwu et al., 2019)

Symbol	Description
$\|\Psi\rangle$	A bra-ket notation is used to represent a quantum state. it can represent a single state as well as multiple states, unlike classical computing where there exists only one state at one time.
$\alpha_0\|0\rangle$	Probability amplitude of states represents the probability to remain the system in any state.
X, Y, Z, H, I	Represents the quantum gates.
$U_f\|0\rangle$	Represent oracle function, responsible for transformations of states. Used in Deutch-Jozsa algorithm.

3.2.6 Tensor Product

The tensor product of two vectors *V* and *V*' are represented by *VV*'.

A tensor product *VV*' can be written as

$$C_0\left(V_0V_0'\right)+C_1\left(V_1V_1'\right)+C_2\left(V_2V_2'\right)+\ldots+C_{n-1}\left(V_{n-1}V'_{n-1}\right).$$

where $C \in \mathbb{C}$; V_0, V_1, $V_{n-1} \in V$ and $V_0^{'}, V_1^{'}, V_{n-1}^{'} \in V'$.

3.3 Quantum Hardware

The physical hardware required for manipulating and processing quantum information is referred as quantum hardware. Such hardware is completely different from classical hardware because it leverages principles of quantum mechanics to perform the computational tasks using the symbols as depicted in table 1. The subsequent sub-sections of this section will depict the common hardware required to accomplish the computational tasks in quantum computers.

3.3.1 Qubit

Like bit for classical computers, a qubit is basic building block for quantum computers A qubit can be defined as a unit vector within a two-dimensional complex Hilbert space. It can be expressed using Dirac notation as $|\psi\rangle$, where $|\psi\rangle$ denotes the state vector of the qubit $|\Psi\rangle = a_0|0\rangle + a_1|1$ α_0 and α_1 are said to be the complex numbers with probability magnitude.

$|a_0|^2 + |a_1|^2 = 1.$

$|0\rangle$ and $|1\rangle$ are the basic states of Qubits, that can be either $|0\rangle$, $|1\rangle$, or both at the same time, which is called superposition (Ying et al., 2010).

3.3.2 Quantum Register

When multiple Qubits are formed together, they are called registers, and can be represented as:

$$|\Psi\rangle = a_t|t\rangle = a_t, a_t \ldots a_t \, |t_1, t_2 \ldots t_n\rangle$$

Several registers can be put together to form a bigger register by using a tensor product of registers of the states.

$|\Psi i\rangle = a_{i,t}|t^{(i)}\rangle$ let 'i' be qubit state for any $1 \leq i \leq k$, then the tensor product can be defined as:

$$|\Psi 1 \ldots |\Psi k = \otimes_i^n = 1$$

$$|\Psi i == \alpha 1, t1, \ldots \alpha k, tk \mid t1, t2 \ldots tk\rangle$$

or, we can simply write, $|\Psi\rangle \otimes k$ for $|\Psi\rangle \quad |\Psi k\rangle$

3.3.3 Quantum Gates

Just as the classical bits manipulation, we need gates such as OR, NOR, AND, so in quantum computing for the manipulation of qubits quantum gates are needed (Ratan et al., 2007).Quantum circuits are essential for describing quantum algorithms. Any quantum circuit is a combination of qubits and gates that operate on that qubit. All the gates in quantum mechanics are reversible, unitary operators that are represented by $2^m \times 2^m$ unitary matrix with 'm' is the number of qubits on which the gates operate. There are two most commonly used gates, single qubit, and two qubits gate as envisaged in table 2, below (EDA, 2017).

Bloch sphere representation: The Bloch sphere serves as a visual depiction of a complex number z, $|z|^2=1$, As a point located on the unit circle within the complex plane. If $Z=x+iy$, and x and y are real numbers then $|z|^2=z^*z$, $|z|^2=(x-iy)(x+iy)$, $x^2+ixy-ixy+i^2y^2$, $|z|^2=x^2+y^2$ and $x^2+y^2=1$, The equation representing a circle with a radius of 1 and cantered at the origin represent a qubit as depicted in figure 3 (Glendinning et al., 2005).

Matrix Representation of a Qubit: The representation of quantum states is frequently accomplished through the utilisation of Dirac notation. In Dirac notation vectors are represented by "kets" ($|0\rangle$) and dual vectors are represented by "bras" ($\langle 1|$). Matrix representation denotes the probability of some qubit in some quantum state. The topmost entry of a matrix represents The possibility of transitioning to state $|0\rangle$, and the lower part represents the possibility of transitioning to state $|1\rangle$.

$$|0 = \begin{bmatrix} 1 \\ 0 \end{bmatrix}; |1 = \begin{bmatrix} 0 \\ 1 \end{bmatrix};$$

Above equations represent both Dirac notation and qubit in matrix form.

$$|\Psi\rangle = a_0|0\rangle + a_1 \mid 1\rangle = \begin{bmatrix} \alpha_0 \\ \alpha_1 \end{bmatrix}$$

Table 2. Common quantum gates for quantum computing

Quantum Gate	Input Type	Equivalent Classical Gate	Matrix	Symbol	Description
Hadamard gate	Single qubit	Not available	$\frac{1}{\sqrt{2}}\begin{bmatrix}1 & 1\\ 1 & -1\end{bmatrix}$	H	For superposition
Pauli X-gate	Single qubit	NOT gate	$\begin{bmatrix}0 & 1\\ 1 & 0\end{bmatrix}$	X	Creates X rotations
Pauli Y-gate	Single qubit	Not available	$\begin{bmatrix}0 & -i\\ i & 0\end{bmatrix}$	Y	Creates Y rotations
Pauli Z-gate	Single qubit	Not available	$\begin{bmatrix}1 & 0\\ 0 & -1\end{bmatrix}$	Z	Creates Z rotations
C-NOT gate	2 or more qubits	NOT gate	$\begin{bmatrix}1 & 0 & 0 & 0\\ 0 & 1 & 0 & 0\\ 0 & 0 & 1 & 0\\ 0 & 0 & 0 & 1\end{bmatrix}$		Measures second qubit

$|a_0|^2 + |a_1|^2 = 1$

Where $\alpha 0$ and $\alpha 1$ are complex numbers and represent the probability amplitude of states vectors and their sum of squares must be equal to 1. If there are 'n' numbers of qubits, then it may have n2 records in the matrix. The below equation shows the matrix representation of two qubits:

$$|\Psi = a_0 |00 + a_1 |01 + a_2 |10 + a_3 |11 = \begin{bmatrix}\alpha 0\\ \alpha 1\\ \alpha 2\\ \alpha 3\end{bmatrix}$$

In Dirac notation and matrix form two Qubits can be represented as:

Figure 3. Bloch sphere representation of single qubit

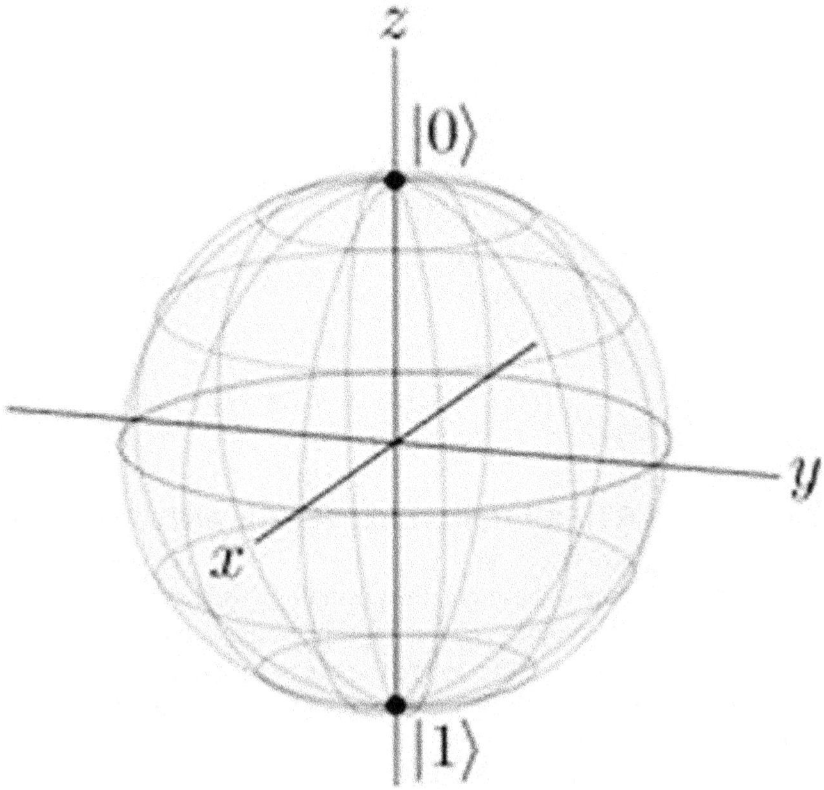

$$|00\rangle = \begin{bmatrix}1\\0\\0\\0\end{bmatrix};\ |01\rangle = \begin{bmatrix}0\\1\\0\\0\end{bmatrix};\ |10\rangle = \begin{bmatrix}0\\0\\1\\0\end{bmatrix};\ |11\rangle = \begin{bmatrix}0\\0\\0\\1\end{bmatrix}$$

Physical Realizations of Quantum Gates: Basically X, Y, Z and H; are quantum single qubit gates. X Gate functioning is similar to classical NOT gate. It toggles between ON and OFF. It transforms $|0$ into $|1\rangle$ and $|1\rangle$ into $|0$. Z Gate is Phase Flip Gate, it changes the Phase when Qubit is ON. It doesn't affect $|0\rangle$. The Y Gate is formed by the conjunction of the X Gate and the Z Gate. It transforms $|0\rangle$ into $i|0\rangle$ and $|1\rangle$ into $-i|0\rangle$. H gate creates simple superpositions. It transforms $|0\rangle$ into $\sqrt{\frac{1}{2}}|0+\sqrt{\frac{1}{2}}|1$ and $|1$ into $\sqrt{\frac{1}{2}}|0-\sqrt{\frac{1}{2}}|1$. The basic operation of single qubit gates on Quirk simulator is depicted in figure 4.

Figure 4. Basic operation of single qubit gates on quirk simulator

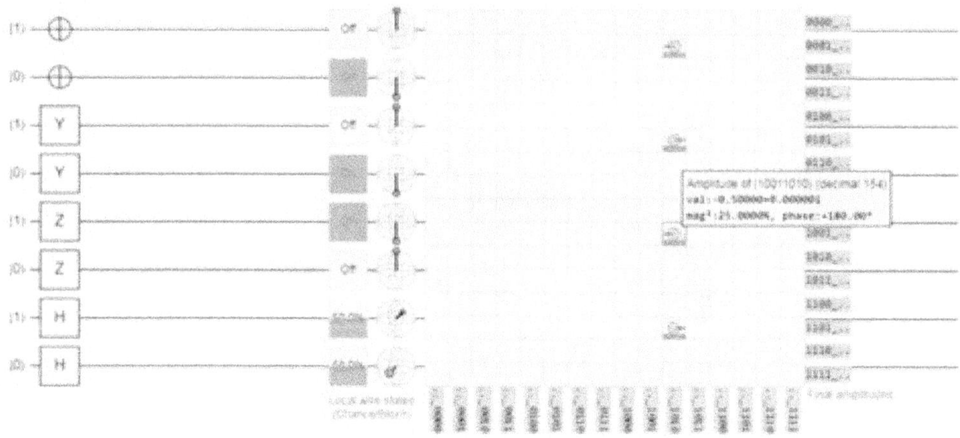

4. SURGERY AIDED BY ARTIFICIAL INTELLIGENCE AND ROBOTICS

The realization of quantum computers is based on the technology used for taming of qubits. A practical qubit can be formulated using the techniques *viz.* Ion traps, photonic qubits, neutral atoms, photons, silicon, superconducting, topological, polarization state of a photon and spin state of an atom, etc. Although the prevailing techniques require sophisticated instruments to generate and keep the qubit operational and such a system will occupy an entire room for extremely limited capabilities such as 100 qubits, a quantity that seems insufficient for the tasks they are capable of performing. The following sub-section will describe the key issues and major challenges encountered by present-day quantum computers (Ushaa, Khang et al., 2023).

4.1 Compatibility Issues

The establishment of a continuous exchange of information between the classical pre-processing stage and the quantum experiments is important to facilitate a hybrid approach required for present-day quantum processors. The challenge of transferring information between different architectures arises from the need to match samples from conventional and quantum models. The challenge can alternatively be resolved by developing technology capable of rapidly generating a set of quantum Gibbs states for a given class. Nevertheless, employing this strategy may give rise to other concerns, a few of which are outlined in the subsequent sections (Perdomo et al., 2018).

4.2 Decoherence and Fidelity Issues

Qubits have a high degree of fragility; any interaction between qubits and its surroundings results in decoherence causing progressive degradation of information that otherwise exists indefinitely if kept in isolation. Establishing isolation in practice can be challenging, particularly given the advanced quantum technology that is presently accessible. Complete isolation is not ideal as it is important for computation and communication to engage with the qubits, particularly during reading and writing operations.

The duration of decoherence, which is mostly determined by the technological selection for qubit implementation is typically used to quantify decoherence. Quantum fidelity is a key measure of merit in this context. The fidelity, which ranges in value from 0 to 1, is a measurement of the capacity to differentiate between two quantum states where lower fidelity results from the imperfections in a bigger physical implementation of any given quantum operation. Since teleportation involves a series of actions on quantum states, the accuracy of the "transmitted" qubit is impacted by the quality of these operations (Teja et al., 2007).

4.3 Entanglement Distribution Issues

The creation of entangled qubit pair is the basis of quantum computation and the most herculean task in the whole process of harnessing the computation from quantum principles. Because of the deterioration of the rate of entanglement distribution with increasing distance, long-distance entanglement distribution still poses a significant problem. The no cloning theorem prevents qubits from being cloned; therefore conventional signal amplification methods are also ineffective. Therefore, we generally use intermediate nodes or quantum repeaters to enable the entanglement of distant nodes without physically transporting an entangled qubit across the full distance. It can be achieved by employing a swapping mechanism to exchange the entanglement established through shorter links which acts as an obvious breakthrough in design of quantum internet. It can be achieved by completely redesigning and reengineering the network function used for both conventional and quantum communications tools (Parhami et al., 2020).

Technical giants like IBM, Google, D-wave systems are continuously working on the betterment of quantum computers but the quantum hardware is still not stable. Most of the experiments, algorithms, and protocols are still in the development phase (Yang et al., 2023). Technology barriers in quantum computation, such as limited qubit connection, insufficient gate fidelities, and the necessity for a substantial quantity of qubits to facilitate error correction are attributed to inadequate software implementation. Another significant obstacle to the development of quantum computing is the lack of cooperation and communication between businesses and academics. Furthermore, operations of such quantum computers require temperatures that are almost exactly zero, and it is always difficult to maintain (Prabu, Khang et al., 2023).

4.4 Fault-Tolerant Quantum Computing and Quantum Error Correction

The true potential of a quantum computer can be realized by mitigating noise and de-coherence. The initial hypothesis of 'no cloning theorem' and EPR Paradox are against quantum error correction but Bell's inequalities open a way forward to distribute the information over multiple entangled qubits to make fault-tolerant quantum computing possible. The Quantum Error Correction map 'k' (2^k - dimensional Hilbert Space) qubits into 'n' qubits (2^n - dimensional Hilbert Space) such that 'n>k'. The 'k' qubits are known as physical qubits and 'n' qubits are known as logical qubits obtained by concatenating the former.

The quantum system is inevitably entangled with the larger non-local environment system. Let |0 be the state of the environment then the unitary evolution (U) will interact with the environment as

$$U\left|\psi\right|0 \rightarrow \alpha\left|0\right|e_{00} + \alpha\left|1\right|e_{01} + \beta\left|0\right|e_{10} + \beta\left|1\right|e_{11}$$

Therefore,

$$U|\psi|0 \rightarrow (\alpha|0+\beta|1)\frac{1}{2}(|e_{00}+|e_{11})$$
$$+(\alpha|0+\beta|1)\frac{1}{2}(|e_{00}-|e_{11})+(\alpha|0+\beta|1)\frac{1}{2}(|e_{01}+|e_{10})+(\alpha|0+\beta|1)\frac{1}{2}(|e_{01}-|e_{10})$$

In symbolic form, it can be written as equation (2),

$$U|\psi|0 \rightarrow I|\psi|e_I + Z|\psi|e_Z + X|\psi|e_X + Y|\psi|e_Y$$

Where I|ψ→ no error, Z|ψ→ bit–flip error, X|ψ→ phase–flip error and Y|ψ→ both bit–flip, and phase–flip errors simultaneously.

The error of n-qubit can be described as, $E_a=M_1M_2\ldots\ldots M_n$ with $M_i\rightarrow$ Pauli operators (I,X,Y,Z). The necessary and sufficient condition to detect an error in two code words |i and |j can be written as $i|E_a|j = C_{aij}$.

5. QUANTUM COMPUTATIONAL JOURNEY

Quantum computing Research was first initiated by Planck, Bohr, Heisenberg, Einstein, and Schrödinger in 1923, but some argument and conflicts were arising continuously about the encapsulation of Quantum mechanics with nature. In 1981 Feynman proposed a framework for Quantum computation and for simulating the states of the quantum computer (Botsinis et al., 2019).

6. IMPACT OF QUANTUM COMPUTING

Quantum computing possesses weird potential which has never been witnessed by any technology in the past. The main technological advantages of quantum computing are listed in below sub-sections.

6.1 Cryptography

Presently, AES is an example of a symmetric cipher that allows the messages to be encrypted by the private keys of 128-bit, 196-bit and 256-bit. 128-bit AES is deployed mostly among the all while securing the Internet of things. A brute force attack is mostly used to break the keys by using all possible combinations of keys. By using Grover's algorithm this process can be sped up dramatically but the key size of AES also needs to be doubled. To prevent from 128-bit attack, in quantum computer, the key size of AES is doubled (Cheng et al., 2017).

Quantum computers have the potential to exert a positive as well as a negative impact on the field of cryptography. The remarkable velocity exhibited by quantum computers will yield seamless management and evaluation of information but the negative impact is that the release of Shor's algorithm, the prime

Table 3. Chronological development of quantum computing

Year	Achievement
1980	Russian mathematician Yuri Manin in proposed a quantum automata and Paul Benioff designed a quantum Hamiltonian to simulate Turing machine (Feynman et al., 1981).
1981	Feynman proposed a framework to simulate the computers with quantum physics (Feynman et al., 1981).
1985	Deutch realised quantum theory of the Turing machine principle and universal quantum computer (Deutsch et al., 1985).
1992	Deutch-Jozsa algorithm to compete with classical computers (Labs et al., 1994) .
1994	Simon, offline cursive word recognition (Simon et al., 1994).
1994	Shor algorithm for discrete logarithm and factoring in Quantum computation (Bhatia et al., 2020).
1997	Grover algorithm for searching problems in quantum computing (Mandviwalla et al., 2018).
1998	A. Seth developed a quantum algorithm for increasing speed exponentially to find eigenvalues and eigenvectors (Abrams et al., 1998). Jones designed quantum algorithm for counting on NMR ensemble quantum computer (Han et al., 2002).
2002	Han encapsulated a quantum based algorithm for optimization problems (Beach et al., 2004).
2003	Beach *et al.* proposed image processing using quantum computation (Beach et al., 2004).
2009	Lau and Chang implemented quantum based algorithm for unit commitment (Lau et al., 2009).
2018	Botsinis come with a quantum based search algorithm for wireless communication (Botsinis et al., 2019).
2019	Egger come with a quantum computer based credit risk analysis algorithm and Google claimed quantum supremacy (Egger et al., 2019).
2020	Eskandarpour come with a quantum algorithm to enhance Grid Security(Eskandarpour et al., 2020).
2021	Two Chinese teams claimed achievement of quantum supremacy (Egger et al., 2019).

factoring problems that were very difficult to solve now can be easily solvable using the Shor's algorithm. That gives vulnerability to brute attacks, thus releasing a negative impact (Njorbuenwu et al., 2019).

6.2 Fast Searching and Optimization

All organizations, government, and companies use their computers for searching and optimization problems; to schedule their work, for searching from large databases, to design chips & aero plane, any organization may need quantum computers to speed up such tasks (De Wolf et al., 2017). The design of quantum computers may revolutionarized the field of computer optimisation.

6.3 Database Management

With the advancement of technology, more and more information is generated every day. Organizations are striving hard to find solutions to problems about the management, analysis, and processing of information. The features of quantum computers such as entanglement, and superposition can tackle and resolve these problems easily and computers based on such principles will be able to perform by minimizing such warranted issues.

6.4 Big Data Analysis

Big data is driven by characteristics such as veracity, velocity, and variety. The algorithms based on artificial intelligence techniques such as neural networks, natural language processing, support vector machines and machine learning can exponentially speed up quantum computing machines for better performance. Machine learning in quantum computing is still in infancy and imposes challenges but treasures progress may be very beneficial for big data analysis especially in the field of healthcare (Khang & Medicine, 2023) and other social sectors (Shaikh et al., 2016).

6.5 Information Hiding

The procedure of integrating concealed data into a multimedia platform such as image, music, video, or text is referred to as information hiding and very important concept for the present-day information processing industry. The preservation of confidentiality and the maintenance of data integrity are the primary objectives of information security hiding. Quantum steganography and quantum watermarking are two methodologies employed to conceal information quantum computers can be of great use. In comparison to their classical computer equivalents, quantum computers are more efficient and secure. It can be asserted that quantum computers will significantly advance technology in the fields of computation, communication, image processing, and other areas by applying the technique of information hiding (Abd El-Latif et al., 2018).

7. QUANTUM COMPUTING ALGORITHMS

The quantum algorithms are significantly different and more efficient from their classical counterparts. The working of algorithms is characterized by steps such as starting the quantum system with the quantum bits in any classical state, superposition of states, applications of various unitary operations on superposition states, and final measurements of quantum bits (Yanofsky et al., 2008). In this section, we will discuss about few common quantum computing algorithms and its significance.

7.1 Deutsch-Jozsa's Algorithm

The initial quantum algorithm was devised in 1992 by David Deutsch and Richard Jozsa, to provide a competitive edge over classical computers. For instance, if two coins are thrown simultaneously their final outcomes can be either head or tail, but cannot remain unknown before trial. In such experiments, the observer is unable to guess whether coins exhibit a consistent outcome of both heads and both tails (referred to as a constant function), or if one coin displays heads while the other shows tails (referred to as a balanced function). In the field of quantum physics, it is conceivable that the characterization of the unknown function can potentially be ascertained with a sole measurement. This is made feasible by quantum parallelism, which is absent in classical systems, and a particular state preparation (Tesch et al., 2004).

$f: \{0,1\} \rightarrow \{0,1\}$

Figure 5. The implementation of this algorithm using IBM's Qiskit for values n=2bits

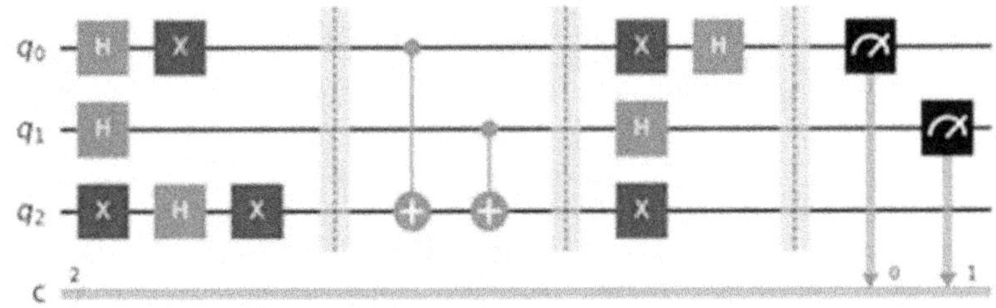

This algorithm guarantees output as an equal number of 0's and 1's; called a constant function and if half of the outputs are 0's another half will be 1's is called balanced function. The basic principle of this algorithm is that after taking the input, the Hadamard operation is applied on all configurations prepared by the oracle function and then grouping is done to achieve the result in either '0' and '1' as depicted in figure 5.

7.2 Shor's Algorithm

Shor's Algorithm was the first successful quantum algorithm developed by Peter Shor in 1994. The algorithm seeks to effectively determine the prime factorization of extremely large numbers by the utilization of quantum computers, sometimes also referred to as Shor's factoring algorithm. This algorithm can efficiently ascertain the prime factors of extremely big numbers with a polynomial time complexity which otherwise posed a significant challenge for classical computers.

Let us consider the scenario in which an integer x, selected at random, satisfies the condition of being less than 'N' and has no non-trivial shared factors with 'N'. The factors of such a number can be determined by Euclid's Algorithm by calculating GCD (x. N). If GCD ^{1} 1, that there is a factor of N, If GCD = 1 than 'x' is co-prime of 'N'. Such factorization involves finding the values of the function: $f_{x,N}(a)= x^a Mod\ N$.

By determining the smallest 'r', such that $fx,\ N(r)= x^r Mod\ N=1$, we find the period of the function (Bhatia et al., 2020). According to a theorem in number theory, any co-prime x<N, the output of $f_{x,\ N}$ will be '1' for some r<N. When it hits '1', the sequence repeats. If $fx,\ N(r)=1$ then, $fx,\ N(r+1)=fx,\ N(1)$. In general, $fx,\ N(r+s)=fx,\ N(s)$.

7.3 Grover's Algorithm

Grover's algorithm was the first quantum search algorithm developed by Grover in 1996. It searches a target element in an unordered list having 'N' elements and can find the targeted element in O(ÖN) operations while a classical computer takes O(N) operations, giving a sped up searching operations over its classical counterpart.

In this algorithm, we find the index of the target element amongst the list of $N = 2^n$ elements, where 'N' is the list of elements from which have to find the elements and 'n' is the number of qubits (Mandviwalla et al., 2018). Grover's algorithm operates the search in the follow steps, also depicted in figure 6:

Figure 6. Grover's iteration about √N times

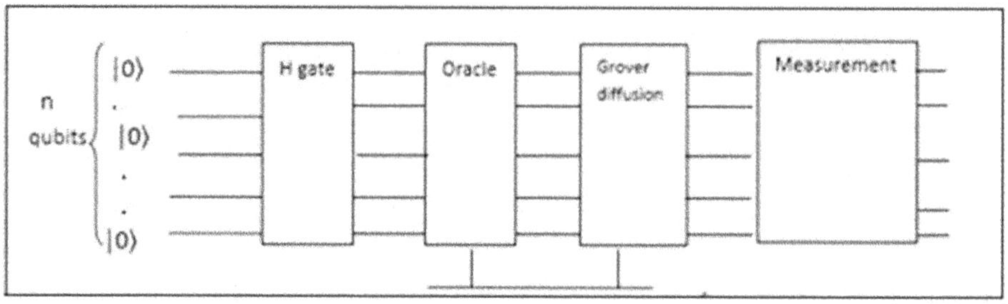

a) Get $|0\rangle \otimes n$ where Ä is the tensor product, and |0ñÄⁿ mean $|0\rangle \otimes |0\rangle \otimes |0\rangle \otimes \ldots |0\rangle$ up to 'n' times.
b) Apply Hadamard for superposition, H Äⁿ.
c) Apply the Oracle function O to mark the searching element by negating its sign $O|x$=-$|x$, where 'x' is the searching element.
d) Grover diffusion operator 'D' is applied to intensify the probability amplitude of the searched element.
e) Repeat steps c) and d) for $\sqrt{N}$ times.
f) Perform measurements.

7.4 Simon's Periodicity Algorithm

Simon's Algorithm, proposed by Daniel Simon in 1994, is used to find the hidden patterns in any set of functions. Shor's algorithm and Simon's algorithm both are abelian hidden sub-group problems. This algorithm contains procedures of both classical and quantum. Suppose a function f: $\{0,1\}^n \rightarrow \{0,1\}^n$ is any black box, and surely contains a secret string 's' *i.e.* for all strings $x.y \in \{0,1\}^n$; where f(x)= f(y) *iff* $x = y \otimes s$.

The objective here is to find 's' and hence the values of 'f' will replicate in some kind of pattern which will be determined by 's' while in classical algorithms determining such a 2-to-1 function is considered the worst-case scenario because it requires $2^{n-1}+1$ while Simon's algorithm requires only 'n' evaluations as shown in table 4 (Siddiqui et al., 2014).

Table 4. Quantum algorithms and protocols

Algorithm	**Description**
Teleportation	Transportation of Quantum states from one Quantum system to another by using entanglement. Almost all Quantum systems are based on teleportation (Harwood et al., 2021)
Deutsch–Jozsa algorithm	Determines black box function, whether it is balanced or constant (Labs et al., 1994)
Shor's algorithm	Solves the prime factoring problems very efficiently (Bhatia et al., 2020)
Grover's algorithm	Optimally searches in an unsorted database index (Mandviwalla et al., 2018)
Simon's algorithm	Finds hidden pattern (Simon et al., 1994)
Quantum KD	Exchanges secret key in both the classical and quantum systems.

8. TOOLS FOR QUANTUM COMPUTING

Quantum tools are the lightweight platforms for creating or embedding real time, responsible, and modular applications in quantum computing(Basak et al., 2022). Presently, the quantum computing tools are either provided as FOSS or have been offered as freely subscribed cloud services application programming interface (API) and graphical the user interface (GUI) where user writes programs on a classical computer and runs on a real quantum simulator. Some of the common simulators offered by various organizations are depicted in table 5.

Table 5. Various tools in quantum computing

Tool	Description
Qiskit	It is a Python library developed by IBM. It allows the matrices created by a user to run to IBM's Qiskit simulator or on IBM's Quantum cloud services (Miceli et al., 2018)
MS Quantum Development Kit (Q#)	It is an open-source SDK, having Q# compiler and a range of resource estimators & Quantum simulators for the development of Quantum systems (Mykhailova et al., 2022)
Google Cirq	Google Cirq is a Python based platform that enables creating, invoking & editing noise intermediate-scale quantum systems(Oumarou et al., 2020)
PyQuil	Quantum computing Forest by Rigette, allows the developers to simulator their quantum circuits on Quil (quantum instruction language) (Carrascal et al., 2021)
Project	An open-source SDK featuring quantum hardware and a C++ QC that backs up a python based platform(Jones et al., 2019)
QSim	QSim is software derived from QEMU, which is an open source emulator capable of emulating entire systems. QEMU utilizes a dynamic binary translator to facilitate the support of diverse instruction sets(Wang et al., 2014)

9. GLOBAL SCENARIO IN QUANTUM COMPUTING

Recently, quantum computing has been a hot topic in both industry and academia across the globe. The contemporary scenario of quantum computing across the world and within India is presented below

Recently, quantum computing has been a hot topic in both industry and academia across the globe. The contemporary scenario of quantum computing across the world and within India is presented below:

9.1 Quantum Computing in World

With current global investments reaching $36 billion (estimated), the global research and innovation fostering in quantum science and technology is steadily growing. By 2027, it is anticipated that the market for quantum technology will have grown to $42.4 billion. The innovation ecosystems of Asia, North America, Europe, and Australia are quite dissimilar. As a result, different national governments take different routes when it comes to quantum computing. The United States implemented the National Quantum Initiative Act, 2019 to facilitate the advancement of various quantum technologies. The establishment of organizations such as the Quantum Economic Development Consortium (QED-C) has been undertaken to foster the development of the quantum industry and facilitating the establishment of a quantum supply chain.

Figure 7. Country-wise government funding for quantum computing

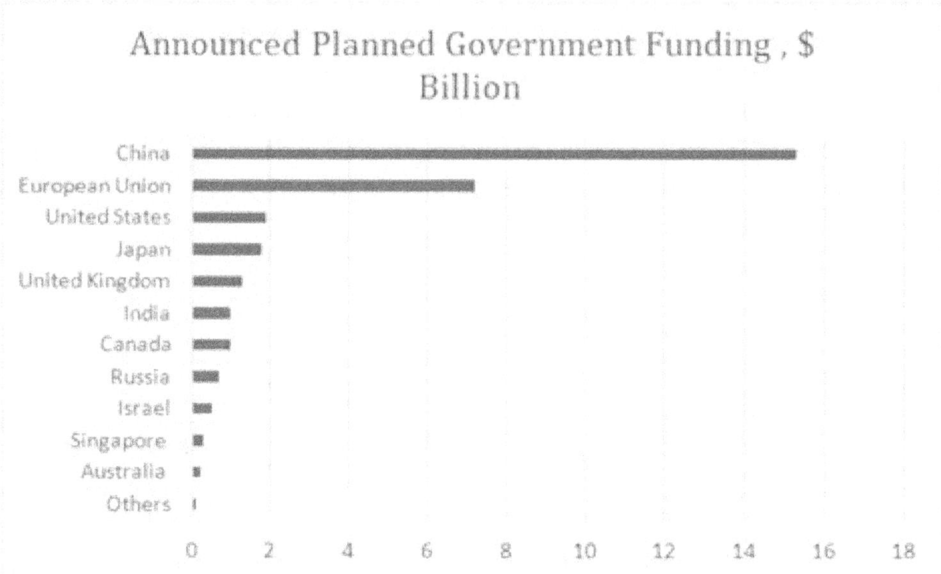

The national and provincial plans of Canada and its provinces are continually being improved. National and regional initiatives have been started throughout Europe to drive the development of quantum computing technologies. With a combined amount of public support of over $7 billion, the UK, Netherlands, Germany, and France have launched national quantum projects and policies. The EU Quantum Flagship is a separate $1.1 billion research and innovation project launched by the European Commission to advance and commercialize quantum technology within the European Union.

China's five-year plan, released in 2016, focused on preserving national technological dominance. China has targeted investments in quantum technology over $1 billion during the last ten years, with an additional $150 million reserved for start-up funds. Longer-term plans call for public investments of up to $15 billion which reportedly seems fairly substantial, although no exact number is given to the public.

Beginning in 2019, Japan and South Korea formalized their quantum policies, making attempts to develop quantum computers and access them via the cloud on third-party hardware. India declared early in 2022 that it would invest more than $1 billion over the following five years to aid in the development of quantum technologies, including a project to create a tiny quantum computer prototype. Figure7, depicts the country-wise details of quantum funding across the globe (Mohr et al., 2022)

9.2 Quantum Computing in India

To speed up advanced training and research in quantum computing; the technology giant IBM announced a partnership with eleven respected academic institutions in May 2021, including the Indian Institute of Science in Bangalore. These institutions are highly engaged in terms of their intellectual contributions to the field of quantum computing. The institutions of IISER Thiruvananthapuram, ISI Kolkata, and IIT Madras have announced their plans to offer Quantum Computing Lab courses for its advanced undergraduate and graduate students. As a culminated effort of fund infusion, a large number of start-ups have been emerged even in tier-2 and tier-3 cities, *viz.* BosonQ in Bhilai and QNu Labs in Bangalore.

A software business called BosonQ Psi develops simulations using the possibilities of quantum computing. The company creates simulation capabilities using a hybrid infrastructure of traditional high-performance computers (HPC) and quantum computers to emphasize near-term value additions to its customers. Additionally, BosonQ is researching brand-new simulation methods for upcoming fault-tolerant quantum computers. The Quantum Computing Applications Lab (QCAL) is a prominent initiative developed by the Ministry of Electronics and Information Technology (MeitY) in collaboration with AWS. By giving researchers and developers access to quantum computers, resources, and tools, QCAL hopes to hasten the uptake of the technology in India.

The Indian government has approved a $730 million plan to transform the country into a quantum communications and computing giant by the year 2031. The objectives of the National Quantum Mission (NQM) are to develop intermediate scale quantum computers including 50-1000 physical qubits. While accomplishing this would be impressive, it would not be a quantum leap. IBM's Osprey quantum computer has 433 qubits, and by 2025, Big Blue intends to have a system with over 4,000 qubits. The objective of the mission is to position India as a prominent player in the advancement of Quantum Technologies & Applications (QTA), even though the goals of the National Quantum Mission (NQM) do not prioritize India's attainment of the highest global ranking in terms of quantum processing capability.

QSIM: With the assistance of MeitY, IISC Bangalore, IIT Roorkee, and C-DAC are jointly working on the QSim project. This native toolset will be a crucial educational and research tool for quantum computing students and researchers. Researchers and students can generate and debug Quantum Code, which is necessary for creating Quantum Algorithms, using QSim. Even required controls and observations disrupt quantum systems because of their extreme sensitivity to environmental disturbances (Khang & Muthmainnah et al., 2023).

10. APPLICATIONS OF QUANTUM COMPUTING

The obvious reason for the popularity of quantum computing is the technological promises enunciated by it. The major applications of quantum computing, expected to drive the future of human societies, are:-

10.1 Intelligent Transportation Systems

Throughout the recent decades, Intelligent Transportation Systems (ITS) have remained a main driving force of the industrial revolution and emerged as a viable approach for enhancing the operational capabilities of road transportation. The objective of Intelligent Transportation Systems (ITS) is to enhance road safety, improve driving comfort, and disseminate up-to-date traffic information considering the observation that the coherent superposition of quantum states enables parallel computation and exponential growth, the entanglement of quantum systems and the 'no cloning theorem' guarantee the absolute security of data during transmission and processing. Additionally, the quantum tunneling effect increases the likelihood of escaping local extremes and obtaining the global optimal solution (Wang et al., 2022).

10.2 Power Grid

As the power system is progressively dependent on variable renewable generation and intermittent loads, the development of forecasts for the condition of the system necessitates varied analytical capabilities. Another viable example is the occurrence of several outages in reaction to the increasing frequency and intensity of extreme weather events and natural disasters, which can be directly attributed to climate change. Hence, the grid of the future necessitates the development of novel models, algorithms, and computational tools to cater to its requirements. Quantum computing is a transformative technology that has the potential to effectively tackle numerous difficulties, while simultaneously establishing a computational framework for a continuously advancing grid system (Khang & Kali et al., 2023).

10. Three Consumer Applications

The ongoing growth of computing can be related to the demands arising from several industries such as commerce, banking, security, and logistics. The current body of research is mostly centered on the development of novel computing capabilities through the utilization of revolutionary methodologies. Quantum computing is a paradigm of computation that harnesses the principles of quantum mechanics to perform computational tasks. The aforementioned approach exhibits significant potential in enhancing secure computation, trusted data storage, and the efficiency of applications (Humble et al., 2018).

10.4 Quantum Machine Learning

Machine Learning (ML) problems often include analyzing and categorization of several vectors in multidimensional domains. Within the domain of machine learning (ML), which presently comprises a range of methodologies relying upon complicated and massive datasets, there exists an increasing urge to employ these techniques in the context of "big data". The domain of Quantum Machine Learning (QML) is now undergoing significant growth due to its integration of quantum computing with machine learning. The basic concept of Quantum Machine Learning (QML) is to exploit the inherent advantages offered by quantum computing, including superposition, entanglement, and quantum parallelism to augment the efficacy of traditional machine learning techniques (Ramezani et al., 2020).

11. CONCLUSION

In comparison to classical techniques, quantum computing promises tremendous speedups in the resolution of many issues. However, running a quantum circuit involves numerous intricate pre and post processing steps that call for a high level of technical and mathematical expertise. Because of the fact that the classical application must complete all of these tasks to interact with the quantum circuit, the integration of classical applications with quantum circuits is a difficult endeavor. Although quantum computers have a huge advantage, their full potential is yet unknown. There are several quantum algorithms that outperform their classical counterparts giving rise to many possibilities simultaneously quantum simulations will continue to draw academics interested in quantum calculations for a longer period .

A significant path for future study is the development of novel and beneficial applications for existing quantum algorithms. For complex machine learning algorithms, researchers have incorporated quantum algorithms as subroutines. The proposed algorithms increase speed theoretically but in practice they might be expensive to run. The physical realization of a quantum computer is a challenge for quantum algorithms. In a nutshell, the present ontological study concluded that to unearth the true potential of quantum computing, it is immensely required to nurture an ecosystem with more talent in this regime and infuse more funds for academics and research.

REFERENCES

Abd El-Latif, A. A., Abd-El-Atty, B., Hossain, M. S., Rahman, M. A., Alamri, A., & Gupta, B. B. (2018, March). Efficient Quantum Information Hiding for Remote Medical Image Sharing. *IEEE Access : Practical Innovations, Open Solutions*, *6*, 21075–21083. doi:10.1109/ACCESS.2018.2820603

Abdi, H. (2007). *The Eigen-Decomposition: Eigenvalues and Eigenvectors*. UTD. https://www.utd.edu/

Abrams, D. S., & Lloyd, S. (1998). *A quantum algorithm providing exponential speed increase for finding eigenvalues and eigenvectors*. APS. https://journals.aps.org/prl/abstract/10.1103/PhysRevLett.83.5162

Azad, U., Behera, B. K., Ahmed, E. A., Panigrahi, P. K., & Farouk, A. (2022). Solving Vehicle Routing Problem Using Quantum Approximate Optimization Algorithm. *IEEE Transactions on Intelligent Transportation Systems*. doi:10.1109/TITS.2022.3172241

Basak, U. P., Faruk, J. H., Nazim, M., Masum, M., Hossain, M. S., Uddin, G., Barzanjeh, S., Sheikh, I. A., & Rahman, A. (2022). *Evolution of Quantum Computing: A Systematic Survey on the Use of Quantum Computing Tools*. IEEE. https://ieeexplore.ieee.org/abstract/document/9842601/

Beach, G., Lomont, C., & Cohen, C. (2003). Quantum image processing (QuIP). Proc. - Appl. Imag. Pattern Recognit. IEEE. doi:10.1109/AIPR.2003.1284246

Bhat, H. A., Khanday, F. A., Kaushik, B. K., Bashir, F., & Shah, K. A. (2022). Quantum Computing: Fundamentals, Implementations and Applications. *IEEE Open Journal of Nanotechnology*, *3*, 61–77. doi:10.1109/OJNANO.2022.3178545

Bhatia, V., & Ramkumar, K. R. (2020). An Efficient Quantum Computing technique for cracking RSA using Shor's Algorithm. *2020 IEEE 5th International Conference on Computing Communication and Automation, ICCCA 2020*. Institute of Electrical and Electronics Engineers Inc. 10.1109/ICCCA49541.2020.9250806

Botsinis, P., Alanis, D., Babar, Z., Nguyen, H. V., Chandra, D., Ng, S. X., & Hanzo, L. (2019, April). Quantum Search Algorithms for Wireless Communications. *IEEE Communications Surveys and Tutorials*, *21*(2), 1209–1242. doi:10.1109/COMST.2018.2882385

Carrascal, G., del Barrio, A. A., & Botella, G. (2021). First experiences of teaching quantum computing. *The Journal of Supercomputing*, *77*(3), 2770–2799. doi:10.100711227-020-03376-x

Cheng, C., Lu, R., Petzoldt, A., & Takagi, T. (2017, February). Securing the Internet of Things in a Quantum World. *IEEE Communications Magazine*, *55*(2), 116–120. doi:10.1109/MCOM.2017.1600522CM

De Wolf, R. (2017). *The Potential Impact of Quantum Computers on Society*. arXiv. https://arxiv.org/abs/1712.05380

Deutsch, D. (1985). *Quantum-Theory, The Church Turing Principle and the Universal Quantum Computer.* Royal Society Publishing. https://royalsocietypublishing.org/doi/abs/10.1098/rspa.1985.0070

EggerD. J.GutiérrezR. G.MestreJ. C.WoernerS. (2019). Credit Risk Analysis using Quantum Computers. arXiv. https://arxiv.org/abs/1907.03044

Eskandarpour, R., Gokhale, P., Khodaei, A., Chong, F. T., Passo, A., & Bahramirad, S. (2020, September). Quantum Computing for Enhancing Grid Security. *IEEE Transactions on Power Systems*, *35*(5), 4135–4137. doi:10.1109/TPWRS.2020.3004073

Eswaran, U. (2023). Role of Quantum Computing in The Era of Artificial Intelligence," Applications and Principles of Quantum Computing. IGI Global Press. https://doi.org/ doi:10.4018/979-8-3693-1168-4-ch003

Fernandez-Carames, T. M., & Fraga-Lamas, P. (2020). Towards Post-Quantum Blockchain: A Review on Blockchain Cryptography Resistant to Quantum Computing Attacks. *IEEE Access : Practical Innovations, Open Solutions*, *8*, 21091–21116. doi:10.1109/ACCESS.2020.2968985

Gill, S. S., Kumar, A., Singh, H., Singh, M., Kaur, K., Usman, M., & Buyya, R. (2022). Quantum computing: A taxonomy, systematic review and future directions. *Software, Practice & Experience*, *52*(1), 66–114. doi:10.1002pe.3039

Glendinning, I. (2005). European Centre For Parallel Computing At Vienna. The Bloch Sphere. http://andcorporation.com/bloch-sphere.pdf

Han, K. H., & Kim, J. H. (2002, December). Quantum-inspired evolutionary algorithm for a class of combinatorial optimization. *IEEE Transactions on Evolutionary Computation*, *6*(6), 580–593. doi:10.1109/TEVC.2002.804320

Harwood, S., Gambella, C., Trenev, D., Simonetto, A., Bernal Neira, D., & Greenberg, D. (2021, January). Formulating and Solving Routing Problems on Quantum Computers. *IEEE Transactions on Quantum Engineering*, *2*, 1–17. doi:10.1109/TQE.2021.3049230

Humble, T. (2018). Consumer applications of quantum computing. *IEEE Consumer Electronics Magazine*, 1–5. https://ieeexplore.ieee.org/abstract/document/8482518/

Jones, J. A., & Mosca, M. (1999). Approximate quantum counting on an NMR ensemble quantum computer. *Physical Review Letters*, *83*(5), 1050–1053. doi:10.1103/PhysRevLett.83.1050

Jones, T., Brown, A., Bush, I., & Benjamin, S. C. (2019, December). QuEST and High Performance Simulation of Quantum Computers. *Scientific Reports*, *9*(1), 10736. doi:10.103841598-019-47174-9 PMID:31341200

Khang, A. (2023). Applications and Principles of Quantum Computing. IGI Global Press. https://doi.org/ doi:10.4018/979-8-3693-1168-4

Khang, A. (2023). Enabling the Future of Manufacturing: Integration of Robotics and IoT to Smart Factory Infrastructure in Industry 4.0. AI-Based Technologies and Applications in the Era of the Metaverse. IGI Global Press. doi:10.4018/978-1-6684-8851-5.ch002

Khang, A. (2023). Applications and Principles of Quantum Computing. IGI Global Press. https://doi.org/ doi:10.4018/979-8-3693-1168-4

Khang, A. (2023). Quantum Mechanics Primer - Fundamentals and Quantum Computing," Applications and Principles of Quantum Computing. IGI Global Press. https://doi.org/ doi:10.4018/979-8-3693-1168-4-ch001

Khang, A. (2023). Introduction to Quantum Computing and Its Integration Applications. Applications and Principles of Quantum Computing. IGI Global Press. https://doi.org/ doi:10.4018/979-8-3693-1168-4-ch002

Khang, A. (2023). AI and IoT-Based Technologies for Precision Medicine. IGI Global Press. doi:10.4018/979-8-3693-0876-9

Khang, A., Chowdhury, S., & Sharma, S. (2022). *The Data-Driven Blockchain Ecosystem: Fundamentals, Applications, and Emerging Technologies* (1st ed.). CRC Press. doi:10.1201/9781003269281

Khang, A., & Muthmainnah, M. (2023). AI-Aided Teaching Model for the Education 5.0 Ecosystem. AI-Based Technologies and Applications in the Era of the Metaverse. (1 Ed.) Page (83-104). IGI Global Press. doi:10.4018/978-1-6684-8851-5.ch004

Khang, A., Shah, V., & Rani, S. (2023). AI-Based Technologies and Applications in the Era of the Metaverse. IGI Global Press. doi:10.4018/978-1-6684-8851-5

Khanh H. H., & Khang A. (2021). The Role of Artificial Intelligence in Blockchain Applications. *Reinventing Manufacturing and Business Processes through Artificial Intelligence*, *2*(20-40). CRC Press. doi:10.1201/9781003145011-2

Krauss, T., & McCollum, J. (2021). Solving the Network Shortest Path Problem on a Quantum Annealer. *IEEE Transactions on Quantum Engineering*, *1*, 1–12. doi:10.1109/TQE.2020.3021921

Labs, B., Avenue, M., & Nj, M. H. (1994). *Quantum Mechanics helps in searching for a needle in a haystack.*, https://journals.aps.org/prl/abstract/10.1103/PhysRevLett.79.325

Lau, T. W., Chung, C. Y., Wong, K. P., Chung, T. S., & Ho, S. L. (2009). Quantum-inspired evolutionary algorithm approach for unit commitment. *IEEE Transactions on Power Systems*, *24*(3), 1503–1512. doi:10.1109/TPWRS.2009.2021220

Le, L., & Nguyen, T. N. (2022). DQRA: Deep Quantum Routing Agent for Entanglement Routing in Quantum Networks. *IEEE Transactions on Quantum Engineering*, *3*, 1–12. doi:10.1109/TQE.2022.3148667

Mandviwalla, A., Ohshiro, K., & Ji, B. (2019). Implementing Grover's Algorithm on the IBM Quantum Computers. *IEEE International Conference on Big Data, Big Data 2018*. Institute of Electrical and Electronics Engineers Inc. 10.1109/BigData.2018.8622457

Miceli, R. & McGuigan, M. (2018). *Quantum Computation and Visualization of Hamiltonians using Discrete Quantum Mechanics and IBM QISKit*. NYSDS. . doi:10.1109/NYSDS.2018.8538959

Mohr, N., Masiowski, M., Zesko, M., & Soller, H. (2022). *Quantum Technology Monitor*. McKinsey. https://pubs.acs.org/doi/abs/10.1021/ct1007235

MykhailovaM. (2022). Developing Programming Assignments for Teaching Quantum Computing and Quantum Programming. arXiv. https://arxiv.org/abs/2211.16347 doi:10.1109/QCE53715.2022.00092

Njorbuenwu, M., Swar, B., & Zavarsky, P. (2019). A Survey on the Impacts of Quantum Computers on Information Security. *Proceedings - 2019 2nd International Conference on Data Intelligence and Security, ICDIS 2019*. Institute of Electrical and Electronics Engineers Inc. 10.1109/ICDIS.2019.00039

Oumarou, O., Paler, A., & Basmadjian, R. (2020). *QUANTIFY: A framework for resource analysis and design verification of quantum circuits*. arXiv. https://arxiv.org/abs/2007.10893 doi:10.1109/ISVLSI49217.2020.00032

Parhami, B. (2020). Reliability Inversion: A Cautionary Tale. *Computer*, *53*(6), 28–33. doi:10.1109/MC.2019.2958907

Perdomo-Ortiz, A., Benedetti, M., Realpe-Gómez, J., & Biswas, R. (2018). Opportunities and challenges for quantum-assisted machine learning in near-term quantum computers. *Quantum Science and Technology*, *3*(3), 1–13. doi:10.1088/2058-9565/aab859

Prabu, B., & Khang, A. (2023). Quantum Computing Smart City: Fostering Sustainability and Green Infrastructure with Smart City. Applications and Principles of Quantum Computing. IGI Global Press. https://doi.org/ doi:10.4018/979-8-3693-1168-4-ch009

Ramezani, S. B., Sommers, A., Manchukonda, H. K., Rahimi, S., & Amirlatifi, A. (2020). Machine Learning Algorithms in Quantum Computing: A Survey. *Proc. Int. Jt. Conf. Neural Networks, 2*. 10.1109/IJCNN48605.2020.9207714

Ratan, R., Shukla, M. K., & Oruç, A. Y. (2007). Quantum Switching Networks with Classical Routing. IEEE. https://ieeexplore.ieee.org/abstract/document/4298416/

Scherer, W. (2019). *Mathematics of Quantum Computing: An Introduction*. Springer International Publishing. doi:10.1007/978-3-030-12358-1

Shaikh, T. A., & Ali, R. (2016). Quantum Computing in Big Data Analytics: A Survey. *2016 IEEE International Conference on Computer and Information Technology (CIT)*. IEEE. 10.1109/CIT.2016.79

Shyam, R. (2023). Effects of Quantum Technology on Metaverse. AI-Based Technologies and Applications in the Era of the Metaverse. IGI Global Press. doi:10.4018/978-1-6684-8851-5.ch009

Siddiqui, S., Islam, M. J., & Shehab, O. (2014). *Five Quantum Algorithms Using Quipper*. arXiv. https://arxiv.org/abs/1406.4481

Solmeyer N., Dixon, R. & Balu, R. (2017). *Quantum routing games*. IOP Science. . doi:10.1088/1751-8121/aae31f

Teja, V., Banerjee, P., Sharma, N. N., & Mittal, R. K. (2007). Quantum cryptography: State-of-art,challenges and future perspectives. *2007 7th IEEE Int. Conf. Nanotechnol. - IEEE-NANO*. IEEE. 10.1109/NANO.2007.4601420

Tesch, C. M., & De Vivie-Riedle, R. (2004, December). Vibrational molecular quantum computing: Basis set independence and theoretical realization of the Deutsch-Jozsa algorithm. *The Journal of Chemical Physics*, *121*(24), 12158–12168. doi:10.1063/1.1818131 PMID:15606234

Thapliyal, H., & Ranganathan, N. (2009). Design of efficient reversible binary subtractors based on a new reversible gate. *Proc. 2009 IEEE Comput. Soc. Annu. Symp.* IEEE. 10.1109/ISVLSI.2009.49

Wang, J. (2014). Manifold: A parallel simulation framework for multicore systems. *ISPASS 2014 - IEEE Int. Symp. Perform. Anal. Syst. Softw*. IEEE. 10.1109/ISPASS.2014.6844466

Wang, S., Pei, Z., Wang, C., & Wu, J. (2022). Shaping the Future of the Application of Quantum Computing in Intelligent Transportation System. *Intell. Converg. Networks*, *2*(4), 259–276. doi:10.23919/ICN.2021.0019

Wang, Y., Yu, X., Zhao, Y., Nag, A., & Zhang, J. (2022, December). Pre-established entanglement distribution algorithm in quantum networks. *Journal of Optical Communications and Networking*, *14*(12), 1020–1033. doi:10.1364/JOCN.465432

Yang, Z., Zolanvari, M., & Jain, R. (2023). A Survey of Important Issues in Quantum Computing and Communications. *IEEE Communications Surveys and Tutorials*, *25*(2), 1059–1094. doi:10.1109/COMST.2023.3254481

Yanofsky, N. S., & Mannucci, M. A. (2008). *Quantum computing for computer scientists*. Cambridge University Press. https://books.google.com.vn/books?hl=en&lr=&id=U1chAwAAQBAJ&redir_esc=y doi:10.1017/CBO9780511813887

Ying, M. (2010, February). Quantum computation, quantum theory and AI. *Artificial Intelligence*, *174*(2), 162–176. doi:10.1016/j.artint.2009.11.009

Zeilinger, A. (1999). Experiment and the foundations of quantum physics. *APS*. https://journals.aps.org/rmp/abstract/10.1103/RevModPhys.71.S288

Chapter 13
Quantum Computing Smart Grids:
A Review on Internet of Things-Enabled Technologies in Smart Grids

Rashmi Sharma
Indus University, Ahmedabad, India

Zalak Patel
Indus University, Ahmedabad, India

Miloni Ganatra
Indus University, Ahmadabad, India

ABSTRACT

Throughout the past few decades, the electricity system's condition has evolved. The power grid has recently experienced a number of difficulties and problems. With power outages, grid failures, and a lack of smart technology on the one hand, and rising security risks on the other, there is a daily rise in the demand for electrical energy. These problems cannot be solved by the current electricity grid. Because of its key characteristics, the internet of things (IoT) has expanded quickly in a very short period of time. We can improve the conventional grid's efficiency, capacity, dependability, sustainability, scalability, and stability by integrating IoT into the electrical grid. The myriad issues that are present in smart grids are solved by using the IoT. Security concerns have been recognized as one of the main drawbacks with IoT enabled smart grid (SG) systems, according to the most recent research. The numerous security applications and elements of integrating the IoT with smart grids are reviewed in this chapter, along with various frameworks in the era of quantum computing.

DOI: 10.4018/979-8-3693-1168-4.ch013

1. INTRODUCTION

The majority of power plants in the world today produce electricity primarily using fossil fuels. Yet, these fossil fuel supplies are being exhausted and are anticipated to exhaust themselves soon. Alternative approaches must be used to guarantee the availability of the electrical energy required for human survival. The use of a smart grid is one such remedy, which overcomes the drawbacks of conventional grids by easing not only the transfer of electricity but also tracking and managing energy consumption. Power grids have often evolved over the course of electricity's development. Despite the electrical grid's robust design, new technologies have appeared. Modern communication and information technologies (ICTs) have the potential to significantly increase the electrical grid's dependability, stability, and efficiency while also meeting a variety of customer needs (Khang & Quantum, 2023).

A noteworthy trend is the growth of domestic energy production as well as sources of clean energy like wind and solar energy. These energy sources, as opposed to the conventional power grid, require less centralized infrastructure and, over time, can produce more predictable outputs. The reliability of the electric power grid is directly impacted by these shifting requirements, necessitating the employment of sophisticated ICTs for intensive monitoring and control. The current power infrastructure has evolved into what is commonly known as the smart grid, a digitally-based and interconnected system. Unlike earlier centralized power systems, the smart grid enables the bidirectional exchange of energy between service providers and users. This transformation has been driven by a confluence of factors, including increased power demands, the proliferation of smart devices, the imperative for stronger safety measures, and the ever-growing need for enhanced dependability and efficiency (Parvin et al., 2022).

Moreover, a promising development in this context is the integration of quantum computing into the smart grid infrastructure. Quantum computing, with its unparalleled computational capabilities, offers the potential to revolutionize the way we manage, optimize, and secure the smart grid. It can facilitate advanced simulations and real-time decision-making, enabling the grid to adapt swiftly to dynamic energy supply and demand patterns. Additionally, quantum computing can play a crucial role in enhancing the security of the smart grid by providing quantum-resistant encryption techniques, thereby safeguarding critical data and communications. As we move towards a more decentralized and environmentally sustainable energy landscape, the fusion of quantum computing with the smart grid represents a cutting-edge solution that addresses the complexities of modern energy systems while ensuring reliability, resilience, and heightened security (Shyam & Khang et al., 2023).

The smart grid uses the Internet of Things, or IoT, to handle communication, automation, and monitoring concerns. It includes a variety of grid monitoring and control devices. Smart meters, actuators, and other IoT devices are key components of the smart grid architecture. The smart grid can avoid and minimize damage during natural catastrophes, minimize economic losses, and improve the dependability of power transmission by utilizing IoT technologies. Three key processes are commonly involved in IoT implementation in the power sector: digitalizing assets, gathering asset data, and creating computational algorithms for control systems. Telecommunication infrastructures must assure quality of service (QoS), be compliant with industrial protocols, and adhere to strict security standards in order for the smart grid to function effectively.

Connecting previously disconnected items is the IoT's main goal. Healthcare (Khang & Rana et al., 2023), food production, transportation (Khang et al., 2023), cities (Khang & Rani et al., 2022) that are intelligent, renewable energy, and electric grids are just a few of the areas of human life that have undergone radical change. The Internet of Things (IoT) enables continuous communication and data sharing by equipping any gadgets and objects with sensors and connecting them to an IoT platform. IoT-enabled technologies create a strong network between people and machines in the setting of the smart grid employing a variety of sensing tools like RFID, or radio frequency infrared detectors, the Global Positioning System intelligent safety mechanisms, and laser scanning.

Power production, distribution, and transmission are all included in the IoT's coverage of the electrical infrastructure. Power generation facilities, distributed power stations, emission levels, utilization of energy, battery storage, transmission networks, substations, network of things control, service that is and devices management, smart meters, multi-network convergence, which refers electric vehicle battery life, environmental control, energy demand management, and more are all made easier by this. In order to transform energy from multiple sources and sustain economic growth, the power grid is essential. It is crucial to create a power infrastructure for the future that promotes sustainable development. A properly operating smart grid can aid in lowering carbon emissions, enhancing the efficiency of energy conversion, streamlining end-use procedures, and encouraging environmentally friendly mobility (Abir et al., 2021).

1.1 Motivation and Contribution

Technology is advancing at a rapid pace across numerous IoT businesses, ushering in a wave of possibilities for enhancing the seamless operation of intelligent electrical grid systems. This technological evolution is driven, in part, by the proliferation of smart devices, which are changing the way we interact with and manage electrical energy. As smart devices become more prevalent, the demand for electricity continues to grow. However, this expansion has also raised significant security concerns, especially as the power grid becomes increasingly interconnected and reliant on digital systems.

To address these concerns, innovative technologies are being deployed to manage electricity supply and demand while bolstering the security of the overall infrastructure. Quantum computing has emerged as a transformative force in this landscape. Quantum computing's immense processing power allows for complex optimization and real-time decision-making, making it an ideal candidate for managing the intricate dynamics of a modernized power grid. Furthermore, quantum computing plays a pivotal role in ensuring the security of the smart grid. Its ability to generate unbreakable encryption methods, known as post-quantum cryptography, fortifies the grid's defenses against potential cyber threats.

In this rapidly evolving landscape, IoT stands out as a potent technology for revitalizing antiquated power systems and transforming them into smarter, more adaptable, and highly efficient power grids. With the integration of quantum computing, the modernization of the power grid gains additional momentum, not only addressing the growing energy demands posed by IoT devices but also fortifying the grid's resilience and security against emerging challenges.

IoT has the ability to offer a comprehensive answer to the problems that traditional power systems are now experiencing. For numerous smart grid applications, its adoption is growing in popularity in both residential and commercial structures. Smart grids can operate more effectively at all phases of energy production, exchange, and distribution by including sensors and smart metering. The electrical industry's many issues are successfully solved by this combination. In this area, extensive study is now being done. The goal of this study is to give scholars a clear grasp of the IoT, intelligent grids, Internet-aided intelligent grid construction, models, different IoT and non-IoT applications, developments, and security concerns (Lewis et al., 2009). Table 18.1 displays the main contributions of this review.

Table 1. Review on latest technologies in on IoT for smart grids and its findings

Author(s)	Year	Findings	Remarks
Ghasempour et al. (2019)	2019	Explores IoT architecture, applications, and key technologies in smart grids. - Discusses challenges associated with IoT in smart grids.	Provides a comprehensive overview of IoT in smart grids.
Abir et al. (2021)	2021	Explores the uses and difficulties of IoT-enabled intelligent power networks.	Focuses on applications specific to smart energy grids.
Mocrii et al. (2018)	2018	IoT-based smart home aspects are reviewed, including system design, programs, communications, privacy, and security.	Primarily focuses on smart homes and related aspects.
Jia et al. (2019)	2019	Reviews enabling technologies and applications for IoT in smart buildings.	Emphasizes the development of IoT in smart buildings.
Reka et al. (2018)	2018	Provides a comprehensive review of IoT and smart grid, emphasizing the role of energy delivery.	Focuses on the future role of energy delivery.
Zeadally et al. (2020)	2020	The design concepts for harvesting energy in the IoT are discussed.	Emphasizes energy harvesting aspects in IoT.
Martín et al. (2020)	2020	Reviews IoT energy platforms aimed at end users.	Focuses on energy platforms from an end-user perspective.
Tao, J. et al. (2014)	2020	Examines the effects of IoT on power systems, assisted by advancing 5G.	Highlights the intersection of IoT and 5G in power systems.
Rose et al. (2014)	2014	Provides an overview and recommendations on the Internet of Things (IoT).	Offers general insights into IoT.
Da Silva et al. (2022)	2022	Focuses on future IoT experimentation and evaluation.	Emphasizes the need for experimentation and evaluation.

The IoT and smart grids are briefly discussed in this article, along with the methods, architectures, prototypes, and methods of communication used in IoT-enabled smart grid systems. An example of Internet of Things (IoT) uses in smart grids that solve security concerns and acceptance obstacles.

1.2 Article Organization

To fully cover the subject at hand, this paper employs a systematic format. The paper's the beginning, motivation, and contributions are covered in Section 1. A thorough analysis of the body of research that has already been done in the topic is provided in Section 2's literature scan. We now go on to Section 3, where many facets of Internet and smart grid technology are briefly covered, including the fundamental ideas, tools, and guidelines. Section 4 examines several energy management architectures and prototypes that are IoT-enabled in order to offer insights into realistic deployment strategies. Section 5 explores the variety of IoT potential non-IoT technologies used in smart grid infrastructure, addressing their functionalities, advantages, and potential drawbacks. The IoT's applications in smart grids, presenting real-world use examples that show increased efficacy and efficiency.

2. LITERATURE SURVEY ON IOT BASED SYSTEMS IN SMART GRID

The review focuses on the concept of the future power internet in the context of consular energy service and required management in smart grids. It discusses the current practices and challenges associated with the adaptation of smart intelligent grids and provides insights into the future directions of this field. The authors highlight the importance of integrating advanced communication technologies, intelligent control systems, and renewable energy resources to develop an efficient and sustainable energy internet. They discuss various aspects related to utility energy services, such as energy flow, control, and intelligent consumption. The paper also addresses the challenges faced in implementing the future energy internet, including technical, economic, and regulatory aspects. To guarantee the reliability as well as the safety of the energy internet infrastructure, it emphasizes the necessity for standardized guidelines, interoperability, and cybersecurity measures. The authors also outline potential future courses and approaches for using the full capabilities of the energy internet, such as sophisticated demand response strategies, energy storage technologies, and grid integration of electric vehicles (Andreadou et al., 2016).

The integration of Internet of Things, also known as IoT, in smart grids is the main topic of the study. It offers a study of the IoT integration into smart grid systems' architecture, applications, services, essential technologies, and obstacles. The author begins by outlining the many parts and levels that make up the framework of IoT enabled smart grids. Sensors, such as networks for communication, processing of information, and decision-making systems are all included in this. In addition, the article investigates how cloud and edge computing enable IoT-based smart grids. The following section of the paper looks at the several smart grid applications made possible by IoT. Smart measurement, response to demand, distribution automation, integration of renewable energy sources, and grid monitoring are some of these uses. To demonstrate how these applications are used in practise, the author offers examples and case studies. Furthermore, the paper discusses the services that can be provided by IoT in smart grids. These services include real-time monitoring, predictive maintenance, fault detection, energy management, and optimization. The author highlights the benefits of these services in improving the efficiency, reliability, and sustainability of the grid. The key technologies enabling IoT in smart grids are also analyzed in the paper. This includes wireless sensor networks, communication protocols, data analytics, machine learning, and cybersecurity. The author discusses how these technologies contribute to the successful implementation of IoT-based smart grids. The paper addresses the challenges and potential issues associated with deploying IoT in smart grids. These difficulties include the requirement for standardization, interoperability, scalability, and data privacy and security. To guarantee the effective implementation of IoT in electrical power systems, the author emphasizes the significance of addressing these difficulties (Chen et al., 2010).

The paper highlights various uses of IoT for smart energy grids. These include smart metering, demand response, renewable energy integration, grid monitoring and control, energy efficiency optimization, and electric vehicle integration. The authors provide insights into how IoT technologies enable these applications and contribute to the efficient operation of the energy grid (Zhao et al., 2013) (Wang et al., 2012).

The review discusses the benefits of implementing IoT for smart energy grids. These benefits include improved energy management, increased grid reliability and resilience, enhanced renewable energy integration, optimized energy consumption, and cost savings for consumers and utilities. The paper emphasizes how IoT technologies enable real-time monitoring, data analytics, and decision-making capabilities for efficient grid operations. The paper addresses the challenges associated with implementing IoT in smart energy grids. These challenges include data uses and security concerns, hacking issues, scalability of IoT systems, regulatory and policy considerations, and the need for standardized protocols.

To enable the successful use of IoT technology in the energy sector, the authors stress the significance of addressing these issues. The incorporation of IoT and other modern technologies in the power sector is covered by the authors. These include artificial intelligence (AI), blockchain (Khang & Chowdhury et al., 2022), edge computing, and cloud computing. The paper highlights the potential synergies and benefits of combining these technologies with IoT in creating smart energy grids. The collection of papers analyzed various aspects of the Internet of Things (IoT) in the context of energy systems, with a particular focus on smart grids. (Billewicz et al., 2016) paper addressed IoT research in energy systems, highlighting the importance of cloud computing and data analytic platforms. Privacy and security concerns were also discussed, along with potential solutions like blockchain technology (Khanh & Khang et al., 2021).

In the (Bikmetov et al., 2015) paper, the importance of a thorough debate on IoT-enabled smart grid networks and their architectures was emphasized. It also provided a recap of previous studies and examined the benefits and uses of IoT in smart grids. The research conducted a survey of the literature, looking at the IoT-enabled smart grid systems' current architectures, uses, and prototypes. A wide range of sectors were covered in the fourth paper's investigation of IoT applications in companies and smart electrical systems, along with their consequences. The fifth paper established the idea of the electricity internet for utilities energy delivery and demand-side management, outlining major elements and difficulties as well as outlining the need for additional study in the field (Alsuwian et al., 2022). Taking into account recent advancements in 5G networks and edge computing, the paper (Ou et al., 2012) thoroughly analyzed environmentally friendly communications and computing approaches in smart grids.

The study (Mavroeidakos et al., 2020) examined recent technological developments in IoT-enabled energy systems, as well as security threats and suggested solutions for addressing them. The article (Aktar et al., 2022), which addressed applications, difficulties, and the expansion of security requirements for power grids, concluded by examining the connection among IoT and smart grids. In general, these papers provide insightful information about the possible uses, difficulties, and solutions for integrating IoT into utilities, particularly within the setting of smart grids.

3. IOT AND SMART GRID IN TODAY'S WORLD

The network of physical items, including machines, automobiles, appliances, and other things, that are equipped with software, hardware, and communication capabilities so they can gather and share data over the internet is known as the Internet of Things (IoT). The aim of IoT is to create a seamless integration between the physical and digital worlds, enabling smarter and more efficient processes, improving decision-making, and enhancing the overall quality of life for individuals and society. IoT characteristics is shown in Figure 1.

Alt-text: Figure 1 depicts the network of physical items, including machines, automobiles, appliances, and other things, that are equipped with software, hardware, and communication capabilities of IoT.

Figure 1. IoT characteristics

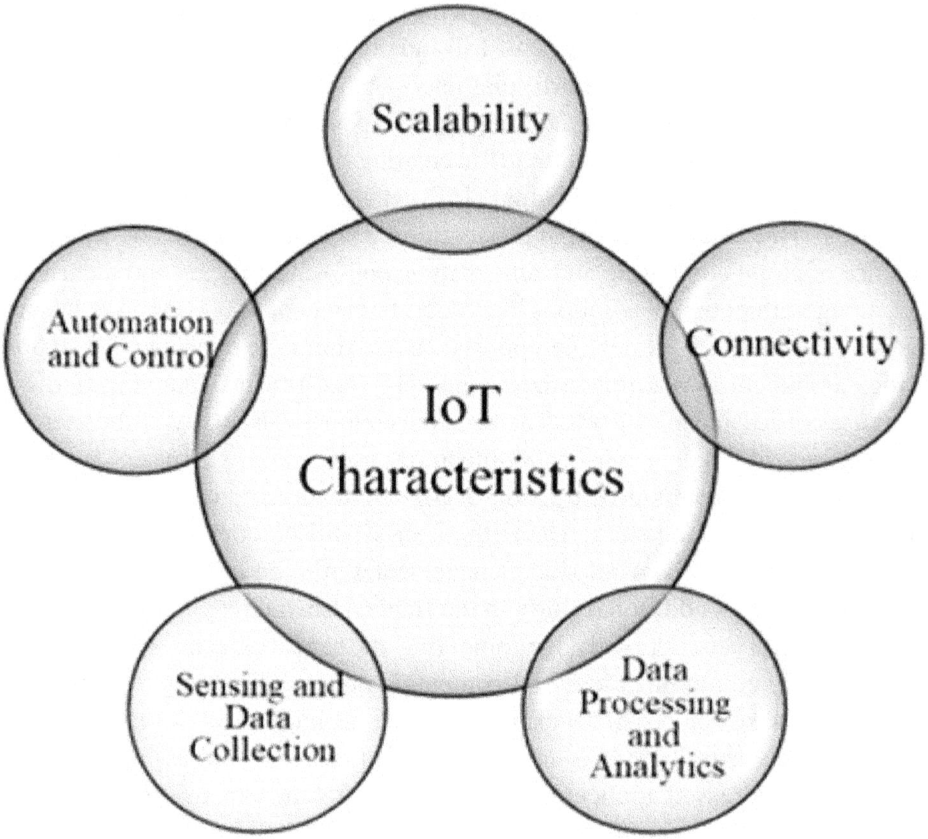

3.1 IoT Introduction

3.1.1 Characteristics of IoT

The concept of the "Internet of Things" (IoT) describes how physical systems, objects, and gadgets are connected to the internet. To enable these devices to gather and exchange data, it entails the integration of numerous various actuators, sensors, and communication technologies (Gunduz et al., 2020). IoT devices have integrated distinctive identifiers that enable remote connection and control. Following are some essential qualities:

IoT devices can communicate with one another and with other platforms or systems thanks to their internet connectivity. Different technologies, including Bluetooth, Wi-Fi, cell phone networks, and lowpower devices wide-area networks (LPWAN), can be used to provide this connectivity. IoT devices are fitted with sensors that enable them to gather information about their surroundings. Depending on the use, these sensors may include ones for humidity, temperature, light, motion, and a wide variety of

Figure 2. IoT technologies

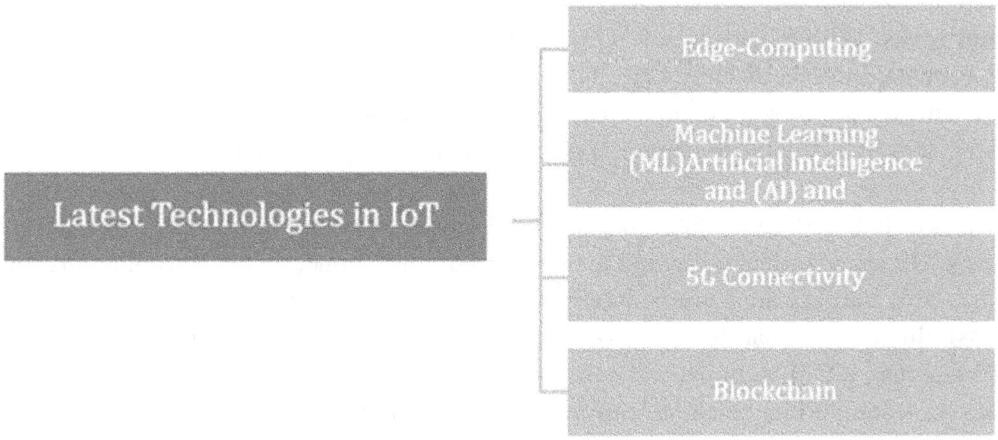

other things. Following collection, the data is sent to the internet or additional platforms for analysis and processing.

IoT creates enormous amounts of data, which are processed and used to produce insightful conclusions using cutting-edge analytics techniques. Predictive analysis, real-time monitoring, and making decisions based on the gathered data are all made possible by data analytics.

Automated and remote control are made possible via the Internet of Things (IoT). IoT systems may autonomously react to changes, initiate activities, and optimize operations without the need for human intervention thanks to connectivity and data exchange. IoT systems are scalable, allowing them to handle an enormous amount of gadgets and information sources. The proliferation of applications for the IoT and the incorporation of various devices and technologies are made possible by this scalability.

3.1.2 Latest Technologies in IoT

The system of IoT is constantly evolving, with new technologies being developed to enhance its capabilities and address emerging challenges (Kimani et al., 2019) as Figure 2.

Some of the latest technologies in IoT include:

- **Edge Computing**: Edge Computing technology reduces latency and enables real-time processing and making choices near the edge of the network's infrastructure by bringing computing power and information storage close to the Internet of Things devices themselves. This technique is very helpful in applications where high dependability and low latency are essential.
- **AI and ML**: The use of artificial intelligence (also called AI) and neural networks: AI and ML methods are applied to IoT data in order to obtain insights, spot patterns, and form wise forecasts. These innovations allow IoT devices to gain knowledge from data and gradually increase their performance (Rana & Khang et al., 2021).
- **Blockchain**: A decentralized and secure means to authenticate and keep track of interactions or transactions between IoT devices is provided by blockchain technology. In IoT networks, it im-

proves trust, safety, and data integrity, especially in situations where numerous devices or entities must cooperate or communicate data.

- **5G networks**: With the advent of 5G networks, connection has been quicker and more dependable, allowing Internet of Things (IoT) devices to exchange data with less latency and more bandwidth. This technology offers novel use cases that call for real-time communication and fast data transfer, as well as the implementation of large-scale IoT applications.

IoT intends to link and give digital intelligence to physical items, building a connected ecosystem that improves efficiency, productivity, and the standard of life. Connectivity, data collecting, the process, automation, and scalability are some of its primary features. IoT is constantly developing and opening up new possibilities for humanity in a variety of fields, such as intelligent homes, medical treatment (Khang & Medicine, 2023), farming, and transportation and industrial automation. These new technologies include edge computing, machine learning and artificial intelligence blockchain, and 5G.

3.2 Smart Grid

In order to increase the efficiency, dependability, long-term viability and adaptability associated with electricity production, distribution, transmission, and consumption, a sophisticated electrical grid infrastructure known as a "smart grid" is used. This infrastructure combines contemporary communication, data, and energy technology. A sophisticated electrical grid system known as a "smart grid" uses contemporary communications and technological advancements to improve the efficiency, dependability, and sustainability of power production, distribution, transmission, and consumption. Several of a smart grid's essential components include:

- **Distributed Energy Resources (DERs)**: Demand response programs, energy storage technologies, and energy from renewable sources including solar and wind power are all incorporated into smart networks. This enables better integration of intermittent renewable energy, enhances grid resilience, and supports energy optimization.
- **Grid Automation and Control**: Automation and management systems are used by smart grids to continuously monitor and regulate the flow of electricity. This includes self-healing capabilities to detect and isolate faults, optimize grid operations, and enable quick restoration of power.
- **Demand Response and Energy Efficiency**: Demand response programs are made possible by smart grids, allowing consumers to modify their electricity usage in response to price signals or grid circumstances. Energy efficiency measures, such as smart appliances and load management systems, help reduce peak demand and overall energy consumption.

4. REQUIREMENTS FOR THE PRESENT ENERGY SCENARIO

The present energy scenario, characterized by an increasing demand for electricity, a growing percentage of renewable energy, and the need to reduce greenhouse effect and its emissions, necessitates the adoption of smart grids (Ullah et al., 2020). The major requirements include:

Integration of Renewable Energy: Smart grids can effectively handle the variability and intermittency of renewable energy sources, ensuring their smooth integration into the grid.

Table 2. Comparison between smart grids and convectional grids

Parameters	Convectional Grids	Smart Grids
Communication type	Limited or single-way with renewables	Bidirectional communication
Energy Generation and Distribution	Mainly centralized	Distributed generation with renewables
Monitoring and Control	Limited visibility and control	Real-time monitoring and control
Grid Resilience	Susceptible to outages, slower restoration	Self-healing capabilities and fault detection and isolation
Energy Efficiency	Limited demand-side management	Programs for managing loads and demand
Renewable Integration	Limited integration of renewable energy	Better integration of intermittent renewables
Data Analytics	Limited data utilization	Advanced analytics for optimization
Cybersecurity	Less focus on cybersecurity	Robust security measures and protocols
Scalability	Limited scalability	Scalable to accommodate a large number of devices
Cost	Lower initial investment	Higher initial investment and infrastructure costs
Environmental Impact	Higher reliance on fossil fuels	Reduced carbon footprint and increased sustainability

Grid Flexibility: Smart grids offer the flexibility to accommodate a diverse range of energy resources, including distributed generation and energy storage, enabling efficient utilization and management of energy.

Demand-Side Management: Smart grids enable active participation of consumers in energy management, allowing them to monitor and control their energy consumption, adopt energy-efficient practices, and participate in demand response programs.

4.1 Difference Between Current Grid and Smart Grid

The current electrical grid primarily operates on a centralized and one-way flow of electricity, with limited visibility into real-time conditions. A comparison is shown in table 2 In contrast, a smart grid introduces the following differences:

- **Bidirectional Communication**: Bidirectional connection between the utility company and the customer is made possible by smart grids, which enable continuous surveillance, control, and information sharing.
- **Increased Automation**: Smart grids incorporate automation and control systems to optimize grid operations, detect and respond to faults, and enable self-healing capabilities.
- **Enhanced Reliability and Resilience**: The formation of advanced monitoring and control management technologies in smart grids improves the efficiency, reliability and resilience of the grid by detecting and isolating faults, minimizing downtime, and enabling rapid restoration.

Traditional grids operate on a one-way communication system with limited monitoring capabilities, while smart grids utilize advanced communication and sensing technologies to enable two-way communication and real-time data exchange between utility providers and consumers. This enhances grid

Figure 3. Smart grid structure

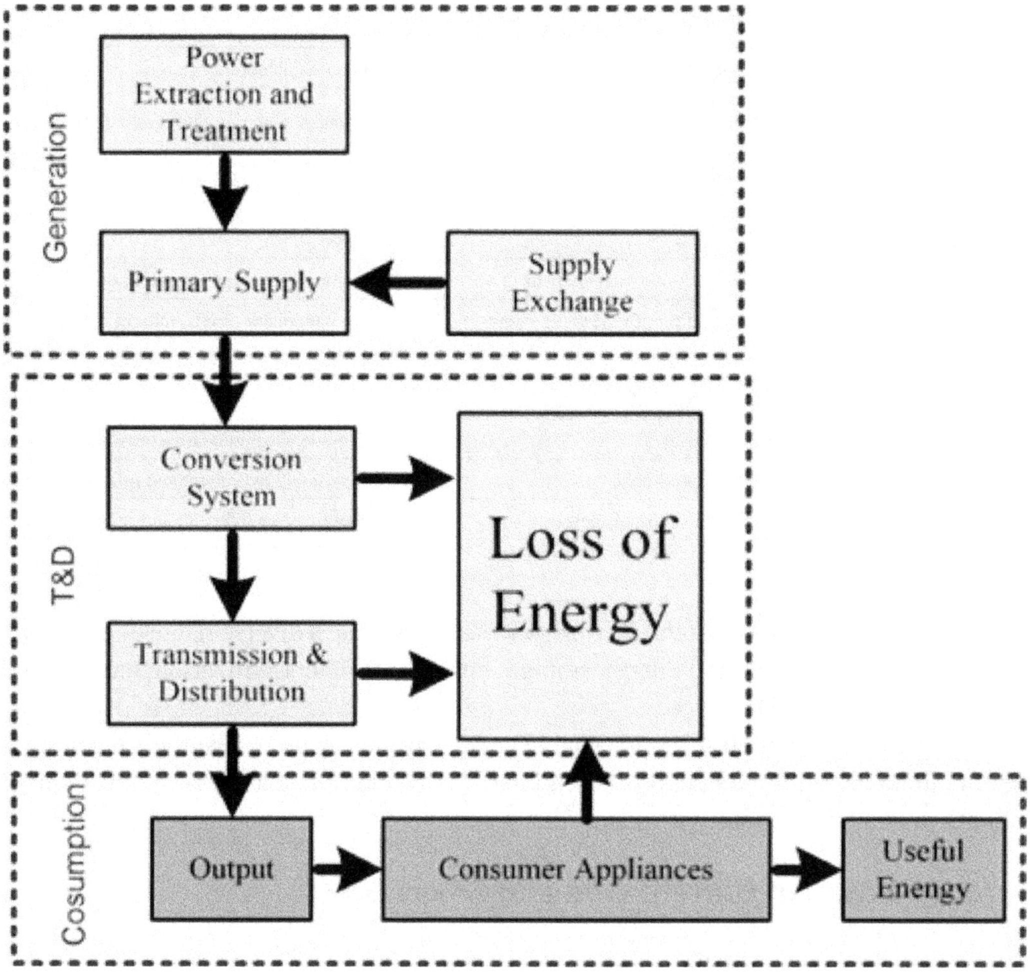

stability and efficiency by enabling better monitoring and control of power generation, transmission, and consumption.

Smart grids incorporate advanced metering infrastructure, which includes smart energy meters that provide detailed and accurate information about electricity usage. In contrast, traditional grids rely on manual meter reading, resulting in delayed and less precise consumption data. The availability of real-time data in smart grids enables dynamic pricing models, demand-response programs, and better load management, empowering consumers to make regular decisions for their energy usage with potentially reduce their power bills (Borlase et al., 2017).

Smart grids also make it easier to incorporate distributed energy resources and renewable energy sources. They provide improved monitoring and control systems that optimize the grid's integration of sporadic renewable energy output, like solar and wind energy. This encourages the utilization of clean energy, lessens dependency upon fossil fuels, and helps create an energy system that is more resilient and sustainable. The early detection and remediation of problems and outages is one of the key benefits of smart grids. Smart grids can locate problems and isolate impacted sections thanks to automatic fault recognition and self-healing features, reducing the impact on the entire system. Traditional grids often require manual intervention, resulting in longer restoration times and extended periods of service interruptions as Figure 3.

Alt-text: Figure 3 shows the structure of smart energy grid.

4.2 Shortcomings of Smart Grids

Cost and Infrastructure: Implementing a comprehensive smart grid infrastructure requires significant investments in upgrading existing systems, deploying new technologies, and building the necessary communication and control infrastructure.

Data Privacy and Security: The increased connectivity and data exchange in smart grids raise concerns about data privacy and cybersecurity. Robust security measures and protocols are necessary to save against potential threats and provide the privacy of consumer data.

Interoperability and Standards: The integration of diverse devices, technologies, and stakeholders in a smart grid ecosystem requires interoperability and standardization to ensure seamless communication, compatibility, and effective collaboration.

Regulatory and Policy Frameworks: The development and deployment of smart grids require supportive regulatory frameworks and policies that incentivize investments, promote innovation, and address regulatory barriers.

Smart grids are an essential part of the transition to a more robust and sustainable energy future despite these difficulties since they significantly improve efficiency, dependability, sustainability, and consumer empowerment.

5. SMART GRID ARCHITECTURE WITH IOT SUPPORT

There are many different architectures & prototypes which have been developed in the application of IoT-enabled smart power plants to improve the functioning and productivity of the power grid (Gope et al., 2021). These systems and prototypes use IoT technology to make it possible to monitor, regulate, and optimize energy production, distribution, and consumption in real-time. The following are some noteworthy IoT-enabled electrical grid architectures and prototypes:

- **Hierarchical Architecture**: This architecture consists of multiple hierarchical levels, including the field level, control level, and management level. Each level is equipped with IoT devices and sensors that enable data collection, analysis, and decision-making at different stages of the grid operation.
- **Peer-to-Peer (P2P) Architecture**: P2P architectures aim to decentralize control and decision-making within the smart grid. It allows devices and nodes to communicate and exchange information directly with each other, promoting efficient and distributed energy management.
- **Cloud-Based Architecture**: This architecture leverages cloud computing technologies to process and monitor vast range of data collected from IoT equipment's within the smart grid. Cloud platforms enable scalable and flexible data storage, analytics, and visualization, supporting advanced grid management and optimization.

Table 3. Different study on IoT usage in smart grid

Study	Year	Architecture/Prototype	Key Findings
Smith, et al. (Smith et al., 2020)	2020	Hierarchical Architecture	Improved grid monitoring and fault detection
Li, et al (Li et al., 2021)	2018	Peer-to-Peer Architecture	Decentralized control for efficient grid management
Chen, et al. (Chen et al., 2019)	2019	Cloud-Based Architecture	Scalable data storage and analytics for grid optimization
Johnson, et al. (Johnson et al., 2017)	2017	Plug-and-Play Architecture	Seamless integration of IoT devices in the grid
Xu, et al. (Xu et al., 2014)	2016	Microgrid Architecture	Enhanced renewable energy resources and its integration
Kim, et al. (Kim et al., 2015)	2015	Advanced Metering Infrastructure	Real time data handling and demand response programs
Zhang, et al.(Zhang et al., 2018)	2018	Hierarchical Architecture	Efficient energy management and cost optimization
Liu, et al. (Liu et al., 2019)	2019	Peer-to-Peer Architecture	Improved grid resilience and fault recovery
Yang, et al. (Yang et al., 2017)	2017	Cloud-Based Architecture	Advanced data analytics for predictive maintenance
Zhou, et al (Zhou et al., 2020)	2020	Plug-and-Play Architecture	Easy installation and replacement of devices
Wang, et al. (Uslar et al., 2019)	2018	Microgrid Architecture	Optimal utilization of distributed energy resources

- **Plug-and-Play Architecture**: This architecture focuses on the interoperability and seamless integration of IoT devices and systems within the smart grid. It allows for easy installation, configuration, and replacement of devices, ensuring smooth operation and scalability.
- **Microgrid Architecture**: Microgrids are small, autonomous electrical systems that can run separately or in conjunction with the larger grid. Integrated distributed sources of energy, energy storage, and overall intelligent controls are used in IoT-enabled microgrids to maximize energy production, intake, and grid resilience.
- **Advanced Metering Infrastructure -AMI**: It is a crucial part of smart grids with IoT capabilities. It contains smart meters that have been installed at customer locations and allow for two-way communication between customers and the utility company. Demand response, billing, and real-time data monitoring are all made possible by AMI.

These architectures are accompanied by various IoT-enabled smart grid prototypes that have been implemented for research and demonstration purposes. These prototypes often involve the deployment of sensors, actuators, and IoT devices in real-world grid scenarios to validate the feasibility and effectiveness of IoT technologies in optimizing grid operations, improving energy efficiency, and enhancing grid reliability (Lyulyov et al., 2021). Table 3 shows the recent study on IoT usage in smart grid.

6. INTEGRATION OF THE SMART GRID BASED ON COMMUNICATION TECHNOLOGY WITH THE IOT

All IoT as well as non-IoT communication technologies are essential for the integration of the smart grid because they provide seamless data transfer between devices. These technologies make it easier for smart meters, Internet of Things (IoT) equipment, and utility control centers to communicate with one another and with other smart grid components (Nian et al., 2019). The following are some of the major IoT plus non-IoT communication techniques used in the integration of the smart grid:

6.1 Wired Technologies

- **Ethernet**: Ethernet technology, such as Ethernet cables, switches, and routers, provides reliable and high speed wired communication within the smart grid infrastructure.
- **Power Line Communication - PLC**: It permits data transmission through current power lines, enabling communication between electrically connected devices.

6.2 Wireless Technologies

Wireless Fidelity called Wi-Fi: This technology provides wireless connectivity within local area networks, allowing devices to communicate among each other and in connection to the internet.

- **Cellular based Networks**: These are 3G, 4G, and upcoming 5G network devices, provide large area wireless coverage, enabling communication between devices over long distances.
- **Zig-bee**: It is a less-power wireless technology for communication, used for short-range communication between devices in a smart grid.
- **Z-Wave**: It is a wireless technology optimized as low-power home devices and automation, often used in smart home applications within the smart grid context.
- **Long Range Wide Area Network**: LoRaWAN is a wide-area network technology with low-power that permits low data rate, long distance communication, suitable for IoT devices in smart grid deployments.
- **IoT over narrowband (NB-IoT)**: It is a cellular network technology with low-power IoT devices, providing efficient and wide coverage for smart grid applications.

These technologies facilitate communication between various components of the smart grid, ensuring reliable and secure data transmission. The distance that must be covered between devices, the necessary data rate, the permitted power consumption, and the particular use case inside the smart electricity infrastructure all play a role in the decision of which communication technology to utilize (Prabu, Khang et al., 2023).

6.3 IoT Applications in Smart Grid

The IoT plays a crucial role in implementing smart grids by leveraging various IoT technologies to enhance the processing, monitoring, alerting, self-healing, disaster recovery, and reliability of the grid. It enables the integration of intelligent devices, censoring instruments, meters, and communication en-

Figure 4. IoT role in smart grid

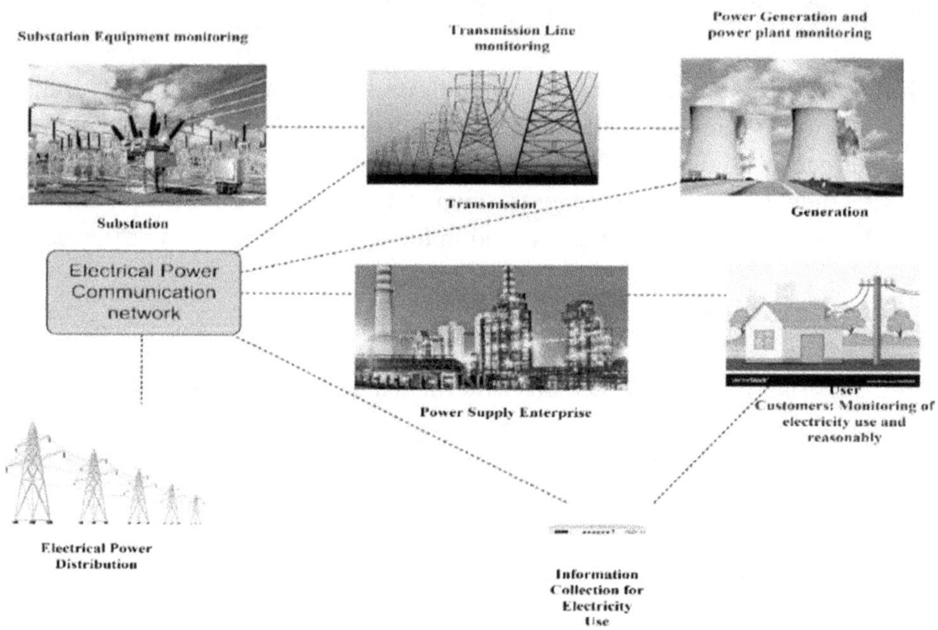

abled devices, thereby facilitating state monitoring, data gathering, control, safety, and security within the smart grid as Figure 4.

Alt-text: Figure 4 illustrates the diverse applications of the smart grid across all range of the IoT. Data transmission, processing, and data collecting make up the three main parts of the IoT.

The ability to acquire data via instruments and real-time access, including information on device locations, necessary electronic devices, and characteristic changes, is one of the major IoT breakthroughs. The fundamental system, network architecture, working mechanism, and transfer protocols of IoT technology advance with time, successfully answering the needs of the smart grid (Yao et al., 2018).

Technically speaking, the smart grid's IoT network function focuses on three main areas: data collecting, transmission, and processing. These areas encompass activities such as collecting data from various grid components, transmitting the data between devices and control centers, and processing the collected information to make informed decisions and enable efficient grid management.

Following is a summary of the IoT's function in several aspects of the smart grid, such as electricity production, transmission, distribution, and consumption:

6.3.1 Power Generation

IoT technologies have a significant contribution in power generation by enabling monitoring, control, and optimization of power generation facilities. These technologies facilitate actual data acquisition, data analysis, and decision power, leading to improved performance and efficiency in energy production.

One example of IoT application in power generation is the use of sensors and IoT-enabled devices to monitor the performance of renewable sources such as solar grids or wind power. Sensors embedded in these devices can collect data on parameters such as solar radiation intensity, wind speed, and turbine rotation speed. This data is passed to a central control system using IoT communication technologies. To operate power producing facilities as efficiently as possible, the acquired data can be real-time analyzed. For instance, if the IoT system detects a decrease in solar radiation intensity, it can automatically adjust the positioning of solar panels to maximize their exposure to sunlight. Similarly, in the case of wind turbines, the IoT system can adjust the blade pitch or turbine speed based on real-time wind speed data to maximize energy production (Khang & Kali et al., 2023). Furthermore, IoT technologies enable remote monitoring and control of power generation facilities. Operators can access real-time performance data and receive alerts or notifications in case of equipment malfunctions or deviations from optimal operating conditions. This allows for proactive maintenance and troubleshooting, minimizing downtime and improving overall plant performance (Sridhar et al., 2017).

In addition to monitoring and control, IoT-enabled power generation facilities can also participate in grid-level optimization and demand response programs. For example, during periods of high electricity demand, an IoT-enabled power plant can automatically adjust its output based on signals from the grid operator, contributing to grid stability and meeting fluctuating energy demands efficiently.

6.3.2 Transmission Lines

IoT devices and sensors deployed along transmission lines have an important role in monitoring and maintaining the condition of these lines. They provide valuable data that helps detect faults, identify potential issues, and enable predictive maintenance activities, ultimately ensuring reliable and secure power transmission. One example of IoT application in transmission line monitoring is the use of sensors to measure parameters such as temperature, humidity, vibration, and electrical current along the line. These sensors can be placed at strategic locations, including pylons, towers, and substations, to continuously collect data on the line's condition. For instance, temperature sensors can detect overheating in the transmission lines, which could indicate an increased risk of failure or fire.

The IoT system can monitor the temperature readings in real-time and trigger alerts or notifications if the temperature exceeds predefined thresholds. This enables operators to take immediate action, such as rerouting power or dispatching maintenance crews to address the issue promptly. Vibration sensors are another essential component of IoT-enabled transmission line monitoring. They can detect excessive vibrations caused by wind, equipment malfunction, or other external factors. By analyzing vibration data, the IoT system can identify potential structural issues or loose connections, enabling timely maintenance or repair actions to prevent line failures or outages. The incorporation of sophisticated machine learning and analytics techniques to process the gathered data and find patterns or anomalies is another benefit of IoT technologies (Khang & Misra et al., 2023). By analyzing historical data, the IoT system can predict potential faults or degradation in transmission lines. Predictive maintenance techniques can then be employed, scheduling maintenance activities during planned downtime or when the grid's load is low, minimizing disruption to the power transmission (Li et al., 2017).

IoT-enabled transmission line monitoring can improve the security of power transmission by detecting unauthorized access or tampering. For example, IoT devices equipped with intrusion detection sensors can alert operators if any unauthorized activity is detected near the transmission lines, enabling swift response and preventing potential sabotage or theft attempts.

6.3.3 Distribution

In the distribution generation enabling renewable sector of the smart grid, IoT-enabled devices and sensors have brought significant advancements in monitoring and controlling distribution networks. These technologies enable real-time data analysis, load balancing, and fault detection, leading to improved distribution efficiency and reliability.

6.4 Recent Trends in IoT-Enabled Distribution Systems

The recent trends in distribution system using IoT is listed below:

6.4.1 AMI, or Advanced Metering Infrastructure

AMI is a key trend in distribution systems, involving the deployment of smart meters equipped with IoT capabilities. These smart meters offer real-time information on power quality, voltage levels, and energy use. They allow for communication in both directions between the utility and the customer, improving load control, enabling distant disconnect/reconnect services, and providing real-time billing information (Khang & Vugar et al., 2024).

6.4.2 Distributed Energy Resources (DER) Integration

With the increasing adoption of renewable energy sources and distributed generation, IoT technologies play a crucial role in integrating DERs into the distribution grid. IoT devices and sensors enable real-time monitoring of DERs such as solar panels, wind turbines, and energy storage systems. This data helps optimize the utilization of these resources, balance the grid's load, and manage energy flows effectively.

6.4.3 Grid Automation and Control

IoT technologies facilitate the automation and control of distribution networks. IoT-enabled devices, such as intelligent switches and sensors, enable real-time monitoring of grid conditions, including voltage levels, power flows, and equipment status. This data allows for remote control and automated actions to improve grid performance, reduce outage durations, and enhance fault detection and isolation.

6.4.4 Demand Response and Load Management

IoT technologies enable demand response programs, where consumers can adjust their energy consumption based on real-time price signals or grid conditions. IoT-enabled devices, such as smart thermostats and appliances, can communicate with the grid and automatically adjust energy usage to optimize demand and reduce peak loads. This helps balance energy supply and demand, enhance grid stability, and support energy conservation efforts.

6.4.5 Data Analytics and Predictive Maintenance

IoT devices produce enormous volumes of data, and current trends focus on using powerful machine learning and analytics algorithms to extract useful insights (Khang & Hajimahmud et al., 2023). Data analytics can identify patterns, anomalies, and predict equipment failures or maintenance needs. This makes it possible for utilities to adopt preventative maintenance plans, cutting downtime and raising distribution network reliability.

The IoT's integration into the smart grid empowers utilities, grid operators, and consumers with enhanced monitoring, control, and optimization capabilities, resulting in enhanced grid performance, energy reliability and efficiency.

6.5 Future Research Directions and Security Challenges

Security threat is a critical concern in IoT based smart grids, as the interconnected nature of devices and systems introduces new vulnerabilities and risks. Addressing security issues is crucial to ensure the reliability, privacy, and integrity of smart grid operations (Song et al., 2015). Some of the key security challenges and future research directions in IoT for smart grids include:

6.5.1 Device Authentication and Access Control

Ensuring the authenticity and secure access of IoT devices in smart grids is essential. Research should focus on developing robust authentication mechanisms, such as cryptographic protocols and digital certificates, to prevent unauthorized access and device spoofing.

6.5.2 Data Privacy and Confidentiality

Smart grids generate massive several of sensitive information in data, including power usage, its consumption cycle and customer private information. Protecting data privacy and confidentiality requires secure data transmission, encryption techniques, and access control mechanisms to prevent unauthorized data disclosure or tampering.

6.5.3 Secure Communication Protocols

IoT devices in smart grids need secure communication protocols to manage the confidentiality, honesty, and authenticity of data exchanged between devices and the utility control center. Research should focus on developing lightweight, energy-efficient cryptographic protocols that can withstand potential attacks and provide end-to-end security.

6.5.4 Intrusion Detection and Prevention

Smart grids are susceptible to cyberattacks from viruses, distributed denial of-service (DDoS) assaults, and persistent advanced threats (APTs). The goal of research should be to create efficient systems for the detection and prevention of intrusions (IDPS) that can identify anomalies, malicious activity, and unauthorized entry attempts in real-time.

6.5.5 Resilience and Disaster Recovery

Smart grids should have robust mechanisms to withstand and recover from cyber-attacks, natural disasters, and physical damage. Research should focus on developing resilient architectures, redundancy mechanisms, and disaster recovery plans to ensure continuity of operations and minimize downtime.

6.5.6 Security Monitoring and Incident Response

Continuous monitoring of smart grid systems is crucial to detect and respond to security incidents promptly. Future research should explore techniques for real-time monitoring, security event correlation, and automated incident response to mitigate the impact of security breaches.

6.5.7 Standardization and Regulatory Frameworks

Establishing standardized security protocols, guidelines, and regulatory frameworks is essential to ensure consistent security practices across smart grid deployments. Research should focus on developing industry standards and regulatory policies to address the unique security challenges of IoT-enabled smart grids.

6.5.8 Security Awareness and Training

Building security awareness among utility operators, consumers, and stakeholders is crucial for maintaining a secure smart grid ecosystem. Future research should emphasize security education and training programs to promote best practices, security hygiene, and threat awareness among all stakeholders (Khang & Muthmainnah et al., 2023).

In addressing the security challenges in IoT for smart grids requires a holistic approach involving device authentication, data privacy, secure communication, intrusion detection, resilience, and standardized security practices. Future research should continue to explore innovative solutions and technologies to ensure the secure and trustworthy operation of IoT-enabled smart grids (Ushaa, Khang et al., 2023).

7. CONCLUSION

IoT technology holds immense potential for revolutionizing smart grid systems and addressing the challenges of the present energy landscape. By integrating IoT devices, sensors, and data analytics, smart grids can achieve enhanced efficiency, reliability, and sustainability. However, several critical areas require attention and further research. These include security and privacy measures to protect sensitive data, standardized protocols to ensure interoperability and scalability, energy-efficient solutions for IoT devices, advanced data analytics techniques, resilience against cyber threats and physical attacks, user-centric design considerations, and regulatory frameworks to address privacy and ethical concerns (Khang & Kali, 2023).

Quantum computing, with its unprecedented processing power and capabilities in data analytics and post-quantum cryptography, adds a game-changing dimension to these challenges. By integrating quantum computing alongside IoT, we can shape the future of smart grids, creating a more intelligent, secure, and sustainable energy ecosystem where data is safeguarded, decisions are optimized, and the grid adapts to the evolving energy landscape with efficiency and precision. By addressing these challenges and exploring potential solutions, the future of IoT for smart grids can be shaped, enabling a more intelligent, secure, and sustainable energy ecosystem (Khang & Shah et al., 2023).

8. FUTURE SCOPE OF RESEARCH IN IOT

Future research in IoT for smart grids should address various challenges and explore potential solutions to enhance the efficiency, security, and scalability of these systems (Wang et al., 2014). One crucial area of focus is security and privacy enhancements, where advanced mechanisms like anomaly detection, machine learning-based intrusion detection, and blockchain technology can be employed to protect sensitive data and prevent unauthorized access (Tailor & Ranu et al., 2022).

Interoperability and scalability are also critical, necessitating standardized protocols and interfaces to ensure seamless integration of IoT devices within the smart grid infrastructure. Moreover, research should aim to develop energy-efficient IoT solutions, including low-power communication protocols, energy harvesting techniques, and optimized resource management algorithms to minimize energy consumption while maintaining functionality and reliability (Khang, Quantum, 2023).

Edge computing and fog computing hold promise for processing and analyzing IoT-generated data locally, reducing latency and bandwidth requirements. Advanced data analytics techniques such as machine learning and big data analytics can extract valuable insights from the vast amount of data collected by IoT devices, enabling optimized energy management and decision-making processes. Resilience against cyber threats, physical attacks, and natural disasters should be a priority, with robust cybersecurity frameworks, incident response strategies, and recovery mechanisms in place (Khang & Vugar et al., 2023).

Additionally, considering user-centric design and human factors can enhance user acceptance and engagement in energy management. Developing regulatory and policy frameworks to address privacy protection, data governance, and ethical considerations is crucial (Panet al., 2015). By focusing on these research areas and proposing viable solutions, IoT for smart grids can unlock its full potential in transforming the energy landscape and meeting the evolving needs of the future (Khang & Kali, 2024).

REFERENCES

Abir, S. M. A. A., Anwar, A., Choi, J., & Kayes, A. S. M. (2021). IoT-Enabled Smart Energy Grid: Applications and Challenges. *IEEE Access : Practical Innovations, Open Solutions*, *9*, 50961–50981. https://ieeexplore.ieee.org/abstract/document/9381850/. doi:10.1109/ACCESS.2021.3067331

Aktar, T. (2022). A Survey on IoT-Enabled Smart Grids: Technologies, Architectures, Applications, and Challenges. *sustainability*, *15*(1), 717. https://www.mdpi.com/2071-1050/15/1/717

Alsuwian, T., Shahid Butt, A., & Amin, A. A. (2022). Smart Grid Cyber Security Enhancement: Challenges and Solutions—A Review. *Sustainability (Basel)*, *14*(21), 14226. https://www.mdpi.com/2071-1050/14/21/14226. doi:10.3390u142114226

Andreadou, N., Guardiola, M. O., & Fulli, G. (2016). Telecommunication Technologies for Smart Grid Projects with Focus on Smart Metering Applications. *Energies*, *9*(5), 375. https://www.mdpi.com/1996-1073/9/5/375. doi:10.3390/en9050375

Bikmetov, R., Raja, M. Y. A., Kazi, K., Chowdhury, B., & Enslin, J. (2015). *Dynamic prediction capabilities of Smart Metering Infrastructure.* Proceedings of the 2015 North American Power Symposium (NAPS), Charlotte, NC, USA. https://ieeexplore.ieee.org/abstract/document/7335235/

Billewicz, K. (2016). Possibility of Internet of things technology implementation in smart power grids. *Energetyka (Warszawa)*, *5*, 264–270. https://www.infona.pl/resource/bwmeta1.element.baztech-8da0dc88-22bc-4963-bf28-d034e1a32d8d

Borlase, S. (2017). *Smart Grids: Advanced Technologies and Solutions.* CRC Press. https://www.google.com/books?hl=en&lr=&id=VD4PEAAAQBAJ&oi=fnd&pg=PP1&dq=Smart+Grids:+Advanced+Technologies+and+Solutions.+CRC+Press&ots=uKXRz1EEFq&sig=lRRoKWYshe0alojZ0RlSvGwU4js

Chen, C. (2019). *A Social Network Based Approach for IoT Device Management and Service Composition.* IEEE. https://ieeexplore.ieee.org/abstract/document/7196496/

Chen, X., Li, X. Z., & Wang, H. (2010). The Application of Smart Grid in IOT. *J. North China Power*, *3*, 50–53. https://www.sciencedirect.com/science/article/pii/S1364032114003761

Da Silva, N., dos Santos, J., Gomes, R. L., Lins, R. A., & da Rocha, R. C. (2022). A Requirements Engineering Process for IoT Systems. *International Journal of Engineering Technologies and Management Research, 9.* https://www.academia.edu/download/63178921/Apr-2020-26-1587884210-A2520200502-113660-rma669.pdf

Eswaran, U. (2023). Role of Quantum Computing in The Era of Artificial Intelligence. Applications and Principles of Quantum Computing. IGI Global Press. https://doi.org/ doi:10.4018/979-8-3693-1168-4-ch003

Ghasempour, A. (2019). Internet of Things in Smart Grid: Architecture, Applications, Services, Key Technologies, and Challenges. *Inventions (Basel, Switzerland)*, *4*(1), 22. https://www.mdpi.com/2411-5134/4/1/22. doi:10.3390/inventions4010022

Gope, P., & Sikdar, B. (2021). A Privacy-Aware Reconfigurable Authenticated Key Exchange Scheme for Secure Communication in Smart Grids. *IEEE Transactions on Smart Grid*, *12*(6), 5335–5348. https://ieeexplore.ieee.org/abstract/document/9518386/. doi:10.1109/TSG.2021.3106105

Gunduz, M. Z., & Das, R. (2020). Cyber-security on smart grid: Threats and potential solutions. *Computer Networks*, *169*, 107094. https://www.sciencedirect.com/science/article/pii/S1389128619311235. doi:10.1016/j.comnet.2019.107094

Hassan, M., Rehmani, M. H., & Chen, J. (2019). Privacy preservation in blockchain-based IoT systems: Integration issues, prospects, challenges, and future research directions. *Future Generation Computer Systems*, *97*, 512–529. https://www.sciencedirect.com/science/article/pii/S0167739X18326542. doi:10.1016/j.future.2019.02.060

Jia, M. (2019). Adopting Internet of Things for the development of smart buildings: A review of enabling technologies and applications. *IEEE Access : Practical Innovations, Open Solutions*, *7*, 113601–113616. https://www.sciencedirect.com/science/article/pii/S0926580518307064

Johnson, D. (2017). *The impact of IoT security labelling on consumer product choice and willingness to pay*. PLOS. https://journals.plos.org/plosone/article?id=10.1371/journal.pone.0227800

Khang, A. (2023). *Advanced Technologies and AI-Equipped IoT Applications in High-Tech Agriculture* (1st ed.). IGI Global Press. doi:10.4018/978-1-6684-9231-4

Khang, A. Applications and Principles of Quantum Computing. (Eds.) (2023). ISBN: 9798369311684. IGI Global Press. https://doi.org/ doi:10.4018/979-8-3693-1168-4

Khang, A. (2023). Quantum Mechanics Primer - Fundamentals and Quantum Computing. Applications and Principles of Quantum Computing. IGI Global Press. https://doi.org/ doi:10.4018/979-8-3693-1168-4-ch001

Khang, A. (2023). Introduction to Quantum Computing and Its Integration Applications. Applications and Principles of Quantum Computing. IGI Global Press. https://doi.org/ doi:10.4018/979-8-3693-1168-4-ch002

Khang, A. (2023). Enabling the Future of Manufacturing: Integration of Robotics and IoT to Smart Factory Infrastructure in Industry 4.0. AI-Based Technologies and Applications in the Era of the Metaverse. IGI Global Press. doi:10.4018/978-1-6684-8851-5.ch002

Khang, A. (2023). AI and IoT-Based Technologies for Precision Medicine. IGI Global Press. doi:10.4018/979-8-3693-0876-9

Khang, A. (2024). Advanced IoT Technologies and Applications in the Industry 4.0 Digital Economy (1 Ed.) CRC Press. https://doi.org/ doi:10.1201/9781003434269

Khang A. (2024). *The Quantum Evolution: Application of AI and Robotics in the Future of Quantum Technology*. CRC Press. https://doi.org/ doi:10.1201/978-1-032-64207-9

Khang, A., Chowdhury, S., & Sharma, S. (2022). *The Data-Driven Blockchain Ecosystem: Fundamentals, Applications, and Emerging Technologies* (1st ed.). CRC Press. doi:10.1201/9781003269281

Khang, A., Hajimahmud, V. A., Gupta, S. K., Babasaheb, J., & Morris, G. (2023). AI-Centric Modelling and Analytics: Concepts, Designs, Technologies, and Applications. CRC Press. https://doi.org/ doi:10.1201/9781003400110

Khang, A., Misra, A., Gupta, S. K., & Shah, V. (2023). AI-aided IoT Technologies and Applications in the Smart Business and Production. CRC Press. https://doi.org/ doi:10.1201/9781003392224

Khang, A., & Muthmainnah, M. (2023). AI-Aided Teaching Model for the Education 5.0 Ecosystem. AI-Based Technologies and Applications in the Era of the Metaverse. IGI Global Press. doi:10.4018/978-1-6684-8851-5.ch004

Khang, A., Rana, G., Tailor, R. K., & Hajimahmud, V. A. (2023). Data-Centric AI Solutions and Emerging Technologies in the Healthcare Ecosystem. CRC Press. doi:10.1201/9781003356189

Khang, A., Rani, S., & Sivaraman, A. K. (2022). *AI-Centric Smart City Ecosystems: Technologies, Design and Implementation* (1st ed.). CRC Press. doi:10.1201/9781003252542

Khang, A., Shah, V., & Rani, S. (2023). AI-Based Technologies and Applications in the Era of the Metaverse. IGI Global Press. doi:10.4018/978-1-6684-8851-5

Khanh H. H., & Khang A. (2021). The Role of Artificial Intelligence in Blockchain Applications. *Reinventing Manufacturing and Business Processes through Artificial Intelligence*. CRC Press. doi:10.1201/9781003145011-2

Kim, J. (2015). *A study on the adoption of IoT smart home service: using Value-based Adoption Model*. Taylor & Francis. https://www.tandfonline.com/doi/abs/10.1080/14783363.2017.1310708

Kimani, K., Oduol, V., & Langat, K. (2019). Cyber security challenges for IoT-based smart grid networks. *International Journal of Critical Infrastructure Protection*, *25*(1), 36–49. https://www.sciencedirect.com/science/article/pii/S1874548217301622. doi:10.1016/j.ijcip.2019.01.001

Lewis, R. P., Igict, P., & Zhou, Z. (2009). Assessment of communication methods for smart electricity metering in the U.K. *Proceedings of the 2009 IEEE PES/IAS Conference on Sustainable Alternative Energy (SAE),* (pp. 1-4). IEEE. https://ieeexplore.ieee.org/abstract/document/5534884/

Li, B. (2021). *The evolution of the Internet of Things (IoT) over the past 20 years*. Science Direct. https://www.sciencedirect.com/science/article/pii/S0360835221000784

Li, F., Liang, X., & Jin, X. (2017). Integration of Internet of Things and Smart Grid in Industrial Systems. In 2017 IEEE International Conference on Internet of Things (iThings) and IEEE Green Computing and Communications (GreenCom) and IEEE Cyber, Physical and Social Computing (CPSCom) and IEEE Smart Data (SmartData). IEEE., https://ieeexplore.ieee.org/abstract/document/8276863/

Lyulyov, O., Vakulenko, I., Pimonenko, T., Kwilinski, A., Dzwigol, H., & Dzwigol-Barosz, M. (2021). Comprehensive Assessment of Smart Grids: Is There a Universal Approach? *Energies*, *14*(11), 3497. https://www.mdpi.com/1996-1073/14/12/3497. doi:10.3390/en14123497

Martín-Lopo, M. M. (2020). A Literature Review of IoT Energy Platforms Aimed at End Users. Proceedings of the 2020 IEEE International Conference on Industrial Internet. ICII. https://www.sciencedirect.com/science/article/pii/S138912861931271X

Mavroeidakos, T., & Chaldeakis, V. (2020). Threat Landscape of Next Generation IoT-Enabled Smart Grids. Advances in Information and Communication Technology, AIAI. Springer. https://link.springer.com/chapter/10.1007/978-3-030-49190-1_11

Mocrii, D. (2018, April). IoT-based smart homes: A review of system architecture, software, communications, privacy and security. *IEEE Internet of Things Journal*, *5*(2), 812–835. https://www.sciencedirect.com/science/article/pii/S2542660518300477

Nian, X., Sun, Y., Li, X., & Wang, Z. (2019). Integration of IoT in Smart Grid: A Survey. *IEEE Access : Practical Innovations, Open Solutions*, *7*, 70563–70577. https://ieeexplore.ieee.org/abstract/document/8625421/

Ou, Q., Zhen, Y., Li, Y., Zhang, Y., & Zeng, L. (2012). *Application of Internet of Things in Smart Grid Power Transmission.* Third FTRA International Conference on Mobile, Ubiquitous, and Intelligent Computing, Vancouver, BC, Canada. https://ieeexplore.ieee.org/abstract/document/6305831/

Pan, J., Jain, R., Paul, S., Vu, T., Saifullah, A., & Sha, M. (2015). An Internet of Things Framework for Smart Energy in Buildings: Designs. *IEEE Internet of things Journal.* https://ieeexplore.ieee.org/abstract/document/7061425/

Parvin, K., Hannan, M. A., Mun, L. H., Lipu, M. S. H., Abdolrasol, M. G. M., Ker, P. J., Muttaqi, K. M., & Dong, Z. Y. (2022). The future energy internet for utility energy service and demand-side management in smart grid: Current practices, challenges and future directions. *Sustainable Energy Technologies and Assessments, 53*, 102648. https://www.sciencedirect.com/science/article/pii/S221313882200697X. doi:10.1016/j.seta.2022.102648

Prabu, B., & Khang, A. (2023). Quantum Computing Smart City: Fostering Sustainability and Green Infrastructure with Smart City. Applications and Principles of Quantum Computing. IGI Global Press. https://doi.org/ doi:10.4018/979-8-3693-1168-4-ch009

Rana, G., Khang, A., Sharma, R., Goel, A. K., & Dubey, A. K. (2021). Reinventing Manufacturing and Business Processes through Artificial Intelligence. CRC Press. doi:10.1201/9781003145011

Rose, K. (2014). The Internet of Things: Understanding the Issues and Challenges of a More Connected World (2014). *Proceedings of the 2014 11th IEEE International Conference on Ubiquitous Intelligence and Computing*. IEEE. https://doi.org/10.1201/9781003434269

Shyam, R. & Khang A. (2023). Effects of Quantum Technology on Metaverse. AI-Based Technologies and Applications in the Era of the Metaverse. IGI Global Press. doi:10.4018/978-1-6684-8851-5.ch009

Smith, A. (2020). *User values and the development of a cybersecurity public policy for the IoT.* Science Direct. https://www.sciencedirect.com/science/article/pii/S0268401219313301

Song, Y. E., Liu, Y., Fang, S., & Zhang, S. Research on Applications of the Internet of Things in the Smart Grid. In Proceedings of the 7th International Conference on Intelligent Human-Machine Systems and Cybernetics, 26–27 August 2015, Hangzhou, China, 2015. https://ieeexplore.ieee.org/abstract/document/7334945/

Sridhar, V., & Bharathi, V. M. (2017).Internet of Things (IoT) Applications in Power Grid: A Review. In *2017 International Conference on Innovations in Green Energy and Healthcare Technologies* (pp. 1-6). IEEE. https://doi.org/10.1201/9781003434269

Tailor, R. K., Pareek, R., & Khang A. (2022). Robot Process Automation in Blockchain. CRC Press. doi:10.1201/9781003269281-8

Tao, J. (2019). IoT-Based Intelligent Perception and Access of Manufacturing Resource Toward Cloud Manufacturing. In Proceedings of the 2019 15th International Conference on Computational Intelligence and Security. CIS. https://dl.acm.org/doi/abs/10.1145/3334480.3375032

Ullah, Z., Al-Turjman, F., Mostarda, L., & Gagliardi, R. (2020). Applications of artificial intelligence and machine learning in smart cities. *Computer Communications*, *154*, 313–323. https://www.sciencedirect.com/science/article/pii/S0140366419320821. doi:10.1016/j.comcom.2020.02.069

Uslar, M., Rohjans, S., Neureiter, C., Andrén, F. P., Velasquez, J., Steinbrink, C., Efthymiou, V., Migliavacca, G., Horsmanheimo, S., & Brunner, H. (2019). Applying the Smart Grid Architecture Model for Designing and Validating System-of-Systems in the Power and Energy Domain: A European Perspective. *Energies*, *12*(2), 258. https://www.mdpi.com/1996-1073/12/2/258. doi:10.3390/en12020258

Wang, C., Li, X., Liu, Y., & Wang, H. (2014). *The research on development direction and points in IoT in China power grid*. In International Conference on Information Science, Electronics and Electrical Engineering, Sapporo, Japan. https://ieeexplore.ieee.org/abstract/document/6948106/

Wang, Y. F., Lin, W. M., Zhang, T., & Ma, Y. Y. (2012). Research on application and security protection of Internet of Things in Smart Grid. In *Proceedings of the IET International Conference on Information Science and Control Engineering (ICISCE 2012)*, Shenzhen, China. 10.1049/cp.2012.2311

Xu, S. (2014). *The internet of things: a survey*. Science Direct. https://www.sciencedirect.com/science/article/pii/S1389128610001568

Yang, H. (2017). *The Internet of Things for Smart Manufacturing: A Review*. Taylor and Francis. https://www.tandfonline.com/doi/abs/10.1080/24725854.2018.1555383

Yao, Y., Li, Y., Han, R., & Wang, X. (2018). Internet of Things for Smart Grid: Technologies, Standards and Challenges. *IEEE Communications Surveys and Tutorials*, *20*(1), 674–707. https://ieeexplore.ieee.org/abstract/document/9139976/

Zeadally, S. (2020). Toward efficient smartification of the Internet of Things (IoT) services. In *Proceedings of the 2020 18th International Conference on Advances in Mobile Computing & Multimedia (MoMM)*, (pp. 392-397). Science Direct. https://www.sciencedirect.com/science/article/pii/S0167739X17317326

Zhang, L. (2018). *IoT-Based Intelligent Perception and Access of Manufacturing Resource Toward Cloud Manufacturing*. IEEE. https://ieeexplore.ieee.org/abstract/document/6742605/

Zhao, L. J., Zhang, F. Z., & He, H. X. (2013). *The Application of Internet of Things in Smart Grid," in Advanced Materials Research*. Trans Tech Publications, Ltd. https://www.mdpi.com/2071-1050/15/1/717

Zhou, W. (2020). *Internet of Things (IoT) in high-risk Environment, Health and Safety (EHS) industries: A comprehensive review*. Science Direct. https://www.sciencedirect.com/science/article/pii/S0167923618300344

Chapter 14
Quantum Computing in Pharmaceutical Science

Bancha Yingngam
https://orcid.org/0000-0001-7215-9123
Ubon Ratchathani University, Thailand

ABSTRACT

In the field of pharmaceutical science, there is an increasing demand for advanced computational techniques to enhance drug discovery and understand complex molecular interactions. Classical models, despite their strengths, often struggle with complex biological systems and accurate drug predictions. Quantum computing, using qubits that exist in multiple states, processes information at unparalleled speeds. This chapter sheds light on the principles of quantum computing, emphasizing algorithms specifically designed for pharmaceutical applications. These methods signal groundbreaking advancements in drug design, molecular simulation, and drug interaction predictions. Quantum computers have the potential to surpass classical computers in simulating intricate molecules, thus offering more precise drug side effect forecasts. While challenges such as noise and error rates remain, the continuous evolution of quantum technology holds significant potential for breakthroughs in drug discovery and development.

1. INTRODUCTION

Pharmaceutical science, an interdisciplinary domain, is central to the discovery, development, and production of drugs. This field consistently relies on computational models to understand molecular behavior, optimize drug designs, and predict drug interactions. Such models have revolutionized the drug discovery process, diminishing both time and costs tied to traditional experimental methods. Notably, the pharmaceutical sector reinvests approximately 15% of its sales into research and development, which accounts for more than 20% of all research and development expenditures in the global economy (Mahesh & Shijo, 2023). Traditionally, the trial-and-error method is used to create drugs, which is not only expensive but also risky and difficult to complete. However, as the pharmaceutical landscape grows more intricate—especially when addressing diseases stemming from multifactorial molecular interactions—classical computational approaches are nearing their limits.

DOI: 10.4018/979-8-3693-1168-4.ch014

One significant challenge confronting the industry is effectively modeling complex biological systems. The multitude of interactions among proteins, enzymes, and potential drug molecules, notably in multifunctional drugs, requires substantial computational power and time. Classical computers, even with their advancements, often struggle to provide accurate predictions in a reasonable timeframe. Such hurdles carry tangible consequences: inaccuracies or delays can lead to extended drug development cycles, increasing costs, and potentially missed therapeutic opportunities.

Additionally, the rising demand for personalized medicine—therapies tailored according to an individual's genetic profile—intensifies these computational challenges. Such a tailored approach necessitates the understanding of individual genetic variations and their influence on drug responses, a computationally intensive task. By enabling businesses to undertake more drug discoveries and develop breakthrough medical treatments, these computer advancements have the potential to significantly boost efficiency and lead to a more productive pharmaceutical industry. Pharmaceutical research and development has recently embraced artificial intelligence (AI) (Khang & Kali, 2024).

The next digital frontier in drug detection is quantum computing. This is where quantum computing enters science. Rooted in quantum mechanics principles, this emerging computational approach holds the potential to simultaneously process vast data amounts, addressing challenges previously considered insurmountable. The integration of quantum computing in pharmaceutical sciences promises not only to expedite drug discovery but also to introduce an era marked by unparalleled precision and customization in drug design and application (Dash et al., 2023; Sihare & Khang, 2023).

In essence, the confluence of quantum computing and pharmaceutical sciences could be the panacea for the current computational bottlenecks in drug discovery and development (Gupta et al., 2023). The significance of addressing these challenges cannot be understated, as it directly influences the speed and efficacy with which life-saving drugs are brought to the market, thus influencing global health outcomes (Izsák et al., 2023). The aims of this chapter are as follows:

- **Illuminate the fundamentals of quantum computing**: Before exploring its applications in the pharmaceutical sector, it is pivotal to grasp the foundational concepts and principles that underpin quantum computing. This knowledge allows for appreciating the transformative potential quantum technology offers to drug discovery and other pharmaceutical processes.
- **Emphasize the synergy between quantum computing and pharmaceutical sciences**: Quantum computing represents more than just an addition to the computational toolkit; it signifies a paradigm shift in problem-solving approaches. This chapter underscores the unique advantages that quantum computing offers, particularly in addressing the complex challenges inherent in pharmaceutical research.
- **Showcase current and potential applications**: From molecular modeling to drug interaction predictions, quantum computing is starting to redefine what is possible in drug discovery. This chapter endeavors to present an in-depth analysis of existing applications and offer a glimpse into the future, where quantum solutions could drastically streamline pharmaceutical research and development (R&D) processes.
- **Discussion challenges and limitations**: Like any emerging technology, quantum computing in pharmaceutical sciences is not without its hurdles. It is essential to offer a balanced view, addressing the current challenges—ranging from hardware constraints to error rectifications—and how they might impact the broader adoption of quantum methodologies in pharmaceutical research.

- **Charting the road ahead**: The final aim is to provide readers with insights into the future trajectory of quantum computing in this field. This encompasses ongoing research, anticipated breakthroughs, and the evolving landscape of quantum-enhanced pharmaceutical research, equipping researchers and professionals with the knowledge to navigate this exciting frontier.

The organization of this chapter is as follows: Section 2 elucidates the foundational concepts intrinsic to quantum computing, laying the groundwork for subsequent exploration. In Section 3, the spotlight shifts to computational methods tailored for pharmaceutical research, elucidating how they can bring transformative changes to the industry. Section 4 delves into the profound impact of quantum technology on molecular interactions, offering insights into more precise predictions of drug side effects. Section 5 addresses the challenges and constraints of integrating quantum methodologies into the pharmaceutical realm, highlighting both hardware and software limitations. Section 6 paints a vision of the anticipated advancements in this interdisciplinary domain, hinting at the boundless possibilities on the horizon. Finally, Section 7 provides a comprehensive synthesis, tying together the key takeaways and outlining the future trajectory of this promising confluence of technologies.

2. FUNDAMENTALS OF QUANTUM COMPUTING

2.1 Quantum Bits (Qubits) vs. Classical Bits

Classical bits, the foundational units of classical computing, operate within a binary framework. Each bit exists in one of two possible states: either 0 or 1. The entirety of classical computing, from basic arithmetic to intricate simulations, relies on manipulating these bits using established logical operations (Ur Rasool et al., 2023). Over the years, improvements in classical computing have primarily centered around miniaturizing hardware components and enhancing processing speeds, but the fundamental binary nature of classical bits has remained unchanged. In contrast, quantum bits, or qubits, introduce an entirely novel paradigm to data representation and computation as shown in Table 1.

Unlike their classical counterparts, qubits operate based on the principles of quantum mechanics. This allows them to exist not only in a definite state of 0 or 1 but also in a superposition of these states, representing both simultaneously. When measured, however, a qubit collapses into a definite state (Khang & Shah et al., 2023). Beyond superposition, qubits also exhibit entanglement, a unique quantum phenomenon where the states of two or more qubits become interdependent, regardless of the distance separating them. This ability to exist in superposition and to become entangled grants quantum computers their unparalleled computational power and parallelism, enabling them to process vast amounts of data and solve complex problems that are currently beyond the reach of classical machines.

2.2 Principles of Quantum Superposition and Entanglement

Quantum superposition is one of the cornerstone principles of quantum mechanics. It posits that any quantum system, such as a qubit, can exist in multiple states simultaneously until it is observed or measured. In the context of qubits, this means that instead of being definitively in a state of 0 or 1, as classical bits are, a qubit can concurrently be in both states. This phenomenon is not simply a blend of the two states but a coexistence, allowing quantum systems to embody a vast array of potential outcomes

Table 1. Comparison of classical and quantum computing

Attributes	Classical computing	Quantum computing	Reference
Computing units	Bit (0 or 1)	Qubits (0, 1, or both)	Ur Rasool et al. (2023)
Computing capacity	Deterministic; linear growth in capacity with additional bits	Exponential growth in capacity; capable of processing vast amounts of data in parallel due to superposition	Sihare and Khang (2023)
Suitability	Best for most traditional tasks; databases, basic arithmetic, general software applications	Suitable for specific problems like factorizing large numbers, simulating quantum systems, and optimization problems	Liu (2023)
Error rates	Relatively low; mature technology with error correction mechanisms	Currently higher; prone to decoherence and noise; active research in quantum error correction technique	Kim et al. (2023)

simultaneously. When a measurement is made, the superposed state collapses into one of its possible definite states, with the outcome governed by probabilistic rules. This attribute of superposition facilitates quantum computers to process a multitude of potential solutions at once, drastically enhancing their computational capabilities.

Quantum entanglement, another quintessential quantum principle, is a phenomenon where particles, such as qubits, become interconnected in such a way that the state of one particle is dependent on the state of another, irrespective of the distance between them. This means that a change in the state of one entangled qubit will instantaneously influence the state of its entangled partner, even if they are light-years apart. Einstein famously referred to this as "spooky action at a distance". In the realm of quantum computing, entanglement is invaluable. It enables complex computations that leverage the coordinated behavior of multiple qubits, allowing quantum algorithms to solve intricate problems with a depth and efficiency that classical systems cannot achieve.

2.3 Quantum Gates and Circuits

At a rudimentary level, classical computers process data using bits, which exist in either a 0 or 1 state. These bits are manipulated using logical gates, such as AND, OR, and NOT gates, to perform computations. Quantum computing, on the other hand, introduces a paradigm shift in this approach. The fundamental unit of quantum information is the qubit. Unlike classical bits that can be either 0 or 1, a qubit can exist in a superposition of these states (Sasi et al., 2023). This means that a qubit can simultaneously represent both 0 and 1, allowing for vast parallelism in computations.

Quantum gates are the quantum analogs of classical logic gates. They operate on qubits to change their state, facilitating quantum computation. Some common quantum gates include:

- Pauli-X, Pauli-Y, and Pauli-Z gates: These gates act as quantum versions of the classical NOT gate but with nuanced differences, given their operation in the quantum realm (Just, 2022).
- Hadamard gate (H-gate): This gate creates superposition. When applied to a qubit, it transforms to a state where it has an equal probability of collapsing to either 0 or 1 upon measurement (Mummadi & Rudra, 2023).

- Controlled-NOT (CNOT) gate: A two-qubit gate where the state of the second qubit is flipped if the first qubit is in a certain state, exemplifying quantum entanglement (Wong, 2024a).

Concerning quantum circuits, much like classical circuits built using a sequence of gates to perform a specific computation, quantum circuits are designed by arranging quantum gates in a particular sequence to manipulate qubits and produce a desired quantum state. The design and optimization of these circuits are pivotal to efficient quantum computations.

- Building a quantum circuit: Starts with an initial set of qubits, usually in the $|0\rangle$ state. Quantum gates then act on these qubits, manipulating their states. Finally, a measurement is made, collapsing the qubits to definite states (either 0 or 1) (Movassagh, 2023).
- Entanglement and superposition: Two quintessential quantum phenomena, entanglement and superposition, play a crucial role in quantum circuits. When qubits are entangled, the state of one qubit is dependent on the state of another, even if they are physically separated. Superposition, as mentioned, allows qubits to exist in multiple states simultaneously, amplifying computational parallelism.

3. QUANTUM ALGORITHMS RELEVANT TO PHARMACEUTICAL SCIENCES

Pharmaceutical sciences began involving computational applications by focusing on molecular modeling and simulation in its initial stages. The 1970s and 1980s introduced molecular mechanics and quantum mechanics methods, enhancing the understanding of molecular structures and energetics. This era also saw the rise of tools such as the Cambridge structural database and computer-aided drug design (CADD), which transformed drug discovery by predicting molecular behavior in biological systems. By the turn of the 21st century, the emphasis shifted toward high-throughput screening (HTS) and the emergence of bioinformatics. With the vast amounts of biological data generated, computing became indispensable for managing and analyzing these datasets (Khang & Quantum, 2023).

HTS employs computational methods to quickly and efficiently test thousands to millions of compounds for their potential therapeutic efficacy. Bioinformatics, on the other hand, integrated computational methods with vast biological datasets, particularly in genomics and proteomics. This merging deepened our understanding of diseases at the molecular level and laid the foundation for personalized medicine through the analysis of individual genetic information. The recent decade has experienced explosive growth in the application of deep learning techniques in pharmaceutical sciences. Advanced machine learning algorithms, particularly neural networks, have enhanced predictive modeling, drug repurposing, and understanding of complex biological systems. The integration of big data with artificial intelligence has further amplified the accuracy and efficiency of drug discovery and development processes. In parallel, quantum computing has started to emerge as a promising tool. Although in its infancy, it holds the potential to revolutionize the pharmaceutical field by tackling complex molecular simulations and optimizing drug design processes, setting the stage for the next large leap in computational pharmaceutical research (Khang & Kali et al., 2023).

Charting the progress, the evolution of computing in pharmaceutical research is emblematic of the broader technological evolution. Figure 1 depicts the timeline of computing evolution in pharmaceutical research. From the early data management systems of the 1960s to the artificial intelligence (AI)-driven models of the 2000s and the quantum horizons of the 2020s, each era has expanded the boundaries of what is possible in drug discovery. Since the 1960s, the pharmaceutical industry has experienced significant transformations in its research methodologies, largely propelled by advances in computing technologies. In the 1960s and 1970s, the industry began to adopt cutting-edge data collection methods. Computer systems, albeit rudimentary by today's standards, facilitated the management and analysis of vast experimental datasets. These systems replaced cumbersome manual structures, helping to predict drug behaviors and identify potential drug candidates more efficiently. Laboratories equipped with these computers could store, retrieve, and process data in ways that were previously unthinkable, laying the groundwork for subsequent technological revolutions in the field.

Entering the new millennium, particularly in the early 2000s, the emergence of advanced computing capabilities heralded the age of AI and machine learning (ML) in pharmaceutical research. These technologies were transformative. AI/ML algorithms could sift through extensive databases of molecular structures, identifying potential drug candidates based on predefined criteria at speeds previously unimaginable. Additionally, ML models started predicting how certain compounds would interact within biological systems, optimizing drug design processes. The potential for accelerated drug discovery and development using these tools became evident, leading to a surge in the identification of new therapeutic agents.

From 2020 onwards, quantum computing began to influence pharmaceutical research. Even though it was in its nascent stages, the capability of quantum computers to process vast amounts of information simultaneously offered promising prospects for drug discovery. These systems could simulate intricate molecular interactions with unprecedented accuracy, giving insight into drug behaviors that classical computers struggled to predict. For instance, understanding protein folding or simulating the nuances of multifactorial molecular interactions—tasks that were hefty for classical systems—became more feasible with quantum computers. However, this progression was not without challenges. Each computational leap brought its own set of technical, ethical, and practical concerns. The integration of each new computing paradigm into existing research infrastructures needed not only technological adaptation but also shifts in mindset, training, and sometimes even the restructuring of entire research processes. With every advance, the pharmaceutical industry had to ensure that the new methodologies maintained the rigor and safety standards essential for drug discovery (Khang & Kali, 2023).

Before exploring the applications of quantum computing in pharmaceutical sciences, it is essential to provide an overview of key quantum algorithms and understand how they intersect with pharmaceutical applications. Table 2 presents a concise overview of six pivotal quantum algorithms and their respective applications in the pharmaceutical domain. Grover's algorithm facilitates faster searching in extensive drug databases, providing a more efficient method for identifying potential drug candidates. Quantum phase estimation (QPE) aids in molecular simulations by determining the energy levels of molecules, a critical component for grasping molecular interactions in drug discovery. The variational quantum eigensolver (VQE) focuses on ascertaining the ground-state energy of molecules, which is instrumental in predicting molecular properties and, consequently, in discovering new drugs (Ushaa, Khang et al., 2023).

Additionally, the quantum approximate optimization algorithm (QAOA) offers solutions for combinatorial problems and can be employed to identify optimal molecular structures, enhancing drug formulations. On the other hand, Shor's algorithm, known predominantly for its ability to quickly factorize large numbers, finds its niche in the pharmaceutical sector by bolstering data security, especially in transmitting sensitive personalized medicine data. Last, the HHL algorithm, named after its developers Harrow, Hassidim, and Lloyd, is proficient in solving linear systems of equations. This capability be-

Figure 1. Timeline of computing evolution in pharmaceutical research

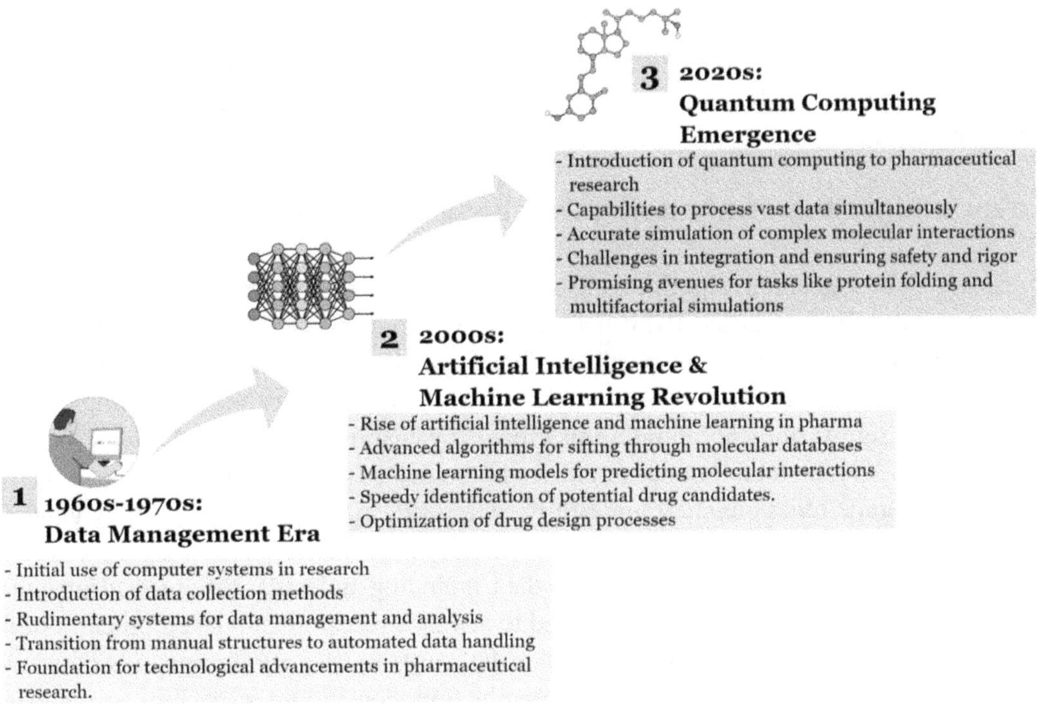

Table 2. Key quantum algorithms and their pharmaceutical applications

Quantum Algorithm	Application in Pharmaceuticals	Description	Reference
Grover's algorithm	Drug database searching	Speeds up the search process in large drug databases, potentially finding drug candidates more quickly.	Bonde et al. (2024)
Quantum phase estimation	Molecular simulations	Used to determine energy levels of molecules, which is vital for understanding molecular interactions in drug discovery.	Chaganti et al. (2024)
Variational quantum Eigen solver	Drug discovery	Helps find the ground state energy of molecules, aiding in predicting molecular properties for new drugs.	Ma et al. (2023)
Quantum approximate optimization algorithm	Optimizing molecular structures	Can be used to find optimal molecular structures or configurations that might be relevant in drug formulations.	Lindh and Fdez Galván (2023)
Shor's algorithm	Cryptography in personalized Medicine data	While traditionally used for factorizing numbers, it can ensure data security in transmitting personalized medicine data.	Sharma et al. (2023)
HHL algorithm (Harrow-Hassidim-Lloyd)	Personalized medicine, drug discovery	Solves linear systems of equations; can be used to model biological processes or predict drug responses in patients.	Koczka et al. (2023)

comes particularly salient in the field of personalized medicine, as it can be employed to model intricate biological processes and, in turn, predict individual drug responses in patients. In the following section, a deeper dive into these algorithms is undertaken (Khang & Vugar et al., 2023).

Alt-text: Figure 1 depicts charting the transformative journey from rudimentary data management systems in the 1960s through the AI and ML-driven revolutions of the early 2000s to quantum computing innovations starting in 2020.

3.1 Grover's Algorithm: Enhancing Database Searches

In the expansive field of quantum algorithms, Grover's algorithm stands out for its promise in enhancing database search capabilities. Classically, if one needed to find a specific item within an unsorted database, one would, on average, need to inspect half of its entries before a successful match. In a worst-case scenario, they might need to inspect every single entry. Grover's algorithm, when implemented on a quantum computer, significantly accelerates this process by enabling a successful search with just about $\sqrt{N}$ queries, where N is the total number of entries in the database. This quadratic speed-up, although less dramatic than some other quantum algorithms, can result in substantial efficiency gains for large databases.

In pharmaceutical sciences, vast repositories of data, including molecular structures, drug interactions, and patient genetic profiles, are frequently queried to advance research and development. For instance, drug discovery often involves searching molecular databases to identify compounds that might bind effectively to a target protein. Classical methods can be time-consuming, especially as databases grow exponentially with new discoveries. By employing Grover's algorithm on quantum computers, scientists have the potential to identify prospective drug candidates or molecular interactions much faster than with traditional computing methods. Moreover, Grover's algorithm is not limited to straightforward database searches. Its framework can be adapted for other optimization and search-related problems that are abundant in pharmaceutical research. Whether it is finding the optimal molecular structure with desired properties or searching for patterns in genetic data for personalized medicine, Grover offers a versatile toolset that can be tailored to a myriad of applications (Hema & Arowolo, 2023).

3.2 Quantum Phase Estimation and its Importance in Molecular Dynamics

QPE is one of the most fundamental algorithms in the quantum computing landscape. At its core, it aims to determine the phase (or eigenvalue) of a given eigenvector for a unitary operator. While this may sound esoteric, it translates to extracting valuable information about a system's behavior, especially when that system is governed by quantum mechanics. In essence, QPE provides insights into the dynamics of quantum systems by estimating the eigenvalues of their corresponding unitary transformations. Molecular dynamics, which investigates the motions and interactions of atoms and molecules, relies heavily on understanding these quantum systems.

The behavior of molecules, particularly in complex reactions or when influenced by external fields, can be described by their quantum states. These states are often eigenstates of specific Hamiltonians, with associated eigenvalues representing the energy levels of the system. Accurately determining these eigenvalues is paramount for predicting molecular behavior, reaction rates, and interaction strengths (Izsák et al., 2023). Classical computational methods often struggle to efficiently calculate these values for large or complex molecular systems, leading to approximations. Enter QPE, which offers a powerful quantum solution to this challenge. When applied to molecular systems, QPE can provide precise energy level (eigenvalue) estimations for molecules. This is particularly crucial in molecular dynamics simulations, where understanding energy variations and transitions can elucidate reaction mechanisms, molecular interactions, and structural transformations. For instance, in drug discovery, accurately determining the energy levels can help predict how a drug molecule might interact with its target protein, thereby influencing its efficacy and potential side effects.

3.3 Variational Quantum Eigensolvers for Drug Design

VQE is one of the most promising quantum algorithms in the quest for practical quantum computing applications. VQE operates by leveraging both quantum and classical processing powers to find the approximate ground state energy of a quantum system, especially when exact solutions are computationally expensive or infeasible. In essence, VQE employs a quantum subroutine to prepare quantum states and estimate energies, while a classical optimizer refines the parameters to minimize the energy further, iteratively. The field of drug design, specifically, stands to benefit immensely from VQE. The mechanism of drug interactions at the molecular and atomic levels is predominantly governed by quantum mechanics (Bonde et al., 2024). One of the central challenges in drug design is predicting the binding affinity of a drug molecule to its target, which hinges on accurately determining the lowest energy configuration (or ground state) of the drug-target complex. Given the vast conformational space and quantum nature of molecular interactions, traditional computational methods can be prohibitively slow or approximate. With VQE, there is an avenue for more precise and efficient determination of these ground state energies. The algorithm's hybrid approach ensures that while the quantum computer captures the intricacies of quantum mechanics, the classical optimizer can quickly fine-tune parameters, ensuring a balance between accuracy and computational efficiency. For pharmaceutical research, this translates to quicker, more accurate predictions of drug efficacy and interactions, allowing for more streamlined drug optimization processes and potentially reducing the time it takes to bring a drug to market.

3.4 Quantum Approximate Optimization Algorithm

The QAOA stands as a state-of-the-art quantum algorithm developed specifically to solve combinatorial optimization problems, where potential solutions are evaluated to find the best one. Within the pharmaceutical sector, combinatorial challenges are rampant, especially when one considers the myriad of ways molecules can be combined or modified to yield a desired therapeutic effect. For instance, in the field of drug design and formulation, researchers are often faced with the task of identifying the optimal molecular structures or configurations that yield maximum therapeutic efficacy with minimal side effects (Bhatia et al., 2023).

QAOA, leveraging the principles of quantum superposition and entanglement, has the potential to significantly accelerate the search for these optimal solutions by exploring multiple possibilities simultaneously rather than sequentially, as in classical methods (Bonde et al., 2024). In practical terms, when applied to pharmaceuticals, QAOA has the potential to revolutionize the drug discovery process. Consider a scenario where scientists aim to design a new drug molecule that interacts in a very specific way with a target protein in the body. Traditional methods often involve extensive trial and error, testing numerous molecular combinations to find a suitable fit. With QAOA, the quantum computer can assess multiple molecular structures simultaneously, pinpointing the most promising candidates in a fraction of the time it would take classically. Thus, QAOA not only accelerates the drug discovery pipeline but also offers the chance to identify molecular combinations that may be missed using classical methods. This quantum advantage holds the promise of faster, more efficient drug development, benefiting both the pharmaceutical industry and patients awaiting novel treatments.

3.5 Shor's Algorithm

Originally devised by mathematician Peter Shor, Shor's algorithm is best known for its capability to efficiently factorize large numbers into their prime factors—a task that classical computers find increasingly difficult as the numbers grow larger (Aggarwal et al., 2023). At first glance, one might question the relevance of this number theory breakthrough to the pharmaceutical world. However, the significance becomes apparent when considering data encryption and security. Modern encryption systems, including the RSA algorithm (Rivest-Shamir-Adleman), which underpins a significant portion of online data security, rely heavily on the difficulty of factorizing large numbers to ensure the confidentiality of information. If Shor's algorithm were to be implemented on a sufficiently advanced quantum computer, it would render many of these classical encryption methods vulnerable, potentially compromising data security.

Within the field of pharmaceuticals, the protection of sensitive data is paramount. This includes patient medical records, clinical trial data, proprietary drug formulas, and personalized medicine data (Khang & Medicine, 2023). The potential of Shor's algorithm to undermine existing encryption methods prompts the pharmaceutical industry, along with many other sectors, to seek quantum-resistant encryption techniques. On the flip side, the algorithm could also be employed defensively by pharmaceutical companies, ensuring that their data encryption methods are robust against potential quantum decryption threats. In essence, while Shor's algorithm might not directly influence drug discovery or molecular interactions, its implications for data security in the pharmaceutical domain are profound, emphasizing the need for proactive measures in data protection as quantum computing matures.

3.6 HHL Algorithm (Harrow-Hassidim-Lloyd)

The HHL algorithm, named after its developers Harrow, Hassidim, and Lloyd, is a quantum algorithm designed to solve linear systems of equation efficiency. These equations are pervasive in numerous scientific and engineering domains. In classical computing, as the size and complexity of these systems increase, the computational cost of finding solutions can become prohibitively expensive. Utilizing the HHL algorithm's quantum approach, these large-scale linear systems can be solved more efficiently, offering solutions in significantly less time than traditional methods when applied to suitable systems (Feng et al., 2023).

In the pharmaceutical context, linear systems often arise in modeling biological processes, pharmacokinetics, and drug interactions. For instance, predicting how a drug will be metabolized in the body, understanding its distribution, or forecasting its interaction with other compounds can often be represented as linear systems (Rani et al., 2023). The HHL algorithm can be instrumental in these scenarios, providing rapid insights into drug behaviors and interactions. Another salient application within pharmaceuticals is personalized medicine. Here, patient-specific data can be used to create models that predict individual responses to drugs. By leveraging the HHL algorithm, researchers can quickly process vast amounts of patient data to tailor treatments more effectively, paving the way for more customized, effective, and safer therapeutic interventions (Khang, Vugar and Abuzarova et al., 2023).

4. QUANTUM-ENHANCED MOLECULE MODELING

4.1 Predicting Drug Interactions and Side Effects

Molecular modeling is a pivotal component of modern pharmaceutical research. Through the simulation of molecular interactions, scientists can gain insights into how potential drug compounds interact with biological targets, enabling predictions about efficacy, side effects, and potential drug–drug interactions (Wang et al., 2023). Traditionally, these simulations have relied on classical computational methods, which, while powerful, possess inherent limitations when diving deep into the quantum realm, where many pivotal drug interactions occur. Quantum-enhanced molecular modeling emerges as a solution. Tapping directly into quantum mechanics, quantum computers offer a platform more attuned to simulating molecular behaviors. They can capture the subtle nuances of quantum phenomena such as tunneling, superposition, and entanglement, which play critical roles in molecular interactions (Bhatia et al., 2023).

For instance, a quantum model might offer a more precise depiction of how electrons in a drug molecule interact with those in a target protein—a detail that may be overly simplified or approximated in classical simulations. This heightened fidelity in quantum simulations is of paramount importance when predicting the potential side effects of drugs. Side effects often arise from unforeseen interactions between drug molecules and unintended biological targets. Quantum-enhanced modeling can provide a richer, more detailed view of these potential off-target interactions, allowing researchers to anticipate and mitigate adverse effects earlier in the drug's lifecycle. Furthermore, in an era where polypharmacy (the concurrent use of multiple medications by one individual) is common, predicting drug–drug interactions becomes crucial. Quantum computers, given their ability to process vast amounts of data simultaneously and model complex systems, can potentially simulate the simultaneous interactions of multiple drug molecules within the biological system. This capability could lead to a more holistic understanding of how drug combinations might work in tandem or, conversely, interfere with each other.

4.2 Optimizing Drug Synthesis Routes

At the heart of the pharmaceutical industry lies the complex and multifaceted process of drug synthesis. Beyond the discovery of a potential drug molecule lies the challenge of producing it efficiently, economically, and at scale. Traditional drug synthesis involves intricate pathways that might require multiple steps, each with its own set of reactants, conditions, and catalysts. The optimization of these pathways to achieve maximal yield, minimal waste, and economic feasibility is a daunting computational task that presents considerable challenges to classical computers (Yingngam, 2023). Quantum computing, with its unique ability to process and analyze vast combinatorial spaces, offers a revolutionary approach to this problem. Consider a multistep synthesis process: at each stage, multiple reactants, solvents, catalysts, and conditions can be chosen. The combination possibilities quickly become astronomical.

Quantum computers can explore these myriad combinations simultaneously, thanks to the principles of superposition. By modeling the quantum nature of molecular interactions directly, they can predict the outcomes of reactions more accurately, guiding chemists toward the most promising synthesis routes. Additionally, certain quantum algorithms are tailor-made for optimization problems. For instance, algorithms such as the quantum approximate optimization algorithm can be adapted to find the most efficient synthesis pathways, minimizing the number of steps, reducing waste, or optimizing for other desired outcomes (Bonde et al., 2024).

Given that even slight enhancements in efficiency can lead to substantial cost reductions and a diminished environmental footprint, the potential advantages of such optimizations are considerable. Quantum computing's prowess does not end there. It can also be instrumental in discovering and modeling novel catalysts. Catalysts, which accelerate reactions without being consumed, are integral to drug synthesis, influencing the pace, selectivity, and yield of reactions. Quantum simulations can shed light on the underlying mechanisms of catalyst actions, potentially leading to the discovery of more effective catalysts and further streamlining the drug synthesis process (Deswal et al., 2023).

4.3 Error Correction and Noise in Quantum Computers

Quantum computing, while offering transformative potential, also faces significant challenges in molecular drug modeling. Among the most pressing is error correction, compounded by the inherent noise that comes with quantum systems. Unlike classical computers, where bits are relatively stable and errors can be easily managed, quantum bits—qubits—are sensitive to their environment. Even minor perturbations from thermal fluctuations, electromagnetic waves, or other sources can introduce errors in quantum computations. In the context of pharmaceutical sciences, where precision is paramount, these errors can significantly impede the reliability of quantum-enhanced research outcomes (Kikuchi et al., 2023). Noise in quantum computers arises from the delicate nature of quantum states. Qubits, when in a state of superposition or entanglement, can easily be disrupted, leading to the collapse of their quantum states or unintended interactions. This "decoherence" can occur rapidly, and as quantum systems scale up in size and complexity, managing and minimizing this noise becomes increasingly challenging (Kim et al., 2023). Given that drug design and molecular simulations require sustained and intricate quantum computations, the mitigation of noise is of utmost importance.

Fortunately, the field of quantum error correction has seen considerable advancement in recent years. Quantum error correction codes, analogous to classical error correction but tailored to the unique challenges of quantum systems, have been developed. These codes work by encoding a logical qubit (the information-carrying unit) across multiple physical qubits. By spreading out the information, localized errors affecting a few qubits can be detected and corrected without compromising the integrity of the overall computation (Garani et al., 2023). In pharmaceutical contexts, employing these error-correcting codes ensures that molecular simulations and drug interactions modeled on quantum computers remain accurate and trustworthy. However, the implementation of quantum error correction does introduce overheads.

The need for multiple physical qubits to represent a single logical qubit means larger, more complex quantum systems. Additionally, error correction routines require additional computations, potentially offsetting some of the speed advantages quantum computers offer. However, in pharmaceutical sciences, where the stakes are high, this trade-off between speed and accuracy is often acceptable, especially when the end goal is the development of safe and effective therapeutic agents.

5. CHALLENGES AND LIMITATIONS

5.1 Error Correction and Noise in Quantum Computers

Despite the potential of quantum error correction in ensuring the reliability of quantum computations, the practical implementation of these techniques in pharmaceutical sciences presents its own set of challenges (Izsák et al., 2023). At its core, quantum error correction requires the dedication of significant additional quantum resources (De Maio & Brandic, 2023). This overhead, in terms of qubit count and quantum operations, can sometimes offset the computational speed advantages that quantum systems intrinsically offer, making certain computations less efficient than their classical counterparts. For example, imagine trying to simulate the binding affinity of a new drug molecule to its target protein using a quantum computer. If the quantum system requires too many qubits dedicated solely to error correction, the remaining qubits may not be sufficient to accurately represent the drug molecule and its target protein, thus limiting the simulation's fidelity.

Physical constraints further complicate matters. Current quantum computers, especially those that harness superconducting qubits, require extremely low temperatures to function. These conditions are necessary to reduce thermal noise and enhance qubit coherence times (Nemoto, 2023). However, as we scale up the number of qubits to accommodate error correction, maintaining the uniformity of these ultracold conditions across the entire quantum system becomes challenging (Softley, 2023). Should there be inconsistencies in qubit performance due to temperature variations across the quantum system, the resultant simulation may inaccurately represent certain atomic interactions, leading to an erroneous prediction of the drug's efficacy.

In the context of intricate molecular simulations or drug pathway optimizations, any inconsistency in qubit performance can introduce errors, potentially leading to misleading or inaccurate results. Another challenge lies in the intricacies of the error correction codes themselves. While quantum error correction techniques can manage a variety of errors, they are often designed with specific error models in mind (Shaib et al., 2023). For instance, if an unanticipated error occurs for which the error correction code has not been designed, it could incorrectly simulate how the drug molecule interacts with water molecules in its environment, leading to false insights about its solubility or stability. In real-world quantum systems, however, the error landscape is complex and multifaceted (Acharya et al., 2023). Unanticipated errors or those outside the model's purview can still compromise computations. Ensuring that quantum error correction codes are robust against all conceivable error types in a pharmaceutical setting, where precision is vital, remains a formidable task (Aggarwal et al., 2023).

Moreover, while error correction addresses issues of qubit stability, it does not necessarily solve all problems related to noise. Quantum gates, which drive the transitions and operations of qubits, are themselves sources of noise (Kikuchi et al., 2023). For example, suppose quantum gates introduce noise when simulating electron movements in drug molecules. In that case, the prediction of the molecule's reactivity, and thus its potential side effects or interactions with other drugs, could be misleading. Optimizing these gates while ensuring that they do not introduce more errors than they rectify is a balancing act. This becomes especially pertinent when simulating intricate molecular systems or predicting nuanced drug interactions, where even slight inaccuracies can have cascading effects.

5.2 Scalability Concerns

The promise of quantum computing for pharmaceutical sciences largely hinges on the ability to scale quantum systems to handle complex molecular simulations, drug interaction predictions, and optimization tasks (Aggarwal et al., 2023). However, scaling up quantum computers presents a myriad of technical and theoretical challenges that directly impact their potential utility in pharmaceutical applications.

First and foremost is the physical challenge of adding more qubits to a quantum system. As the number of qubits increases, so does the complexity of the system's architecture (Golestan et al., 2023). Interconnecting these qubits, ensuring their stability, and maintaining the necessary conditions for their optimal operation become exponentially difficult. For instance, superconducting qubits, one of the leading platforms for quantum computing, necessitate cryogenic temperatures (Joshi & Moazeni, 2023). Ensuring uniform and stable conditions across a vast array of qubits is technologically challenging (Sood & Chauhan, 2023). In the realm of pharmaceuticals, where simulations often demand large, complex quantum systems, these scaling issues become particularly pertinent.

Beyond the physical challenges, there is the issue of error accumulation. The larger the quantum systems are, the higher the chance of errors manifesting (Bonde et al., 2024). As quantum systems grow, the probability of errors occurring within the system naturally increases. While quantum error correction methods can mitigate some of these errors, the overhead introduced by these techniques can diminish the advantages of a larger quantum system (Flöther, 2023). In the context of pharmaceutical sciences, where precision is of the essence, even a marginal increase in error rates can have significant ramifications, potentially leading to incorrect conclusions or missed drug discovery opportunities (Bano et al., 2023).

Additionally, the software and algorithms designed for quantum systems also face scalability concerns. As quantum systems expand, the algorithms need to be adapted and optimized to harness the increased computational power effectively (Gharehchopogh, 2023). This requires a continuous cycle of development and refinement, ensuring that quantum software can keep pace with hardware advancements. For drug discovery processes, which already demand extensive computational resources, the need for constantly evolving algorithms adds another layer of complexity (Avramouli et al., 2023).

There are also concerns related to the integration of quantum computing with existing pharmaceutical research methodologies. A scalable quantum system needs to work seamlessly with classical computational systems, data storage solutions, and analysis tools (Gupta et al., 2023). Ensuring this smooth integration, while also scaling the quantum system, demands a holistic approach, considering both the quantum and classical components of the research pipeline.

5.3 Integrating Quantum With Classical Systems

The integration of quantum computers into the existing infrastructure of pharmaceutical research, which is largely built upon classical computing paradigms, poses a set of unique challenges and limitations (Aithal, 2023). The symbiotic relationship between quantum and classical systems is crucial for effective computational research, but harmonizing these distinct computational entities is no trivial task. One primary challenge arises from the inherently different nature of quantum and classical computations. Quantum computers excel at specific tasks such as optimization problems, quantum simulations, and factorization, but they are not designed to replace classical computers in all aspects (Singh & Bhangu, 2023). For a smooth workflow in pharmaceutical sciences, which often involves a mix of quantum-suitable tasks and classical tasks (such as data analysis, storage, and retrieval), orchestrating a seamless handoff between quantum and classical systems becomes imperative (Ur Rasool et al., 2023). Achieving this requires sophisticated middleware and interfaces that can translate and route tasks to the appropriate computational platform. Furthermore, data translation and compatibility have become significant concerns. Quantum computers process and produce information in quantum states, which do not directly align with the binary data of classical systems. Converting quantum results into a format that classical systems and, consequently, researchers can interpret and utilize is a nontrivial challenge. Moreover, ensuring that this conversion does not result in the loss of crucial information or introduce errors is vital, especially in pharmaceutical applications where accuracy is paramount.

There's also the matter of infrastructure. Quantum computers, especially those that rely on superconducting qubits or trapped ions, have stringent environmental requirements, including cryogenic cooling and electromagnetic shielding (Brijwani et al., 2023). Integrating these systems physically within the existing infrastructure of pharmaceutical research facilities and ensuring that they can effectively communicate with classical systems presents logistical and engineering challenges. The need for specialized technicians and quantum experts to maintain and operate these machines adds another layer to the integration dilemma. Finally, there is a cultural and educational aspect to consider. The principles of quantum computing are fundamentally different from classical computing. For researchers and scientists accustomed to classical methodologies, there is a steep learning curve associated with understanding and harnessing quantum capabilities (Bethel et al., 2023). Bridging this knowledge gap, ensuring that pharmaceutical researchers can effectively utilize integrated quantum-classical systems, will necessitate extensive training and perhaps a rethinking of educational paradigms in the field (Khang & Muthmainnah et al., 2023).

Overall, when integrating quantum computing with pharmaceutical applications, several challenges arise, categorized broadly into technical, operational, and ethical challenges (Table 3). From a technical standpoint, the infancy of quantum computing means that issues such as noise, error management, and the limited coherence times of qubits remain persistent obstacles. Scalability remains a pressing concern, with current quantum hardware not yet sufficiently advanced to handle the large-scale computations often needed in pharmaceutical research. The integration of quantum and classical systems is nontrivial, and the industry still lacks standardized quantum programming languages. Moreover, maintaining delicate quantum states requires specialized infrastructure, often involving ultralow temperatures, further complicating the implementation process. From an operational perspective, the nascent nature of the field has led to a shortage of experts who are well versed in both quantum computing and its pharmaceutical applications. The initial setup and maintenance costs of quantum systems can be prohibitively high for many institutions.

Moreover, access to cutting-edge quantum platforms remains limited, and integrating these systems into existing IT infrastructures presents its own set of challenges. The development and validation of quantum algorithms, specifically tailored for pharmaceutical needs, is a slow-moving process. While the preceding text did not delve into all potential ethical challenges, the importance of ethical considerations in this integration is clear. Concerns about data privacy, especially in light of patient confidentiality, are paramount. The potential misuse of quantum technology, particularly in designing harmful biological agents, cannot be overlooked. As quantum computing promises to accelerate drug discovery, there is a risk of widening disparities in health care research, with only a few having access to these advanced tools. Additionally, the rapid design of drugs using quantum insights brings about ethical dilemmas, particularly if exhaustive testing is bypassed. Last, as organizations delve deeper into quantum-enhanced drug discovery, intellectual property concerns and the sharing of methodologies will become pivotal points of discussion.

Table 3. Challenges in quantum computing integration for pharmaceutical applications

Category	Challenges
Technical	- Noise and error management in quantum computations
	- Scalability issues with current quantum hardware
	- Limited qubit coherence times
	- Difficulty in interfacing quantum systems with classical systems
	- Lack of standardized quantum programming languages
	- Need for specialized infrastructure to maintain quantum systems (e.g., ultralow temperatures)
Operational	- Shortage of experts trained in quantum computing for pharmaceutical applications
	- High initial costs of quantum hardware and maintenance
	- Limited access to advanced quantum computing platforms for many research institutions
	- Integration challenges with existing IT systems and data management structures in pharmaceutical companies
	- Slow development and validation of quantum algorithms tailored for specific pharmaceutical needs
Ethical	- Concerns about data privacy in quantum systems, especially with patient data
	- The potential misuse of quantum technology for designing malicious biological agents
	- Unequal access to quantum-enhanced drug discovery tools, potentially widening disparities in healthcare research and outcomes
	- Ethical implications of rapidly designed drugs without exhaustive testing, due to accelerated quantum-aided drug discovery processes
	- Questions about intellectual property and the sharing of quantum-enhanced drug discovery methodologies between organizations

6. FUTURE PERSPECTIVES

6.1 Advances in Quantum Hardware and Algorithms

6.1.1 Advances in Quantum Hardware

6.1.1.1. Scalability

Scalability remains a pivotal consideration in quantum computing. To execute more complex computations and simulations—especially those pertinent to pharmaceutical sciences—a quantum computer with an expanded number of qubits is essential (Izsák et al., 2023). At the core of quantum computing are qubits, quantum analogs, and classical bits. Unlike classical bits, which can be either 0 or 1, qubits can exist in a superposition, allowing them to simultaneously represent both 0 and 1. This feature enables quantum computers to process a vast amount of information concurrently (Liu, 2023). However, qubits are highly sensitive to external factors. Even minor disturbances can disrupt their superposition, a phenomenon known as "decoherence". As more qubits are added to a system (scaling), ensuring that all these qubits maintain their coherence becomes progressively challenging.

Consequently, researchers are striving to find ways to increase the number of qubits in a quantum system without compromising their coherence (Ezratty, 2023). Several promising approaches, such as topological qubits that inherently resist local disturbances and quantum error correction codes (designed to detect and correct errors that occur during computation without disturbing the qubits), are under investigation (Dejpasand & Sasani Ghamsari, 2023). The significance of scalability in the field of drug discovery is immense. To understand and predict the behavior of complex molecules and their interactions within the body, vast amounts of data need to be processed. Quantum computers, with their potential for parallelism, are well suited to this task. However, for precise simulations of complex molecular interactions, quantum systems with a large number of qubits that can maintain coherence are needed. As an illustration, the simulation of caffeine—a comparatively simple molecule—demands a quantum computer equipped with no fewer than 160 qubits (Shinde & Bendale, 2023).

6.1.1.2. Error correction

Quantum computing offers vast potential, yet it is not without susceptibility to errors. The intrinsic property of qubits—their ability to exist in superpositions—renders them vulnerable to mistakes from various external influences, including environmental disturbances (Aggarwal et al., 2023). This is where quantum error correction (QEC) comes into play. QEC is a set of techniques designed to protect quantum information from errors due to decoherence and other quantum noise. QEC is essential because even a small error in a quantum calculation can result in significant inaccuracies, especially in sensitive fields such as pharmaceuticals where precision is crucial (Bonde et al., 2024). For example, in classical computing, redundant bits are used to detect and correct errors. Analogously, in quantum computing, a single logical qubit can be represented using multiple physical qubits. This presentation aids in the detection and correction of errors without collapsing the qubit's state.

However, the challenge here is implementing error correction without adding significant overhead or complexity to the quantum computation. Recent advancements have seen the development of various quantum error-correcting codes. One of the most promising is the surface code, which uses a two-dimensional lattice of qubits and has become a frontrunner due to its relatively high threshold for errors (Dejpasand & Sasani Ghamsari, 2023). In the context of pharmaceutical applications, consider the simulation of a complex protein folding process. If there is an error in the quantum computation, it might predict an incorrect folding pattern, leading to misconceptions about the protein's function or its interaction with drugs. Enhanced QEC techniques would ensure that such simulations are more accurate, paving the way for more precise drug interactions and treatments.

6.1.1.3. Hybrid systems

Hybrid systems that combine quantum and classical computing aim to take advantage of the unique strengths of both computational paradigms. While quantum computers offer the potential to address certain problems more efficiently than classical systems, they are not universally superior for all computational tasks. Thus, a synergistic combination can present optimal outcomes (De Maio & Brandic, 2023). Quantum computers excel in areas such as factorization, quantum system simulations, and tackling optimization problems. Within pharmaceutical sciences, they can be invaluable for simulating molecular interactions, optimizing drug design parameters, and more. Classical computers, on the other hand, are adept at handling large-scale data processing, storage, and tasks that do not inherently benefit from quantum parallelism. For instance, in drug discovery, a quantum computer might be tasked with simulating a specific molecular interaction, with the results subsequently transferred to a classical system for deeper analysis, such as visualizing the data or comparing it with expansive databases of related simulations. When delving into optimization problems in pharmacology, there may be a need to both optimize a drug molecule's structure and determine the logistics of its delivery.

Here, a quantum computer's strength in leveraging quantum parallelism could be harnessed to identify the optimal structure of the drug molecule, while a classical computer efficiently manages logistics, scheduling, and distribution. Furthermore, in the field of machine learning, quantum computers exhibit the capability to expedite certain tasks, such as training specific neural network models (Bonde et al., 2024). Within a hybrid framework, the quantum system might focus on accelerating the training phase, while its classical counterpart adeptly handles tasks such as data preprocessing, storage, and posttraining evaluations.

6.1.1.4. Cooling and noise reduction

Noise is a primary challenge in quantum computing. In quantum mechanics, "noise" denotes any kind of disturbance that affects the fragile quantum states of qubits, thus influencing the accuracy and reliability of quantum operations. This noise typically results from the external environment or imperfections in the quantum device (Baiardi et al., 2023). Given the extreme sensitivity of qubits to their surroundings, even minute disturbances can lead to errors in quantum computations (Golestan et al., 2023). One primary method for reducing quantum noise is through advanced cooling technologies. Quantum computers, especially those based on superconducting qubits, operate at extremely low temperatures, often just fractions of a degree above absolute zero. These ultracold environments are necessary because they minimize thermal noise (Ezratty, 2023). For instance, dilution refrigerators are commonly used to achieve these temperatures, and advancements in cooling technologies could lead to more efficient and compact refrigerators in the future (Agarwal & Mohanta, 2023).

A pertinent paper in this domain is reviewed by Bonde et al. (2024), which highlights the challenges and prospects of noise in near-term quantum devices. Beyond cooling, there are active research efforts geared toward developing techniques specifically tailored to mitigate quantum noise. Techniques such as dynamical decoupling and quantum error correction are being refined to reduce the impact of noise on quantum operations. An illustrative example can be found in a study by Singh et al. (2023). This work explores the application of error correction techniques in quantum hardware to combat noise. Additionally, as quantum hardware evolves, the materials and architectures employed will likely also evolve to inherently produce less noise. For example, research into topological qubits, which are thought to be more resistant to environmental noise due to their unique structure, is an active area of exploration (Zou et al., 2023).

6.1.2 Evolution of Quantum Algorithms

6.1.2.1. Efficient drug simulations

One of the most promising applications of quantum computing lies in its potential to simulate molecular and chemical reactions with high precision and efficiency. Classical computers face significant limitations in this regard, especially when simulating larger and more complex molecules. Quantum computers, on the other hand, have the potential to manage these simulations, thanks to their inherent compatibility with quantum mechanics—the fundamental principle governing molecular interactions (Pal, Bhattacharya, Dash, et al., 2023). For example, the quantum phase estimation algorithm can be applied to determine the energy levels of molecules, which is central to understanding their behavior and interactions. At its core, quantum phase estimation estimates the phase (or eigenvalue) of a unitary operator when given a quantum state that is an eigenvector of the operator.

In simpler terms, for certain mathematical operations, there are special states (eigenvectors) that, when the operation is applied to them, behave very predictably. They are simply scaled by a factor (eigenvalue). This algorithm can be used to find this scaling factor efficiently using quantum mechanics. Diving into the field of molecules, the term "energy levels" typically denotes the potential energy states that electrons in a molecule can occupy. These states are derived from the Schrödinger equation, which stands as a cornerstone in quantum mechanics. When solving this equation, it reveals the distinct energy states (also known as eigenvalues) a molecule possesses. From the perspective of quantum mechanics intertwined with linear algebra, the Schrödinger equation manifests as an eigenvalue challenge. This sets the stage for the role of quantum phase estimation. To find the energy of the molecule, one can map the problem onto a quantum computer in the form of a unitary operator.

Using quantum phase estimation on this operator, the algorithm returns the eigenvalues, which correspond to the energy levels of the molecule. Why is this important? The behavior, reactions, and interactions of molecules are heavily influenced by these energy levels. For instance, when two molecules react, they often do so in a way that allows them to reach a lower energy state. Knowing the energy levels in detail allows researchers to predict how molecules will behave and interact with one another. The challenge is that as molecules become larger and more complex, solving the Schrödinger equation becomes computationally very challenging for classical computers. This is due to the exponential growth of possible quantum states the electrons can be in. Quantum computers, by their very nature, are designed to handle such quantum states, making them potentially more efficient for this kind of problem (Bonde et al., 2024).

While Shor's algorithm is primarily known for its ability to factorize large numbers, the techniques it introduced are foundational for many quantum algorithms, including those used for molecular simulations. Shor's algorithm is one of the most famous quantum algorithms. Its claim to fame is its ability to factor large numbers into their prime components exponentially faster than the best-known classical algorithms. If implemented on a large-scale, fault-tolerant quantum computer, Shor's algorithm could break many of the cryptographic systems currently in use, as they rely on the difficulty of factoring large numbers (Bonde et al., 2024). Beyond its factoring application, the underlying techniques and principles introduced by Shor's algorithm are what make it foundational for other quantum algorithms. Some of the techniques include:

- Quantum Fourier transform: This is a quantum analog of the classical discrete Fourier transform. It is at the heart of Shor's algorithm and allows quantum systems to be transformed into a different representation, making certain computational problems more tractable (Wong, 2024b).
- Phase estimation: This technique, which is closely related to the quantum phase estimation discussed earlier, is about determining the phase of an eigenstate of a unitary operator. This process is integral to Shor's algorithm and has found applications in numerous other quantum algorithms (Golestan et al., 2023).
- Amplitude amplification: While more associated with Grover's algorithm, manipulating the probability amplitudes of quantum states to make certain outcomes more probable is a concept that, along with techniques from Shor's algorithm, has influenced the landscape of quantum algorithm development (Khanal et al., 2023).

When molecular simulations and number factorizations seem unrelated, they share underlying mathematical and computational challenges. Molecular simulations, at their core, involve solving complex eigenvalue problems (determining the energy levels of a system). The phase estimation techniques from Shor's algorithm, which help in identifying eigenvalues, are directly applicable here. This is why the foundational principle of Shor's algorithm can be adapted and extended to quantum algorithms designed for molecular simulations.

Another relevant quantum algorithm is the variational quantum eigensolver. This is a hybrid quantum-classical algorithm, which means it combines the strengths of quantum and classical computers. The primary aim of this algorithm is to determine the lowest energy state (ground state) of a quantum system, such as a molecule (Jattana et al., 2023). Why is ground-state energy important? The ground-state energy of a molecule defines its most stable configuration. In pharmaceutical chemistry, understanding this stable configuration is paramount for various applications. For drug design, in particular, knowing the ground state energy of a molecule can provide insights into how that molecule may interact with others, which is key for developing new medications.

Accurate prediction of the energy of molecular interactions allows for the prediction of their behavior. The variational quantum eigensolver uses a variational approach. First, a quantum computer is used to prepare a quantum state (a guess for the ground state) of the system. This state is then measured to estimate its energy. The way this energy estimation is performed is heavily reliant on the principles of quantum mechanics. The estimated energy is then sent to a classical computer, which adjusts the parameters of the quantum state based on the outcome to obtain a better guess for the ground state. This process is iteratively refined, with the quantum computer preparing states and the classical computer optimizing them, until the best approximation for the ground state energy is found. Current quantum computers, often called "noisy intermediate-scale quantum" (NISQ) devices, are not yet fault-tolerant and can only run relatively short quantum circuits before errors become significant. The variational quantum eigensolver is designed to work with these limitations. Its hybrid nature, relying partly on classical computers, and its iterative approach, which involves shorter quantum circuits, make it one of the more practical algorithms for quantum computers available today (Bonde et al., 2024).

6.1.2.2. Machine learning and AI

Machine learning and AI are transforming a multitude of fields, and when combined with the capabilities of quantum computing, they can further revolutionize drug discovery. Quantum machine learning (QML) merges quantum algorithms with machine learning techniques (Jadhav et al., 2023). Quantum systems, thanks to superposition, can process vast amounts of data simultaneously. This allows quantum algorithms to potentially tackle certain computational problems more efficiently than their classical counterparts. For example, one application of QML in drug discovery is optimizing molecular structures for desired properties. With the advantage of quantum parallelism—where quantum systems can represent and process vast amounts of data simultaneously due to superposition—QML algorithms can, in theory, achieve unprecedented speeds in data processing and analysis.

A significant part of drug discovery involves searching for molecules with desired properties, whether their binding affinity, solubility, or metabolic stability. Classical machine learning has already been employed to predict these properties by training on large databases of molecular data. With QML, this process can be turbocharged. Quantum computers can handle and process high-dimensional data spaces more efficiently. This means that optimizing molecules more quickly and accurately than classical methods leads to more efficient identification of drug candidates with desired properties (Avramouli et al., 2023).

One notable quantum algorithm with implications for machine learning is the quantum support vector machine (QSVM). QSVM can be used for both classification and regression tasks, making it suitable for tasks such as drug classification based on molecular properties (Bonde et al., 2024). This means it has the potential to classify drugs based on their molecular properties or predict specific outcomes, such as binding affinities, providing a more efficient approach to analyzing and categorizing pharmaceutical compounds. Another intriguing avenue is the quantum Boltzmann machine (QBM), a quantum version of the classical Boltzmann machine that is used in deep learning. QBMs can be trained faster than their classical counterparts, potentially accelerating drug discovery processes that rely on deep learning. This accelerated performance could be pivotal in drug discovery, where deep learning is employed to analyze vast datasets, potentially streamlining the identification and optimization of new therapeutic compounds (Avramouli et al., 2023). Moreover, pattern recognition in complex molecular datasets is a significant challenge in drug discovery.

Quantum-enhanced machine learning algorithms can identify patterns in these datasets more efficiently, potentially predicting drug behavior, side effects, or interactions that might not be apparent through classical methods. However, current quantum computers, especially NISQ devices, are noisy. Future developments in QML will likely focus on algorithms that are more resilient to such noise, enabling more accurate predictions even on imperfect hardware. As with many quantum algorithms today, the future may see more hybrid quantum-classical machine learning algorithms. Such models can leverage the strengths of both quantum and classical systems, leading to more practical and scalable solutions.

6.1.2.3. Personalized medicine

The integration of quantum computing into personalized medicine has the potential to bring about transformative healthcare solutions. Personalized medicine aims to tailor medical treatments to individual patient characteristics, often leveraging genetic or molecular profiling. The vastness and complexity of genetic data present significant challenges for classical algorithms, especially when processing entire genomes. Quantum algorithms, however, could significantly expedite this analysis.

A case in point is the quantum version of the Fourier transform, a core component of many genetic algorithms. This could dramatically speed up tasks such as DNA sequence alignment (Chakraborty et al., 2023). Genetic variations can cause drugs to be effective in some patients and ineffective in others. Quantum algorithms, with their ability to swiftly analyze genetic profiles, can pinpoint which drug compounds are likely to be most beneficial for specific individuals. This has the potential to reduce the trial-and-error nature of prescribing medication, leading to more effective treatments and fewer side effects. In the field of oncology, for example, understanding genetic mutations in tumors is pivotal to selecting the best treatment strategies. Quantum-enhanced algorithms could expedite the process of sequencing tumor DNA and comparing it to healthy DNA. This allows for quicker and more efficient identification of crucial mutations (Flöther, 2023).

6.1.2.4. Optimization problems

Optimization challenges are ubiquitous in the pharmaceutical industry, and they can span multiple domains from research and development to distribution. Quantum algorithms have shown the potential to address certain optimization problems more efficiently than classical methods. Concerning supply chain optimization, the pharmaceutical supply chain is a complex web of manufacturing, distribution, and retail networks. Identifying the most efficient way to distribute drugs while minimizing costs and ensuring timely delivery can be computationally intensive. Quantum algorithms, through techniques such as quantum annealing, can find optimal solutions faster than classical methods (Fasnacht & Straube, 2023).

For example, consider a scenario where a pharmaceutical company may need to determine the best route to distribute a temperature-sensitive vaccine to various hospitals and clinics across the country while ensuring that it remains within a particular temperature range. A quantum computer could quickly analyze all potential routes and conditions, selecting the best route. Additionally, pinpointing the precise drug dosage that optimizes efficacy and minimizes adverse effects is crucial. This involves analyzing vast amounts of clinical trial data and patient feedback. Quantum algorithms can assist by rapidly processing these data and identifying the best dosage ranges. In the field of drug design, discovering a molecule that combines the desired therapeutic effect and minimal side effects is a challenge. Quantum algorithms can search the vast space of molecular structures more efficiently to find potential drug candidates.

6.2 The Synergy of Quantum Computing and Artificial Intelligence in Pharmaceuticals

The future of pharmaceuticals is set to be profoundly shaped by the convergence of quantum computing and AI. Quantum computing, which is renowned for its capacity to handle and process vast amounts of complex data through quantum parallelism, presents the potential to solve problems that are currently computationally infeasible for classical systems (Liu, 2023). On the other hand, AI, with its advanced machine learning algorithms, can analyze patterns, make predictions, and optimize processes based on its data analytics (Askr et al., 2023). As these two technologies merge, the pharmaceutical industry can look forward to unprecedented advancements in drug discovery, molecular simulations, and personalized medicine (Wang et al., 2023).

Quantum-enhanced machine learning algorithms could streamline drug discovery by rapidly identifying potential drug candidates and predicting their behavior more effectively than traditional models (Bonde et al., 2024). In the sphere of personalized medicine, the synergy of quantum computing and AI has the potential to revolutionize treatments tailored to individual genetic makeups. Quantum algorithms are poised to excel in analyzing vast and intricate genetic datasets, while AI illuminates how genetic variations contribute to disease susceptibility and drug responses (Sagingalieva et al., 2023). This integrative tactic would enable the creation of drug treatments optimized for individual patients, increasing efficacy and reducing potential side effects. Additionally, with both quantum computing and AI technologies advancing in tandem, there is promise for even more sophisticated and efficient tools. The resultant tools and methodologies could dramatically hasten the drug discovery process, optimize pharmaceutical supply chains, and lead to the development of novel, effective therapeutic strategies tailored to individual needs.

6.3 Potential Quantum Breakthroughs in Personalized Medicine

The field of personalized medicine is on the brink of significant transformation due to the potential integration of quantum computing (Ur Rasool et al., 2023). Traditional medical approaches typically involve broad treatments that cater to the average patient. However, with advances in genetic sequencing, the vision of treatments tailored to an individual's unique genetic makeup is becoming increasingly tangible (Chakraborty et al., 2023). Quantum computing, with its exceptional computational power, can process and analyze vast genetic datasets at speeds unimaginable with classical systems (Liu, 2023).

This power presents opportunities for breakthroughs in understanding the numerous genetic factors influencing disease susceptibility, progression, and drug response (Pal, Bhattacharya, Lee, et al., 2023). Such insights could lead to the development of highly personalized therapeutic strategies, where treatments and medications are optimized for individual patients, greatly enhancing efficacy and minimizing potential adverse reactions (Niazi, 2023).

Furthermore, as quantum algorithms become more sophisticated, they can probe the complex relationships between genes, proteins, and metabolic pathways, revealing detailed information about individual health profiles (Pal, Bhattacharya, Lee, et al., 2023). These insights could provide the foundation for preventive medicine, where potential health issues are identified and addressed even before they manifest. Beyond just treatments, personalized medicine powered by quantum computing could also guide individualized wellness and lifestyle recommendations, promoting holistic health management (Aggarwal et al., 2023). In essence, the merger of quantum computing and personalized medicine promises a future where healthcare is not just about treating illnesses but proactively maintaining and optimizing health based on one's unique genetic composition.

Overall, based on the author's perspective, potential quantum innovations in the next few decades will be transformative for pharmaceutical science applications. As outlined in Table 4, technological advancements are expected to begin with the introduction of noise-reduced quantum chips, focusing on refining quantum accuracy and scalability. As quantum technology matures, the integration of artificial

Table 4. Potential quantum innovations in the next few decades for pharmaceutical applications

Technological Advancements	Research Initiatives	Innovations in Pharmaceutical Sciences
Introduction of noise-reduced quantum chips	Large-scale studies on quantum noise mitigation	Enhanced precision in drug molecule simulations
Development of scalable quantum error correction techniques	Quantum simulations of simple proteins	Prediction of simple protein–ligand interactions using quantum algorithms
Improved hybrid quantum-classical algorithms	Expansion of quantum-computing-based bioinformatics tools	Quantum-aided drug repurposing initiatives
Quantum computers reaching beyond 1000 qubits	Research into quantum machine learning for biological data	Faster and more accurate virtual drug screenings
Enhanced quantum annealing techniques	Large-scale molecular dynamics simulations on quantum systems	Optimized synthesis pathways for complex drug molecules
Advancements in quantum encryption for biomedical data	Quantum-enhanced genomics and personalized medicine studies	Quantum algorithms for personalized drug dosing and scheduling
Integration of AI and quantum computing	Exploratory research into quantum effects in biological systems	Understanding and designing drugs for multifactorial diseases using quantum insights
Quantum computers with 10,000+ qubits	Pilot projects for full protein folding on quantum platforms	Breakthroughs in understanding complex biological systems at quantum resolution
New quantum algorithms tailored for multidrug interactions.	Expansion of quantum pharmacology initiatives.	Quantum-enhanced prediction of drug–drug and drug-disease interactions.
Advanced quantum hardware tailored for pharmaceutical research	Quantum studies of advanced drug delivery systems	Development of novel drug delivery mechanisms informed by quantum simulations
Establishment of industry-wide quantum pharmaceutical research standards	Large-scale quantum collaboration initiatives across pharmaceutical giants	Introduction of the first quantum-designed blockbuster drugs to the market

intelligence and quantum computing will become a reality, ultimately leading to quantum computers boasting over 10,000 qubits. These advancements will run parallel to research initiatives that encompass large-scale studies on quantum noise mitigation, delving deep into quantum effects in biological systems, and culminating in broad quantum collaboration initiatives spanning major pharmaceutical giants.

On the application front, the initial enhancements in precision for drug molecule simulations will set the stage for various innovations. The industry will witness quantum algorithms' capability to predict protein–ligand interactions, repurpose drugs, and facilitate faster drug screenings. Quantum technology will also underpin the optimized synthesis of complex drug molecules and the design of drugs for multifactorial diseases. Furthermore, as quantum research deepens its roots in pharmaceutical applications, the possibility of introducing the first quantum-designed blockbuster drugs to the market by the end of the projected period looks promising. This synergy between quantum advancements, research, and pharmaceutical innovations presents an exciting path for the future of medicine.

7. CONCLUSION

In the ever-evolving landscape of pharmaceutical sciences, the integration of quantum computing emerges as a pivotal tool of transformative potential. This chapter has shed light on the fundamental principles of quantum computing, and its synergy with pharmaceutical processes has been elucidated. This exploration paints a picture of a future where intricate molecular interactions can be understood with unprecedented depth and precision. The capabilities of quantum computers, especially their ability to process vast amounts of data simultaneously, offer a promising solution to some of the most complex challenges faced by the pharmaceutical industry today. While the current applications of quantum computing in pharmaceuticals are still budding, they already demonstrate the potential to revolutionize drug design, molecular simulations, and interaction predictions.

The accuracy and speed afforded by quantum algorithms have potentially shortened drug discovery timelines, bringing life-saving medications to the market with greater speed and efficiency. However, the journey is not devoid of hurdles. As with any emerging technology, quantum computing faces unique challenges, primarily related to hardware limitations and error corrections. Addressing these challenges is crucial for the wider integration and reliability of quantum methodologies in pharmaceutical research.

Peering into the future, it is evident that the horizon is rife with potential. As quantum hardware and algorithms advance, it is anticipated that they will play an even more integral role in the pharmaceutical sector. Current research endeavors are poised to bring forth innovations that will further refine the applications of quantum computing, potentially unlocking solutions to longstanding challenges in drug discovery.

Furthermore, as quantum mechanics principles increasingly align with the biological intricacies of molecular interactions, the convergence of quantum computing and pharmaceutical sciences could drive innovations in personalized medicine, fostering treatments aligned with individual genetic profiles. In essence, while the integration of quantum computing into the pharmaceutical sciences is still in its early stages, its potential suggests a significant influence on healthcare, redefining the scope of drug discovery and development in the coming years.

8. ABBREVIATIONS

AI Artificial intelligence
HTS High-throughput screening
ML Machine learning
NISQ Noisy intermediate-scale quantum
QBN Quantum Boltzmann machine
QEC Quantum error correction
QML Quantum machine learning
QSVM Quantum support vector machine
R&D Research and development
VQE Variational quantum eigensolver

9. ACKNOWLEDGMENT

The author extends sincere gratitude to the expert reviewers for their invaluable feedback and insights that greatly enriched this chapter. Their expertise and dedication were instrumental in shaping the narrative and ensuring the accuracy of the content presented. While this chapter was not supported by any grants, the collaborative spirit and commitment of the reviewers played a pivotal role in its realization.

REFERENCES

Acharya, R., Aleiner, I., Allen, R., Andersen, T. I., Ansmann, M., Arute, F., & Google Quantum, A. I. (2023). Suppressing quantum errors by scaling a surface code logical qubit. *Nature, 614*(7949), 676–681. doi:10.103841586-022-05434-1 PMID:36813892

Agarwal, A., & Mohanta, C. (2023). Development, Working and Standardization of Quantum Computing. *Journal of The Institution of Engineers (India): Series B, 104*(2), 523-531. doi:10.1007/s40031-023-00868-1

Aggarwal, L., Sachdeva, S., & Goswami, P. (2023). Quantum healthcare computing using precision based granular approach. *Applied Soft Computing, 144*, 110458. doi:10.1016/j.asoc.2023.110458

Aithal, P. (2023). Advances and new research opportunities in quantum computing technology by integrating it with other ICCT underlying technologies. *International Journal of Case Studies in Business* [IJCSBE]. *IT and Education, 7*(3), 314–358. doi:10.47992/IJCSBE.2581.6942.0304

Askr, H., Elgeldawi, E., Aboul Ella, H., Elshaier, Y. A. M. M., Gomaa, M. M., & Hassanien, A. E. (2023). Deep learning in drug discovery: An integrative review and future challenges. *Artificial Intelligence Review, 56*(7), 5975–6037. doi:10.100710462-022-10306-1 PMID:36415536

Avramouli, M., Savvas, I. K., Vasilaki, A., & Garani, G. (2023). Unlocking the potential of quantum machine learning to advance drug discovery. *Electronics (Basel), 12*(11), 2402. doi:10.3390/electronics12112402

Baiardi, A., Christandl, M., & Reiher, M. (2023). Quantum computing for molecular biology. *ChemBioChem, 24*(13), e202300120. doi:10.1002/cbic.202300120 PMID:37151197

Bano, I., Butt, U. D., & Mohsan, S. A. H. (2023). New challenges in drug discovery. In S. Das, S. Thomas, & P. P. Das (Eds.), *Novel Platforms for Drug Delivery Applications* (pp. 619–643). Woodhead Publishing. doi:10.1016/B978-0-323-91376-8.00021-5

Bethel, E., Amankwah, M. G., Balewski, J., Van Beeumen, R., Camps, D., Huang, D., & Perciano, T. (2023). Quantum computing and visualization: A disruptive technological change ahead. *arXiv preprint arXiv:2310.04937, 43*(6), 1-11. doi:/arXiv.2310.04937 doi:10.48550

Bhatia, A. S., Saggi, M. K., & Kais, S. (2023). Quantum machine learning predicting ADME-Tox properties in drug discovery. *Journal of Chemical Information and Modeling*, acs.jcim.3c01079. doi:10.1021/acs.jcim.3c01079 PMID:37603536

Bonde, B., Patil, P., & Choubey, B. (2024). The Future of Drug Development with Quantum Computing. In A. Heifetz (Ed.), *High Performance Computing for Drug Discovery and Biomedicine* (pp. 153–179). Springer US., doi:10.1007/978-1-0716-3449-3_7

Brijwani, G. N., Ajmire, P. E., & Thawani, P. V. (2023). Future of Quantum Computing in Cyber Security. In *Handbook of Research on Quantum Computing for Smart Environments* (pp. 267–298). IGI Global., doi:10.4018/978-1-6684-6697-1.ch016

Chaganti, S. K., Kesari, A. S., & Chowdhury, C. (2024). Execution of a quantum algorithm for calculating energies of some quantum mechanical systems. *Chemical Physics, 576*, 112082. doi:10.1016/j.chemphys.2023.112082

Chakraborty, C., Bhattacharya, M., Dhama, K., & Lee, S.-S. (2023). Quantum computing on nucleic acid research: Approaching toward next-generation computing. *Molecular Therapy. Nucleic Acids, 33*, 53–56. doi:10.1016/j.omtn.2023.06.007 PMID:37449046

Dash, S., Priyadarshini, S., Mohanty, S. N., Priyadarshini, S., & Dulla, N. (2023). Quantum computing in netnomy: A networking paradigm in e-pharmaceutical setting. In Evolution and Applications of Quantum Computing (pp. 145-161). doi:10.1002/9781119905172.ch9

De Maio, V., & Brandic, I. (2023). Accelerating Scientific Applications with the Quantum Edge: A Drug Design Use Case. In A. Bienz, M. Weiland, M. Baoulin, & C. Kruse (Eds.), *High Performance Computing, ISC High Performance 2023*. Springer Nature Singapore. doi:10.1007/978-3-031-40843-4_11

Dejpasand, M. T., & Sasani Ghamsari, M. (2023). Research trends in quantum computers by focusing on qubits as their building blocks. *Quantum Reports, 5*(3), 597–608. doi:10.3390/quantum5030039

Deswal, Y., Asija, S., Tufail, A., Dubey, A., Deswal, L., Kumar, N., Saroya, S., Kirar, J. S., & Gupta, N. M. (2023). Instigating the in vitro antidiabetic activity of new tridentate Schiff base ligand appended M (II) complexes: From synthesis, structural characterization, quantum computational calculations to molecular docking, and molecular dynamics simulation studies. *Applied Organometallic Chemistry, 37*(4), e7050. doi:10.1002/aoc.7050

Eswaran, U., & Khang A. (2023). Role of Quantum Computing in The Era of Artificial Intelligence. Applications and Principles of Quantum Computing. IGI Global Press. https://doi.org/ doi:10.4018/979-8-3693-1168-4-ch003

Ezratty, O. (2023). Perspective on superconducting qubit quantum computing. *The European Physical Journal A, 59*(5), 94. doi:10.1140/epja10050-023-01006-7

Fasnacht, D., & Straube, C. (2023). *Quantencomputing als Basistechnologie für den nächsten Konjunkturzyklus*. HMD Praxis der Wirtschaftsinformatik. doi:10.136540702-023-00969-x

Feng, C., Chen, L., & Zhao, L.-J. (2023). Coherence and entanglement in Grover and Harrow–Hassidim–Lloyd algorithm. *Physica A, 626*, 129048. doi:10.1016/j.physa.2023.129048

Flöther, F. F. (2023). The state of quantum computing applications in health and medicine. *arXiv preprint arXiv:2301.09106*. doi:/arXiv.2301.09106 doi:10.48550

Garani, S. S., Nadkarni, P. J., & Raina, A. (2023). Theory behind quantum error correcting codes: An overview. *Journal of the Indian Institute of Science, 103*(2), 449–495. doi:10.100741745-023-00392-7

Gharehchopogh, F. S. (2023). Quantum-inspired metaheuristic algorithms: Comprehensive survey and classification. *Artificial Intelligence Review, 56*(6), 5479–5543. doi:10.100710462-022-10280-8

Golestan, S., Habibi, M. R., Mousazadeh Mousavi, S. Y., Guerrero, J. M., & Vasquez, J. C. (2023). Quantum computation in power systems: An overview of recent advances. *Energy Reports, 9*, 584–596. doi:10.1016/j.egyr.2022.11.185

Gupta, S., Modgil, S., Bhatt, P. C., Chiappetta Jabbour, C. J., & Kamble, S. (2023). Quantum computing led innovation for achieving a more sustainable Covid-19 healthcare industry. *Technovation, 120*, 102544. doi:10.1016/j.technovation.2022.102544

Hema, T., & Arowolo, M. O. (2023). Quantum Computing Significance on Multidimensional Data: R-Tree Search Based on Grover's Search Algorithm. In A. K. Tyagi (Ed.), *Handbook of Research on Quantum Computing for Smart Environments* (pp. 217–230). IGI Global. doi:10.4018/978-1-6684-6697-1.ch012

Izsák, R., Riplinger, C., Blunt, N. S., de Souza, B., Holzmann, N., Crawford, O., Camps, J., Neese, F., & Schopf, P. (2023). Quantum computing in pharma: A multilayer embedding approach for near future applications. *Journal of Computational Chemistry, 44*(3), 406–421. doi:10.1002/jcc.26958 PMID:35789492

Jadhav, A., Rasool, A., & Gyanchandani, M. (2023). Quantum Machine Learning: Scope for real-world problems. *Procedia Computer Science, 218*, 2612–2625. doi:10.1016/j.procs.2023.01.235

Jattana, M. S., Jin, F., De Raedt, H., & Michielsen, K. (2023). Improved variational quantum eigensolver via quasidynamical evolution. *Physical Review Applied, 19*(2), 024047. doi:10.1103/PhysRevApplied.19.024047

Joshi, S., & Moazeni, S. (2023). Scaling up superconducting quantum computers with cryogenic rf-photonics. *Journal of Lightwave Technology*, 1-10. (In press) doi:10.1109/JLT.2023.3311806

Just, B. (2022). Quantum Gates on One Qubit. In B. Just (Ed.), *Quantum Computing Compact: Spooky Action at a Distance and Teleportation Easy to Understand* (pp. 69–82). Springer Berlin Heidelberg. doi:10.1007/978-3-662-65008-0_9

Khanal, B., Orduz, J., Rivas, P., & Baker, E. (2023). Supercomputing leverages quantum machine learning and Grover's algorithm. *The Journal of Supercomputing*, *79*(6), 6918–6940. doi:10.100711227-022-04923-4

Khang, A. (2023). Applications and Principles of Quantum Computing. IGI Global Press. https://doi.org/ doi:10.4018/979-8-3693-1168-4

Khang, A. (2023). Quantum Mechanics Primer - Fundamentals and Quantum Computing. Applications and Principles of Quantum Computing. IGI Global Press. https://doi.org/ doi:10.4018/979-8-3693-1168-4-ch001

Khang, A. (2023). Introduction to Quantum Computing and Integration of Quantum Computing. Applications and Principles of Quantum Computing. IGI Global Press. https://doi.org/ doi:10.4018/979-8-3693-1168-4-ch002

Khang, A. (2023). AI and IoT-Based Technologies for Precision Medicine. IGI Global Press. doi:10.4018/979-8-3693-0876-9

Khang, A. (2023). AI-Aided Data Analytics Tools and Applications for the Healthcare Sector. AI and IoT-Based Technologies for Precision Medicine. IGI Global Press. doi:10.4018/979-8-3693-0876-9.ch018

Khang, A., & Kali, C. R. (2023). Enabling the Future of Manufacturing: Integration of Robotics and IoT to Smart Factory Infrastructure in Industry 4.0. AI-Based Technologies and Applications in the Era of the Metaverse. IGI Global Press. doi:10.4018/978-1-6684-8851-5.ch002

Khang, A., & Muthmainnah, M. (2023). AI-Aided Teaching Model for the Education 5.0 Ecosystem. AI-Based Technologies and Applications in the Era of the Metaverse. IGI Global Press. doi:10.4018/978-1-6684-8851-5.ch004

Khang, A., Shah, V., & Rani, S. (2023). AI-Based Technologies and Applications in the Era of the Metaverse. IGI Global Press. doi:10.4018/978-1-6684-8851-5

Kikuchi, Y., Mc Keever, C., Coopmans, L., Lubasch, M., & Benedetti, M. (2023). Realization of quantum signal processing on a noisy quantum computer. *NPJ Quantum Information*, *9*(1), 93. doi:10.103841534-023-00762-0

Kim, Y., Eddins, A., Anand, S., Wei, K. X., van den Berg, E., Rosenblatt, S., Nayfeh, H., Wu, Y., Zaletel, M., Temme, K., & Kandala, A. (2023). Evidence for the utility of quantum computing before fault tolerance. *Nature*, *618*(7965), 500–505. doi:10.103841586-023-06096-3 PMID:37316724

Koczka, F., Prantner, C., & Bíro, C. (2023). Educational issues raised by the availability of the quantum computer. *Central-European Journal of New Technologies in Research. Education and Practice*, *5*(1), 1–17. doi:10.36427/CEJNTREP.5.1.5051

Lindh, R., & Fdez Galván, I. (2023). Molecular structure optimizations with Gaussian process regression. In P. O. Dral (Ed.), *Quantum Chemistry in the Age of Machine Learning* (pp. 391–428). Elsevier. doi:10.1016/B978-0-323-90049-2.00017-2

Liu, Q. (2023). Comparisons of conventional computing and quantum computing approaches. *Highlights in Science. Engineering and Technology*, *38*, 502–507. doi:10.54097/hset.v38i.5875

Ma, H., Liu, J., Shang, H., Fan, Y., Li, Z., & Yang, J. (2023). Multiscale quantum algorithms for quantum chemistry. *Chemical Science (Cambridge)*, *14*(12), 3190–3205. doi:10.1039/D2SC06875C PMID:36970085

Mahesh, V., & Shijo, S. (2023). Accelerating drug discovery with quantum computing. In S. N. Mohanty, R. Alvalu, & S. Mohanty (Eds.), *Evolution and Applications of Quantum Computing* (pp. 175–181). Wiley. doi:10.1002/9781119905172.ch11

Movassagh, R. (2023). (in press). The hardness of random quantum circuits. *Nature Physics*. Advance online publication. doi:10.103841567-023-02131-2

Mummadi, S., & Rudra, B. (2023). Fundamentals of Quantum Computation and Basic Quantum Gates. In A. K. Tyagi (Ed.), *Handbook of Research on Quantum Computing for Smart Environments* (pp. 1–24). IGI Global. doi:10.4018/978-1-6684-6697-1.ch001

Nemoto, K. (2023). Our future with quantum computers. *JSAP Review*, *230212*. doi:10.11470/jsaprev.230212

Niazi, S. K. (2023). The coming of age of ai/ml in drug discovery, development, clinical testing, and manufacturing: The FDA perspectives. *Drug Design, Development and Therapy*, *17*(null), 2691-2725. doi:10.2147/DDDT.S424991

Pal, S., Bhattacharya, M., Dash, S., Lee, S.-S., & Chakraborty, C. (2023). Future potential of quantum computing and simulations in biological science. *Molecular Biotechnology*. doi:10.100712033-023-00863-3 PMID:37717248

Pal, S., Bhattacharya, M., Lee, S.-S., & Chakraborty, C. (2023). (in press). Quantum computing in the next-generation computational biology landscape: From protein folding to molecular dynamics. *Molecular Biotechnology*. doi:10.100712033-023-00765-4 PMID:37244882

Rani, S., Kumar Pareek, P., Kaur, J., Chauhan, M., & Bhambri, P. (2023). Quantum machine learning in healthcare: Developments and challenges. *2023 IEEE International Conference on Integrated Circuits and Communication Systems (ICICACS)*. IEEE. 10.1109/ICICACS57338.2023.10100075

Sagingalieva, A., Kordzanganeh, M., Kenbayev, N., Kosichkina, D., Tomashuk, T., & Melnikov, A. (2023). Hybrid quantum neural network for drug response prediction. *Cancers (Basel)*, *15*(10), 2705. doi:10.3390/cancers15102705 PMID:37345042

Sasi, P., Gulshan, S., Tyagi, A. K., Kakulapati, V., Shyam Mohan, J. S., & Singh, R. k. (2023). Quantum Computing and the Qubit: The Future of Artificial Intelligence. In A. K. Tyagi (Ed.), *Handbook of Research on Quantum Computing for Smart Environments* (pp. 231–244). IGI Global. doi:10.4018/978-1-6684-6697-1.ch013

Shaib, A., Naim, M. H., Fouda, M. E., Kanj, R., & Kurdahi, F. (2023). Efficient noise mitigation technique for quantum computing. *Scientific Reports*, *13*(1), 3912. doi:10.103841598-023-30510-5 PMID:36890156

Sharma, S., Ramkumar, K. R., Kaur, A., Hasija, T., Mittal, S., & Singh, B. (2023). Postquantum Cryptography: A Solution to the Challenges of Classical Encryption Algorithms. In R. Agrawal, C. Kishore Singh, A. Goyal, & D. K. Singh (Eds.), *Modern Electronics Devices and Communication Systems*. Springer Nature Singapore., doi:10.1007/978-981-19-6383-4_3

Shinde, A. R., & Bendale, S. P. (2023). Evolution of Quantum Machine Learning and an Attempt of Its Application for SDN Intrusion Detection. In R. Pandey, N. Srivastava, N. K. Singh, & K. Tyagi (Eds.), *Quantum Computing: A Shift from Bits to Qubits* (pp. 437–456). Springer Nature Singapore. doi:10.1007/978-981-19-9530-9_22

Sihare, S., & Khang, A. (2023). Effects of Quantum Technology on the Metaverse. In A. Khang, V. Shah, & S. Rani (Eds.), *Handbook of Research on AI-Based Technologies and Applications in the Era of the Metaverse* (pp. 174–203). IGI Global. doi:10.4018/978-1-6684-8851-5.ch009

Singh, J., & Bhangu, K. S. (2023). Contemporary quantum computing use cases: Taxonomy, review and challenges. *Archives of Computational Methods in Engineering*, *30*(1), 615–638. doi:10.100711831-022-09809-5

Singh, K., Bradley, C. E., Anand, S., Ramesh, V., White, R., & Bernien, H. (2023). Mid-circuit correction of correlated phase errors using an array of spectator qubits. *Science*, *380*(6651), 1265–1269. doi:10.1126cience.ade5337 PMID:37228222

Softley, T. P. (2023). Cold and ultracold molecules in the twenties. *Proceedings - Royal Society. Mathematical, Physical and Engineering Sciences*, *479*(2274), 20220806. doi:10.1098/rspa.2022.0806

Sood, V., & Chauhan, R. P. (2023). (in press). Archives of quantum computing: Research progress and challenges. *Archives of Computational Methods in Engineering*. doi:10.100711831-023-09973-2

Ur Rasool, R., Ahmad, H. F., Rafique, W., Qayyum, A., Qadir, J., & Anwar, Z. (2023). Quantum computing for healthcare: A review. *Future Internet*, *15*(3), 94. doi:10.3390/fi15030094

Wang, P. H., Chen, J. H., Yang, Y. Y., Lee, C., & Tseng, Y. J. (2023). Recent advances in quantum computing for drug discovery and development. *IEEE Nanotechnology Magazine*, *17*(2), 26–30. doi:10.1109/MNANO.2023.3249499

Wong, H. Y. (2024a). Quantum Gate Introduction: NOT and CNOT Gates. In H. Y. Wong (Ed.), *Introduction to Quantum Computing: From a Layperson to a Programmer in 30 Steps* (pp. 133–141). Springer International Publishing. doi:10.1007/978-3-031-36985-8_15

Wong, H. Y. (2024b). Shor's Algorithm. In H. Y. Wong (Ed.), *Introduction to Quantum Computing: From a Layperson to a Programmer in 30 Steps* (pp. 289–298). Springer International Publishing., doi:10.1007/978-3-031-36985-8_29

Yingngam, B. (2023). New drug discovery. In M. S. Aslam & M. M. Ahmad (Eds.), *Multidisciplinary Applications of Natural Science for Drug Discovery and Integrative Medicine* (pp. 134–184). IGI Global. doi:10.4018/978-1-6684-9463-9.ch005

Zou, D., Pan, N., Chen, T., Sun, H., & Zhang, X. (2023). Experimental simulation of topological quantum computing with classical circuits. *Advanced Intelligent Systems*, 2300354. doi:10.1002/aisy.202300354

Chapter 15
Quantum Computing in Healthcare:
A Review

N. Ambika
https://orcid.org/0000-0003-4452-5514
St. Francis College, India

ABSTRACT

This research examined national elections' impact on crash risks of individual stocks listed in the Indonesia Stock Exchange. The study used national elections as a proxy for political uncertainty and negative skewness of stock returns as a proxy for crash risk. On the industry level, researchers found nuances in which national elections' impact varies according to industry, with some industries tending to have a greater risk of crash than others during national elections and post-national election periods.

1. INTRODUCTION

Quantum processing (Gruska 1999) (Ambika 2020) is a quickly arising innovation that bridles the laws of quantum mechanics to tackle issues excessively complex for traditional computers. Quantum mechanics is the numerical construction that embraces basic physics on a fundamental level. Quantum mechanics underlies the way of behaving of all systems. Quantum calculations adopt another strategy to such complex issues making multi-faceted spaces where the examples connecting individual information focus arise.

Quantum registering utilizes actual quantum peculiarities such as quantum superposition and quantum ensnarement. Quantum PC exploits a surprising perception in quantum physical science, which addresses a solitary piece in both '1' furthermore '0' that is known as a quantum bit or a qubit. Utilizing this peculiarity, quantum registering makes a solid registering foundation for handling various bits of information simultaneously. It empowers the handling of monstrous measures of data progressively. Quantum processing has, as of late, seen a flood of interest by analysts who are hoping to take figuring ability to the next level as we move past the time of Moore's regulation. The quantum PC will involve three significant advances (Egger et al., 2020).

DOI: 10.4018/979-8-3693-1168-4.ch015

- The arrangement should stack its information from a traditional PC into the quantum PC in a few stages. This outcome is in a superposition, consolidating the central states depicted above in a manner that mirrors the stacked dataset. While piling information into a PC is frequently disregarded as a technicality in old-style registering, quantum information stacking can be a significant part of an arrangement plan.
- Whenever information is stacked, the arrangement should quickly play out a calculation on the stacked information inside the quantum PC. It includes control of the qubits in a way that changes the primary states in a manner that mirrors the result of an ideal calculation.
- The estimation step mentions an objective fact of the registered "yield" superposition and reports it back to an old-style PC. Quantum calculations are often rehashed numerous times, with every redundancy called a shot. Each gave reports a result among the dissemination of potential impacts from the figure step. A probabilistic picture arises of which development has gotten the most elevated likelihood.

E-medical services (Ambika, 2022) (Lv and Qiao, 2020) innovation in brilliant urban communities consolidates savvy advances, smart wearable gadgets, and multi-specialist sensors to help intelligent e-medical services applications, which can construct a smart medical services city. The smart wearable gadgets help to remotely survey individual wellbeing status or wellness system with no expert help. Smart-city medical services innovation cooperates with savvy devices, gathers information delivered by these gadgets, and lastly moves these to specialists, scientists, and medical care specialists for better-customized findings and arrangements.

The chapter is divided into 5 sections. The second segment briefs the various applications of quantum computing. The third division details literature of the domain. Future directions are explained in segment four. The work is concluded in section five as shown in Figure 1.

Figure 1. Quantum bit
(Mosteanu and Faccia 2021)

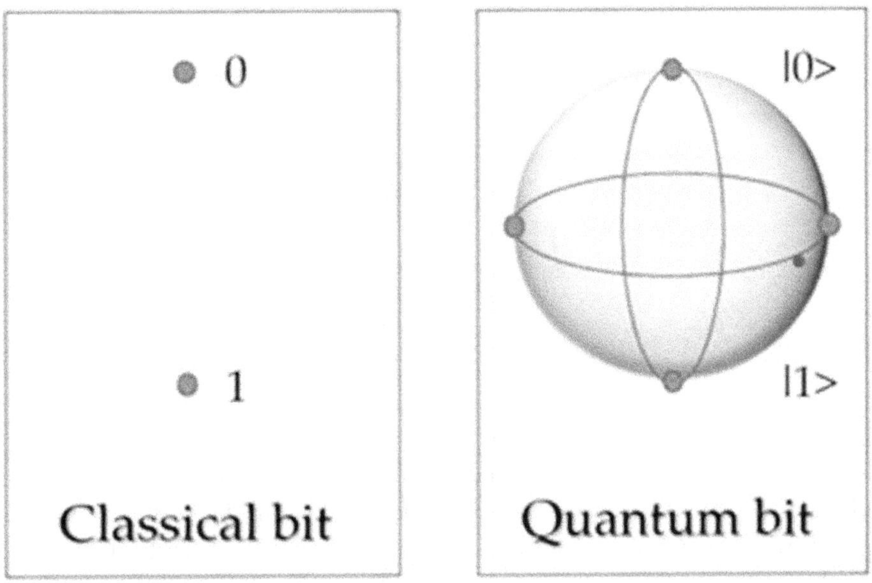

2. APPLICATIONS

2.1 Banking and Finance

Combinatorics challenges (Heffernan, 2005) (Ramalingam and Venkatesan, 2019) are expected in banking and money, from exchange to credit scoring to subsidiaries advancement. Compelling the arrangement of potential arrangements implies that occasionally the best performance is never found. There are two kinds of mistakes that banks could pursue while settling on a credit decision. One sort of mistake emerges when a bank's credit scoring model recommends loaning cash to a client, and afterward, the client, in this way, defaults. One more kind of mistake emerges when a bank's model recommends not loaning a client cash; instead, the client could not have possibly defaulted had the bank given credit. It is exorbitant to lend some money to individuals who default. Declining beneficial customers is likewise expensive as shown in Figure 2.

Figure 2. Private quantum money-based verification protocol
(Kumar, 2019)

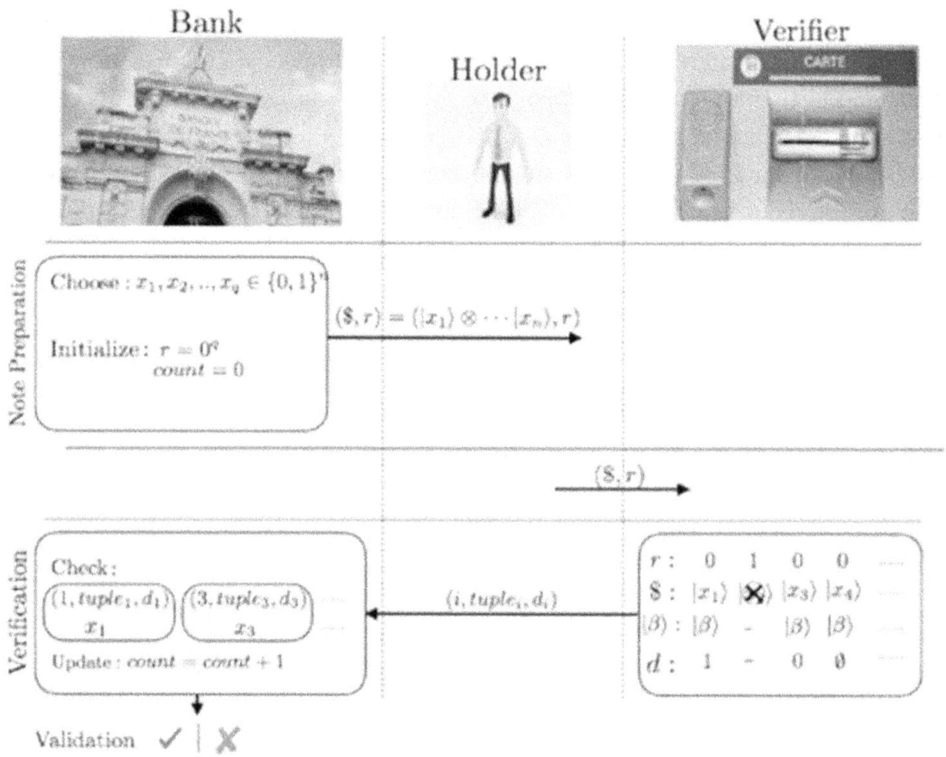

The plan (Kumar, 2019) includes a bank that delivers and circulates quantum notes. These noteholders are untrusted and believed nearby verifiers of the bank to whom the holders send their messages to do transactions. A private quantum cash plot includes a calculation utilized by a confided substance, a bank, to create various notes and a convention run between a holder of the message and the bank to check the realness of the notice. The plan includes two stages. First is the note readiness stage, where the bank picks various n-digit paired strings autonomously and consistently arbitrarily. These strings are then encoded into single-photon states in superposition over modes.

The quantum note is the mix of single-photon conditions relating to every one of the picked strings. It is circulated among the untrusted holders. In the confirmation stage, the holder's note is shipped off the nearby verifier for testing. The verifier haphazardly chooses a few duplicates of the note state. He runs the confirmation convention utilizing the conspire and does a nearby check given the measurements of the traditional result. On the off chance that the insights are not the same as the ones created by the note of a fair holder, he nullifies the letter. The results are imparted to the bank if the message finishes this assessment. The bank contrasts these results and his confidential n-digit strings.

2.2 Advanced Manufacturing

Brilliant assembling (Williams and Clearwater 1998) works on long haul seriousness by streamlining work, energy, and material to create an excellent item and find a fast reaction to variety in market requests and conveyance time. The added substance-producing strategy is applied to manufacture different designs and complex parts. Its processes are fundamentally the cycles that add a few materials to the past surface through various testimony procedures that lead to various part quality, thickness, and mathematical precision. The traditional bikes are typically subtractive or blend a few cycles in the event

Figure 3. Smart manufacturing in Industry 4.0
(Mehrpouya, et al. 2019)

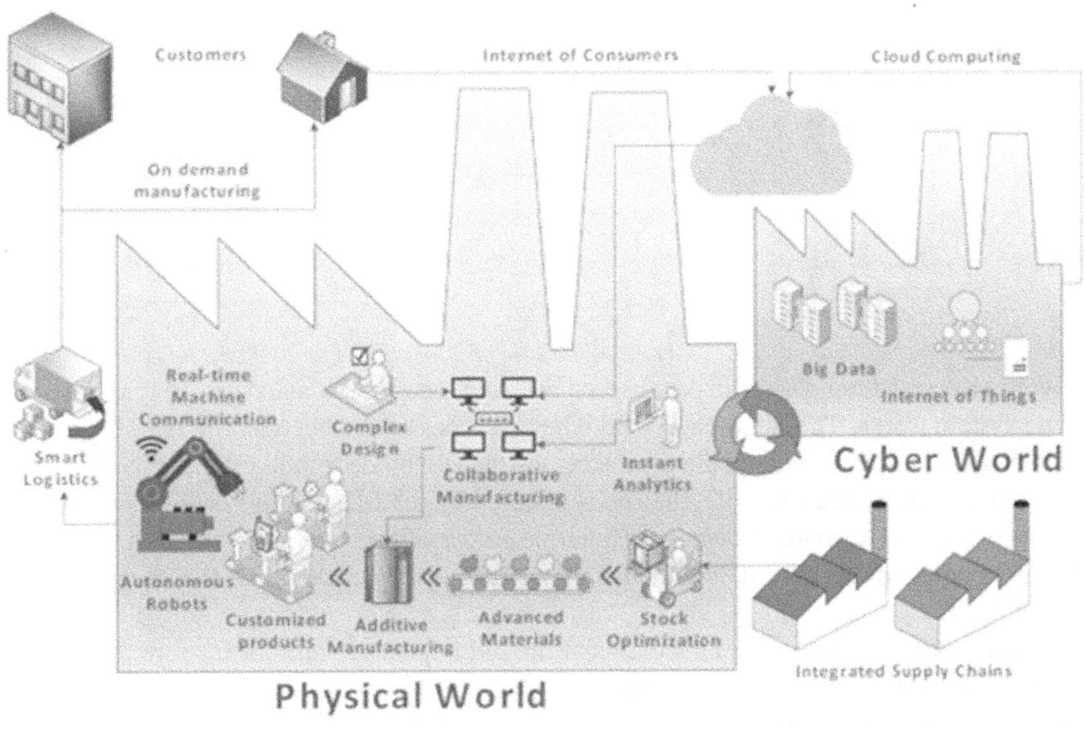

of confounded parts. The general method of all AM processes is that a layer of material is kept, and this cycle proceeds to the point that the last 3D item is finished as shown in Figure 3.

2.3 Cyber Security

There is much potential to create a new generation of cybersecurity algorithms thanks to quantum, a novel technology. Quantum cryptography makes use of quantum mechanics to protect user-to-user communication via networks. Clients can share keys while preventing unauthorized users from obtaining them via a quantum key exchange, a revolutionary critical distribution method based on quantum physics. The first communicating party begins the first stage by communicating with the second one by randomly switching between the filters while sending photons across a quantum channel. However, the second party also utilizes random filters because he is unsure which filter to use to determine the photon's polarization. In the second step, the second one told Ellie about the filter he employed, not the spin or the value as shown in Figure 4.

Figure 4. File encryption system
(Ko and Jung 2021)

The work (Ko and Jung, 2021) makes use of the AES encryption standard. In real-time, the file monitoring system works. The file is encrypted when an access event occurs. The AES is an algorithm with symmetric keys. Consequently, the user accesses the file by decrypting the encryption key. Even if a disaster or a hacker damages a file, the file is restored by the file restore system. Even if the file is leaked, it also prevents secondary damage if internal information is revealed through the encryption system. It uses quantum computing to enhance the efficiency of conventional file encryption and decryption algorithms.

A classical register to store each bit's classical bit and a quantum resistor to store the qubit. A list of all qubits is kept in the Quantum Register. Classical bits are stored in the Classical Register. The quantum circuit that runs each register can only be used if the classical bits are measured. Based on the configured logs, a quantum circuit is built. The quantum gate for each classic logic gate operation is used in the quantum circuit. The original AES is implemented in work with Python. The proposed method is implemented through the quantum random walk with the help of a linear congruential generator, and the Mersenne Twister made possible by IBM Quantum accounts.

2.4 Chemical Engineering

Chemical product design involves the generation of numerous candidates, their evaluation, and screening to satisfy consumer, service, and process needs. Chemical process design has long benefited from computer-aided methods and tools to develop new processes and services that can meet the needs of society. It is emphasized that the challenges and opportunities in the market for multi-scale modeling focusing on property models suitable for computer-aided applications and solution strategies that can solve a wide range of chemical and engineering problems. Computer-aided product and process design have been the focus of many scholarly works, primarily within the process systems engineering discipline. It is emphasized that the need for multi-scale modeling can address a wide range of chemical product design issues and a framework for chemical product design. There have been attempts to use molecular scale modeling and computational chemistry methodologies for product design to control the complexity involved with solution strategies to create unique products.

3. LITERATURE REVIEW

The work (Naresh, et al. 2020) is a medical services framework design named smart medical services city. It partitions a given city into different zones. It utilizes a Quantum GKA strategy given eminent Quantum Diffie-Hellman. The hub is a Gathering Regulator and structures 2-party bunches with great gateways, laying out a QDH-style shared key per every two-party. It then, at that point, joins these keys into a solitary gathering key through an XOR activity, going about as a standard gathering node. It is a schematic plan of the multi-specialist framework in the zonal medical services framework, which works with progress like medical services. It is a quantum bunch key understanding reasonable for certain correspondence among specialists to accomplish security in sharing patient data as shown in Figure 5.

The framework Layer is the patient-information maker layer. It gathers wellbeing information from wearable gadgets, WBANs, houses/structures, and the sensors connected to the vehicles. All collected data moves to the middleware layer through correspondence conventions. Middleware Layer liabilities incorporate information assortment from the framework layer, information handling, information capacity on neighborhood information capacity region, and imparting the data to the concerned partner for additional handling of the patient information. Partner Layer comprises all the medical services specialists who can be ready to chip away at patient wellbeing records.

A crossover move learning model (Shahwar et al., 2022) has twofold identification, in which a quantum circuit works on the exhibition of pre-prepared ResNet34 engineering. It is a blend of traditional and quantum brain organizations. It comprises a component vector of the old-style web, a variational quantum circuit, and a wholly associated layer of the conventional organization that groups Alzheimer's sickness utilizing hysterical and non-insane X-ray pictures. A convolution brain network design called ResNet34 is the component extractor. It works on the given info picture into a 512-highlight vector. Then, at that point, dimensionality decrease was performed utilizing the quantum variational circuit. The

Figure 5. System Architecture
(Naresh et al., 2020)

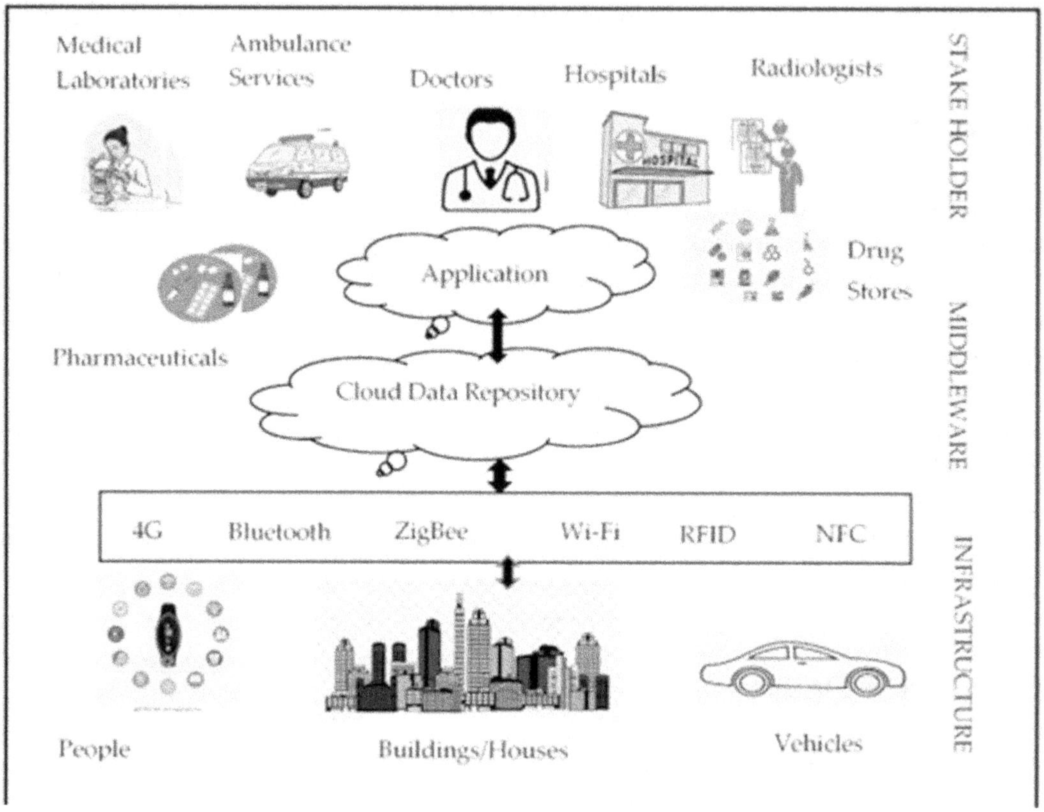

512-highlight vector changes to 4 aspects using the variational quantum circuit. The quantum circuit gives a four-layered vector as a result. The associated layer lessens the four-layered vector to a two-layered one.

The organization used the variational quantum circuit and the ResNet3 four-highlight vector to characterize hysterical and non-deranged X-ray pictures. Its design relies on the necessities of the dataset. The variational equation is regularly deteriorated into two phases: old style to-traditional exchange learning and old style to-quantum move learning. ResNet34 is the pre-prepared model by saving all convolutional and pooling layers, and another teachable classifier is added to the organization to characterize the X-ray pictures. The dataset was taken from the Kaggle platform. It has 6400 MRI scans. The size is 176 × 208.

The study (Gupta et al., 2022) seeks to understand how quantum computing is changing the healthcare industry and building a more long-lasting healthcare system. A semi-structured interview approach is used. A systematic coding strategy is applied, employing an open, axial, and selective process. The professionals in the healthcare sector—representing those working in the pharmaceutical industry, hospitals, clinics, and nursing homes—have been contacted. They must have at least three years of relevant experience. From all the areas that were directly relevant to the healthcare industry, they reached 250 professionals.

Following a brief explanation of the interview's goal, participants are given 25–30 minutes to respond thoughtfully to each question. Only 21 out of 31 have discussed their hopes and innovations that could use quantum computing to transform the healthcare sector completely. Only 20 of the responses are deemed to be valid after serious evaluation. The triangulation method evaluated the research questions for validity and in-depth examination from several points of view. The results show the potential uses of quantum computing for pharmaceuticals, hospitals, health insurance, and patient organizations to have accurate and speedy answers to issues that may accomplish increased precision and speed.

The proposed methodology (Al-Saggaf et al., 2022) uses smart card and biometrics to authenticate users. A new post-quantum fuzzy commitment scheme-based lightweight two-factor user authentication protocol for the IoT-enabled healthcare sector. The biometric template protected the indirect biometric matching process. An aging population, chronic diseases, automated patient care, and collection and analysis of health records are all made possible by IoT-enabled healthcare. During the COVID-19 outbreak, IoT-enabled healthcare offered a better environment for doctors and patients.

The linking of intelligent devices and applications over the Internet is the IoT-enabled healthcare ecosystem. Measurements of patients' conditions can be collected, followed, and analyzed thanks to the IoT-enabled healthcare network. The system gathers and tracks the patient's biological conditions for later processing. The patient's body houses wearable or implantable IoT medical devices, which measure and collect data on the patient's physical problems. These biological conditions are transmitted via an app from an IoT medical device to a smartphone. These physical conditions are sent via an app from an IoT medical device to a smartphone. The healthcare server receives the biological conditions from the smartphone for additional analysis and decision-making.

The study (Jayanthi and Muralikrishna, 2022) focuses on the various quantum neural networks used on COVID-19 and photos of diabetes retinopathy. The suggested model uses a 4 4 input picture fed into a unitary matrix to provide features at various output channels. With these features, a quant circuit model was developed and assembled using a loss function and an optimizer from the TensorFlow and Keras model utility package. The age groups of the CT scan samples range from 20 to 60. The system uses positive and negative examples to train the suggested model. The experimental procedure entails selecting a base model for the image's training and then developing it using the data gathered.

QNN was chosen for continued benchmarking throughout the trials after evaluating the various performance criteria between QNN, QCNN, Hybrid CNN with a single filter, and Hybrid CNN with multiple filters from the circuit design and performance measurement. The system uses TensorFlow Quantum (TFQ) and a Python framework for developing QML. The experiment uses D-wave Leap and TensorFlow Quantum Framework for training and evaluation. On a system with 2041 qubits and 13.5 qubit temperatures, the expected wait time for problem submission was 1–10 s.

The project (Bhavin, et al. 2021) is a healthcare-friendly blockchain-based architecture. The patient, the doctor, the lab, the researchers, the health insurance, and the pharmacy's information are all included in the blockchain data. Access is granted to various intelligent contracts following their roles. The SDK or a client application is used to register the participant. The certificate authority now provides the enrollment certificate through a membership service provider. Every participant now has a new ID and private key as a result. The blockchain network's transactions are stored on the Hyperledger Fabric blockchain. Record addition to the blockchain is done through a client application. The chain code is called upon during the process of adding blocks.

Updated transactions are distributed across the network following the transaction's addition to the blockchain. This administrator ensures that all participants receive transactions. The previous hash and timestamp are used to add a new trade. On the blockchain network, every user has access to the most recent records. The clinician and laboratory staff need permission to view and update the patient's data. The patient is the only person who can grant clinicians, lab workers, or any other participant's access.

4. FUTURE DIRECTIONS

4.1 Quantum Computing for Big Data Processing

Recent months have seen a lot of interest in big data analytics across several businesses. Regarding business decision-making, its data gathering and analytics automation capabilities offer a comprehensive picture and improved insight into the manufacturing process, risk management, and customer relationship management as shown in Figure 6.

Figure 6. Computational model of turning machine
(Sarkar and Bertels, 2021)

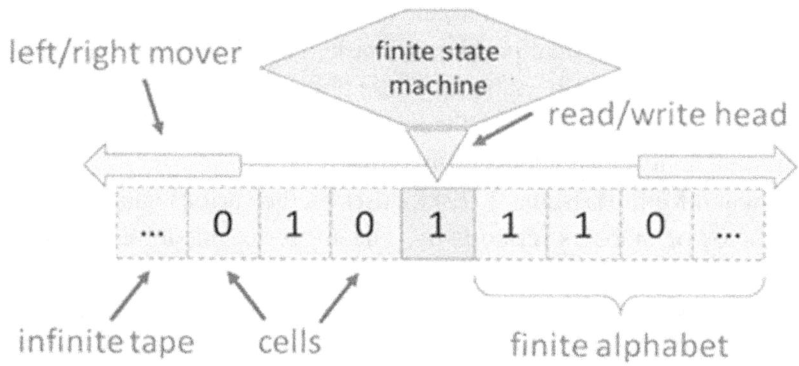

Big data analytics is gaining knowledge from enormous amounts of data, obtaining business intelligence, searching for patterns, drawing conclusions, and making specific forecasts. A quantum computer is a computer that runs calculations using the principles of quantum physics. It can solve more quickly than the newest, quickest computer. It promises computing power beyond what a regular computer could ever have as shown in Figure 7.

The work machine (Sarkar and Bertels, 2021) estimates algorithmic information metrics using a quantum computing framework based on the circuit model. To make the target metrics calculable under realistic assumptions, the canonical Turing machine computation model has limited time and space resources. It provides simulation results considering practical resource assumptions regarding the runtime and qubits. It is a complete and scalable circuit description of the entire construction in two well-known quantum programming languages.

Programs that self-replicate are the first use case. There are two steps to the constructor mechanism: First, a new machine encoded in the description is built using the universal constructor, and then a copy of that description is made using the versatile copier in the new device. The cell's operating system is its dynamics, which, when it is not reproducing, also performs metabolic functions as additional functions. Based on the five parameters, the implementation provides a scalable quantum circuit implementation of the QPULBA model.

Figure 7. Application development framework
(Sarkar and Bertels, 2021)

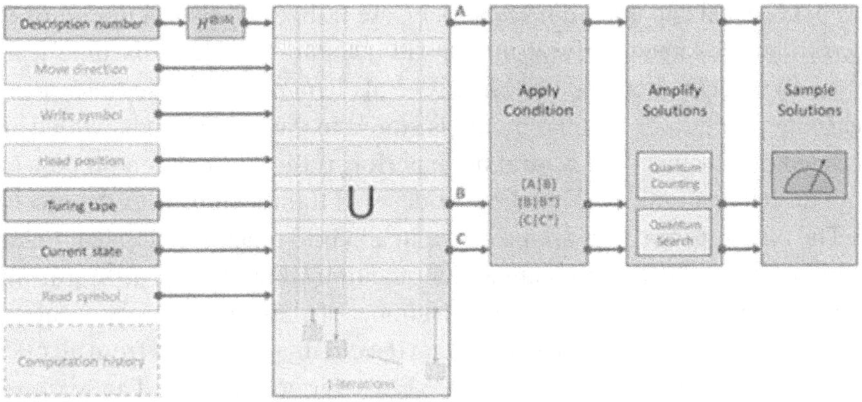

4.2 Quantum AI/MI Applications

The development of models that implicitly learn from previous experiences is the goal of machine learning. The bit is the binary basis for storing, handling, and calculating data in traditional computing. It has the probability that an electron will "spin up" or "spin down" when it passes through a magnetic field using quantum computers. The trend and the analysis and classification of many vectors in multi-dimensional spaces are frequently involved in machine learning tasks. The main idea behind QML is to use quantum computing's inherent advantages to boost the performance of traditional machine learning algorithms. While using quantum physical concepts like entanglement and supper position, the vast

Figure 8. Federated quantum machine learning
(Chen and Yoo, 2021)

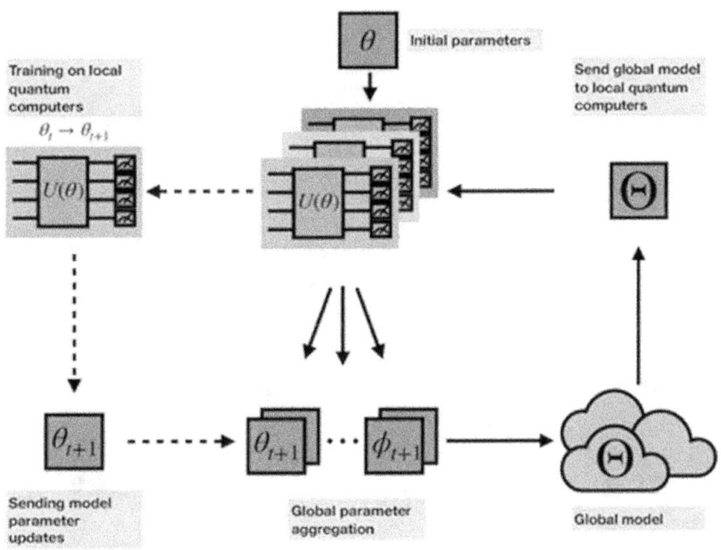

majority of quantum machine learning algorithms are based on ideas and architectures borrowed from classical algorithms. To circumvent limitations in NISQ-era quantum computers, hybrid systems, which are combinations of classical and quantum computers, are utilized as shown in Figure 8.

The study focuses on Federated training on hybrid quantum-classical machine learning (Ambika, 2022) (Ayodele, 2010) models. A central node and several client nodes are the fundamental components of a federated learning process. The global model is stored in the central node, which also receives the client devices' trained parameters. The central node performs the aggregation process to share the new global model with all its client nodes. The client nodes will train locally with the received model with their small data. The circuit parameters are prepared in a hybrid quantum-classical manner by the local clients of our proposed framework, which are either quantum computers or quantum simulators.

Each training round will choose a predetermined number of client nodes to carry out the local training. After the client training, the central node will combine the circuit parameters from all client nodes. Same is portrayed in Figure 8. The model can be combined in a variety of ways. The work selects the client model mean. The position serves as the foundation for the hybrid quantum-classical transfer learning method. The study uses the traditional cats versus dogs' dataset to perform binary classification. The dimensions of each image in this dataset vary slightly.

4.3 Quantum Computers for Simulation

The initial working hypothesis (Villalba-Diez, et al. 2021) is that quantum simulations can better model a machine's internal sensor network than traditional models based on decision networks. A case study is used as a first step to evaluate the impact of integrating quantum simulations into I4.0 environments and quantitatively test this hypothesis. This case study generates a digital quantum twin to provide the 4.0 operator with a shopfloor management tool that enables him to visualize the machine's state in real-time.

Before submitting the work, the study integrated and tested the proposed solution in a factory for 12 weeks. The device in question has five sensors that measure the rotational speed of two motors. These

Figure 9. Case study
(Villalba-Diez et al., 2021)

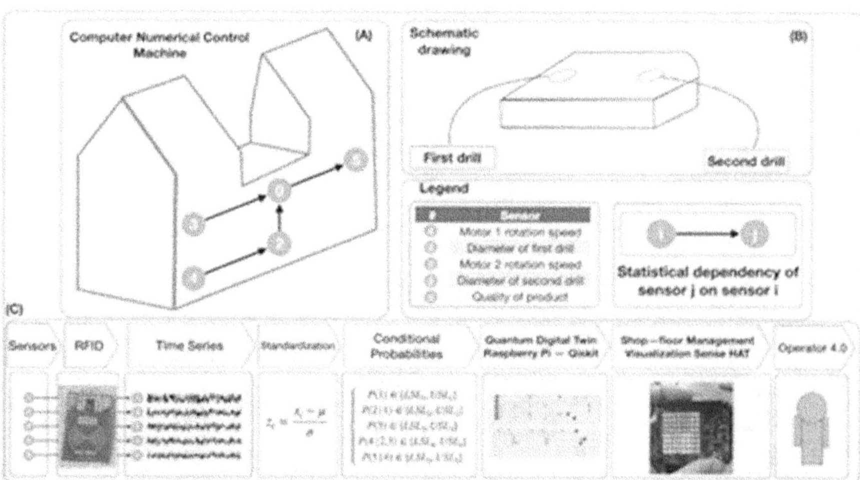

motors drive two drills that drill two holes. The relative positions of those drills and engines determine the quality of the finished product as shown in Figure 9.

4.4 Quantum Web and Cloud Services

Web services that operate independently to fulfill users' requests are called single web services. Composition is the process of putting together a collection of different services to meet a problematic functional or non-functional requirement that a single service cannot meet. The concept of clustering, which has been demonstrated to be effective in numerous domains, is inherited when services providing the same functionality are grouped into communities. The primary objective of the trust and reputation models that are being proposed for this architecture is to assist users in selecting the appropriate Web service that will best fulfill their requests as shown in Figure 10.

The data layer, the application layer, and the presentation layer are separated in GeoVITe's three-tier architecture (Iosifescu-Enescu, et al. 2017). Back-end systems like PostgreSQL geodatabases and Network Attached Storage shares, where databases and folder structures are organized according to a defined data management schema, the data management layer of GeoVITe hosts available vector and raster datasets. Each available vector product has its database in PostgreSQL geodatabases. The servers hosting geoprocessing and view services were the foundation for developing the server-side application layer, which was based on well-known open-source software and libraries. On top of the GeoAdmin and OpenLayers frameworks, the GeoVITe Graphical User Interface is built with renowned Web standards like HTML, CSS, and JavaScript. It organizes the available layers, products, and services in an object-oriented way. An Apache Tomcat web server serves the entire GeoVITe GUI to the Internet.

Figure 10. system Architecture
(Iosifescu-Enescu et al., 2017)

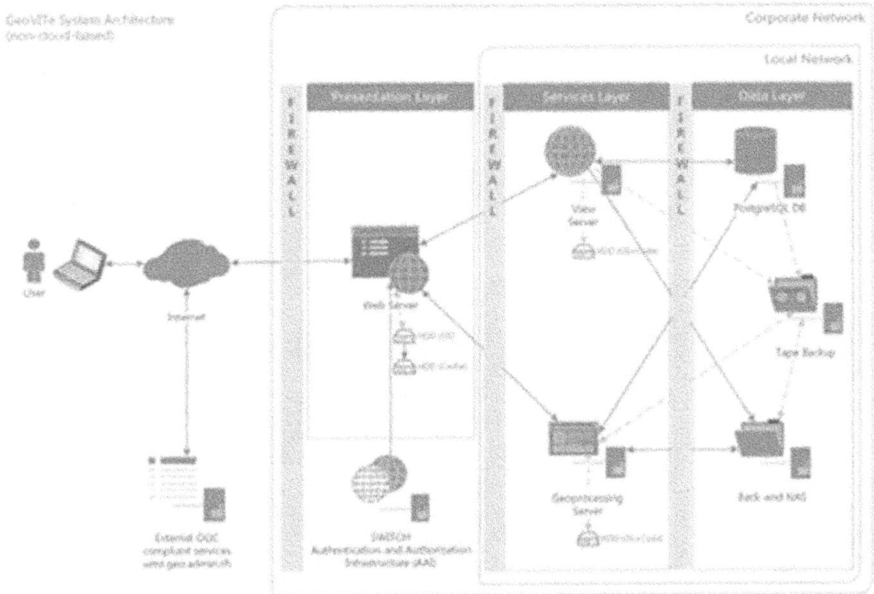

4.5 Quantum Game Theory

Decision sciences and decision support systems both rely heavily on game theory. The information exchange processes among players and the realization of rewards in quantum games are explicitly depicted. A quantum game is described using a more general form than a triple. The combination of evolutionary biology and game theory, with an emphasis on players' strategy dynamics, led to the development of evolutionary game theory. An evolutionarily stable strategy is one that, once established in the population, cannot be invaded by a mutant process, as defined by the associated equilibrium concept. Information theory based on quantum effects is the focus of quantum information theory as shown in Figure 11.

A typical mobile database system design (Madbouly, Mokhtar and Darwish 2021) creates a small fragment of the primary database on the mobile host. Mobile host and satellite services have eased accessibility restrictions, so this design addresses them. It is possible for the mobile host to interact directly with another mobile host in the vicinity if it is present in the cell serviced by the mobile support station. Many mobile hosts and a base station make up each cell, allowing the mobile host to move between them freely. The base station set up the stations as a wireless gateway, making it possible for them to communicate with the mobile host and send data over the wireless network. Static games with complete information serve as the model for the suggested play. The players simultaneously select strategies and receive rewards based on their actions. It can use a normal-form representation to formalize games of this kind. Log monitoring, a mobile agent, and a hybrid method that combines movement-based checkpointing and message recording are its three recovery protocols.

Figure 11. Four player games
(Madbouly and Darwish, 2021)

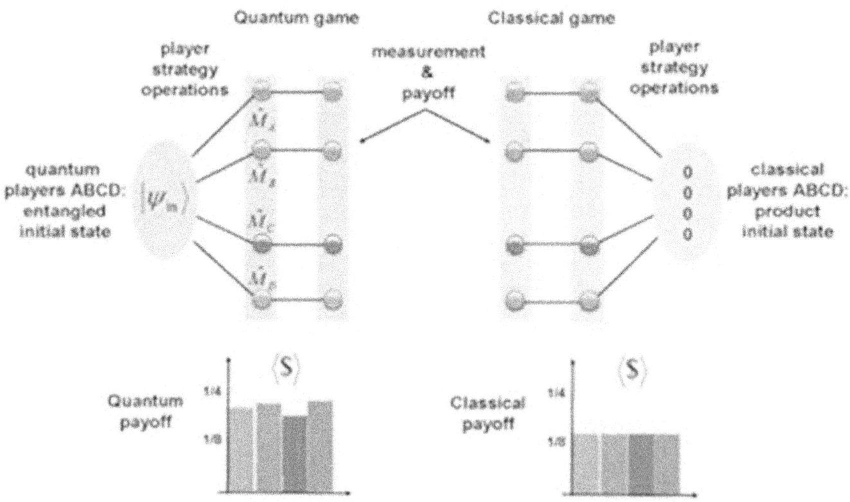

Figure 12. Suggested Framework
(Harun et al., 2020)

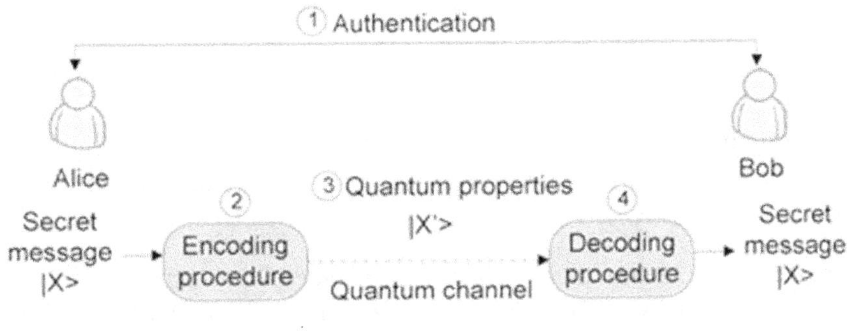

4.6 Quantum Security Applications

In conflict situations, game theory is a tool for making rational decisions. Modeling decision-making situations in which the outcomes are dependent on the interacting strategies of two or more agents with conflicting or, at best, self-interested motives has long been every day in economics, the social sciences, and biology. The game theory aims to model a situation in which agents interact mathematically. Players are the game's agents, and a prescription—also known as a strategy—specifies the specific move that it should make in all possible game situations is called a prescription. A system is a plan of action that accounts for all possible outcomes based on the game's rules. A pure strategy always plays a specific move in response to a particular game situation.

On the other hand, a mixed strategy chooses between different moves by utilizing a randomizing device. A player's desire for a particular outcome can be expressed numerically as its utility to them. For each game outcome, a payoff matrix provides numerical values for the players' utility. The players have their abilities within the constraints of the game's rules. Games of perfect information are those in which the choices made by the players are immediately known as shown in Figure 12.

Sharing an authentication key and a secret angle between Alice and Bob over a multi-stage quantum channel is the primary focus of the SSAK protocol (Harun et al., 2020). The proposed protocol uses a data-associated Information Travel Time or Time Stamp method. In the scenario where Eve is situated in the middle of the distance between Alice and Bob, the ITT is unmistakably different because Alice and Bob are further apart than Eve and Bob. A Python-based simulation has been used to evaluate the proposed SSAK implementation.

9. CONCLUSION

The quantum mechanism has received much attention and is rapidly evolving continuously. The quantum field is equipped with numerous application scenarios and brand-new vitality thanks to the quantum mechanism's mighty computing power and high parallel ability. Intelligent algorithms have always been one of the most active areas of research. Biology, mathematics, and other subjects are perfectly integrated into this cutting-edge interdisciplinary field. When it comes to solving specific problems, quantum computing outperforms conventional computers. The chapter details usage of quantum computing in healthcare sector.

REFERENCES

Al-Saggaf, A. A., Sheltami, T., Alkhzaimi, H., & Ahmed, G. (2022). Lightweight Two-Factor-Based User Authentication Protocol for IoT-Enabled Healthcare Ecosystem in Quantum Computing. *Arabian Journal for Science and Engineering*, 1–11. PMID:36164325

Ambika, N. (2020). Improved Methodology to Detect Advanced Persistent Threat Attacks. In Quantum Cryptography and the Future of Cyber Security. US: IGI Global. doi:10.4018/978-1-7998-2253-0.ch009

Ambika, N. (2022). Aiding IoT and Cloud to Control COVID-19: A Systematic Approach. In Husain M.S., Adnan M.H.B.M., Khan M.Z., Shukla S. and Khan F.U. (eds), Pervasive Healthcare, 349-365. Cham.: Springer. doi:10.1007/978-3-030-77746-3_21

Ambika, N. (2022). An Economical Machine Learning Approach for Anomaly Detection in IoT Environment. In S. A. Suresh, Y. Vimal, Harold Robinson, Dhinesh Kumar, R. Ramaswami and Udendhran (eds.), Bioinformatics and Medical Applications: Big Data Using Deep Learning Algorithms, 215-234. Hoboken, New Jersey: Wiley Publications. doi:10.1002/9781119792673.ch11

Ayodele, T. O. (2010). Types of machine learning algorithms. *New advances in machine learning, 3*, 19-48.

Bhavin, M., Tanwar, S., Sharma, N., Tyagi, S., & Kumar, N. (2021). Blockchain and quantum blind signature-based hybrid scheme for healthcare 5.0 applications. *Journal of Information Security and Applications*, *56*, 102673. doi:10.1016/j.jisa.2020.102673

Chen, S. Y.-C., & Yoo, S. (2021). Federated Quantum Machine Learning. *Entropy (Basel, Switzerland)*, *23*(4), 460. doi:10.3390/e23040460 PMID:33924721

Egger, D. J., Gambella, C., Marecek, J., McFaddin, S., Mevissen, M., Raymond, R., Simonetto, A., Woerner, S., & Yndurain, E. (2020). Quantum computing for finance: State-of-the-art and future prospects. *IEEE Transactions on Quantum Engineering*, *1*, 1–24. doi:10.1109/TQE.2020.3030314

Gruska, J. (2005). *1999. Quantum computing. Vol.* McGraw-Hill.

Gupta, S., Modgil, S., Bhatt, P. C., Jabbour, C. J. C., & Kamble, S. (2022). Quantum computing led innovation for achieving a more sustainable Covid-19 healthcare industry. *Technovation*, 102544.

Heffernan, S. (2005). *Modern banking*. John Wiley & Sons.

Iosifescu-Enescu, I., Matthys, C., Gkonos, C., Iosifescu-Enescu, C. M., & Hurni, L. (2017). Cloud-Based Architectures for Auto-Scalable Web Geoportals towards the Cloudification of the GeoVITe Swiss Academic Geoportal. *ISPRS International Journal of Geo-Information*, *6*(7), 192. doi:10.3390/ijgi6070192

Jayanthi, P., Rai, B. K., & Muralikrishna, I. (2022). The Potential of Quantum Computing in Healthcare. In *Technology Road Mapping for Quantum Computing and Engineering* (pp. 81–101). IGI Global. doi:10.4018/978-1-7998-9183-3.ch006

Ko, K.-K., & Jung, E.-S. (2021). Development of Cybersecurity Technology and Algorithm Based on Quantum Computing. *Applied Sciences (Basel, Switzerland), 11*(19), 9085. doi:10.3390/app11199085

Kumar, N. (2019). Practically Feasible Robust Quantum Money with Classical Verification. *Cryptography, 3*(4), 26. doi:10.3390/cryptography3040026

Lv, Z., & Qiao, L. (2020). Analysis of healthcare big data. *Future Generation Computer Systems, 109*, 103–110. doi:10.1016/j.future.2020.03.039

Madbouly, M. M., Mokhtar, Y. F., & Darwish, S. M. (2021). Quantum Game Application to Recovery Problem in Mobile Database. *Symmetry, 13*(11), 1984. doi:10.3390ym13111984

Mehrpouya, M., Dehghanghadikolaei, A., Fotovvati, B., Vosooghnia, A., Emamian, S. S., & Gisario, A. (2019). The Potential of Additive Manufacturing in the Smart Factory Industrial 4.0: A Review. *Applied Sciences (Basel, Switzerland), 9*(18), 3865. doi:10.3390/app9183865

Mosteanu, N. R., & Faccia, A. (2021). Fintech Frontiers in Quantum Computing, Fractals, and Blockchain Distributed Ledger: Paradigm Shifts and Open Innovation. *Journal of Open Innovation, 7*(1), 19. doi:10.3390/joitmc7010019

Naresh, V. S., Nasralla, M. M., Reddi, S., & García-Magariño, I. (2020). Quantum Diffie–Hellman Extended to Dynamic Quantum Group Key Agreement for e-Healthcare Multi-Agent Systems in Smart Cities. *Sensors (Basel), 20*(14), 3940. doi:10.339020143940 PMID:32679823

Ramalingam, H., & Venkatesan, V. P. (2019). Conceptual analysis of Internet of Things use cases in Banking domain. In *IEEE Region 10 Conference (TENCON)* (pp. 2034–2039). IEEE. doi:10.1109/TENCON.2019.8929473

Sarkar, A., Al-Ars, Z., & Bertels, K. (2021). Estimating Algorithmic Information Using Quantum Computing for Genomics Applications. *Applied Sciences (Basel, Switzerland), 11*(6), 2696. doi:10.3390/app11062696

Shahwar, T., Zafar, J., Almogren, A., Zafar, H., Rehman, A. U., Shafiq, M., & Hamam, H. (2022). Automated Detection of Alzheimer's via Hybrid Classical Quantum Neural Networks. *Electronics (Basel), 11*(5), 721. doi:10.3390/electronics11050721

Svítek, M. (2022). Emergent Intelligence in Generalized Pure Quantum Systems. *Computation (Basel, Switzerland), 10*(6), 88. doi:10.3390/computation10060088

Villalba-Diez, J., Gutierrez, M., Grijalvo Martín, M., Sterkenburgh, T., Losada, J. C., & Benito, R. M. (2021). Quantum JIDOKA. Integration of Quantum Simulation on a CNC Machine for In–Process Control Visualization. *Sensors (Basel), 21*(15), 5031. doi:10.339021155031 PMID:34372267

Williams, C. P., & Clearwater, S. H. (1998). *Explorations in quantum computing*. Springer.

Chapter 16
Quantum Computing in FinTech:
Recent Progress and Perspectives on Quantum Computing for the FinTech Sector

Chinnadurai Kathiravan
https://orcid.org/0000-0002-2064-1927
Vellore Institute of Technology, India

Padma Lakshmi Govindarajan
https://orcid.org/0000-0003-1701-1681
Vellore Institute of Technology, India

Rajesh Mamilla
Vellore Institute of Technology, India

N. Manoraj
Vellore Institute of Technology, India

Kasilingam Lingaraja
https://orcid.org/0000-0002-7701-2507
Thiagarajar College, India

ABSTRACT

Among the hot research topics, Fintech is leading the trend in terms of the newest technology applications. The relatively new emerging paradigms in various sciences, such as geometry (fractals), physics (quantum), and database systems (distributed ledger—blockchain), seem to potentially contribute to a greater shift in the framework of the finance industry, bringing also some concerns (cyber-threats). Consistent and extensive investigation of the reasonable potential impact of these new models (and their underlying technologies) is performed, and then tested through a SWOT analysis, as the main objective

DOI: 10.4018/979-8-3693-1168-4.ch016

of this research. This research confirms that information availability and the increasing interconnection of crosswise applications of each discovery to the different fields of science is determining the rapid succession of revolutions identified by evident large shifts in economic paradigms. The growing computing capacity and the development of increasingly powerful predictive software are leading to a competitive, extremely dynamic, and challenging system.

1. INTRODUCTION

In this chapter, although limited to a general qualitative overview, the authors carried out an analysis aimed at highlighting not only the exposure of the financial sector to the technologies (FinTech) that are driving its exponential development, but also the inferences with different scientific fields, such as geometry (fractal), physics (quantum), and database systems (blockchain distributed ledger). The potential impact of each scientific progression on Fintech is then assessed through a SWOT analysis to verify and systematically confirm the assumptions.

Considering that many sciences are witnessing exponential progress thanks to globalization and the easier dissemination of information, and given that social and technological revolutions have always been the result of a mix of innovations, it would be limiting to focus only on a single driving sector for the Fintech development. An analysis of the possible and potential interactions of different fields of science with the Fintech industry is, therefore, undoubtedly relevant and interesting (Khang & Quantum, 2023).

1.1 Fintech Origins and Definition

Fintech is undoubtedly a very popular and trending topic, especially when it comes to considering its impact in the finance industry. It is, therefore, considered useful to define its origins to better identify the period in which to contextualize its evolution, in parallel with the scientific and social revolutions that are shaping it. It is possible to reasonably confirm that the digitization of finance has sharply accelerated in the last decade of the twentieth century, with the start-up of online banks, it is, therefore, not surprising that the term "FinTech" or "Fintech" was used for the first time by the chairman of Citicorp, John Reed, in 1990, as the contraction of Finance (Fin) and Technology (Tech).

The field of research relating to Fintech is, however, controversial depending on (a) characteristics of the technologies to be included; (b) identification of technologies that can be considered "new", therefore including only "the new breed of companies that specialize in providing financial services primarily through technologically-enabled mobile and online platform"; (c) the willingness to include only technology-enabled business model innovations. A broader meaning is referred to any use of new digital technologies applied in the financial sector to perform forecasts, analyses, or to facilitate financial transactions.

As per the Oxford Dictionary, Fintech is identified as any "Computer programs and other technology used to support or enable banking and financial services". Given this broad definition, alternative payment systems are included in the analyses, therefore overcoming the academic controversy as to which "cryptocurrencies" or "virtual currencies" (based on the blockchain) may or may not be framed within the Fintech scope, although not backed by any government (since decentralized) and, therefore, lacking any intrinsic value.

Figure 1. Fintech and shifts in related sciences

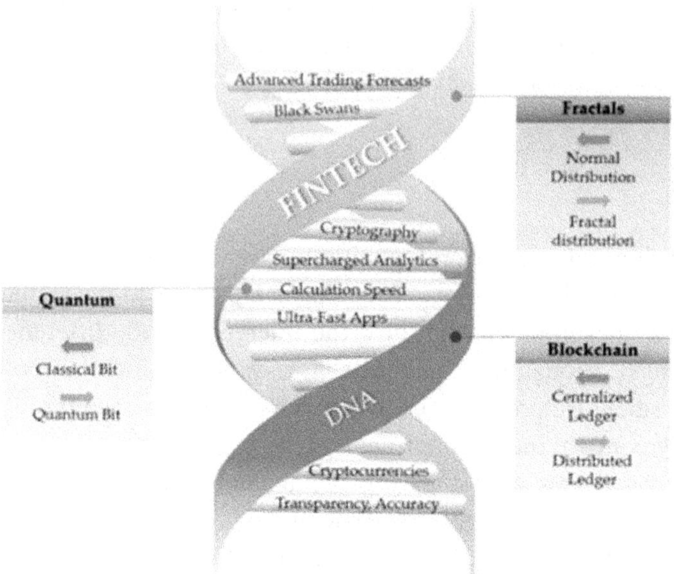

1.2 Fintech and Interactions With Other Fields

In general, Fintech companies aim to provide the most innovative financial services which, thanks to digital technologies, can be developed and evolve to bring new benefits to end-users, individual consumers, large companies, or SMEs. The future of finance has indeed a digital DNA as shown in Figure 1, and shifts in other sciences will have a great impact on the Fintech industry. As will be demonstrated below, three sciences, in particular, are identified as those that will have the greatest impact, given the specific applications and the main characteristics that make them particularly suitable for driving the development of the sector (Khang & Kali, 2023).

Banking, financial and insurance services have been witnessing profound change over the last two decades. The impact of digital technologies in finance is indisputable; FinTech is, therefore, the most challenging frontier. It is not limited to the banking sector, since it can be extended to open banking, API (Application Programming Interface), start-ups, robot-advisor, process automation, and crowdfunding online platforms. In recent years, the interdependence among different areas of science has proven to be the best driver of the main advancements (Khang & Kali et al., 2023). Fields that did not even exist a couple of centuries ago, such as data science (algorithms, data analytics), quantum computing, fractal geometry, distributed ledger, and cloud systems represent nowadays not only hot research topics but the driver of most strategic public and private projects.

The need for ever-greater computing power, availability of big data and complex mathematical algorithms, decentralization of ledgers, seem to be the ideal context for the development and application of quantum computing. Financial and insurance innovations are relentless and driven by the customers' satisfaction, increasingly convinced that digital services make every process more agile. Therefore, the willingness to use new technologies is becoming crucial. In this sense, great attention is focused on

Blockchain, Big Data Analytics, Artificial Intelligence, Cloud Systems, and new 5G Communication Technologies (Khang & Agriculture et al., 2023).

Given that "Information" is the key driver that binds both technology (in the broader meaning of IT-Information Technology) and Finance, this research, therefore, identifies three main fields of science that may have the greatest potential impact (both in terms of opportunities and threats) on the Fintech industry, by enabling the availability of better information, namely: (a) quantum computing, through increased computing capacity; (b) fractal geometry, through a comprehensive approach that does not exclude the tails of the Gaussian bell; (c) blockchain distributed ledger, through decentralization of information. It is reasonable to argue that the possible further integration of these related sciences could lead to enormous advances, but it also needs great attention from the regulatory point of view to mitigating the systemic risk always linked to the financial system (Khang & Vugar et al., 2023).

2. FRONTIERS' ORIGINS AND CONNECTIONS WITH FINTECH

Among its various meanings, a paradigm can be understood as a "model"; it has recently been introduced in the sociology and philosophy of science to identify a complex of methodological rules, explanatory models and problem-solving criteria that characterize a community of scientists in a given phase of the historical evolution of their discipline: the so-called "scientific revolutions" can be traced back to paradigm changes.

Many scientific disciplines in recent decades have witnessed significant shifts, among which, however, three, in particular, have been more frequently brought to the world of finance. In the following paragraphs, three scientific sectors and their paradigm shifts are, therefore, briefly but exhaustively analyzed: geometry, physics, and database systems. All the possible known or potential applications that the new technologies introduced by these scientific revolutions can impact the Fintech sector will also be identified in the findings of this research, therefore expanding its frontiers (Ushaa, Khang et al., 2023).

2.1 Quantum Computing

At the beginning of the twentieth century, scientists were convinced that they understood the basic principles of nature. Atoms were the "building blocks" with which the natural world was built; Newton's motion laws explained that most of the physics problems seemed to be solved. However, starting from Einstein's theory of relativity, which replaced Newtonian mechanics, scientists have gradually understood that their knowledge was far from complete.

Quantum physics, which completely altered the fundamental axioms of physics, is particularly relevant. The transition to this new approach involves fields of nanotechnology for the production of hardware with very strong computing power and, at the same time, it generates cybersecurity risks that cannot be ignored as current cryptography systems will soon become completely obsolete (since vulnerable). The Fintech industry will certainly be among the first to be affected both from the side of predictive calculation (performance of financial instruments), and from that of trust and security in financial systems (Prabu, Khang et al., 2023).

Quantum mechanics, or quantum theory, or quantum physics is a theory that its creators did not fully understand, but that has proved to be the only one capable of explaining the behavior of matter in the microscopic world. At the heart of the matter, there is an immense world, made up of billions and

Figure 2. Classical bit vs. quantum bit

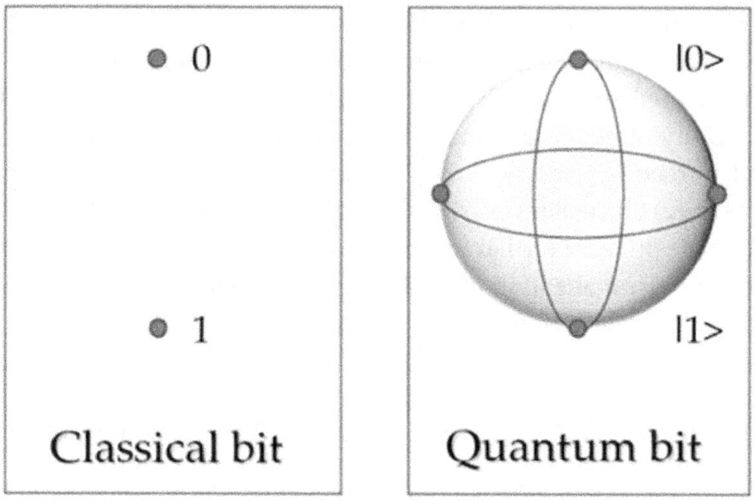

billions of particles, which escapes our senses and our intuition. Quantum computers are new types of devices that allow us to represent and manipulate information not through the classic bits, “0” and “1”, but through the quantum bits or qubits, more complex objects that take advantage of some properties that are peculiar to quantum physics such as the superposition of states, entanglement, and quantum interference. Just as in classical information it is encoded on two bits (0, 1); in quantum information, it is encoded on a two-level system (for example, the polarization of a photon), identified by “|0>” and “|1>” as shown in Figure 2.

In the simplest possible way, the three main properties of quantum physics that are the basis of its extraordinary applications are explained below:

- **Superposition of states.** Under this principle, two or more quantum states can be added (“superimposed”), and the result will be another valid quantum state; conversely, each quantum state can be represented as the sum of two or more other distinct states. A single “particle” can, therefore, be prepared in a superposition of states, and for this reason, it is no longer localizable.
- **Entanglement**. It is a quantum phenomenon that describes an important feature of the non-classical world: if two states of different systems are entangled, there is no way to characterize one without referring to the other. When a measurement is made on one of the particles of the state, it also instantly determines the values of the observable quantities of all the other particles, however distant; this is also defined as the so-called quantum non-locality.
- **Interference**. In the analysis of particles in a quantum state, in the case of two waves, it is observed that in certain points these add up, and in others cancel at the same time. This effect cannot be anything other than a direct consequence of the first postulate of superposition (where the two states overlap).

Quantum technologies aim to overcome the limits of current technologies by exploiting the properties of quantum systems. The first and most developed of these disciplines is quantum information, which deals with the study of the coding, transmission, and processing of information by exploiting these properties, in particular those of photons.

The main advantage of quantum computers is that this category of computers could potentially solve some complexity classes, which require excessive temporal, technical, and economic resources to be solved. This is a fascinating field, but one with considerable criticalities: both from a scientific perspective (there are still difficulties in demonstrating the effective superiority of quantum computing compared to classic approaches) and at the engineering level, given the fragility of quantum systems and the need to shield them from radiation, keep them in temperatures close to zero, and correct errors

2.2 Fractal Geometry

Classical statistics generally require that price changes are normally distributed. One reason is that the Normal (or Gaussian) distribution can be described using only two parameters: the mean and the mean square deviation (understood as the square root of the arithmetic mean of the squares of the deviations of the observed values with respect to the mean, also called Deviation Standard). The latter is the usual tool for representing the dispersion around the mean of a random variable. Indeed, the probability of capturing an observation below or above the mean depends only on the standard deviation.

In the real historical series in the economic and financial field—from ratios to equity returns and interest rates—but also in the physical field, however, the distributions of returns take different forms. The study of fractals has influenced the finance industry since Mandelbrot, analyzing the daily fluctuations in the price of cotton sold in New York in the previous hundred years, found a recurrence of similar trends of the aforementioned price in different periods.

A clear example of the presence of fractal geometry in finance can also be observed by analyzing the daily money transfer network between banks. The identification of patterns in price trends of financial instruments is certainly of great interest as it suggests the possibility of predicting price trends and changes in the long term. The need to identify the so-called black swans, unlikely events that can, however, produce devastating effects, is becoming increasingly widespread. This model contrasts with that of the Gaussian curve which instead theorizes the presumed "normality" of events, but which miserably fails to support decisions in uncertain and volatile contexts, as in the current time as shown in Figure 3.

Figure 3. Normal distribution vs. fractal approach: (a) Gaussian Bell; (b) Mandelbrot set

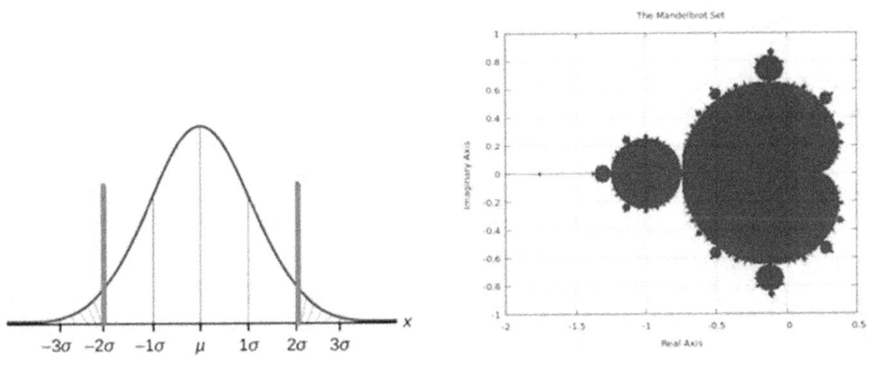

(a) (b)

2.3 Database Systems: Blockchain Distributed Ledger

It is quite difficult to frame the blockchain in a single definition. It can be considered and presented from different perspectives as it is a sub-category of technologies in which the ledger is structured as a chain of blocks containing transactions and whose validation is entrusted to a consensus mechanism, distributed on all nodes of the network in the case of permission-less or public blockchains or on all nodes the nodes that are authorized to participate in the transaction validation process to be included in the register in the case of "permissioned" or private blockchains. The main characteristics of blockchain technologies are the immutability of the register, transparency, traceability of transactions, and security based on cryptographic techniques.

The blockchain is based on a network, and from the point of view of functionality, it allows us to manage a database in a distributed way. It is an alternative to centralized databases, and it allows us to manage the updating process of data with the collaboration of the network participants and with the possibility of having shared, accessible, distributed data among all participants. It, therefore, allows data management in terms of verification and authorization without the need for a central authority. It is understood as a communication protocol that identifies a technology based on the logic of the distributed database (a database in which data are not stored on a single computer but on multiple machines connected, called nodes).

Blockchain is often confused or identified with Bitcoins, or one of the uses of the Blockchain and in particular to the one that underlies the digital currency or crypto-currency Bitcoin. The Blockchain was, therefore, immediately associated with e-money, digital currency, and payment. The blockchain seems to have been conceived by Satoshi Nakamoto (pseudonym of the inventor of the blockchain and its source code), and made famous by its best-known protocol, the virtual currency Bitcoin. Satoshi Nakamoto revealed his project and his vision in October 2008 with the publication of a white paper that describes the possibility of developing a digital currency independent of any central body or institution in the form of Bitcoin.

The white paper called Bitcoin is a Peer-to-Peer Electronic Cash System immediately met with great interest, in particular, because it opens a perspective for decentralized monetary and financial exchanges. Ledger technology distributed in finance is not limited to cryptocurrencies, but numerous studies are extending its application to accounting for auditing purposes, to take advantage of those intrinsic characteristics of transparency and reliability as shown in Figure 4.

Figure 4. Centralized ledger vs. distributed ledger

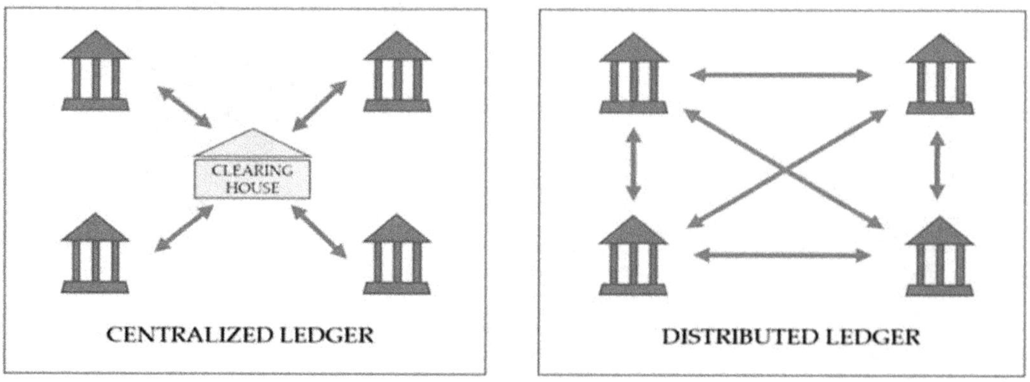

3. QUANTUM COMPUTING APPLICATION IN FINTECH

Frontier of computational evolution, quantum computing is now taking its first concrete steps but is still far from meeting expectations. However, thanks also to important public and private investments, research is growing and in the short term, the first commercial applications will be available. The main advantage of quantum computers is that potentially this category of computers could solve some families of problems, in technical jargon, certain "complexity classes", which today are very difficult and require excessive time and technical and economic resources to be able to deal with. This is a fascinating field, but with considerable criticalities: both from a scientific point of view (there are still difficulties in demonstrating the effective superiority of quantum computing over classical approaches) and at the engineering level, given the fragility of quantum systems and the need to shield them from radiation, it is necessary to keep them in temperatures close to freezing and correct errors.

In the first decade of the 2000s, many physical implementations of qubits were conducted, however, from a technology development perspective, the most recent advancements were determined by Google (Sycamore) in October 2019, when its dedicated research team declared the achievement of the quantum supremacy. Subsequently, Honeywell Venture (System Model H1), in June 2020, announced it had built the most powerful quantum computer in the world that reached supremacy, three months after disclosing (in March 2020), "investment in Cambridge Quantum Computing and Honeywell Quantum Systems confirms CQC as the first beta user of their powerful quantum computer". From the perspective of business opportunities and market domination, it is certainly desirable to obtain the possibility of making calculations in a few seconds that supercomputers based on traditional digital technology would be able to carry out in hundreds of years (Khang & Medicine, 2023).

The evolution of research now addresses numerous areas, from systems scalability to error correction, from the exploration of early industrial applications to the creation of frameworks and languages to develop software able to exploit the potential of quantum technologies. In the short term, however, given that they require relatively few qubits, the most probable applications of quantum computing concern simulations, in particular as regards systems themselves based on quantum properties. Further applications related to simulations and big data can also be hypothesized, such as in finance or other scientific research fields. In the long term, quantum attacks could likely break some of the cryptographic algorithms regularly used today in financial transactions (including many cryptographic systems on which blockchains are based) and in securing military communications.

Very large quantities of qubits are needed to make these applications feasible; however, there are research lines related to postquantum cryptography (or "quantum encryption" or "quantum cyber-security"). A few days ago (30 November 2020), IBM announced the launch of quantum-safe cloud services (through IBM key protect encryption), therefore starting a new market. From a cybersecurity perspective, if quantum physics-based technology is commercialized, all current cryptographic systems, including those that protect financial systems, would become extremely vulnerable and useless.

A further potential application of quantum computing is related to artificial intelligence and machine learning. Although it is still considered a hypothetical scenario to date, research on quantum machine learning can be soon made available on cloud quantum platforms that will create suitable testing environments, therefore massively accelerating learning in a neural network or improving statistical learning systems. It is also reasonable to consider that the first applications in the Fintech field will concern the detection of fraud (reducing the cases of false positives and false negatives) or supporting Monte Carlo simulations for a more accurate calculation of options. Moreover, the applications will be countless and would affect all sectors, not only Fintech, since artificial intelligence is already applied almost everywhere, and it will be boosted by extraordinary computing power.

4. MISCELLANEOUS FURTHER FINTECH APPLICATIONS

In addition to the possible applications determined by the paradigmatic shifts listed above, some further miscellaneous drivers constitute forthcoming applications. It is worth listing the major applications that, through the use of new technologies, are already impacting the Fintech sector, shaping the future of the industry, although unrelated to any of the three previous drivers.

- **Payment gateways**: Software that facilitates a transaction by communicating information about transactions. A payment gateway authorizes credit card payment for online retailers, traditional stores, and e-commerce ventures. The payment gateway protects the details on a credit card by encrypting the sensitive information it holds. This process ensures that personal private details are passed securely between the customer and the merchant. A payment gateway is part of the process that occurs in the background when a credit or debit card transaction occurs. By sending information securely between the website and the credit card network for processing and then returning the transaction details from the payment network to the website, this is a core component that enables e-commerce.
- **Digital wallets (E-wallets)**: A virtual wallet that allows users to make payments, online or in physical stores, using electronic devices. In other words, an e-wallet is a secure tool that can store credit, debit, prepaid, or bank account numbers to make payments quickly and easily. The virtual wallet is the safest solution for digital payments: it allows users to make payments simply by creating a free account and entering an email address. After creating the account, the user can enable payments without sharing their personal data. Indeed, to authenticate the transaction, the consumer will only have to enter the email connected to his wallet and the relative password. In this way, there will be no need to share sensitive data or information, avoiding the risks of fraud or theft. All the user's personal information is stored in dedicated protected environments outside the online pages, making the payment particularly secure. In the current context, increasingly oriented towards payments by phone and on the move, the most used version of e-wallets is undoubtedly the mobile one.
- **Digital banking/digital insurance**: Represents the use of technology to provide banking/insurance products. However, the concept of digital banking/digital insurance (or "InsurTech") cannot be understood only using an online or mobile platform. Going digital means embracing the latest technologies at all functional levels and across all service delivery platforms. A digital bank would behave in the same way at the branch, at the head office, on an online service delivery platform, at ATMs, and vending machines. Many other banking/insurances functions can be managed online, such as (a) risk management, (b) treasury, (c) product development, (d) marketing, and (e) relationship-based sales teams.

5. SWOT ANALYSIS

Quantum Computing, Fractal Geometry, and Blockchain Distributed Ledger impacts on Fintech Industry as shown in Figure 5.

Figure 5. SWOT analysis

	Quantum Computing	Fractal Geometry	Blockchain Distributed Ledger
S	*Unrivaled processing speed and power to be used in calculations for financial forecasts and analyses.*	*Great capacity to identify patterns in forecasts and serving financial analyses, through the use of different scales.*	*Integrity and tamper proof technology. Transparency and decentralized structure. Improvements in public trust.*
W	*Still expensive and instable. It requires a particular environment and constant monitoring.*	*Not all events and systems can be analyzed. Although Fractals are commons they are not always easy to be identified.*	*Expensive implementation. IT necessary skills and training. High investments necessary in e-procurement departments.*
O	*Potential to solve problems with multiple variables and to improve forecast capacity.*	*Potential to identify Black Swans. Improbable, but highly disruptive events can be identified. Systemic Risk reduction.*	*Smart Contracts. Efficiency improvements in the intermediarization with crowdfunding.*
T	*High potential cyber-security risk. Current cryptography not adequate. It needs constant upgrading.*	*Wrong identification of apparent fractals can lead to wrong assumptions. Deep analyses and continuous testing are required.*	*Lack of awareness, slow response, involved entities need to approve each transaction in time. It is impossible to correct mistakes.*

6. CONCLUSION

Given that the financial sector is undoubtedly the most regulated, as a pillar of the functioning of the economy, based on public trust, the development of all technologies with application in this industry appears to be of considerable interest to policymakers and regulators. The incredible ability to speed up transactions, to connect users (potential lenders and borrowers) without adequate reliability checks, along with the decentralization of systems, represent essential challenges and as many concerns for the future of Fintech. The need to reduce (or at least mitigate) systemic risk certainly contrasts with an increasingly easily interconnected system, where "concentrated markets are not necessarily more susceptible to systemic risk than dispersed or disaggregated ones" (Shyam & Khang et al., 2023).

The economy, in general, and finance, in particular, have always been impacted by technological advances. Although some attempts have been already carried out, the identification of the drivers that can most lead (and at the same time threaten) the development of the Fintech industry appears to be a very controversial topic, being to date the object of unsystematic research, only focused on specific features, rather than providing a comprehensive framework that could help to: (a) Improve the regulations, therefore ensuring public trust and reducing systemic risks; (b) Gather relevant information on the opportunities that technologies can bring through a re-antimedicalization.

Given the need for new Open Innovation business models, this article intends to highlight the impact that paradigmatic shifts in three interdependent sciences are generating on the Fintech industry: geometry (from normal to fractal distribution), physics (from classical bit to quantum bit), and database systems (from centralized to the distributed ledger—blockchain). After briefly, but rather comprehensively, analyzing the shifts in those paradigms, it was possible to demonstrate that most of the forthcoming Fintech applications are based and attributable to at least one of them or to a combination of them (i.e., forecasting algorithms to be run on a cloud quantum environment).

The process of introducing these applications has also been accelerated by the current outbreak of the COVID-19 pandemic, which has substantially changed consumer expectations, focusing in particular on the need to ensure consumer trust and ease of use. While the main limitation of this research is related to the impossibility of predicting which of these technologies will generate the greatest impact on the Fintech industry, its main usefulness is due to the ability to shed light on the Fintech sector. The systematic survey, through a classification that can be traced back to major paradigms, allows readers (scholars, regulators, entrepreneurs, and Fintech stakeholders in general) to understand and foresee most of the future technological trends (Khang & Shah et al., 2023).

REFERENCES

Arner, D. W., Barberis, J., & Buckley, R. P. (2015). The evolution of Fintech: A new post-crisis paradigm. *SSRN, 47*, 1271–1285. https://www.zbw.eu/econis-archiv/handle/11159/270063. doi:10.2139srn.2676553

Bagozzi, R. P. (2007). The legacy of the technology Acceptance model and a proposal for a paradigm shift. *Journal of the Association for Information Systems, 8*(4), 3. https://aisel.aisnet.org/jais/vol8/iss4/3/. doi:10.17705/1jais.00122

Bhattacharyya, R., Banerjee, M., Heiblum, M., Mahalu, D., & Umansky, V. (2019). Melting of interference in the fractional quantum Hall Effect: Appearance of neutral modes. *Physical Review Letters, 122*(24), 246801. https://journals.aps.org/prl/abstract/10.1103/PhysRevLett.122.246801. doi:10.1103/PhysRevLett.122.246801 PMID:31322402

Campbell, P., & Abhyankar, S. (1978). *Fractals, Form, Chance and Dimension*. Springer. https://link.springer.com/chapter/10.1007/978-3-642-70880-0_9 doi:10.1007/BF03023043

Corrado, C. A., & Hulten, C. R. (2010). How do you measure a "technological revolution"? *The American Economic Review, 100*(2), 99–104. https://pubs.aeaweb.org/doi/pdf/10.1257/aer.100.2.99. doi:10.1257/aer.100.2.99

De Barros, J. A., & Suppes, P. (2009). Quantum mechanics, interference, and the brain. *Journal of Mathematical Psychology, 53*(5), 306–313. https://www.sciencedirect.com/science/article/pii/S002224960900039X. doi:10.1016/j.jmp.2009.03.005

Egger, D. J., Gambella, C., Marecek, J., McFaddin, S., Mevissen, M., Raymond, R., & Yndurain, E. (2020). Quantum computing for Finance: State of the art and future prospects. *IEEE Trans. Quantum Eng*. IEEE. https://ieeexplore.ieee.org/abstract/document/9222275/

Eswaran, U., Khang A., & Eswaran, V. (2023). Role of Quantum Computing in The Era of Artificial Intelligence. Applications and Principles of Quantum Computing. IGI Global Press. https://doi.org/doi:10.4018/979-8-3693-1168-4-ch003

Fernández-Martínez, M., Guirao, J. L. G., Sánchez-Granero, M. Á., & Segovia, J. E. T. (2019). *Fractal Dimension for Fractal Structures: With Applications to Finance* (Vol. 19). Springer. https://link.springer.com/article/10.1007/s00161-022-01116-5 doi:10.1007/978-3-030-16645-8

Grinin, L., & Grinin, A. (2020). The Cybernetic Revolution and the Future of Technologies. In *The 21st Century Singularity and Global Futures* (pp. 377–396). Springer. https://link.springer.com/chapter/10.1007/978-3-030-33730-8_17 doi:10.1007/978-3-030-33730-8_17

Jones, E., & Knaack, P. (2019). Global financial regulation: Shortcomings and reform options. *Global Policy, 10*(2), 193–206. https://onlinelibrary.wiley.com/doi/abs/10.1111/1758-5899.12656. doi:10.1111/1758-5899.12656

Khang, A. (2023). AI and IoT-Based Technologies for Precision Medicine. IGI Global Press. doi:10.4018/979-8-3693-0876-9

Khang, A. (2023). *Advanced Technologies and AI-Equipped IoT Applications in High-Tech Agriculture* (1st ed.). IGI Global Press. doi:10.4018/978-1-6684-9231-4

Khang, A. (2023). Applications and Principles of Quantum Computing. IGI Global Press. https://doi.org/ doi:10.4018/979-8-3693-1168-4

Khang, A. & Kali, K. (2023). Quantum Mechanics Primer - Fundamentals and Quantum Computing. Applications and Principles of Quantum Computing. IGI Global Press. https://doi.org/ doi:10.4018/979-8-3693-1168-4-ch001

Khang, A., Alyar, A., Khalilov, M., Murad, B., & Litvinova, E. (2023). Introduction to Quantum Computing and Integration of Quantum Computing. Applications and Principles of Quantum Computing. IGI Global Press. https://doi.org/ doi:10.4018/979-8-3693-1168-4-ch002

Khang, A., & Kali, C. R. (2023). Enabling the Future of Manufacturing: Integration of Robotics and IoT to Smart Factory Infrastructure in Industry 4.0. AI-Based Technologies and Applications in the Era of the Metaverse. IGI Global Press. doi:10.4018/978-1-6684-8851-5.ch002

Khang, A., Shah, V., & Rani, S. (2023). AI-Based Technologies and Applications in the Era of the Metaverse. IGI Global Press. doi:10.4018/978-1-6684-8851-5

Knewtson, H. S., & Rosenbaum, Z. A. (2020). *Toward understanding FinTech and its industry*. Emerald. https://www.emerald.com/insight/content/doi/10.1108/MF-01-2020-0024/full/bitcoin.org

Makletsov, S. V., Opokina, N. A., & Shafigullin, I. K. (2019). Application of fractal analysis method for studying stock market. *Int. Trans. J. Eng. Manag. Appl. Sci. Technol., 11*. https://tuengr.com/V11/11A01E.pdf

Mandelbrot, B. B. (1975). *Les Objets Fractals*. Forme, Hasard et Dimension; Champs Flammarion: Paris, France. https://www.jstor.org/stable/24949950

Mandelbrot, B. B. (1982). *The Fractal Geometry of Nature*. W.H. Freeman and Co. https://www.cambridge.org/core/journals/mathematical-gazette/article/fractal-geometry-of-nature-by-benoit-b-mandelbrot-pp-460-2275-1982-isbn-0716711869-wh-freeman/D0714C165F4986ECE608B61347EBB625

Mandelbrot, B. B. (2013). *Fractals and Chaos: The Mandelbrot Set and Beyond*. Springer Science & Business Media. https://link.springer.com/book/10.1007/978-1-4757-4017-2

Mandelbrot, B. B. (2013). *Fractals and Scaling in Finance: Discontinuity, Concentration, Risk. Selecta* (Vol. E). Springer Science & Business Media. https://www.google.com/books?hl=en&lr=&id=H6jqBwAAQBAJ

Mandelbrot, B. B., & Hudson, R. L. (2010). *The (mis) Behavior of Markets: A Fractal View of Risk, Ruin and Reward*. Profile Books. https://www.google.com/books?hl=en&lr=&id=zg91TAIs6bgC

Mishra, P., & Reshef, A. (2019). How do central bank governors' matter? Regulation and the financial sector. *Journal of Money, Credit and Banking*, *51*(2-3), 369–402. https://onlinelibrary.wiley.com/doi/abs/10.1111/jmcb.12578. doi:10.1111/jmcb.12578

Mori, H. (1980). Fractal dimensions of chaotic flows of autonomous dissipative systems. *Progress of Theoretical Physics*, *63*(3), 1044–1047. https://academic.oup.com/ptp/article-abstract/63/3/1044/1920090. doi:10.1143/PTP.63.1044

Omarini, A. (2020). FinTech: A New Hedge for a Financial Re-intermediation. Strategy and Risk Perspectives. *Frontiers in Artificial Intelligence*, *3*, 63. https://www.frontiersin.org/articles/10.3389/frai.2020.00063/full. doi:10.3389/frai.2020.00063 PMID:33733180

Philippon, T. The fintech opportunity (No. w22476). Natl. Bur. Econ. Res. 2016, 1–25. https://www.nber.org/papers/w22476

Plenio, M. B., & Virmani, S. S. (2014). An introduction to entanglement theory. In *Quantum Information and Coherence* (pp. 173–209). Springer. https://link.springer.com/chapter/10.1007/978-3-319-04063-9_8 doi:10.1007/978-3-319-04063-9_8

Prabu, B., & Khang, A. (2023). Quantum Computing Smart City: Fostering Sustainability and Green Infrastructure with Smart City. Applications and Principles of Quantum Computing. IGI Global Press. https://doi.org/ doi:10.4018/979-8-3693-1168-4-ch009

Prasad, K. K., & Aithal, P. S. (2015). Massive Growth of Banking Technology with the Aid of 5G Technologies. [IJMIE]. *Int. J. Manag. IT Eng.*, *5*, 616–627. https://www.academia.edu/download/46356634/58.pdf

Raychev, N. (2016). Interactive Quantum Development Environment (IQDE). *J. Quantum Inform. Sci.*, *6*(2), 105–120. https://www.scirp.org/html/4-1300196_67808.htm. doi:10.4236/jqis.2016.62010

Selvam, A. M. (2009). Fractal fluctuations and statistical normal distribution. *Fractals*, *17*(3), 333–349. https://www.worldscientific.com/doi/abs/10.1142/S0218348X09004272. doi:10.1142/S0218348X09004272

Slussarenko, S., & Pryde, G. J. (2019). Photonic quantum information processing: A concise review. *Applied Physics Reviews*, *6*(4), 41303. https://pubs.aip.org/aip/apr/article/6/4/041303/997349. doi:10.1063/1.5115814

Sultana, F. (2020). Paradigm shift and diversity in finance. *Journal of Financial Accounting Research*, *2*, 94–113. https://www.mdpi.com/2199-8531/7/1/19

Treleaven, P. (2015). Financial regulation of FinTech. *J. Financ. Perspect.*, *3*, 1–17. https://www.emerald.com/insight/content/doi/10.1108/978-1-78973-545-120191018/

Utzerath, C., & Fernández, G. (2017). Shaping science for increasing interdependence and specialization. *Trends in Neurosciences*, *40*(3), 121–124. https://www.cell.com/trends/neurosciences/fulltext/S0166-2236(16)30192-8. doi:10.1016/j.tins.2016.12.005 PMID:28126248

Vasiljeva, T., & Lukanova, K. (2016). Commercial banks and FINTECH companies in the digital transformation: Challenges for the future. *J. Bus. Manag*. https://journals.riseba.eu/index.php/jbm/article/view/91

Wodkiewicz, K., Knight, P. L., Buckle, S. J., & Barnett, S. M. (1987). Squeezing and superposition states. *Physical Review A*, *35*(6), 2567–2577. https://journals.aps.org/pra/abstract/10.1103/PhysRevA.35.2567. doi:10.1103/PhysRevA.35.2567 PMID:9898446

Zeng, B., Chen, X., Zhou, D. L., & Wen, X. G. (2019). *Quantum Information Meets Quantum Matter*. Springer.

Chapter 17
Quantum Internet of Things:
Unlocking New Frontiers for Connectivity, Security, and Efficiency in Data Processing

Debosree Ghosh
https://orcid.org/0009-0005-5585-5588
Shree Ramkrishna Institute of Science and Technology, Kolkata, India

ABSTRACT

This book chapter explores the integration of quantum computing and the internet of things (IoT) and its potential to revolutionize connectivity, security, and efficiency in data processing. The authors delve into the transformative power of quantum computing, the challenges and opportunities of integrating quantum technologies into the IoT landscape, and the implications for various industries. Additionally, they explore quantum-based security mechanisms and highlight how quantum computing enhances data processing efficiency. Through a comprehensive examination of cutting-edge research and real-world applications, this chapter aims to provide a deep understanding of quantum IoT and its transformative impact on connectivity, security, and data processing efficiency.

1. INTRODUCTION

The integration of quantum computing and the Internet of Things (IoT) has opened up a new frontier in technology, paving the way for unprecedented advancements in connectivity, security, and efficiency in data processing. This book chapter explores the exciting domain of Quantum IoT, where the power of quantum computing is harnessed to revolutionize the capabilities of IoT applications.

The Internet of Things has rapidly transformed the way we interact with the world around us. From smart homes and wearable devices to industrial automation and smart cities, IoT has enabled the seamless connection and communication of billions of devices, generating massive amounts of data. However, as the volume and complexity of IoT data continue to grow, traditional computing approaches are facing limitations in terms of processing power, scalability, and data security (Khang & Quantum, 2023).

DOI: 10.4018/979-8-3693-1168-4.ch017

Quantum computing, on the other hand, leverages the principles of quantum mechanics to perform computations at an exponential speed compared to classical computers. This remarkable computational power has the potential to address the challenges faced by IoT applications, enabling real-time data processing, advanced analytics, and optimized resource allocation. Quantum IoT promises to unlock new frontiers in connectivity, allowing devices to communicate and share information in ways that were previously unimaginable (Khang & Kali, 2023).

Quantum IoT represents a paradigm shift in connectivity. By integrating quantum computing into IoT applications, we can achieve enhanced connectivity between devices, enabling real-time communication, seamless data exchange, and efficient resource utilization. Quantum-enabled IoT networks can handle vast amounts of data in real-time, facilitating advanced analytics, intelligent decision-making, and optimized operational processes. This connectivity revolutionizes industries such as healthcare, transportation, energy, and manufacturing, driving efficiency, productivity, and innovation (Ushaa, Khang et al., 2023).

Security is another critical aspect of Quantum IoT. Traditional cryptographic algorithms face the risk of being rendered obsolete by quantum computing's exponential computational power. However, quantum encryption techniques, such as quantum key distribution (QKD), offer unbreakable cryptographic solutions, ensuring secure data transmission and protection against quantum hacking. Quantum-enabled security mechanisms leverage the properties of quantum systems, such as entanglement and superposition, to fortify IoT ecosystems against emerging threats (Prabu, Khang et al., 2023).

Moreover, Quantum IoT opens up avenues for improved efficiency in data processing. Quantum-assisted algorithms and machine learning techniques enable faster and more accurate data analysis, leading to better decision-making and resource optimization. With quantum computing, IoT applications can unlock hidden patterns and insights from vast amounts of data, leading to enhanced operational efficiency and cost savings across various industries.

In this chapter, we explore the fascinating domain of Quantum IoT, where the integration of quantum computing and IoT unlocks new frontiers for connectivity, security, and efficiency in data processing. By delving into the foundations of quantum computing, its application in IoT connectivity, security mechanisms, and data processing efficiency, readers gain an in-depth understanding of the transformative potential of Quantum IoT. We address challenges, discuss future directions, and provide recommendations for researchers, practitioners, and policymakers, emphasizing the need for collaborative efforts to drive Quantum IoT forward and unlock its full potential in reshaping industries and enhancing our connected world (Teja, Shaik et al., 2023).

2. FUNDAMENTALS OF QUANTUM COMPUTING

2.1 Principles of Quantum Mechanics

2.1.1 Superposition

One of the fundamental principles of quantum mechanics is superposition. It states that quantum systems can exist in multiple states simultaneously. Unlike classical bits, which can only be in a state of 0 or 1, quantum bits (qubits) can be in a superposition of both states at the same time. This property allows quantum computers to perform parallel computations and explore multiple possibilities simultaneously.

2.1.2 Quantum Measurement and Collapse of Quantum States

When a measurement is made on a quantum system, it collapses from its superposition into a specific state. Quantum measurement is the process of extracting information from a quantum system. The measurement outcome is probabilistic, with the probabilities determined by the amplitudes associated with the superposition. The act of measurement affects the state of the system, making it inherently probabilistic and non-deterministic.

2.1.3 Quantum Gates and Quantum Circuits

Quantum gates are analogous to logic gates in classical computing. They operate on qubits, transforming their quantum states in a controlled manner. Single-qubit gates manipulate the state of an individual qubit, while multi-qubit gates enable interactions between qubits. Quantum circuits are constructed by arranging quantum gates in a sequence to perform specific computations or transformations on the qubits.

2.1.4 Wave-Particle Duality

Quantum mechanics introduced the concept of wave-particle duality, which states that particles can exhibit both wave-like and particle-like properties. This is encapsulated in the wave function, which describes the probability distribution of finding a particle in different states. The wave nature of particles is observed through interference and diffraction phenomena, highlighting the inherent probabilistic nature of quantum mechanics.

2.1.5 Entanglement

Entanglement is a phenomenon where two or more particles become correlated in such a way that the state of one particle is dependent on the state of the other(s), regardless of the distance between them. This non-local correlation enables instantaneous communication between entangled particles, even if they are separated by vast distances. Entanglement is a key resource in quantum information processing and plays a vital role in quantum algorithms and quantum communication protocols.

2.1.6 Quantum Uncertainty

Heisenberg's uncertainty principle is a cornerstone of quantum mechanics, stating that certain pairs of physical properties, such as position and momentum, cannot be precisely known simultaneously. The more precisely one property is measured, the less precisely the other property can be determined. This principle sets a fundamental limit on the accuracy of measurements in the quantum world and introduces inherent uncertainty into quantum systems.

2.1.7 Quantum Interference

Quantum interference arises when different quantum states interfere constructively or destructively, leading to variations in the probabilities of different measurement outcomes. Interference effects can be observed when the superposition of states interferes with itself, leading to patterns of constructive or destructive interference. Quantum algorithms exploit interference phenomena to enhance desired outcomes and suppress undesired ones, enabling computational advantages.

2.1.8 No-Cloning Theorem

The no-cloning theorem states that it is impossible to create an exact copy of an arbitrary unknown quantum state. This principle is essential for quantum information security and plays a crucial role in quantum cryptography protocols such as quantum key distribution (QKD). The no-cloning theorem ensures that any attempt to intercept or eavesdrop on quantum information will be detectable.

2.2 Quantum Algorithms: Shor's Algorithm, Grover's Algorithm, and More

2.2.1. Shor's Algorithm

Shor's algorithm, proposed by Shor Peter in 1994, is a groundbreaking quantum algorithm that efficiently factors large numbers. Factoring large numbers is a classically challenging problem, and Shor's algorithm demonstrates that a quantum computer can solve it exponentially faster. By utilizing quantum Fourier transforms and modular exponentiation, Shor's algorithm can find the prime factors of a composite number efficiently. The ability to factor large numbers efficiently has significant implications for cryptography, as many encryption schemes rely on the difficulty of factoring large numbers.

2.2.2. Grover's Algorithm

Grover's algorithm, developed by Grover Lov in 1996, addresses the problem of unstructured search. Given an unsorted database of N elements, Grover's algorithm enables the efficient identification of a specific target element. In classical computing, a brute-force search would require an average of N/2 steps, but Grover's algorithm can accomplish this in roughly $\sqrt{N}$ steps. By leveraging principles such as amplitude amplification and phase inversion, Grover's algorithm amplifies the probability of finding the target element, reducing the search time significantly. This quadratic speedup has applications in searching large databases efficiently.

2.2.3. Other Quantum Algorithms

In addition to Shor's algorithm and Grover's algorithm, several other quantum algorithms have been developed for various applications:

- **Quantum Approximate Optimization Algorithm (QAOA)**: QAOA is a quantum algorithm designed for combinatorial optimization problems. It employs a variational approach, using entangling gates and parameterized quantum gates to search for optimal solutions iteratively.
- **Quantum Machine Learning Algorithms**: Quantum machine learning combines the power of quantum computing with machine learning techniques. Quantum Support Vector Machines (QSVM), Quantum Neural Networks (QNN), and Quantum Generative Adversarial Networks (QGAN) are examples of quantum machine learning algorithms that aim to enhance pattern recognition, classification, and optimization tasks using quantum computation.

- **Quantum Simulation Algorithms**: Quantum simulation algorithms facilitate the modeling and simulation of complex quantum systems, including chemical reactions, materials science, and quantum physics. Variational Quantum Eigensolver (VQE), Quantum Monte Carlo methods, and Quantum Fourier Transform are among the techniques employed for quantum simulation.
- **Quantum Phase Estimation (QPE)**: QPE is a quantum algorithm used to determine the eigenvalues of a unitary operator efficiently. It plays a crucial role in quantum simulation and the calculation of quantum properties such as energy levels and molecular properties.

2.3 Quantum Hardware: Challenges and Progress

2.3.1 Qubits and Quantum Gates

Qubits are the building blocks of quantum computing and represent the quantum analogue of classical bits. The main challenge in quantum hardware is creating and maintaining stable qubits with long coherence times. Different physical implementations have been explored, including superconducting circuits, trapped ions, topological qubits, and others. Each implementation has its own advantages and challenges in terms of qubit stability, gate fidelity, and scalability.

2.3.2 Coherence and Decoherence

Coherence refers to the ability of qubits to maintain their quantum state without being disturbed by the environment. However, qubits are highly sensitive to noise and interactions with the surrounding environment, leading to decoherence, where the quantum information degrades and becomes susceptible to errors. Minimizing decoherence and extending coherence times are critical challenges in quantum hardware development.

2.3.3 Quantum Gate Operations

Quantum gates are the operations performed on qubits to manipulate their quantum states. Implementing high-fidelity quantum gates is essential for accurate quantum computations. Challenges include reducing gate errors, improving gate fidelity, and optimizing gate speed. Researchers are developing innovative techniques to enhance gate operations, such as using composite pulses, error mitigation strategies, and optimized control techniques.

2.3.4 Scaling and Connectivity

Quantum hardware must enable the scalability necessary for building larger quantum systems. Increasing the number of qubits and improving connectivity between them is crucial for performing complex computations and implementing error correction codes. Overcoming challenges related to scaling and connectivity involves designing architectures that allow for reliable interactions between qubits and minimizing the impact of crosstalk and noise.

2.3.5 Error Correction and Fault-Tolerant Computing

Quantum hardware must address the challenge of errors that arise due to decoherence and other sources of noise. Error correction codes, such as stabilizer codes and surface codes, provide methods for detecting and correcting errors to enable fault-tolerant quantum computation. Implementing error correction codes requires high-quality qubits, reliable gate operations, and sophisticated error correction protocols.

2.3.6 Progress and Advances

Despite the challenges, significant progress has been made in quantum hardware development. Researchers and technology companies have achieved notable milestones, including increasing the number of qubits, improving qubit coherence times, and demonstrating multi-qubit entanglement. Quantum hardware platforms have become more accessible, with cloud-based quantum computing services offering researchers and developers the opportunity to experiment and innovate. The field continues to witness advancements in qubit quality, gate operations, and error mitigation techniques, bringing practical quantum computing closer to reality.

2.4 Quantum Error Correction and Fault-Tolerant Computing

2.4.1 Quantum Error Correction

Quantum error correction involves encoding and protecting quantum information against errors that arise from decoherence and other sources of noise. Unlike classical error correction, quantum error correction relies on encoding information across multiple physical qubits to preserve the quantum state. This redundancy allows for the detection and correction of errors without directly measuring the fragile quantum information.

2.4.2 Stabilizer Codes

Stabilizer codes are commonly used for quantum error correction. They involve encoding logical qubits into multiple physical qubits and utilizing stabilizer operators to detect and correct errors. These stabilizer operators provide information about the presence and location of errors without measuring the encoded quantum state directly. Examples of stabilizer codes include the Steane code (Andrew, 1996), the surface code (Bravyi et al., 2018), and the color code (Mario, 2022).

2.4.3 Surface Code

The surface code is a prominent and widely studied quantum error correction code. It encodes logical qubits on a two-dimensional lattice of physical qubits. By measuring stabilizer operators associated with the lattice, errors can be detected and corrected. The surface code offers promising scalability and fault-tolerant capabilities, making it a prominent choice for quantum error correction.

2.4.4 Fault-Tolerant Computing

Fault-tolerant computing aims to enable reliable quantum computations in the presence of errors. It builds upon the principles of quantum error correction to design quantum gates and algorithms that can operate accurately despite the presence of errors. Fault-tolerant quantum computing ensures that errors do not accumulate and propagate throughout the computation, thereby allowing for accurate and reliable results.

2.4.5 Threshold Theorem

The threshold theorem establishes the conditions under which fault-tolerant quantum computation becomes possible. It states that if the error rate of individual qubits and gates is below a certain threshold, and if appropriate error correction codes are employed, then arbitrarily long and accurate quantum computations can be performed. The threshold theorem provides a theoretical foundation for fault-tolerant quantum computing.

2.4.6 Challenges

Implementing quantum error correction and fault-tolerant computing faces various challenges. These include maintaining qubit coherence and stability over extended periods, implementing high-fidelity quantum gates, reducing errors during error correction, and managing the impact of noise and decoherence in practical quantum systems. Overcoming these challenges requires advancements in qubit technologies, error correction codes, gate operations, and error mitigation strategies (Debosree, Khang et al., 2023).

2.4.7 Advancements and Future Directions

Quantum error correction and fault-tolerant computing are active areas of research and development. Significant advancements have been made in developing more efficient error correction codes, improving gate fidelity, and exploring novel error mitigation techniques. Ongoing efforts aim to address the challenges associated with scaling up quantum systems, reducing errors, and developing fault-tolerant quantum computing architectures.

3. QUANTUM CONNECTIVITY IN IOT

3.1. Quantum Communication

Quantum communication in IoT involves utilizing quantum properties to establish secure and efficient communication channels between IoT devices. Quantum key distribution (QKD) is a key aspect of quantum communication, enabling the secure exchange of cryptographic keys. By leveraging quantum properties such as entanglement and superposition, quantum communication can enhance the security and privacy of IoT data transmission.

3.2. Quantum Sensing and Measurement

Quantum sensors and measurement devices can significantly enhance the capabilities of IoT systems. Quantum sensors leverage quantum phenomena to achieve highly accurate and sensitive measurements. For example, quantum magnetometers can enable precise detection of magnetic fields, while quantum-based environmental sensors can provide more accurate readings of temperature, humidity, and other parameters. Integrating quantum sensors into IoT devices enables improved data collection and analysis.

3.3. Quantum Cryptography

Quantum cryptography ensures secure communication and data integrity in IoT networks. Quantum encryption algorithms provide protection against attacks by leveraging quantum properties to encrypt and decrypt data. Quantum-resistant cryptographic algorithms, such as those based on post-quantum cryptography, are designed to withstand attacks from quantum computers and can be employed to secure IoT communications against potential future threats.

3.4. Quantum Optimization and Machine Learning

Quantum connectivity can leverage quantum algorithms, such as quantum approximate optimization algorithms (QAOA), to optimize IoT network parameters, resource allocation, and decision-making processes. Quantum machine learning algorithms, such as quantum support vector machines (QSVM) and quantum neural networks, can enable advanced data analysis and pattern recognition in IoT applications, enhancing overall efficiency and performance.

3.5. Quantum-Based IoT Infrastructure

Quantum connectivity in IoT also involves the development of quantum-based infrastructure to support the integration of quantum technologies. This includes the development of quantum processors, quantum networks, and quantum-enabled edge computing systems. Quantum-enabled gateways and routers can facilitate communication between quantum and classical devices, enabling seamless integration within the IoT ecosystem.

3.6. Challenges and Future Directions

Quantum connectivity in IoT faces several challenges, including the need for scalable and practical quantum technologies, the integration of quantum devices into existing IoT infrastructures, and the management of quantum resources. Additionally, the interoperability and standardization of quantum-enabled IoT devices and protocols need to be addressed. Ongoing research and development efforts are focused on overcoming these challenges and realizing the full potential of quantum connectivity in IoT.

3.7. Quantum Machine-to-Machine (M2M) Communication

Quantum M2M communication involves quantum-enabled devices communicating directly with each other, leveraging the principles of superposition and entanglement. Quantum M2M communication can enhance the efficiency of data exchange, enable distributed processing, and facilitate complex decision-making within IoT networks. It has the potential to revolutionize applications such as autonomous vehicles, industrial automation, and smart grids.

4. QUANTUM SECURITY FOR IOT

4.1. Quantum Key Distribution (QKD)

Quantum Key Distribution (QKD) is a cornerstone of quantum security for IoT. QKD utilizes the principles of quantum mechanics to establish secure and unbreakable encryption keys between communicating parties. By exploiting quantum phenomena like the no-cloning theorem and the uncertainty principle, QKD ensures that any attempt to eavesdrop or intercept the key exchange is detected, guaranteeing the confidentiality of data transmitted between IoT devices.

4.2. Quantum Random Number Generators (QRNG)

Random numbers are essential for cryptographic operations. Traditional random number generators rely on algorithms and are not entirely unpredictable, making them potentially vulnerable to certain attacks. Quantum Random Number Generators (QRNG) use the inherent unpredictability of quantum phenomena, such as the measurement outcomes of quantum systems, to generate true random numbers. QRNGs provide a robust source of randomness, enhancing the security of cryptographic keys and protocols in IoT systems.

4.3. Quantum-Resistant Cryptography

Quantum computers have the potential to break traditional cryptographic methods used in classical computing, such as RSA and ECC. Quantum-resistant cryptography, also known as post-quantum cryptography, focuses on developing algorithms that remain secure even in the face of powerful quantum computers. Post-quantum cryptographic algorithms, like lattice-based cryptography, hash-based signatures, and code-based cryptography, protect IoT data and communications from future quantum attacks.

4.4. Quantum Encryption

Quantum encryption offers a new paradigm for secure communication between IoT devices. By leveraging quantum properties for secure key exchange and encryption, quantum encryption protocols provide a high level of security against eavesdropping and unauthorized access. Quantum encryption ensures that data transmitted within IoT networks remains confidential and protected from potential adversaries.

4.5. Quantum Authentication

Quantum authentication methods enhance the security of IoT devices and user verification. Quantum authentication protocols utilize quantum principles to establish trust and verify the identity of communicating parties. This helps prevent unauthorized access, ensure data integrity, and protect IoT networks from various cyber threats.

4.6. Challenges and Future Prospects

While quantum security for IoT shows great promise, several challenges must be addressed for widespread deployment. These include the development of practical and scalable quantum security solutions, the integration of quantum technologies into existing IoT infrastructures, and the standardization of quantum security protocols. Additionally, advancements in quantum hardware and the establishment of quantum networks will play a crucial role in the successful implementation of quantum security measures in IoT systems.

5. QUANTUM EFFICIENCY IN DATA PROCESSING

5.1. Quantum Parallelism

Quantum computers can process multiple possibilities simultaneously through quantum superposition. While classical computers require sequential operations to explore various solutions, quantum parallelism allows quantum computers to consider multiple states at once, leading to significant speedup in certain computations. This inherent parallelism is particularly advantageous for data-intensive applications, such as optimization, search, and simulation tasks.

5.2. Quantum Algorithms

Quantum algorithms offer significant efficiency gains over classical algorithms for specific problems. Notable examples include Shor's algorithm, which efficiently factors large numbers, and Grover's algorithm, which accelerates searching in unsorted databases. Quantum algorithms utilize quantum mechanics to outperform classical counterparts in terms of computational speed, providing a competitive edge in data processing for certain applications.

5.3. Quantum Data Analysis

Quantum computing has the potential to revolutionize data analysis tasks, especially for large and complex datasets. Quantum data analysis techniques, such as quantum machine learning algorithms, leverage quantum parallelism to efficiently analyze patterns and correlations within data. Quantum-inspired algorithms can also be applied to classical data processing tasks, offering potential speedup and improved accuracy.

5.4. Quantum Simulation

Quantum computing excels at simulating quantum systems, making it particularly valuable for applications in physics, chemistry, and materials science. Quantum simulations can accurately model complex quantum behaviors, enabling faster data processing for quantum-related phenomena. Quantum simulators provide insights into quantum interactions that classical computers would struggle to achieve efficiently.

5.5. Cryptography and Data Security

Quantum computing has implications for data security and cryptography. While it poses a threat to classical cryptographic methods, it also provides the potential for quantum-resistant encryption algorithms that ensure data security in the post-quantum era. Quantum key distribution (QKD) enables secure key exchange, ensuring data confidentiality in communication systems.

5.6. Quantum Data Encryption and Decryption

Quantum computing offers potential benefits in cryptography by enabling quantum-resistant encryption algorithms, as well as faster decryption through the use of quantum algorithms. Quantum efficiency in data encryption can enhance data security in IoT, cloud computing, and other data-intensive applications.

5.7. Quantum Data Compression

Quantum data compression techniques leverage quantum superposition to represent and process data more efficiently. Quantum data compression methods have the potential to store a large amount of information in a compact quantum state, reducing storage requirements and enhancing data processing efficiency.

5.8. Challenges and Future Directions

Despite the immense potential, quantum efficiency in data processing faces challenges in error correction, qubit stability, and scalability. Current quantum computing technologies, known as Noisy Intermediate-Scale Quantum (NISQ) devices, have limitations that restrict their applicability to specific problems. Ongoing research aims to address these challenges and pave the way for practical and scalable quantum computers.

5.9. Quantum Fourier Transform

The Quantum Fourier Transform (QFT) is a fundamental quantum computation that plays a crucial role in many quantum algorithms, including Shor's algorithm. The QFT enables efficient transformation between the time and frequency domains, facilitating faster data analysis and signal processing tasks. Quantum Fourier Transform's exponential speedup over classical discrete Fourier transform offers significant advantages in data processing applications.

6. INDUSTRY APPLICATIONS OF QUANTUM IOT

6.1. Healthcare

Quantum IoT can enhance healthcare by providing secure and efficient communication between medical devices and healthcare providers. Quantum encryption ensures the confidentiality of patient data during transmission, while quantum computing can process vast amounts of medical data, leading to faster diagnosis, personalized treatment plans, and drug discovery (Khang & Medicine, 2023).

6.2. Finance and Banking

Quantum IoT can transform the finance and banking industry by enhancing data security and accelerating financial transactions. Quantum-resistant cryptography protects financial data from potential quantum attacks, and quantum computing can optimize portfolio management, risk analysis, and fraud detection.

6.3. Energy and Utilities

Quantum IoT can optimize energy distribution and management in the energy and utilities sector. Quantum algorithms can analyze energy consumption patterns, predict demand, and optimize energy distribution. Quantum sensors can monitor energy usage, grid stability, and equipment health, enabling efficient energy management (Khang, Vugar and Abuzarova et al., 2023).

6.4. Transportation and Logistics

Quantum IoT can improve transportation and logistics by optimizing route planning, vehicle scheduling, and supply chain management. Quantum optimization algorithms can handle complex logistics problems, leading to cost-effective transportation operations. Quantum sensors can monitor vehicle health and track goods in real-time (Keerthika et al., 2023).

6.5. Manufacturing

Quantum IoT can enhance manufacturing processes by enabling predictive maintenance and optimizing production schedules. Quantum sensors can monitor machine health, leading to reduced downtime and improved efficiency. Quantum computing can optimize supply chain management and production workflows.

6.6. Environmental Monitoring

Quantum IoT can play a crucial role in environmental monitoring by providing highly accurate measurements of environmental parameters. Quantum sensors can detect pollutants, monitor air and water quality, and provide real-time data for climate analysis and disaster management (Nayak and Atmika et al., 2023).

6.7. Smart Cities

Quantum IoT can contribute to the development of smart cities by optimizing traffic management, waste disposal, and energy usage. Quantum algorithms can analyze data from various urban systems, leading to better resource allocation and sustainable city planning.

6.8. Agriculture

In agriculture, Quantum IoT can improve crop management and irrigation practices. Quantum sensors can monitor soil conditions, water usage, and crop health, leading to optimized irrigation schedules and improved agricultural productivity (Khang & Agriculture et al., 2023).

6.9. Aerospace and Defense

Quantum IoT has applications in aerospace and defense for secure communication and data exchange between military assets and ground stations. Quantum encryption ensures secure military communication, and quantum computing can optimize flight routes and mission planning.

6.10. Communication and Networking

Quantum IoT can revolutionize communication and networking by providing secure quantum key distribution for encrypted communication. Quantum repeaters and quantum routers may enable long-distance quantum communication and enhance the efficiency of quantum networks.

7. CHALLENGES OF QUANTUM IOT

7.1. Quantum Hardware Limitations

Current quantum computing technologies, especially Noisy Intermediate-Scale Quantum (NISQ) devices, have limitations in qubit stability, error rates, and scalability. Overcoming these challenges and building reliable and scalable quantum hardware is essential for realizing the full potential of Quantum IoT.

7.2. Qubit Decoherence

Quantum systems are sensitive to environmental noise, causing qubits to lose their quantum state through decoherence. Managing and mitigating qubit decoherence is critical to maintaining the accuracy and reliability of quantum computations.

7.3. Quantum Error Correction

Quantum error correction is essential to preserve the integrity of quantum computations in the presence of errors and noise. Developing efficient error correction codes and fault-tolerant quantum computing methods is a complex task that requires ongoing research.

7.4. Quantum Network Infrastructure

Building a quantum network infrastructure for secure long-distance quantum communication poses significant challenges. Quantum repeaters and quantum memories are required to extend the range of quantum communication over large distances.

7.5. Integration with Classical IoT Systems

Integrating quantum technologies with existing classical IoT systems can be complex. Ensuring seamless communication and data exchange between quantum devices and classical devices necessitates the development of compatible protocols and interfaces.

7.6. Standardization and Interoperability

The lack of standardization and interoperability in quantum technologies hinders the broader adoption of Quantum IoT. Establishing industry-wide standards for quantum communication, cryptography, and computing is crucial for successful implementation.

7.7. Quantum-Safe Cryptography Transition

As quantum computers advance, classical cryptographic methods may become vulnerable to attacks. Transitioning to quantum-resistant cryptographic algorithms requires careful planning and coordination to ensure data security in the future quantum era.

8. FUTURE DIRECTIONS OF QUANTUM IOT

8.1. Quantum Error Correction and Fault Tolerance

Ongoing research focuses on developing efficient quantum error correction codes and fault-tolerant quantum computing techniques. Advancements in error correction will enhance the reliability and scalability of quantum computations.

8.2. Quantum Hardware Improvements

Quantum hardware research aims to improve qubit coherence, reduce noise, and increase qubit connectivity. Innovations in quantum processor design, quantum error correction, and materials may lead to more stable and powerful quantum computing devices.

8.3. Quantum Network Development

Research on quantum network technologies, such as quantum repeaters and quantum memories, will enable long-distance quantum communication and quantum cloud computing.

8.4. Quantum Algorithms and Applications

Continued research on quantum algorithms and applications will lead to the discovery of new quantum algorithms for data processing, optimization, and machine learning. These algorithms will unlock the true potential of Quantum IoT in various industries.

8.5. Quantum Security Solutions

Advancements in quantum-resistant cryptographic algorithms will ensure the security of data transmitted and stored in Quantum IoT systems. Research on quantum-safe cryptographic methods is crucial to protect sensitive information from future quantum attacks.

8.6. Quantum Software Development

Improving quantum software tools, languages, and compilers will simplify the programming of quantum algorithms and enable a broader range of developers to work with quantum computing technologies.

8.7. Hybrid Quantum-Classical Architectures

Hybrid quantum-classical computing architectures combine the strengths of quantum and classical computing. Leveraging classical computers for certain tasks can compensate for the limitations of current quantum devices and facilitate hybrid quantum algorithms.

9. CONCLUSION

In this book chapter, we explore the fascinating domain of Quantum IoT, where the integration of quantum computing and IoT unlocks new frontiers for connectivity, security, and efficiency in data processing. By delving into the foundations of quantum computing, its application in IoT connectivity, security mechanisms, and data processing efficiency, readers gain an in-depth understanding of the transformative potential of Quantum IoT. We address challenges, discuss future directions, and provide recommendations for researchers, practitioners, and policymakers, emphasizing the need for collaborative efforts to drive Quantum IoT forward and unlock its full potential in reshaping industries and enhancing our connected world.

In conclusion, Quantum Internet of Things (IoT) holds the promise of unlocking new frontiers for connectivity, security, and efficiency in data processing. Throughout this exploration of Quantum IoT, we have delved into its fundamentals, algorithms, hardware challenges, and potential applications across various industries. The principles of quantum mechanics, such as superposition and entanglement, provide the foundation for quantum computing's exponential computational advantages. Quantum algorithms, like Shor's algorithm and Grover's algorithm, showcase the potential for solving complex problems more efficiently than classical counterparts, making Quantum IoT a compelling prospect for data-intensive tasks. However, Quantum IoT is not without its challenges. Quantum hardware limitations, qubit decoherence, and the need for robust error correction and fault-tolerant computing pose significant hurdles.

Additionally, integrating quantum technologies with existing IoT systems and establishing industry-wide standards demand further research and development efforts (Khang & Muthmainnah et al., 2023).

Looking forward, the future of Quantum IoT holds great promise. Advancements in quantum hardware, error correction, and network infrastructure will pave the way for more powerful and reliable quantum computing devices. Hybrid quantum-classical architectures and quantum-resistant cryptographic solutions will bolster the security and efficiency of Quantum IoT systems (Shyam & Khang et al., 2023).

The potential applications of Quantum IoT span various industries, including healthcare, finance, transportation, energy, and manufacturing. Quantum IoT can revolutionize data analysis, optimize supply chains, improve environmental monitoring, and enhance cybersecurity, among other transformative use cases. To realize the full potential of Quantum IoT, collaboration among researchers, technologists, and industry stakeholders is essential. Continual research and development efforts will drive innovation, leading to practical and scalable quantum computing solutions (Khang & Kali et al., 2023).

In conclusion, Quantum IoT represents a groundbreaking paradigm that has the potential to reshape our digital landscape and pave the way for unprecedented advancements in technology and data processing. As we tackle challenges and push the boundaries of quantum computing, Quantum IoT holds the key to unlocking new possibilities for a connected, secure, and efficient future. Embracing this transformative technology will undoubtedly shape the way we interact with data, driving us toward a more profound understanding of the world and accelerating progress across multiple sectors. The journey towards fully harnessing Quantum IoT has just begun, and with dedication and ingenuity, we are poised to witness a quantum leap into a new era of connectivity, security, and efficiency (Khang & Shah et al., 2023).

REFERENCES

Arpita, N., Patnaik, A., & Satpathy, I. B. (2023). Incorporating Artificial Intelligence (AI) for Precision Medicine: A Narrative Analysis. AI and IoT-Based Technologies for Precision Medicine. IGI Global Press. doi:10.4018/979-8-3693-0876-9.ch002

Eswaran, U., Khang A., & Eswaran, V. (2023). Role of Quantum Computing in The Era of Artificial Intelligence. Applications and Principles of Quantum Computing. IGI Global Press. https://doi.org/ doi:10.4018/979-8-3693-1168-4-ch003

Ghosh, D. & Khang, A. (2023). Quantum Computing Internet of Things: Unlocking New Frontiers for Connectivity, Security, and Efficiency in Data Processing. Applications and Principles of Quantum Computing. IGI Global Press. https://doi.org/ doi:10.4018/979-8-3693-1168-4-ch016

Keerthika, K., Kannan, M., & Khang, A. (2023). Medical Data Analytics: Roles, Challenges, and Analytical Tools. AI and IoT-Based Technologies for Precision Medicine. IGI Global Press. doi:10.4018/979-8-3693-0876-9.ch001

Khang, A. (2023). Applications and Principles of Quantum Computing. IGI Global Press. https://doi.org/ doi:10.4018/979-8-3693-1168-4

Khang, A. (2023). Quantum Mechanics Primer - Fundamentals and Quantum Computing. Applications and Principles of Quantum Computing. IGI Global Press. https://doi.org/ doi:10.4018/979-8-3693-1168-4-ch001

Khang, A. (2023). Introduction to Quantum Computing and Integration of Quantum Computing. Applications and Principles of Quantum Computing. IGI Global Press. https://doi.org/ doi:10.4018/979-8-3693-1168-4-ch002

Khang, A. (2023). AI and IoT-Based Technologies for Precision Medicine. IGI Global Press. doi:10.4018/979-8-3693-0876-9

Khang, A. (2023). AI-Aided Data Analytics Tools and Applications for the Healthcare Sector. AI and IoT-Based Technologies for Precision Medicine. IGI Global Press. doi:10.4018/979-8-3693-0876-9.ch018

Khang, A. (2023). *Advanced Technologies and AI-Equipped IoT Applications in High-Tech Agriculture* (1st ed.). IGI Global Press. doi:10.4018/978-1-6684-9231-4

Khang, A., & Kali, C. R. (2023). Enabling the Future of Manufacturing: Integration of Robotics and IoT to Smart Factory Infrastructure in Industry 4.0. AI-Based Technologies and Applications in the Era of the Metaverse. IGI Global Press. doi:10.4018/978-1-6684-8851-5.ch002

Khang, A., & Muthmainnah, M. (2023). AI-Aided Teaching Model for the Education 5.0 Ecosystem. AI-Based Technologies and Applications in the Era of the Metaverse. IGI Global Press. doi:10.4018/978-1-6684-8851-5.ch004

Khang, A., Shah, V., & Rani, S. (2023). AI-Based Technologies and Applications in the Era of the Metaverse. IGI Global Press. doi:10.4018/978-1-6684-8851-5

Krenn, M. (2002). *Quantum Error Correction: Surface code vs. color code*. StaxExchange. https://physics.stackexchange.com/questions/169176/quantum-error-correction-surface-code-vs-color-code

Prabu, B., & Khang, A. (2023). Quantum Computing Smart City: Fostering Sustainability and Green Infrastructure with Smart City. Applications and Principles of Quantum Computing. IGI Global Press. https://doi.org/ doi:10.4018/979-8-3693-1168-4-ch009

Sergey, B., Englbrecht, M., König, R., & Peard, N. (2018). Correcting coherent errors with surface codes. *Nature*. https://www.nature.com/articles/s41534-018-0106-y#citeas

Shyam, R. &, Khang, A. (2023). Effects of Quantum Technology on Metaverse. AI-Based Technologies and Applications in the Era of the Metaverse. IGI Global Press. doi:10.4018/978-1-6684-8851-5.ch009

Steane, A. 1996, Steane code, https://errorcorrectionzoo.org/c/steane

Teja, S. B. B., & Saheb, S. H. (2023). Quantum Deep Learning: Detection of Drowsiness in Drivers Using Tensorflow Framework. Applications and Principles of Quantum Computing. IGI Global Press. https://doi.org/ doi:10.4018/979-8-3693-1168-4-ch015

Chapter 18
Quantum-Inspired Optimization for Enterprises

R. Thilagavathy
Department of Computing Technologies, SRM Institute of Science and Technology, India

M. Gayathri
Department of Computing Technologies, SRM Institute of Science and Technology, India

G. K. Sandhia
Department of Computing Technologies, SRM Institute of Science and Technology, India

M. Pushpalatha
Department of Computing Technologies, SRM Institute of Science and Technology, India

ABSTRACT

Optimization involves finding the best solution from a set of possible options. Traditional computers employ classical algorithms like gradient descent and simulated annealing to solve these problems. However, as the complexity of problems increases, classical computers encounter limitations in terms of time and resources. Quantum optimization stands as a revolutionary paradigm that harnesses the unique properties of quantum mechanics to tackle complex optimization problems that are beyond the capabilities of classical computers. Quantum computers utilize qubits, the fundamental units of quantum information, which can exist in superpositions of states, enabling them to process a vast number of possibilities simultaneously. Additionally, the phenomenon of entanglement enables qubits to be correlated in ways that classical bits cannot, further enhancing their computational power. In this chapter, the authors will delve into the principles, applications, and current developments in the realm of quantum optimization.

1. INTRODUCTION

In today's rapidly evolving business landscape, enterprises face an ever-increasing demand for optimization solutions that can streamline operations, enhance decision-making processes, and ultimately drive competitiveness. In the quest to solve optimization problems that were once deemed insurmountable,

DOI: 10.4018/979-8-3693-1168-4.ch018

quantum optimization emerges as a game-changing field. By harnessing the principles of quantum mechanics, researchers and industries are poised to unlock new frontiers of computational power (Khang & Vugar et al., 2023).

Quantum mechanics, a branch of physics that studies the behavior of matter and energy at the smallest scales, has long been the realm of theoretical physicists. However, recent advancements in quantum computing and quantum-inspired optimization have brought these complex principles into practical applications for businesses. This marks a pivotal moment in the history of technology, akin to the emergence of classical computing in the mid-20th century (Khang & Shah et al., 2023).

As quantum technology continues to advance, the impact of quantum optimization on various sectors of society is bound to be transformative, ushering in a new era of problem-solving and innovation. Quantum-inspired optimization represents a cutting-edge approach that harnesses the principles of quantum mechanics to tackle complex business problems (Ajagekar, 2020).

Quantum-inspired optimization leverages the principles of superposition and entanglement, fundamental to quantum mechanics, to enhance optimization algorithms. These algorithms can efficiently explore vast solution spaces, providing enterprises with the capability to find optimal solutions to complex problems that were once computationally infeasible (Khang & Kali et al., 2023).

Unlike classical optimization techniques, which often rely on exhaustive searches or heuristic methods, quantum-inspired optimization algorithms have the potential to significantly accelerate problem-solving. They offer the promise of solving intricate problems across diverse domains, including supply chain management, financial modeling, logistics, and more (Ewald 2019).

2. QUANTUM MECHANICS PRIMER

Before we dive into quantum-inspired optimization, it's essential to have a basic understanding of quantum mechanics. At its core, quantum mechanics describes the behavior of particles at the atomic and subatomic levels. Several fundamental principles of quantum mechanics underpin quantum-inspired optimization (Gill, 2022):

- ***Superposition:*** In quantum mechanics, particles can exist in multiple states simultaneously, a phenomenon known as superposition. Quantum-inspired optimization algorithms leverage this principle to explore multiple solutions to a problem simultaneously.
- ***Entanglement:*** Entanglement is a quantum phenomenon where the properties of two or more particles become correlated, even when separated by vast distances. This concept is used to enhance the exploration of solution spaces in quantum-inspired optimization.
- ***Quantum Bits (Qubits):*** Qubits are the quantum analogs of classical bits. Unlike classical bits, which can be either 0 or 1, qubits can exist in a superposition of states, representing both 0 and 1 simultaneously. This property enables quantum-inspired algorithms to process information in new and powerful ways.

3. THE QUANTUM-INSPIRED ADVANTAGE

Quantum-inspired optimization algorithms leverage the principles of quantum mechanics to tackle complex optimization problems efficiently. While true quantum computers are still in their infancy, quantum-inspired algorithms are readily available for enterprises to harness the power of quantum principles on classical hardware (Li, Y., 2020). Here's why enterprises should take notice:

- ***Faster Decision-Making:*** Enterprises often deal with a multitude of optimization problems, from supply chain logistics to financial portfolio management. Quantum-inspired algorithms can significantly speed up the process of finding optimal solutions, reducing decision-making timeframes from weeks to hours or even minutes.
- ***Cost Reduction:*** Quantum-inspired optimization can help enterprises optimize resource allocation, minimize costs, and maximize profits. For instance, in the manufacturing sector, it can optimize production schedules to reduce waste and operational expenses.
- ***Enhanced Predictive Analytics:*** Quantum-inspired algorithms can supercharge machine learning models. They enable better feature selection, hyperparameter tuning, and model optimization, leading to more accurate predictions in areas like customer behavior analysis and demand forecasting.

4. CHALLENGES AND CONSIDERATIONS

While quantum-inspired optimization holds immense promise, enterprises should consider some factors:

- ***Quantum Hardware:*** True quantum computers are still developing. Enterprises may need to rely on quantum-inspired algorithms running on classical hardware until quantum processors mature.
- ***Expertise:*** Quantum computing requires a specialized skill set. Enterprises may need to invest in talent or partner with experts in the field.
- ***Data Security:*** As quantum computing advances, so does the threat to existing encryption methods. Enterprises must stay vigilant about data security.

Though challenges exist, the potential rewards are substantial, making quantum-inspired optimization a compelling choice for forward-thinking enterprises looking to navigate the future with confidence. So, as quantum technology advances, enterprises that invest in understanding and implementing quantum optimization are poised to gain a competitive edge, driving innovation, efficiency, and progress in their respective domains. By embracing the principles of quantum mechanics, enterprises can pave the way for a new era of problem-solving that holds the potential to reshape industries and redefine success (Moussa, 2020).

5. QUANTUM ALGORITHMS FOR OPTIMIZATION

5.1 Understanding Quantum Optimization

5.1.1 Quantum Superposition and Entanglement

At the heart of quantum optimization lies the principles of superposition and entanglement. In a classical computer, bits exist in a state of either 0 or 1. However, in a quantum computer, quantum bits or qubits can exist in multiple states simultaneously due to superposition. This property exponentially expands the computational capacity.

Entanglement further enhances the power of quantum computing. When qubits become entangled, the state of one qubit becomes directly correlated with the state of another, regardless of the distance between them. This intrinsic connection allows quantum computers to process information in parallel, leading to unparalleled computational efficiency.

5.1.2 Quantum Gates and Quantum Circuits

Quantum algorithms leverage quantum gates to manipulate qubits. These gates, analogous to classical logic gates, perform operations on qubits. Quantum circuits, composed of interconnected quantum gates, orchestrate the flow of information within a quantum algorithm. Remarkably, quantum circuits can process complex computations much faster than their classical counterparts.

Figure 1. Categories of quantum optimization algorithms

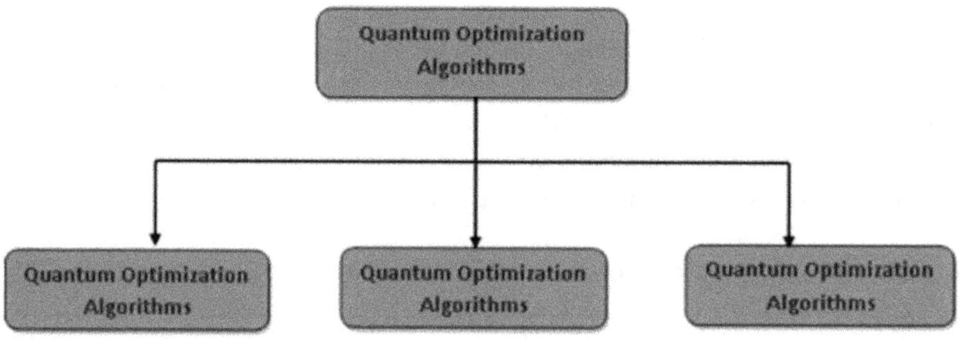

The optimization algorithm in the field of quantum falls under any of the following categories,

- ***Quantum Annealing:*** Quantum annealers like D-Wave's systems harness quantum effects to solve optimization problems. By exploiting quantum tunneling and superposition, quantum annealers can efficiently explore the solution space of complex optimization problems. These machines have found applications in areas such as logistics, finance, and materials science.
- ***Quantum-Inspired Annealing:*** Quantum-inspired annealing algorithms, like the Quantum Approximate Optimization Algorithm (QAOA), use classical hardware to mimic the behavior

of quantum annealers. They employ quantum-inspired techniques to explore problem landscapes more efficiently than traditional optimization algorithms (Denchev, 2012).

- ***Variational Quantum Algorithms:*** Variational quantum algorithms, such as the Variational Quantum Eigensolver (VQE), combine classical and quantum computing to find optimal solutions for complex problems. They use quantum circuits to manipulate quantum states, making them well-suited for problems in quantum chemistry, material science, and machine learning (Barkoutsos, 2020).

Quantum-inspired algorithms for optimization are classical algorithms that draw inspiration from quantum computing principles to solve optimization problems more efficiently than traditional classical methods. These algorithms mimic some of the behaviors of quantum systems, such as superposition and entanglement, to improve their performance on optimization tasks. While they don't leverage actual quantum hardware, they often provide significant advantages over classical approaches (Li, 2021). Here are some key quantum-inspired optimization algorithms:

- ***Quantum Annealing Algorithms:*** Simulated Quantum Annealing (SQA): This algorithm simulates the annealing process of a quantum system. It starts with a random configuration and gradually anneals the system to find the global minimum of an objective function. D-Wave Systems provides quantum annealing hardware (Neven, H., 2012).
- ***Quantum Monte Carlo (QMC) Methods:*** QMC is a quantum-inspired version of the classical Monte Carlo method. It uses random sampling and probability distributions to explore the solution space efficiently. Variational Quantum Monte Carlo (VQMC) combines QMC with variational optimization techniques (Nikolaj, 2018).
- ***Quantum-inspired Genetic Algorithms:*** Quantum-inspired Genetic Algorithms (QGA): QGAs combine genetic algorithms with quantum-inspired operators. Quantum-inspired crossover and mutation operators mimic quantum phenomena like superposition and interference to guide the search process.
- ***Quantum-inspired Simulated Annealing:*** Quantum-inspired Simulated Annealing (QISA): QISA modifies the classical simulated annealing algorithm by introducing quantum-inspired components. These components can help the algorithm escape local optima more effectively.
- **Quantum-inspired Evolution Strategies (QES):**
- QES combines classical evolution strategies with quantum-inspired operators. These strategies use a population-based approach and introduce quantum-inspired components to enhance exploration and exploitation of the search space.
- ***Quantum-inspired Particle Swarm Optimization (QPSO):*** QPSO extends classical particle swarm optimization by incorporating quantum behaviors. Particles exhibit quantum features like superposition and entanglement to improve the convergence speed (Cao, 2020).
- ***Quantum-inspired Ant Colony Optimization (QACO):*** QACO enhances the classical ant colony optimization algorithm by introducing quantum-inspired features. Ants simulate quantum behavior to optimize paths in search spaces.
- ***Quantum-inspired Annealing Optimization (QAO):*** QAO algorithms modify traditional annealing techniques by incorporating quantum-inspired elements to explore the search space more efficiently.

- ***Quantum-inspired Genetic Programming (QGP):*** QGP algorithms use quantum-inspired operators in genetic programming, allowing for the evolution of programs that can optimize complex problems.

These quantum-inspired algorithms are designed to take advantage of quantum-inspired principles to tackle optimization problems in various domains, such as finance, logistics, machine learning, and more. While they don't offer the exponential speedup of true quantum computing, they can outperform classical algorithms in specific scenarios and provide valuable insights into the potential benefits of quantum algorithms ((Moussa, 2020).

5.2 Quantum-Inspired Use Cases for Optimization

Optimization is a critical task in various fields, ranging from logistics and finance to machine learning and materials science. While quantum computing holds the promise of revolutionizing optimization with its potential for exponential speedup, practical quantum computers are still in their infancy. In the meantime, quantum-inspired algorithms have emerged as a powerful alternative for solving complex optimization problems (Bova, 2021). Below are some notable quantum-inspired use cases across different domains:

- ***Portfolio Optimization in Finance:*** Quantum-inspired algorithms can efficiently tackle portfolio optimization, a fundamental problem in finance. By considering factors like risk, return, and constraints, these algorithms help investors build well-balanced portfolios that maximize returns while minimizing risks (Egger, 2020).
- ***Supply Chain Management:*** Quantum-inspired optimization techniques are employed to optimize supply chain logistics, ensuring efficient distribution of goods while minimizing costs. These algorithms can consider multiple variables, such as transportation costs, demand fluctuations, and inventory levels, to find optimal solutions (Egger, 2020).
- ***Drug Discovery and Molecular Modeling:*** In the field of chemistry and pharmaceuticals, quantum-inspired algorithms assist in the search for new drugs and the exploration of molecular structures. They optimize molecular configurations and predict properties with high accuracy, significantly speeding up the drug discovery process (Zhou, 2020).
- ***Machine Learning Model- Hyperparameter Tuning:*** Quantum-inspired optimization algorithms are used to fine-tune hyperparameters of machine learning models. They help data scientists and researchers optimize model performance by efficiently searching through a large hyperparameter space (Elías, 2023).
- ***Energy Grid Management:*** Quantum-inspired optimization plays a vital role in managing energy grids. These algorithms assist in scheduling energy production, distribution, and storage, ensuring a stable and cost-effective energy supply while reducing environmental impact.
- ***Traffic Management and Routing:*** Quantum-inspired optimization can alleviate traffic congestion by optimizing traffic light timings, route planning for vehicles, and public transportation schedules. This leads to reduced commute times and fuel consumption.
- ***Manufacturing and Operations:*** Quantum-inspired algorithms are employed in manufacturing for production scheduling, quality control, and resource allocation. They optimize production processes to increase efficiency and reduce production costs (Zhou, 2020).

- ***Space Exploration and Satellite Orbits:*** In aerospace engineering, quantum-inspired optimization is used to plan satellite orbits and mission trajectories. These algorithms ensure efficient use of fuel and resources, enabling extended missions and deeper space exploration.
- ***Network Design and Telecommunications:*** Quantum-inspired optimization aids in designing robust and efficient telecommunication networks. It optimizes data routing, resource allocation, and network layout to enhance communication quality and reduce latency.
- ***Natural Language Processing:*** Quantum-inspired algorithms are applied to natural language processing tasks, such as text summarization, machine translation, and sentiment analysis. They optimize language models and improve the accuracy of language-related applications.

The quantum-inspired optimization algorithms are making a significant impact across various industries, providing practical solutions to complex optimization challenges. While awaiting the widespread availability of quantum hardware, these classical algorithms offer a glimpse into the potential quantum advantage and continue to drive advancements in optimization techniques (Perdomo-Ortiz, 2019).

5.3 Practical Implementation Strategies

Implementing quantum-inspired optimization strategies effectively requires careful planning and execution. Below are some practical implementation strategies to maximize the benefits of these algorithms:

5.3.1 Problem Formulation and Preprocessing

- ***Define Clear Objectives:*** Begin by precisely defining the optimization problem, including objectives, constraints, and relevant parameters. A well-defined problem statement is crucial for the success of any optimization algorithm.
- ***Data Preprocessing:*** Clean and preprocess data to ensure it is suitable for the chosen quantum-inspired algorithm. Data quality greatly affects optimization results.

5.3.2 Algorithm Selection

Choose the Right Algorithm: Select the quantum-inspired algorithm that aligns best with the problem at hand. Options include simulated quantum annealing, quantum-inspired genetic algorithms, and more. Consider the algorithm's strengths, limitations, and computational requirements (Khang & Kali, 2023).

5.3.3 Platform Selection (Hardware vs. Software)

Decide whether to implement the algorithm on specialized hardware (if available) or through software libraries and frameworks. Factors such as cost, scalability, and compatibility play a role in this decision.

5.3.4 Parameter Tuning (Hyperparameter Optimization)

Quantum-inspired algorithms often have hyperparameters that need tuning. Utilize techniques like grid search or Bayesian optimization to find optimal hyperparameter settings for your specific problem.

5.3.5 Parallelization and Scaling

- ***Parallel Processing:*** Leverage parallelization to expedite the optimization process. Distributing the workload across multiple processors or nodes can significantly reduce execution time, especially for large-scale problems.
- ***Scaling Considerations:*** Assess the scalability of the chosen algorithm. Some quantum-inspired algorithms may require adjustments to handle larger problem sizes efficiently.

5.3.6 Integration With Existing Systems (APIs and Libraries)

Integrate quantum-inspired optimization into your existing software stack using appropriate APIs and libraries. This ensures seamless interaction with other components of your system.

5.3.7 Performance Monitoring and Debugging

- **Real-time Monitoring:** Implement tools for real-time performance monitoring to track algorithm progress and detect potential issues promptly.
- **Debugging Capabilities:** Develop debugging mechanisms to diagnose and resolve problems that may arise during execution.

5.3.8 Resource Management

- ***Resource Allocation:*** Allocate computational resources judiciously. Consider factors such as memory, CPU/GPU availability, and storage to optimize resource usage.
- ***Budgeting and Costs:*** Monitor computational costs and budget accordingly, especially when using cloud-based services.

5.3.9 Robustness and Error Handling

Implement robust error handling mechanisms to manage unexpected situations gracefully. Quantum-inspired algorithms may encounter convergence issues or other errors that need to be addressed.

5.3.10 Documentation and Knowledge Transfer

- ***Documentation:*** Maintain thorough documentation of the implementation, including algorithms used parameter settings, and performance results. This documentation aids in knowledge transfer and troubleshooting.
- ***Knowledge Transfer:*** Ensure knowledge transfer within your team or organization to promote a deeper understanding of the quantum-inspired optimization implementation.

5.3.11 Continuous Improvement (Benchmarking and Optimization)

Regularly benchmark the algorithm's performance against alternatives and explore opportunities for optimization. New research and improvements in quantum-inspired techniques may lead to enhanced results.

On the whole, the successful implementation of quantum-inspired optimization strategies requires a systematic approach that encompasses problem formulation, algorithm and platform selection, robust resource management, and continuous improvement. By following these practical strategies, organizations can harness the power of quantum-inspired algorithms to solve complex optimization problems effectively and drive innovation in various domains (Poggel, B., 2023).

6. NEED AND CHALLENGES

6.1 Why Quantum-Inspired Optimization Matters

The adoption of quantum-inspired optimization offers several compelling advantages for enterprises:

- ***Efficiency:*** Faster problem-solving and better resource allocation lead to increased operational efficiency and cost savings.
- ***Competitive Advantage:*** Enterprises that embrace quantum-inspired optimization can gain a competitive edge by making smarter, data-driven decisions.
- ***Innovation:*** Quantum-inspired algorithms open up new possibilities for innovation and problem-solving in various domains, fostering growth and adaptability.
- ***Sustainability:*** Optimizing resource utilization can contribute to a more sustainable business model, reducing waste and environmental impact.

6.2 Current Challenges and Future Prospects

While quantum optimization holds immense potential, it also faces several challenges. Quantum computers are still in their infancy, characterized by high error rates and limited qubit counts. Error mitigation techniques and advancements in qubit stability are critical for realizing the full potential of quantum optimization (Stilck, 2021).

Looking ahead, the evolution of quantum hardware and the refinement of quantum error correction codes hold the promise of overcoming these challenges. As quantum computers become more robust, they will tackle increasingly complex optimization problems with greater efficiency, revolutionizing industries and scientific research in unprecedented ways (Khang & Muthmainnah et al., 2023).

7. CONCLUSION

In conclusion, quantum optimization stands as a transformative force for enterprises seeking to elevate their problem-solving capabilities. As quantum technology advances, enterprises that invest in understanding and implementing quantum optimization are poised to gain a competitive edge, driving innovation, efficiency, and progress in their respective domains. By embracing the principles of quantum mechanics, enterprises can pave the way for a new era of problem-solving that holds the potential to reshape industries and redefine success (Khang & Quantum, 2023).

However, the integration of quantum optimization into enterprises is not without challenges. Quantum computers are still in their developmental stages, characterized by error rates and limitations in qubit stability. Enterprises must navigate these challenges by collaborating with quantum computing experts, leveraging hybrid quantum-classical approaches, and exploring error mitigation techniques (Khang & Kali, 2024).

REFERENCES

Ajagekar, A., Humble, T., & You, F. (2020). Quantum computing based hybrid solution strategies for large-scale discrete-continuous optimization problems. *Computers & Chemical Engineering*, *132*, 106630. doi:10.1016/j.compchemeng.2019.106630

Barkoutsos, P. K., Nannicini, G., Robert, A., Tavernelli, I., & Woerner, S. (2020). Improving variational quantum optimization using CVaR. *Quantum*, *4*, 256. doi:10.22331/q-2020-04-20-256

Bova, F., Goldfarb, A., & Melko, R. G. (2021). Commercial applications of quantum computing. *EPJ Quantum Technology*, *8*(1), 2. doi:10.1140/epjqt40507-021-00091-1 PMID:33569545

Cao, B., Fan, S., Zhao, J., Yang, P., Muhammad, K., & Tanveer, M. (2020). Quantum-enhanced multi-objective large-scale optimization via parallelism. *Swarm and Evolutionary Computation*, *57*, 100697. doi:10.1016/j.swevo.2020.100697

Denchev, V. S., Ding, N., Vishwanathan, S. V. N., & Neven, H. (2012). *Robust classification with adiabatic quantum optimization.* arXiv preprint arXiv:1205.1148.

Egger, D. J., Gambella, C., Marecek, J., McFaddin, S., Mevissen, M., Raymond, R., Simonetto, A., Woerner, S., & Yndurain, E. (2020). Quantum computing for finance: State-of-the-art and future prospects. *IEEE Transactions on Quantum Engineering*, *1*, 1–24. doi:10.1109/TQE.2020.3030314

Elías, F. (2023). A Practical Guide to Quantum Machine Learning and Quantum Optimization. Packt.

Ewald, R. H. (2019). An Introduction to Quantum Computing and Its Application. In S. Feld & C. Linnhoff-Popien (Eds.), Lecture Notes in Computer Science: Vol. 11413. *Quantum Technology and Optimization Problems. QTOP 2019.* Springer. doi:10.1007/978-3-030-14082-3_1

Gill, S. S., Kumar, A., Singh, H., Singh, M., Kaur, K., Usman, M., & Buyya, R. (2022). Quantum computing: A taxonomy, systematic review and future directions. *Software, Practice & Experience*, *52*(1), 66–114. doi:10.1002pe.3039

Khang, A. (2023). Applications and Principles of Quantum Computing. IGI Global Press. https://doi.org/ doi:10.4018/979-8-3693-1168-4

Khang, A. (2023). Quantum Mechanics Primer - Fundamentals and Quantum Computing. Applications and Principles of Quantum Computing. IGI Global Press. https://doi.org/ doi:10.4018/979-8-3693-1168-4-ch001

Khang, A. (2023). Introduction to Quantum Computing and Integration of Quantum Computing. Applications and Principles of Quantum Computing. IGI Global Press. https://doi.org/ doi:10.4018/979-8-3693-1168-4-ch002

Khang A. (2024). *The Quantum Evolution: Application of AI and Robotics in the Future of Quantum Technology*. CRC Press. https://doi.org/ doi:10.1201/978-1-032-64207-9

Khang, A., & Kali, C. R. (2023). Enabling the Future of Manufacturing: Integration of Robotics and IoT to Smart Factory Infrastructure in Industry 4.0. AI-Based Technologies and Applications in the Era of the Metaverse. IGI Global Press. doi:10.4018/978-1-6684-8851-5.ch002

Khang, A., & Muthmainnah, M. (2023). AI-Aided Teaching Model for the Education 5.0 Ecosystem. AI-Based Technologies and Applications in the Era of the Metaverse. IGI Global Press. doi:10.4018/978-1-6684-8851-5.ch004

Khang, A., Shah, V., & Rani, S. (2023). AI-Based Technologies and Applications in the Era of the Metaverse. IGI Global Press. doi:10.4018/978-1-6684-8851-5

Li, M. W., Wang, Y. T., Geng, J., & Hong, W. C. (2021). Chaos cloud quantum bat hybrid optimization algorithm. *Nonlinear Dynamics*, *103*(1), 1167–1193. doi:10.100711071-020-06111-6

Li, Y., Tian, M., Liu, G., Peng, C., & Jiao, L. (2020). Quantum Optimization and Quantum Learning: A Survey. *IEEE Access : Practical Innovations, Open Solutions*, *8*, 23568–23593. doi:10.1109/ACCESS.2020.2970105

Moll, N., Barkoutsos, P., Bishop, L. S., Chow, J. M., Cross, A., Egger, D. J., Filipp, S., Fuhrer, A., Gambetta, J. M., Ganzhorn, M., Kandala, A., Mezzacapo, A., Müller, P., Riess, W., Salis, G., Smolin, J., Tavernelli, I., & Temme, K. (2018). Quantum optimization using variational algorithms on near-term quantum devices. *Quantum Science and Technology*, *3*(3), 030503. doi:10.1088/2058-9565/aab822

Moussa, C., Calandra, H., & Dunjko, V. (2020). To quantum or not to quantum: Towards algorithm selection in near-term quantum optimization. *Quantum Science and Technology*, *5*(4), 044009. doi:10.1088/2058-9565/abb8e5

Neven, H., Denchev, V. S., Rose, G., & Macready, W. G. (2012, November). Qboost: Large scale classifier training withadiabatic quantum optimization. In *Asian Conference on Machine Learning* (pp. 333-348). PMLR.

Perdomo-Ortiz, A., Feldman, A., Ozaeta, A., Isakov, S. V., Zhu, Z., O'Gorman, B., Katzgraber, H. G., Diedrich, A., Neven, H., de Kleer, J., Lackey, B., & Biswas, R. (2019). Readiness of quantum optimization machines for industrial applications. *Physical Review Applied*, *12*(1), 014004. doi:10.1103/PhysRevApplied.12.014004

Poggel, B., Quetschlich, N., Burgholzer, L., Wille, R., & Lorenz, J. M. (2023, July). Recommending solution paths for solving optimization problems with quantum computing. In *2023 IEEE International Conference on Quantum Software (QSW)* (pp. 60-67). IEEE. 10.1109/QSW59989.2023.00017

Stilck França, D., & Garcia-Patron, R. (2021). Limitations of optimization algorithms on noisy quantum devices. *Nature Physics*, *17*(11), 1221–1227. doi:10.103841567-021-01356-3

Zhou, L., Wang, S. T., Choi, S., Pichler, H., & Lukin, M. D. (2020). Quantum approximate optimization algorithm: Performance, mechanism, and implementation on near-term devices. *Physical Review X*, *10*(2), 021067. doi:10.1103/PhysRevX.10.021067

Chapter 19
Quantum Cryptography:
Reviewing Post-Quantum Cryptography

Sonam Chauhan
SRM University, India

Shipra Srivastava
https://orcid.org/0000-0002-2460-5341
Greater Noida Institute of Technology, India

Ramveer Singh
Galgotias College of Engineering and Technology, India

Ajay Sharma
https://orcid.org/0000-0001-6618-3143
SRM University, India

Deo Brat Ojha
University of Rajasthan, India

ABSTRACT

The world in which we live is heavily reliant on digital communications. Traditional communication methodologies have been replaced with digital communications methods such as e-mail, instant messages, social media, and many more. Every day, terabytes of data are generated and shared around the world. Securing data stored and shared across these networks have evolved into a thriving discipline. Cryptography is an excellent way to secure terabytes of data. The core of cryptography is to transform data into unreadable format. With the introduction of quantum systems, it has been proved that many traditional cryptographic schemes are unsecure and can be broken with quantum computers. This chapter presents a brief review of pre- and post-quantum systems.

DOI: 10.4018/979-8-3693-1168-4.ch019

1. INTRODUCTION

Cryptography is the technology that triumphs in this computing era to keep information hidden from unapproved excess. The field is responsible for securing data from e-mail to e-transactions, online messaging to online shopping, and data transmission to data storage. Access to the World Wide Web (Khang & Shah et al., 2023), digital media, video, use of application and storage servers, and usage of email and instant messaging programs, are all supported via computer networks (Shyam & Khang et al., 2023). Cryptography is the art or science of transforming messages into nonreadable form. The cryptographic schemes ensure that the message is encoded before transmission and can only be decoded by the receiver. The original message is termed plaintext whereas an encrypted message is termed ciphertext. Thus, encryption convert plaintext to ciphertext and decryption performs the reverse (Bruss et al., 2007).

The message before transformation or the original message is termed plaintext. This plaintext is turned into ciphertext via an encryption algorithm. To extract plaintext from ciphertext, a decryption algorithm is used. The encryption and decryption algorithms use keys. Thus, keys can be viewed as the numbers on which encryption-decryption algorithms operate (Rothe, 2005).

The majority of cryptographic schemes are grounded on the hardness of specific mathematical problems, including discrete logarithm. Classical cryptography suffered from two issues (Rothe, 2005; Stinson, 2005; Stallings, 2011, Diffie & Hellman, 1976):

- These cryptosystems are grounded on the difficulty of mathematical problems like discrete logarithm and factoring. These cryptosystems are not completely secure as these problems are not hard. For example, if in RSA public-key cryptosystem (Rivest et al., 1978) large integers are easy to factor then it is not secure.
- The quantum computing has generated new methods that can solve mathematical problems in efficient and quicker way. The quantum computers are powerful enough to break many classical ciphers such as RSA. Although quantum computers fail several classical cryptosystems but they also open a completely new domain to secure the messages using quantum mechanics (Bruss et al., 2007).

In history, scytale was the first recorded used encryption method. In fifth century B.C., Spartans exchange confidential war data between generals using scytale. To encrypt the message, Spartan use a wooden cylinder covered with a strip of leather. The ciphertext was written on the strip from left side to right side. The unspooling strip produces random letters with no meaning. To decrypt the ciphertext, a scytale with the identical radius as the cylinder was used. This is a transposition cipher where the order of the alphabet is changed. There are substitution ciphers also where one alphabet is replaced by another alphabet.

The encryption and decryption combinedly form a system known as cryptosystem where the key is used for both the transformation of plaintext to ciphertext and vice versa. If an eavesdropper had an idea about the encryption scheme or key then s/he can try to break the cryptosystem. In the nineteenth century, Auguste Kerckhoffs asserted that the reliability of cryptosystems is solely dependent on the confidentiality and privacy of the key.

With time, the need to secure information arises and with that need for secure and fast cryptosystems also. Initially, mechanical devices were used to increase the keys complexity and to increase encryption-decryption speed. One such cryptosystem is Enigma. In World War II, Enigma was used by the German military for secure communication. They consider Enigma as secure and unbreakable whereas allied cryptanalysts in Bletchley Park were decrypting their confidential messages. The invention of transistors results in more secure cryptosystems such as DES, AES and RSA.

This chapter is divided into six sections. The second section gives a quick review of traditional cryptography (cryptography before quantum systems). Post-quantum cryptosystems and significant post-quantum protocols are discussed in Section 3. The key distribution strategy for post-quantum cryptography was detailed in section 4. Section 5 discusses lattice-based post quantum cryptosystems. This chapter conclusion is found in Section 6.

2. CLASSICAL CRYPTOGRAPHY

Various textbooks describe classical cryptography such as Rothe (Rothe, 2005), Stinson (Stinson, 2005), and Stallings (Stallings, 2011). In this section, the cryptosystem is defined in terms of tuples. Definition 2.1: A symmetric and deterministic cryptosystem is a five-tuple (P,C,K,E,D) (Stallings, 2011) system that meets the following requirements:

i) The finite set of potential plaintexts is symbolized by P.
ii) The finite set of potential ciphertexts is symbolized by C.
iii) The finite set of potential keys is denoted by K.
iv) We can define rule for encryption as $e_k \in E$ for each key $k \in K$ and its equivalent decryption rule $d_k \in D$, where e_k: $P \rightarrow C$ and d_k: $C \rightarrow P$ are transformation functions that satisfy $d_k(e_k(x))=x$ for each plaintext element $x \in P$.

Assume two communicating parties would like to interconnect via channel. Because their communication is taking place over an insecure channel, an eavesdropper such as Eve can intercept their messages. Alice and Bob can communicate securely by concurring on a key k.

Consider Alice use encryption rule e_k with key k to secure message m. The decrypted text can be obtained as $c=e_k(m)$. Alice after encrypting the message, send that to Bob over an insecure network. Bob after receiving the cipher text, recovers the plaintext using the decryption rule $d_k(c)=d_k(e_k(m))=m$. Without the key, Eve cannot determine the original message m even thought he had ciphertext c. In present, there is a large number of cryptosystems for secure communication. Example: Consider, one-time pad or Vernam Cipher. For plaintext P elements are encoded ranging from 0 to 25 as shown in table 1.

Table 1. One-Time pad encryption example

Plaintext	H	E	L	L	O
	7	4	11	11	14
Key	5	11	7	10	8
Encoded	12	15	18	21	22
Ciphertext	M	P	S	V	W

The aforementioned examples are symmetric cryptosystems in which identical key perform encryption and decryption process. In such a scenario, a secure channel is required for key exchange. How can the key be securely shared between two parties? This is referred to as the key exchange issue. A very simple and secure solution was suggested by Diffie and Hellman (1976) in 1976. They solve the key agreement problem by using hardness of discrete logarithm. After Diffie and Hellman, other researchers also proposed different secret-key agreement protocols. In 1978, Rivest, Shamir and Adleman (1978) proposed a scheme where two identical keys were used in cryptosystem.

The biggest drawback of symmetric ciphers and key distribution appears when large number of parties want to share a secret key. Theoretically, if n parties are there then $n(n+1)/2$ different secret keys are needed. This key distribution problem of public-key cryptosystem is solved by asymmetric cryptosystems. Instead of using one key for each party, asymmetric cyphers use one key pair per party. Diffie and Helman introduced the first public-key cryptosystem (1976). They proposed an idea that uses two distinct keys to perform encryption and decryption.

The RSA cryptosystem was the first of its kind, and it was named after its creators, Ron Rivest, Adi Shamir, and Leonard Adleman. On one-way functions, the RSA algorithm is based. The inverted functions are easier to calculate but hard to invert. In these cryptosystems, Alice generates k_{public} and $k_{private}$ keys. She uses k_{public} for transformation and keeps $k_{private}$ as secret. When Bob demands to share information with Alice, he employs k_{public} for encryption. Because Alice is the only who own the private key, she can retrieve the original message.

In 1985, another cryptosystem dependent on discrete log was introduced by ElGamal (1985). The biggest disadvantage of this cryptosystem is that they use a directory for storing public keys. These directories need security against unauthorized access and attacks. Consider, if an eavesdropper Eve is able to substitute's public key with another key, then she can read Alice's communications.

Public key cryptosystems based on discrete logarithms and factors were assumed to be secure before quantum computers, but Shor showed in 1997 that these could be solved in polynomial time (Shor, 1997).

3. POST QUANTUM CRYPTOGRAPHY

The advent of quantum computers opens a completely new door in science and technology. The front-page article of New York Times opens a debate for severe security flaws of public-key algorithms in Quantum computers. The quantum computers are capable to break the most widely cryptosystems such as RSA, DSA and ECDSA. Is it right to say that quantum computers had destroyed the cryptography? How information can be stored and transmitted securely with quantum computers? Is it correct to say that with quantum computer, cryptography fails? (Khang & Quantum, 2023).

Although quantum computers destroyed RSA, DSA, and ECDSA, cryptography is a broad discipline with many other schemes and algorithms, including hash-based cryptography, code-based cryptography, and Multivariate-quadratic-equations cryptography. These schemes mentioned above are resistant to quantum computers (Bernstein, 2009). The Shor's and Grover's algorithms are capable to break public-key algorithms (Khang & Kali, 2023).

The cryptanalysts continuously invest their efforts, time and energy to find devastating attacks or some smaller loopholes. Sometimes cryptanalysts are not able to suggest any attack that makes us confident that the best imaginable attack has been recognized and the real-world attackers are not able to break that. Consider, factor-based RSA cryptosystem. In this section, we discussed three attacks against the hardness of factorization. For example, if you are writing about cyber warfare, you might mention Russia using cyber-attacks to influence the 2016 US presidential election.

- In 1978, Rivest, Shamir and Adleman proposed a novel scheme (1978) Schroeppel's "linear sieve." This algorithm is able to factor any RSA modulus n and can break RSA with the help of $2^{(1+o(1))(\lg n)^{\frac{1}{2}}(\lg\lg n)^{\frac{1}{2}}}$ operations (Bernstein, 2009). Let's restrict the linear sieve to practice at least 2^b operation, it simply means n should be chosen to have at least $(0.5+o(1))b^2/\lg b$ bits.
- In 1988, Pollard proposed factorization algorithm "number-field sieve." The algorithms factorize any modulus n using $2^{(1.9+\ldots o(1))(\lg n)^{1/3}(\lg\lg n)^{2/3}}$ procedures. Let's restrict algorithm to practice at least 2^boperation, it simply means n should be chosen to have at least $(0.016\ldots+o(1))b^3/\lg b^2$ bits.
- In 1994, Peter Shor proposed a revolutionary algorithm (Shor, 1997) that factor any modulus n with $(\lg n)^{2+o(1)}$ basic operations of size $(\lg n)^{1+o(1)}$ quantum computer. Let's restrict the algorithm to practice at least 2^b operation, it simply signifies n should be chosen to have at least $2^{(0.5+o(1))b}$ bits.

3.1 BB84 Protocol

Charles H. Bennet and Gilles Brassard (Bennet and Brassard, 1984) proposed the first quantum protocol in 1984. This quantum key distribution protocol was named BB84 on its inventor initial and invention year. The protocol works with quantum states that are four in number and these states further comprises of bases that are two in count. These states are as follows: a) up $|\uparrow$, b) down $|\downarrow$, c) left $|\leftarrow$ and d) right $|\rightarrow$. We have maximally conjugate bases, which means if we have two vectors from the same basis then they will overlap. Attribute to bit 0 is states $|\uparrow$ and $|\rightarrow$ and to bit 1 is $|\downarrow$ and $|\leftarrow$. These can also be referred as state qubits.

Alice randomly chooses a state among the four and sends individual spins to Bob. The individual spins can be sent sequentially or all at the same time provided there must be a one-to-one correspondence between Alice and Bob. Bob measured spins with either of the two bases. Bob obtained perfectly correlated results if he used the same basis; otherwise, Bob obtained uncorrelated results. On average Bob gets the information with a bit error rate (raw key) of 25% which is quite high. None of the error correction schemes is capable to withstand this error rate.

In BB84 protocol, a simple straight error correction scheme can be applied. Bob only declares the basis on which he measured the equivalent bit for each qubit. He hides the results that he obtained. Later, Alice revealed if her encoded state is comparable with Bob's declared basis. If the state is compatible, they accepted it; otherwise, they rejected it. By using the process, usually 50%-bit string is rejected. Thus, we obtained a shorter key and this key is termed as shifted key (Ekert and Huttner, 1994). In this case, an eavesdropper Eve can listen in on their conversation but cannot change their information.

Let's review the security of the protocol. Consider, an eavesdropper Eve intercepts the qubit sent by Alice to Bob. In case Bob obtain the unanticipated qubit, he will notify Alice and reject that qubit. Thus, Eve does not get any relevant information. She reduces the bit rate. To really be successful, Eve must transmit a qubit to receiver while remaining anonymous. There are different researches shown in Table 2 that prove the security of BB84 quantum-key-distribution protocol.

Table 2. Security proof of BB84 quantum key-distribution protocol

Reference	Journal/ Conference
(Lo & Chau, 1999)	Science
(Shor & Preskill, 2000)	Physical Review Letters
(Mayers, 2001)	Journal of the ACM
(Biham, 2006)	Journal od Cryptology
(Tsurumarun & Tamaki, 2008)	Physical Review A
(Koashi, 2009)	New Journal of Physics
(Rusca et al., 2018)	Physical Review A
(Su, 2020)	Quantum Information Processing

3.2 No Cloning Theorem

The quantum world is free from perfect copying (Wootters & Zurek, 1982; Wigner, 1961; Dieks, 1982; Milonni & Hardies, 1982; Lindblad, 1999). Let the state of qubit be represented by ψ, |b represents blank copy. The Eve's initial state is represented by $| 0 \epsilon \mathcal{H}_{QCM}$ where Hilbert space $\mathcal{H}_{QCM}$ of quantum cloning machine is taken randomly. With the above conditions, an ideal machine result

$$\psi \otimes \left|b\right\rangle \otimes \left|0 \rightarrow \psi \otimes \psi \otimes \mid f_{\psi}\right. \tag{1}$$

where $|f_{\psi}$ represents Eve's machine final state that depends on ψ. We have

$$\left|\uparrow, \boldsymbol{b}, 0 \rightarrow\right| \uparrow, \uparrow, \boldsymbol{f}_{\uparrow} \tag{2}$$

and

$$\left|\downarrow, \boldsymbol{b}, 0 \rightarrow\right| \downarrow, \downarrow, \boldsymbol{f}_{\downarrow} \tag{3}$$

According to linearity of quantum dynamics

$$\left|\rightarrow, \boldsymbol{b}, 0 = \frac{1}{\sqrt{2}}\left(\left|\uparrow + \right|\downarrow\right) \otimes \right|\boldsymbol{b}, 0 \tag{4}$$

$$\rightarrow \frac{1}{\sqrt{2}}\left(\left|\uparrow, \uparrow, f_{\uparrow} + \right| \downarrow, \downarrow, f_{\downarrow}\right) \tag{5}$$

Thus, an adversary cannot have the exact copy of the quantum. The fact that quantum state or quantum information cannot be copied is the striking fact of quantum information. This attribute plays a key role in quantum cryptography. Two distinct No-Cloning theorem proofs (Wootters & Zurek, 1982; Dieks, 1982) were introduced in 1982. Entanglement permits information transmission at superluminal speeds, according to both pieces of evidence.

The ability to create one or more anticipated copies of an unidentified quantum state exists in quantum cloning machines (Bužek & Hillery, 1996). These devices examine whether pure or mixed states are to be copied, the number of identically prepared input states, the number of desired output copies, and if the cloner device is required to work for an arbitrary input state or for a specific set of input states.

3.3 Two-State Protocols

In 1992, Bennet suggested that only two nonorthogonal states are required (Bennett, 1992). He claimed that quantum cryptography security relies on the perturbation between states, hence only two states are needed. In practice, this is not an ideal solution because two states can only be recognized by perturbation (Ivanovic, 1987; Peres, 1988). This is further practically proved by many researchers (Huttner et al., 1996; Clarke et al., 2000). The interference among a dim and a macroscopic bright pulse with an average of less than single photon can also be used to implement the two-state protocol (Bruss, 1998). The bright pulse made this resistant to an adversary. Thus, Bob can identify if Eve has removed any information by observing the bright pulses. Because bright pulse interference with vacuum causes errors, Eve's ability to eliminate dim pulse without revealing her identity is nearly impossible.

3.4 Six-State Protocols

Initially, four state protocols (Bennet & Brassard, 1984) were introduced that are used as the standards. Later, in 1992, Bennet suggested that only two states are enough (Bennett, 1992). Six state protocols (Bechmann-Pasquinucci & Gisin, 1999; Ekert, 1991; Shaari et al., 2006; Enzer et al., 2002) with improved symmetry of qubit state space were introduced in the late 1990s. The six states protocol, the possibility that sender and receiver pick the identical basis is 1/3. The protocol has simplified security analysis because of symmetry and this decreases Eve's optimal error rate information gain QBER. As compared to the BB84 protocol QBER of 25%, this protocol has 33% QBER.

Take into consideration the following matrices as shown in Formula 6 and Formula 7:

$$\hat{z} = Z = \begin{pmatrix} 1 & 0 \\ 0 & -1 \end{pmatrix},\ \hat{x} = X = \begin{pmatrix} 0 & 1 \\ 1 & 0 \end{pmatrix} \text{ and } \hat{y} = Y = \begin{pmatrix} 0 & -i \\ i & 0 \end{pmatrix} \tag{6}$$

and the corresponding eigen vectors of each are

$$|z_+ = |0 = \begin{pmatrix} 1 \\ 0 \end{pmatrix},\ |z_- = |1 = \begin{pmatrix} 0 \\ 1 \end{pmatrix},\ |x_\pm = \frac{(|z_+ \pm |z_-)}{\sqrt{2}} \text{ and } |y_\pm = \frac{(|z_+ \pm i|z_-)}{\sqrt{2}}. \tag{7}$$

The eigen states of each basis are indicated by the letters in the notation enclosed in brackets. The states |1 and |0 specify the computational basis. The equatorial plane of the Poincaré sphere is defined by the axes $\hat{x}$ and $\hat{z}$.

3.5 Einstein-Podolsky-Rosen (EPR) Protocols

The idea was given by Arthur Ekert (Ekert, 1991) which is the elaborated of David Deutsh work (1985). David Deutsh worked on Quantum cryptography independent of BB84 paper. The idea is highly related to EPR (Einstein, Podolski and Rosen) paradox (Einstein, 1935) and hence is known as Einstein-Podolsky-Rosen protocol. A channel that carries 2 qubits is replaced in this protocol by a channel carrying qubits from a common source. It is possible for the source to selects and emits same state qubits at random. Later, both identify qubit in a randomly and independently chosen states.

Finally, the source discloses its bases, and if Alice and Bob measured their values in a compatible base, they keep the data. The protocol is the identical to the BB84 protocol in case of reliable source. Because the source could be handled by Eve, the Ekert protocol states that the ejection of qubit in the most entangled possible state as shown in Formula 8.

$$\varnothing^{+} = \frac{1}{\sqrt{2}}(|\uparrow,\uparrow + |\downarrow,\downarrow \quad (8)$$

If both communicating parties use identical basis then around half of the cases lead to identical results.

Figure 1. EPR protocol
Alt-text: Figure 1 represents two bits Ekert protocol.

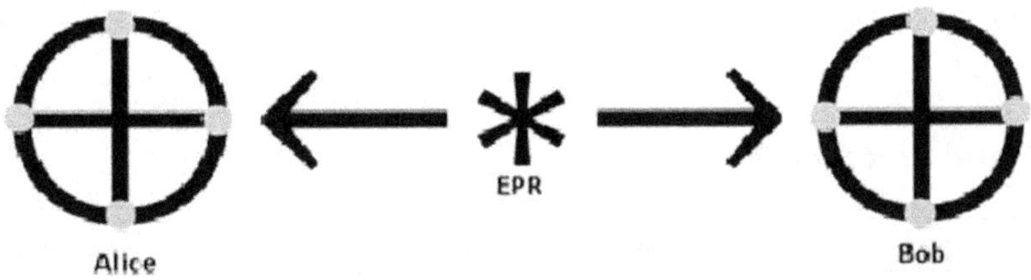

The following equation 9 holds for all unitary evolutions U_1 and U_2.

$$U_1 \otimes U_2 \Phi^{(+)} = \mathbb{I} \otimes U_2 U_1^t \Phi^{(+)} \quad (9)$$

where U_1^t represents the transpose.

In 1991, Ekert in his work (1991) suggested two-qubit protocol security that is based on Bell's inequality. According to Bell's inequality, there exists some correlations that are projected by quantum mechanics and these correlations cannot be recreated by local theory (Bell, 1987). Alice and Bob use

Figure 2. Protocol using EPR quantum systems

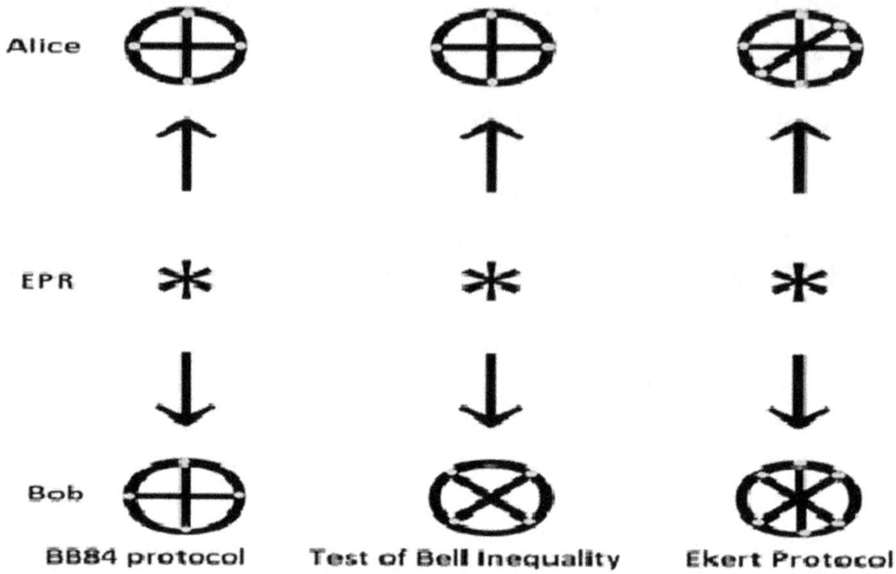

third bases as depicted in fig. 2. The probability of choosing the same basis is reduced from $\frac{1}{2}$ to $\frac{2}{9}$ and they create a key and assemble adequate data to exam Bell's inequality.

In (Bennett et al., 1992) disapprove Ekert's letter and claim that the defilement of Bell's inequality is non-mandatory for Quantum security. Bennett et al. (1992) disapprove of Ekert's letter in 1992, claiming that violating Bell's inequality is not required for the quantum. Numerous studies that make use of the EPR procedure are listed in Table 3.

Table 3. Studies that use EPR protocol

Reference	Year	Journal/ Conference	Application
(Deng et al., 2003)	2003	Physical Review A	Quantum Secure Direct Communication Protocol
(Zheng & Long, 2014)	2014	Science China Physics, Mechanics & Astronomy	EPR pair block-based two-step quantum secure direct dialogue protocol
(Xia et al., 2022)	2022	Journal of Physics A: Mathematical and Theoretical	EPR based semi-quantum signature protocol.
(Pan et al., 2020)	2020	IEEE Access	An entangled pair of photons-based single-photon-memory Quantum Secure Direct Communication protocol.
(Xiang et al., 2019)	2019	Physical Review A	Protocols that provide the distribution of one-way Gaussian steering, which may then be applied to the distribution of one-sided, device-independent (1sDI), quantum keys.
(He et al., 2006)	2006	Physical Review A	Employing continuous variable EPR pair correlations to create a secure communication protocol.

Many researchers proposed variants that use dimension greater than 2 (Bechmann-Pasquinucci & Peres, 2000; Bechmann-Pasquinucci & Tittel, 2001; Bourennane et al., 2001). This idea lacks undetermined practical values. Another variant is proposed by Goldenberg and Vaidman in 1995 (Goldenberg & Vaidman, 1995). In their work, they suggested qubits to be in the superposition of spatially divided state. Alice sends superposition element and waits for Bob to receive it before sending the next element. Conceptually, the idea is good and does not require minimal states to be mutually orthogonal.

4. POST QUANTUM CRYPTOGRAPHY AND KEY ESTABLISHMENT MECHANISM

NIST began standardizing post-quantum cryptosystems in 2017. In the starting they had 69 candidate algorithms for key establishment and digital signatures. 26 out of 69 algorithms reached to the second round after being analyzed and evaluated (Goldenberg & Vaidman, 1995). In 2020, 15 out of the 26 algorithms reached to the next round, where seven are considered finalists and eight are on the alternative list as shown in Figure 3.

Figure 3 depicts the categorization of these algorithms.

The algorithms that are in the finalist list are Kyber crystals (Avanzi et al., 2017; Bos et al., 2018; Avanzi et al., 2018; Bisheh-Niasar et al., 2021; Xing & Li, 2021; Dang et al., 2023; Wan et al., 2022; Nguyen & Gaj, 2021), SABER (Dang et al., 2023; Nguyen & Gaj, 2021), NTRU (Hoffstein et al., 1998; Hoffstein et al., 2003; Hoffstein et al., 2001; Ducas et al., 2014; Gentry & Szydlo, 2002; Bernstein et

Figure 3. Post quantum cryptography key establishment algorithms that reached to the third round (adapted from Avanzi et al., 2015)

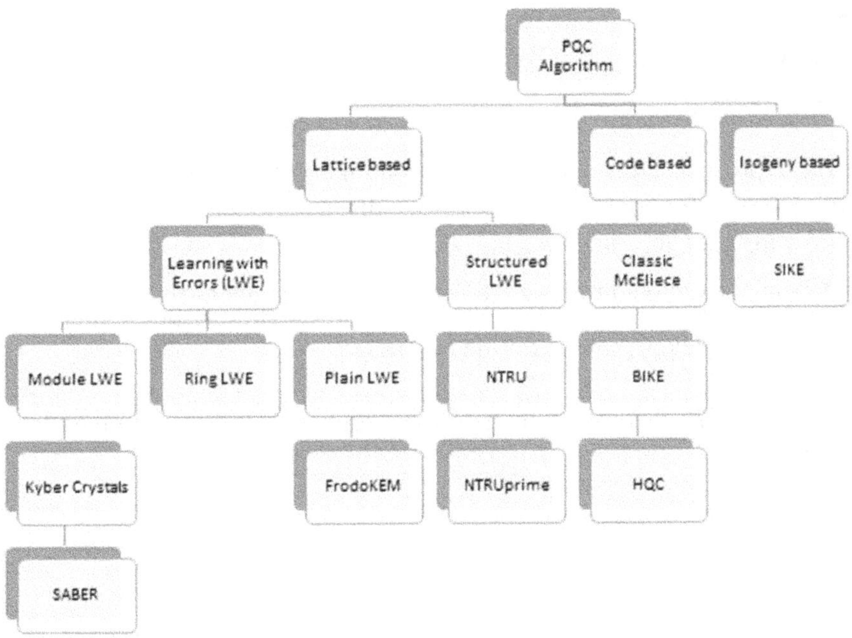

al., 2018) and Classic McEliece (McEliece, 1978; Baldi et al., 2016; Repka & Zajac, 2014; Misoczki & Barreto, 2009; Bernstein et al., 2008).

The algorithms that are placed in the alternative list are FrodoKEM (Guo et al., 2020; Gupta et al., 2021; Bos et al., 2021; Lee et al., 2019; Karl et al., 2022), NTRUprime (Bernstein et al., 2016; Bernstein et al., 2018; Bernstein et al., 2019; Peng et al., 2023; Cheng et al., 2020), BIKE (Guo et al., 2022), HQC (Guo et al., 2022; Schöffel et al., 2022; Goy et al., 2022; Deshpande et al. 2022) and SIKE (Elkhatib et al., 2022; Seo et al., 2020; Alagic et al., 2020; Pandey et al., 2019).

Currently, NIST is working to establish a single algorithm as the standard. The advantages and disadvantages along with the key outcomes of this standardization and evaluation process is summarized in (Alagic et al., 2020). These alternative algorithms can replace the finalist algorithms in case of any vulnerability found in them. These can also become standards after modifications and improvement. In article (Pandey et al., 2019), modified fuzzy commitment scheme has been introduced as a post quantum compatible scheme.

5. LATTICE BASED POST QUANTUM CRYPTOSYSTEMS

5.1 Hard Lattice Problem

In 1996, Ajtai (1996) point out random instance hard problems for n dimensional lattices $\mathcal{L}(\mathcal{B}) \subseteq \mathbb{R}^n$. These lattices are the sum of weighted integers of basic vectors $\mathcal{B} = \left[\overline{b_0}, \overline{b_1}, \ldots, \overline{b_{n-1}}\right] \int \mathbb{R}^n$.

$$\mathcal{L}(\mathcal{B}) = \sum_{i=0}^{n-1} x_i \overline{b_i} : x_i \in \mathbb{Z} \tag{10}$$

One of the most striking properties of these problems is that solving the random instance simply means solving in the worst-case scenario. As a result, they are also known as worst-case scenarios. These three difficult issues are:

i) Determine the approximate distance of the minimum non-zero vector in polynomial factor.
ii) SVP: Identify the smallest non-zero vector v, such that v is unique and all vectors smaller then $n^c\|v\|$ are parallel to v for large absolute constant c.
iii) Shortest Independent vectors problems (SIVP): Identify lattice basis $\left|\overline{b_0}, \overline{b_1}, \ldots, \overline{b_{n-1}}\right|$ of smallest possible length. Here, length can be defined as $\max_i \overline{b_i}$.

According to Minkowski's theorem, "the length of shortest vector in the given lattice has an upper bound of equation 9.

$$\sqrt{n}\left(\det[\overline{b_0}, \overline{b_1}, \ldots, \overline{b_{n-1}}]\right)^{\frac{1}{n}} \text{."} \tag{11}$$

The worst-case lattice glitches that still do not have polynomial-time solution and therefore they are declared no-polynomial time solution. For so long, these problems are considered to have no polynomial-time solution. There exist variants of these problems that are equally hard (Peikert, 2016). For instance, exact vs. approximate variant. The SVP approximate variant identify a lattice vector in lesser then γ(n) time the shortest vector. This problem can be denoted as $SVP\gamma$ and it is equally hard as the SVP problem. The other variant that can be consider is search vs. decision variant. SVP decision variants determines if the shortest vector value is smaller than 1 or more than γ(n). This issue is as hard as the SVP problem and is denoted by GAP$SVP\gamma$.

5.2 Average Lattice Problem

Because of the performance implications, it is not recommended to use the lattice problems mentioned above in lattice-based cryptography. There is a significant difference between hard problems-based schemes and hard lattice-based schemes. The vast majority of modern cryptographic schemes are average-case problems such as learning with errors (LWE) and short integer solution (SIS). Usually, average case problems are connected to some worst-case problem. As a result, proving the cryptographic robustness of lattice-based cryptosystems becomes easy and straightforward. In (Peikert, 2016), lattice-based problems $\mathcal{L}(\mathcal{B})$ is mapped to the hard lattice. problems $\Lambda T(A,q)$. If for a short vector problem in q-ary lattice $\Lambda T(A,q)$ exists a polynomial time algorithm then collection of defined points $h \in \mathbb{Z}^m$ satisfy the following

$$\Lambda^T\left(A,q\right) \triangleq \sum_{i=0}^{m} h_i \overline{a_i} = \overline{0}^n \left(\bmod q\right) : a_i \int \mathbb{Z}_q^n \tag{11}$$

then an algorithm must exist that can solve all hard problems defined above for $\mathcal{L}(\mathcal{B})$. This problem is SIS and is as hard as SIVP on $\Lambda^T(A,q)$. The cryptographic terms on the dual of q-ary lattice and can be represented as

$$\Lambda\left(A,q\right) \triangleq \overline{y} \int \mathbb{Z}^m : y = A^T \overline{s} \text{ for some } \overline{s} \epsilon \mathbb{Z}^n \tag{12}$$

LWE is another common case hard problem (Bos et al., 2016). The problem states that points in a lattice $\mathbb{Z}_q^n$ is generated by basis A of homogeneously sampled n-dimensional vector, error vector e, secret key s. This is hard to identify the difference between $t=As+e$ and a random value t' (Bos et al., 2016)

6. CONCLUSION

This chapter briefly outlines the advancements in quantum cryptography. Here, the article focuses on defining classical and post quantum cryptography. This chapter explains how cryptography changed with quantum computer and how quantum computers are able to break many classical cryptography algorithms such as RSA. This chapter focuses on defining several quantum key distribution protocols such as BB84 protocol, two-state protocol and six-state protocol. Here, the article briefly explained quantum key agreement mechanisms and which protocol is in the finalist list in NIST (Khang & Vugar et al., 2023).

REFERENCES

Ajtai, M. (1996, July). Generating hard instances of lattice problems. In *Proceedings of the twenty-eighth annual ACM symposium on Theory of computing* (pp. 99-108). ACM. https://dl.acm.org/doi/pdf/10.1145/237814.237838

Alagic, G., Alperin-Sheriff, J., Apon, D., Cooper, D., Dang, Q., Kelsey, J., & Smith-Tone, D. (2020). Status report on the second round of the NIST post-quantum cryptography standardization process. *US Department of Commerce, NIST, 2*. https://rnbee.kiev.ua/goto/https://nvlpubs.nist.gov/nistpubs/ir/2020/NIST.IR.8309.pdf

Ardehali, M., Chau, H. F., & Lo, H. K. (1998). Efficient quantum key distribution. *arXiv preprint quant-ph/9803007*. https://link.springer.com/article/10.1007/s00145-004-0142-y

Avanzi, R., Bos, J., Ducas, L., Kiltz, E., Lepoint, T., Lyubashevsky, V., & Stehlé, D. (2017). Crystals-kyber. *NIST, Tech. Rep*. https://csrc.nist.gov/csrc/media/Presentations/2022/crystals-kyber-update/images-media/session-1-schwabe-crystals-kyber-pqc2022.pdf

Avanzi, R., Bos, J., Ducas, L., Kiltz, E., Lepoint, T., Lyubashevsky, V., & Stehlé, D. (2019). CRYSTALS-Kyber algorithm specifications and supporting documentation. *NIST PQC Round, 2*(4), 1-43. https://pq-crystals.org/kyber/data/kyber-specification-round3-20210131.pdf

Baldi, M., Bianchi, M., Chiaraluce, F., Rosenthal, J., & Schipani, D. (2016). Enhanced public key security for the McEliece cryptosystem. *Journal of Cryptology, 29*(1), 1–27. https://link.springer.com/article/10.1007/s00145-014-9187-8. doi:10.100700145-014-9187-8

Bechmann-Pasquinucci, H., & Gisin, N. (1999). Incoherent and coherent eavesdropping in the six-state protocol of quantum cryptography. *Physical Review A, 59*(6), 4238–4248. https://journals.aps.org/pra/abstract/10.1103/PhysRevA.59.4238. doi:10.1103/PhysRevA.59.4238

Bechmann-Pasquinucci, H., & Peres, A. (2000). Quantum cryptography with 3-state systems. *Physical Review Letters, 85*(15), 3313–3316. https://journals.aps.org/prl/abstract/10.1103/PhysRevLett.85.3313. doi:10.1103/PhysRevLett.85.3313 PMID:11019329

Bechmann-Pasquinucci, H., & Tittel, W. (2000). Quantum cryptography using larger alphabets. *Physical Review A, 61*(6), 062308. https://journals.aps.org/pra/abstract/10.1103/PhysRevA.61.062308. doi:10.1103/PhysRevA.61.062308

Bell, J. S. (1966). On the problem of hidden variables in quantum mechanics. *Reviews of Modern Physics, 38*(3), 447–452. https://link.springer.com/chapter/10.1007/978-3-0348-9259-9_21. doi:10.1103/RevModPhys.38.447

Bennett, C. H. (1992). Quantum cryptography using any two nonorthogonal states. *Physical Review Letters, 68*(21), 3121–3124. https://journals.aps.org/prl/abstract/10.1103/PhysRevLett.68.3121. doi:10.1103/PhysRevLett.68.3121 PMID:10045619

Bennett, C. H., & Brassard, G. (1984). IEEE Int. Conf. on Computers, Systems and Signal Processing. *IEEE*. https://iopscience.iop.org/article/10.1088/1367-2630/4/1/343/meta

Bennett, C. H., Brassard, G., & Mermin, N. D. (1992). Quantum cryptography without Bell's theorem. *Physical Review Letters, 68*(5), 557–559. https://journals.aps.org/prl/abstract/10.1103/PhysRevLett.68.557. doi:10.1103/PhysRevLett.68.557 PMID:10045931

Bernstein, D. J. (2009). Introduction to post-quantum cryptography. In *Post-quantum cryptography* (pp. 1–14). Springer Berlin Heidelberg. https://link.springer.com/chapter/10.1007/978-3-540-88702-7_1 doi:10.1007/978-3-540-88702-7_1

Bernstein, D. J., Chuengsatiansup, C., Lange, T., & Van Vredendaal, C. (2016). NTRU Prime. *IACR Cryptol. ePrint Arch., 2016*, 461. http://hyperelliptic.org/tanja/vortraege/caen-ntruprime.pdf

Bernstein, D. J., Chuengsatiansup, C., Lange, T., & van Vredendaal, C. (2018). NTRU prime: reducing attack surface at low cost. In *Selected Areas in Cryptography–SAC 2017: 24th International Conference, Ottawa, ON, Canada*. Springer International Publishing. https://link.springer.com/chapter/10.1007/978-3-319-72565-9_12

Bernstein, D. J., Chuengsatiansup, C., Lange, T., & van Vredendaal, C. (2019). NTRU Prime: round 2. *Submission to the NIST PQC standardization process*. University of Adelaide. https://ntruprime.cr.yp.to

Bernstein, D. J., Lange, T., & Peters, C. (2008). Attacking and defending the McEliece cryptosystem. In *Post-Quantum Cryptography: Second International Workshop,* (pp. 31-46). Springer Berlin Heidelberg. https://link.springer.com/chapter/10.1007/978-3-540-88403-3_3

Biham, E., Boyer, M., Boykin, P. O., Mor, T., & Roychowdhury, V. (2000, May). A proof of the security of quantum key distribution. In *Proceedings of the thirty-second annual ACM symposium on Theory of computing* (pp. 715-724). ACM. https://dl.acm.org/doi/pdf/10.1145/335305.335406

Bisheh-Niasar, M., Azarderakhsh, R., & Mozaffari-Kermani, M. (2021). Instruction-set accelerated implementation of CRYSTALS-Kyber. *IEEE Transactions on Circuits and Systems. I, Regular Papers, 68*(11), 4648–4659. https://ieeexplore.ieee.org/abstract/document/9525069/. doi:10.1109/TCSI.2021.3106639

Bos, J., Costello, C., Ducas, L., Mironov, I., Naehrig, M., Nikolaenko, V., & Stebila, D. (2016, October). Frodo: Take off the ring! Practical, quantum-secure key exchange from LWE. In *Proceedings of the 2016 ACM SIGSAC conference on computer and communications security* (pp. 1006-1018). ACM. https://dl.acm.org/doi/abs/10.1145/2976749.2978425

Bos, J., Ducas, L., Kiltz, E., Lepoint, T., Lyubashevsky, V., Schanck, J. M., & Stehlé, D. (2018, April). CRYSTALS-Kyber: a CCA-secure module-lattice-based KEM. In *2018 IEEE European Symposium on Security and Privacy (EuroS&P)* (pp. 353-367). IEEE. https://ieeexplore.ieee.org/abstract/document/8406610/

Bos, J. W., Ofner, M., Renes, J., Schneider, T., & van Vredendaal, C. (2021). The matrix reloaded: Multiplication strategies in FrodoKEM. In *Cryptology and Network Security: 20th International Conference, CANS 2021, Vienna, Austria, December 13-15, 2021, Proceedings 20* (pp. 72-91). Springer International Publishing. https://link.springer.com/chapter/10.1007/978-3-030-92548-2_5

Bourennane, M., Karlsson, A., & Björk, G. (2001). Quantum key distribution using multilevel encoding. *Physical Review A, 64*(1), 012306. https://journals.aps.org/pra/abstract/10.1103/PhysRevA.64.012306. doi:10.1103/PhysRevA.64.012306

Bruß, D. (1998). Optimal eavesdropping in quantum cryptography with six states. *Physical Review Letters, 81*(14), 3018–3021. https://search.proquest.com/openview/63a2b9c913676ba69db4d45e5ba03e08/1?pq-origsite=gscholar&cbl=2026366&diss=y. doi:10.1103/PhysRevLett.81.3018

Bruss, D., Erdélyi, G., Meyer, T., Riege, T., & Rothe, J. (2007). Quantum cryptography: A survey. *ACM Computing Surveys (CSUR), 39*(2), 6-es. https://dl.acm.org/doi/abs/10.1145/1242471.1242474

Bužek, V., & Hillery, M. (1996). Quantum copying: Beyond the no-cloning theorem. *Physical Review A, 54*(3), 1844–1852. https://journals.aps.org/pra/abstract/10.1103/PhysRevA.54.1844. doi:10.1103/PhysRevA.54.1844 PMID:9913670

Chen, A., Peng, W., & Gui, T. (2013). *A variant of BB84 Protocol based on quantum phase.* Digital Library. https://digital-library.theiet.org/content/conferences/10.1049/cp.2013.2465

Cheng, H., Dinu, D., Großschädl, J., Rønne, P. B., & Ryan, P. Y. (2020). A lightweight implementation of NTRU Prime for the post-quantum internet of things. In *Information Security Theory and Practice: 13th IFIP WG 11.2 International Conference, WISTP 2019, Paris, France, December 11–12, 2019, Proceedings 13* (pp. 103-119). Springer International Publishing. https://link.springer.com/chapter/10.1007/978-3-030-41702-4_7

Clarke, R. B., Chefles, A., Barnett, S. M., & Riis, E. (2001). Experimental demonstration of optimal unambiguous state discrimination. *Physical Review A, 63*(4), 040305. https://journals.aps.org/pra/abstract/10.1103/PhysRevA.63.040305. doi:10.1103/PhysRevA.63.040305

Dang, V. B., Mohajerani, K., & Gaj, K. (2022). High-speed hardware architectures and FPGA benchmarking of crystals-kyber, ntru, and saber. *IEEE Transactions on Computers, 72*(2), 306–320. https://ieeexplore.ieee.org/abstract/document/9954215/. doi:10.1109/TC.2022.3222954

Deng, F. G., Long, G. L., & Liu, X. S. (2003). Two-step quantum direct communication protocol using the Einstein-Podolsky-Rosen pair block. *Physical Review A, 68*(4), 042317. https://journals.aps.org/pra/abstract/10.1103/PhysRevA.68.042317. doi:10.1103/PhysRevA.68.042317

Deshpande, S., Xu, C., Nawan, M., Nawaz, K., & Szefer, J. (2022). Fast and Efficient Hardware Implementation of HQC. *Cryptology ePrint Archive.* https://eprint.iacr.org/2022/1183

Deutsch, D. (1985). Quantum theory, the Church–Turing principle and the universal quantum computer. *Proceedings of the Royal Society of London. A. Mathematical and Physical Sciences, 400*(1818), 97-117. https://royalsocietypublishing.org/doi/abs/10.1098/rspa.1985.0070

Dieks, D. G. B. J. (1982). Communication by EPR devices. *Physics Letters. [Part A], 92*(6), 271–272. https://www.sciencedirect.com/science/article/pii/0375960182900846. doi:10.1016/0375-9601(82)90084-6

Diffie, W., & Hellman, M. E. (2022). New directions in cryptography. In *Democratizing Cryptography: The Work of Whitfield Diffie and Martin Hellman* (pp. 365-390). https://dl.acm.org/doi/pdf/10.1145/3549993.3550007

Ducas, L., Lyubashevsky, V., & Prest, T. (2014). Efficient identity-based encryption over NTRU lattices. In *Advances in Cryptology–ASIACRYPT 2014: 20th International Conference on the Theory and Application of Cryptology and Information Security, Kaoshiung, Taiwan, ROC, December 7-11, 2014, Proceedings, Part II 20* (pp. 22-41). Springer Berlin Heidelberg. https://link.springer.com/chapter/10.1007/978-3-662-45608-8_2

Einstein, A., Podolsky, B., & Rosen, N. (1935). Can quantum-mechanical description of physical reality be considered complete? *Physical Review, 47*(10), 777–780. https://journals.aps.org/pr/abstract/10.1103/PhysRev.47.777. doi:10.1103/PhysRev.47.777

Ekert, A. K. (1991). Quantum cryptography based on Bell's theorem. *Physical Review Letters, 67*(6), 661–663. https://journals.aps.org/prl/abstract/10.1103/PhysRevLett.67.661. doi:10.1103/PhysRevLett.67.661 PMID:10044956

Ekert, A. K., & Huttner, B. (1994). Eavesdropping techniques in quantum cryptosystems. *Journal of Modern Optics, 41*, 2455–2466. https://journals.aps.org/prl/abstract/10.1103/PhysRevLett.84.4733

ElGamal, T. (1985). A public key cryptosystem and a signature scheme based on discrete logarithms. *IEEE Transactions on Information Theory, 31*(4), 469–472. https://elibrary.ru/item.asp?id=29749900. doi:10.1109/TIT.1985.1057074

Elkhatib, R., Koziel, B., & Azarderakhsh, R. (2022, January). Faster Isogenies for Post-quantum Cryptography: SIKE. In *Cryptographers' Track at the RSA Conference* (pp. 49–72). Springer International Publishing. https://link.springer.com/chapter/10.1007/978-3-030-95312-6_3

Enzer, D. G., Hadley, P. G., Hughes, R. J., Peterson, C. G., & Kwiat, P. G. (2002). Entangled-photon six-state quantum cryptography. *New Journal of Physics, 4*(1), 45. https://iopscience.iop.org/article/10.1088/1367-2630/4/1/345/meta. doi:10.1088/1367-2630/4/1/345

Gentry, C., & Szydlo, M. (2002). Cryptanalysis of the revised NTRU signature scheme. In *Advances in Cryptology—EUROCRYPT 2002: International Conference on the Theory and Applications of Cryptographic Techniques Amsterdam, The Netherlands, April 28–May 2, 2002 Proceedings 21* (pp. 299-320). Springer Berlin Heidelberg. https://link.springer.com/chapter/10.1007/3-540-46035-7_20

Goldenberg, L., & Vaidman, L. (1995). Quantum cryptography based on orthogonal states. *Physical Review Letters, 75*(7), 1239–1243. https://journals.aps.org/prl/abstract/10.1103/PhysRevLett.75.1239. doi:10.1103/PhysRevLett.75.1239 PMID:10060243

Goy, G., Loiseau, A., & Gaborit, P. (2022, September). A new key recovery side-channel attack on HQC with chosen ciphertext. In *International Conference on Post-Quantum Cryptography* (pp. 353–371). Springer International Publishing. https://link.springer.com/chapter/10.1007/978-3-031-17234-2_17 doi:10.1007/978-3-031-17234-2_17

Guo, Q., Hlauschek, C., Johansson, T., Lahr, N., Nilsson, A., & Schröder, R. L. (2022). Don't reject this: Key-recovery timing attacks due to rejection-sampling in HQC and BIKE. *IACR Transactions on Cryptographic Hardware and Embedded Systems*, 223-263. https://tches.iacr.org/index.php/TCHES/article/view/9700

Guo, Q., Johansson, T., & Nilsson, A. (2020, August). A key-recovery timing attack on post-quantum primitives using the Fujisaki-Okamoto transformation and its application on FrodoKEM. In *Annual International Cryptology Conference* (pp. 359–386). Springer International Publishing. https://link.springer.com/chapter/10.1007/978-3-030-56880-1_13 doi:10.1007/978-3-030-56880-1_13

Gupta, N., Jati, A., Chauhan, A. K., & Chattopadhyay, A. (2020). Pqc acceleration using gpus: Frodokem, newhope, and kyber. *IEEE Transactions on Parallel and Distributed Systems*, *32*(3), 575–586. https://ieeexplore.ieee.org/abstract/document/9201530/. doi:10.1109/TPDS.2020.3025691

He, G., Zhu, J., & Zeng, G. (2006). Quantum secure communication using continuous variable Einstein-Podolsky-Rosen correlations. *Physical Review A*, *73*(1), 012314. https://journals.aps.org/pra/abstract/10.1103/PhysRevA.73.012314. doi:10.1103/PhysRevA.73.012314

Hellman, M. (1976). New directions in cryptography. *IEEE Transactions on Information Theory*, *22*(6), 644–654. https://dl.acm.org/doi/pdf/10.1145/3549993.3550007. doi:10.1109/TIT.1976.1055638

Hoffstein, J., Howgrave-Graham, N., Pipher, J., Silverman, J. H., & Whyte, W. (2003, February). NTRUSIGN: Digital signatures using the NTRU lattice. In *Cryptographers' track at the RSA conference* (pp. 122-140). Springer Berlin Heidelberg. https://link.springer.com/chapter/10.1007/3-540-36563-X_9

Hoffstein, J., Pipher, J., & Silverman, J. H. (1998, June). NTRU: A ring-based public key cryptosystem. In *International algorithmic number theory symposium* (pp. 267-288). Berlin, Heidelberg: Springer Berlin Heidelberg. https://link.springer.com/chapter/10.1007/BFb0054868

Hoffstein, J., Pipher, J., & Silverman, J. H. (2001). NSS: An NTRU lattice-based signature scheme. In *Advances in Cryptology—EUROCRYPT 2001: International Conference on the Theory and Application of Cryptographic Techniques Innsbruck,* (pp. 211-228). Springer. https://www.google.com/books?hl=en&lr=&id=Ie-qCAAAQBAJ

Huttner, B., Muller, A., Gautier, J. D., Zbinden, H., & Gisin, N. (1996). Unambiguous quantum measurement of nonorthogonal states. *Physical Review A*, *54*(5), 3783–3789. https://journals.aps.org/pra/abstract/10.1103/PhysRevA.54.3783. doi:10.1103/PhysRevA.54.3783 PMID:9913923

Ivanovic, I. D. (1987). How to differentiate between non-orthogonal states. *Physics Letters. [Part A]*, *123*(6), 257–259. https://www.sciencedirect.com/science/article/pii/0375960187902222. doi:10.1016/0375-9601(87)90222-2

Jennewein, T., Simon, C., Weihs, G., Weinfurter, H., & Zeilinger, A. (2000). Quantum cryptography with entangled photons. *Physical Review Letters*, *84*(20), 4729–4732. https://journals.aps.org/prl/abstract/10.1103/PhysRevLett.84.4729. doi:10.1103/PhysRevLett.84.4729 PMID:10990782

Karl, P., Fritzmann, T., & Sigl, G. (2022, April). Hardware Accelerated FrodoKEM on RISC-V. In *2022 25th International Symposium on Design and Diagnostics of Electronic Circuits and Systems (DDECS)* (pp. 154-159). IEEE. https://ieeexplore.ieee.org/abstract/document/9770148/

Khang, A. (2023). Applications and Principles of Quantum Computing. IGI Global Press. https://doi.org/ doi:10.4018/979-8-3693-1168-4

Khang, A. (2023). Quantum Mechanics Primer - Fundamentals and Quantum Computing. Applications and Principles of Quantum Computing. IGI Global Press. https://doi.org/ doi:10.4018/979-8-3693-1168-4-ch001

Khang, A. (2023). Introduction to Quantum Computing and Integration of Quantum Computing. Applications and Principles of Quantum Computing. IGI Global Press. https://doi.org/ doi:10.4018/979-8-3693-1168-4-ch002

Khang, A., Shah, V., & Rani, S. (2023). AI-Based Technologies and Applications in the Era of the Metaverse. (1 Ed.) IGI Global Press. doi:10.4018/978-1-6684-8851-5

Koashi, M. (2009). Simple security proof of quantum key distribution based on complementarity. *New Journal of Physics*, *11*(4), 045018. https://iopscience.iop.org/article/10.1088/1367-2630/11/4/045018/meta. doi:10.1088/1367-2630/11/4/045018

Kumar, M., & Pattnaik, P. (2020, September). Post quantum cryptography (PQC)-An overview. In *2020 IEEE High Performance Extreme Computing Conference (HPEC)* (pp. 1-9). IEEE. https://ieeexplore.ieee.org/abstract/document/9286147/

Lee, E., Kim, Y. S., No, J. S., Song, M., & Shin, D. J. (2019). Modification of frodokem using gray and error-correcting codes. *IEEE Access : Practical Innovations, Open Solutions*, *7*, 179564–179574. https://ieeexplore.ieee.org/abstract/document/8931793/. doi:10.1109/ACCESS.2019.2959042

Lindblad, G. (1999). A general no-cloning theorem. *Letters in Mathematical Physics*, *47*(2), 189–196. https://link.springer.com/article/10.1023/A:1007581027660. doi:10.1023/A:1007581027660

Lo, H. K., & Chau, H. F. (1999). Unconditional security of quantum key distribution over arbitrarily long distances. *science*, *283*(5410), 2050-2056. https://www.science.org/doi/abs/10.1126/science.283.5410.2050

Mayers, D. (2001). Unconditional security in quantum cryptography. [JACM]. *Journal of the Association for Computing Machinery*, *48*(3), 351–406. https://dl.acm.org/doi/abs/10.1145/382780.382781. doi:10.1145/382780.382781

McEliece, R. J. (1978). A public-key cryptosystem based on algebraic. *Coding Thv*, *4244*, 114-116. https://ntrs.nasa.gov/api/citations/19780016269/downloads/19780016269.pdf#page=123

Milonni, P. W., & Hardies, M. L. (1982). Photons cannot always be replicated. *Phys. Lett.;(Netherlands)*, *92*(7). https://www.osti.gov/etdeweb/biblio/5926625

Misoczki, R., & Barreto, P. S. (2009, August). Compact McEliece keys from Goppa codes. In *International Workshop on Selected Areas in Cryptography* (pp. 376–392). Springer Berlin Heidelberg. https://link.springer.com/chapter/10.1007/978-3-642-05445-7_24 doi:10.1007/978-3-642-05445-7_24

Nguyen, D. T., & Gaj, K. (2021, June). Optimized software implementations of CRYSTALS-Kyber, NTRU, and Saber using NEON-based special instructions of ARMv8. In *Proceedings of the NIST 3rd PQC Standardization Conference (NIST PQC 2021)*. https://csrc.nist.gov/CSRC/media/Events/third-pqc-standardization-conference/documents/accepted-papers/nguyen-optimized-software-gmu-pqc2021.pdf

Pan, D., Li, K., Ruan, D., Ng, S. X., & Hanzo, L. (2020). Single-photon-memory two-step quantum secure direct communication relying on Einstein-Podolsky-Rosen pairs. *IEEE Access : Practical Innovations, Open Solutions, 8*, 121146–121161. https://ieeexplore.ieee.org/abstract/document/9129730/. doi:10.1109/ACCESS.2020.3006136

Pandey, J. P., Ojha, D. B., & Sharma, A. (2005). Enhance fuzzy commitment scheme: an approach for post quantum cryptosystem. *Journal of Applied and Theoretical Information Technology*, 16-19. https://www.academia.edu/download/7297632/3vol9no1.pdf

Peikert, C. (2016). A decade of lattice cryptography. *Foundations and trends® in theoretical computer science, 10*(4), 283-424. https://www.nowpublishers.com/article/Details/TCS-074

Peng, B. Y., Marotzke, A., Tsai, M. H., Yang, B. Y., & Chen, H. L. (2023). Streamlined NTRU prime on FPGA. *Journal of Cryptographic Engineering, 13*(2), 167–186. https://link.springer.com/article/10.1007/s13389-022-00303-z. doi:10.100713389-022-00303-z

Peres, A. (1988). How to differentiate between non-orthogonal states. *Physics Letters. [Part A], 128*(1-2), 19. https://www.sciencedirect.com/science/article/pii/0375960187902222. doi:10.1016/0375-9601(88)91034-1

Polanyi Festschrift Committee (Ed.). (2015). *The Logic of Personal Knowledge: Essays Presented to M. Polanyi on His Seventieth Birthday.* Routledge. https://www.google.com/books?hl=en&lr=&id=BzHLCQAAQBAJ

Repka, M., & Zajac, P. (2014). Overview of the McEliece cryptosystem and its security. *Tatra Mountains Mathematical Publications, 60*(1), 57-83. https://sciendo.com/article/10.2478/tmmp-2014-0025

Rivest, R. L., Shamir, A., & Adleman, L. (1978). A method for obtaining digital signatures and public-key cryptosystems. *Communications of the ACM, 21*(2), 120–126. https://dl.acm.org/doi/abs/10.1145/357980.358017. doi:10.1145/359340.359342

Rothe, J. (2005). *Complexity theory and cryptology: an introduction to cryptocomplexity* (Vol. 91). Springer. https://link.springer.com/book/10.1007/3-540-28520-2

Rusca, D., Boaron, A., Curty, M., Martin, A., & Zbinden, H. (2018). Security proof for a simplified Bennett-Brassard 1984 quantum-key-distribution protocol. *Physical Review. A, 98*(5), 052336. https://journals.aps.org/pra/abstract/10.1103/PhysRevA.98.052336. doi:10.1103/PhysRevA.98.052336

Schöffel, M., Feldmann, J., & Wehn, N. (2022, October). Code-based Cryptography in IoT: A HW/SW Co-Design of HQC. In *2022 IEEE 8th World Forum on Internet of Things (WF-IoT)* (pp. 1-7). IEEE. https://ieeexplore.ieee.org/abstract/document/10152031/

Seo, H., Sanal, P., Jalali, A., & Azarderakhsh, R. (2020). Optimized implementation of SIKE round 2 on 64-bit ARM Cortex-A processors. *IEEE Transactions on Circuits and Systems. I, Regular Papers, 67*(8), 2659–2671. https://ieeexplore.ieee.org/abstract/document/9036884/. doi:10.1109/TCSI.2020.2979410

Shaari, J. S., Lucamarini, M., & Wahiddin, M. R. B. (2006). Deterministic six states protocol for quantum communication. *Physics Letters. [Part A], 358*(2), 85–90. https://www.sciencedirect.com/science/article/pii/S0375960106007006. doi:10.1016/j.physleta.2006.05.007

Shor, P. W. (1999). Polynomial-time algorithms for prime factorization and discrete logarithms on a quantum computer. *SIAM Review*, *41*(2), 303–332. https://epubs.siam.org/doi/abs/10.1137/S0036144598347011. doi:10.1137/S0036144598347011

Shor, P. W., & Preskill, J. (2000). Simple proof of security of the BB84 quantum key distribution protocol. *Physical Review Letters*, *85*(2), 441–444. https://journals.aps.org/prl/abstract/10.1103/PhysRevLett.85.441. doi:10.1103/PhysRevLett.85.441 PMID:10991303

Shyam, R. & Sihare, K. A. (2023). Effects of Quantum Technology on Metaverse. AI-Based Technologies and Applications in the Era of the Metaverse. (1 Ed.) Page (104-203). IGI Global Press. doi:10.4018/978-1-6684-8851-5.ch009

Stallings, W. (2006). *Cryptography and network security, 4/E.* Pearson Education India. https://iopscience.iop.org/article/10.1088/1757-899X/981/2/022028/meta

Stinson, D. R., & Paterson, M. (2018). *Cryptography: theory and practice.* CRC press. https://api.taylorfrancis.com/content/books/mono/download?identifierName=doi&identifierValue=10.1201/9781420057133&type=googlepdf

Su, H. Y. (2020). Simple analysis of security of the BB84 quantum key distribution protocol. *Quantum Information Processing*, *19*(6), 169. https://link.springer.com/article/10.1007/s11128-020-02663-z. doi:10.100711128-020-02663-z

Tsurumaru, T., & Tamaki, K. (2008). Security proof for quantum-key-distribution systems with threshold detectors. *Physical Review A*, *78*(3), 032302. https://journals.aps.org/pra/abstract/10.1103/PhysRevA.78.032302. doi:10.1103/PhysRevA.78.032302

Wan, L., Zheng, F., Fan, G., Wei, R., Gao, L., Wang, Y., & Dong, J. (2022, September). A novel high-performance implementation of crystals-kyber with ai accelerator. In *European Symposium on Research in Computer Security* (pp. 514–534). Springer Nature Switzerland. https://link.springer.com/chapter/10.1007/978-3-031-17143-7_25 doi:10.1007/978-3-031-17143-7_25

Weigert, S. (2009). No-cloning theorem. In *Compendium of quantum physics* (pp. 404–405). Springer Berlin Heidelberg. https://link.springer.com/article/10.1007/s00500-022-07043-6 doi:10.1007/978-3-540-70626-7_124

Wootters, W. K., & Zurek, W. H. (1982). A single quantum cannot be cloned. *Nature*, *299*(5886), 802–803. https://www.nature.com/articles/299802a0. doi:10.1038/299802a0

Xia, C., Li, H., & Hu, J. (2022). Semi-quantum digital signature protocol based on Einstein–Podolsky–Rosen steering. *Journal of Physics. A, Mathematical and Theoretical*, *55*(32), 325302. https://iopscience.iop.org/article/10.1088/1751-8121/ac7f6d/meta. doi:10.1088/1751-8121/ac7f6d

Xiang, Y., Su, X., Mišta, L. Jr, Adesso, G., & He, Q. (2019). Multipartite Einstein-Podolsky-Rosen steering sharing with separable states. *Physical Review. A*, *99*(1), 010104. https://journals.aps.org/pra/abstract/10.1103/PhysRevA.99.010104. doi:10.1103/PhysRevA.99.010104

Xing, Y., & Li, S. (2021). A compact hardware implementation of CCA-secure key exchange mechanism CRYSTALS-KYBER on FPGA. *IACR Transactions on Cryptographic Hardware and Embedded Systems*, 328-356. Ruhr Universitat. https://tches.iacr.org/index.php/TCHES/article/view/8797

Zheng, C., & Long, G. (2014). Quantum secure direct dialogue using Einstein-Podolsky-Rosen pairs. *Science China. Physics, Mechanics & Astronomy*, *57*(7), 1238–1243. https://link.springer.com/article/10.1007/s11433-014-5461-x. doi:10.100711433-014-5461-x

Chapter 20

Quantum Computing Security: Application of Quantum Computing in Computer Security

Manish Kaushik
Amity Institute of Information Technology, Amity University Noida, India

Nidhi Sindhwani
Amity Institute of Information Technology, Amity University Noida, India

Komal Saxena
Amity Institute of Information Technology, Amity University Noida, India

ABSTRACT

With the approach of quantum processing, customary cryptographic calculations and conventions utilized for PC security face critical difficulties. This examination paper investigates the utilization of quantum registering in PC security and its possible effect on the field. The authors give an outline of quantum registering, featuring its remarkable properties and computational benefits. Consequently, they talk about the weaknesses of traditional cryptographic frameworks to quantum assaults and the requirement for post-quantum cryptography. Moreover, they dive into explicit utilizations of quantum figuring in PC security, including quantum key dispersion, quantum-safe calculations, and secure multiparty calculation. At last, they look at the present status of examination, difficulties, and future headings in utilizing quantum processing for improving PC security.

1. INTRODUCTION

Quantum processing bridles the standards of quantum mechanics to perform complex calculations dramatically quicker than old style of personal computers (PC). This amazing computational power can possibly change different fields, and PC security is no special case. The use of quantum registering in PC security offers the commitment of improving existing security conventions and growing new cryptographic procedures that can endure assaults from future quantum PCs.

DOI: 10.4018/979-8-3693-1168-4.ch020

Figure 1. Block diagram of quantum computing

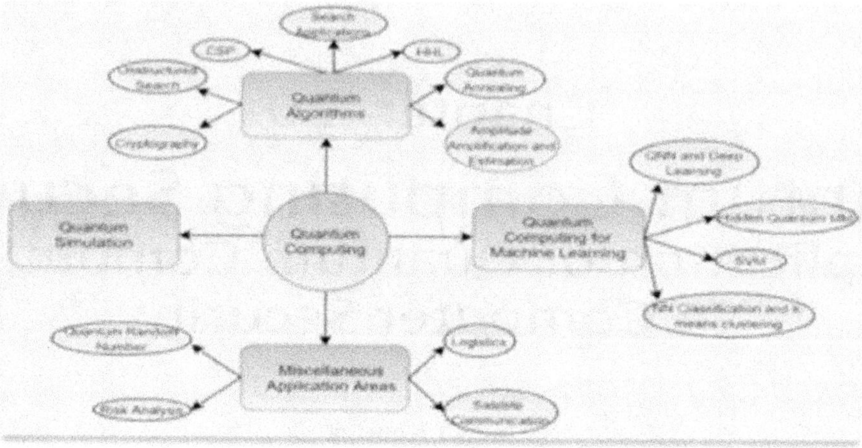

One of the main commitments of quantum registering to PC security lies in its effect on open key cryptography, the underpinning of secure correspondence on the web. At present, broadly utilized public-key encryption calculations, for example, RSA and Diffie-Hellman depend on the computational trouble of considering enormous numbers or taking care of the discrete logarithm issue (Diffie-Hellman, 1976). Notwithstanding, quantum PCs can possibly tackle these issues productively, delivering many existing encryption strategies old.

To address this weakness, analysts are effectively investigating new cryptographic calculations known as post-quantum cryptography (PQC) that can oppose assaults from both old style and quantum PCs. PQC calculations are intended to use the remarkable properties of quantum mechanics to give secure correspondence even within the sight of strong quantum PCs. These calculations, for example, grid based, code-based, and multivariate-based cryptography, offer a promising road for getting delicate data in the post-quantum time.

Past encryption, quantum processing likewise holds the possibility to improve different parts of PC security. Quantum key conveyance (QKD) is an area of exploration that empowers the protected appropriation of cryptographic keys utilizing the standards of quantum mechanics. QKD depends on the major properties of quantum material science, like the vulnerability rule and the no-cloning hypothesis, to lay out secure correspondence stations. This innovation guarantees that any endeavor to capture or alter the communicated keys would be discernible, giving an extraordinary degree of safety.

Also, quantum figuring can add to the field of quantum-safe validation, guaranteeing the honesty and legitimacy of computerized characters. By using quantum-safe calculations, verification instruments can endure assaults from both old style and quantum PCs, moderating the gamble of data fraud and unapproved access as shown in Figure 1.

2. LITERATURE REVIEW

Khang & Quantum et al. (2023) recommended Quantum secure direct correspondence is one of the fundamental areas of quantum cryptography. In quantum secure direct correspondence, classified data can be sent straight over a quantum channel without the requirement for a confidential key. Quantum secure direct correspondence (QSDC) offers more prominent security and speedy correspondence. Since a confidential key isn't used, key administration isn't needed. Since coding just happens in the quantum terminal, rather than quantum key dispersion, when the key is sent from the quantum terminal to the old style terminal there, this further fortifies security at the client's area.

As a hearty essential cryptographic crude, QSDC can be utilized to build further quantum correspondence conventions, including quantum offering, quantum mark, and quantum talk, among others. Here, we momentarily make sense of the basic ideas of QSDC, go over the fundamental conventions, examine the norm, and proposition some point of view Khang & Kali (2023) recommended The most recent four years have seen a quick development of united learning (FL). New protection concerns arose, by the by, as the dispersed halfway outcomes were merged.

The future security safeguarding FL (PPFL) has been declared as an answer for general protection saving AI. Nonetheless, it is as yet testing to keep up with AI's information utility while maintaining information protection. In this review, we present a methodical and complete assessment of the PPFL in view of our proposed 5W situation based scientific classification. We examine momentum techniques, evaluate the protection spillage issues in the FL from five points, and distinguish regions that require more exploration. Park et.al (2022) recommended serious review into the creation of nanocrystals has been driven by both essential logical interest and different modern purposes.

The blend of monodisperse nanocrystals (size variety 5%) is fundamental since the attributes of these nanocrystals are unequivocally dependent on their aspects. For instance, monodisperse attractive nanocrystals are important for the cutting edge multi-terabit attractive capacity media, and the consistency of the nanocrystals considerably affects the variety sharpness of semiconductor nanocrystal-based optical frameworks.

To efficiently manufacture these monodisperse nanocrystals, an innovation that is both productive and reasonable should be created. Tragically, most of the combinations that have been accounted for so far have just delivered little measures of monodisperse nanocrystals — under grams. Uniform-sized nanocrystals of CdSe (Boskovic et al., 2006), (Arya et al., 2019) and Au (Shor, 1996) have been created utilizing colloidal substance combination systems. Monodisperse attractive nanocrystals have additionally been created artificially, including Fe (Abomhara et al., 2015), (Ma & Zeng 2018), Co (Zhou et al., 2018; Azouit et al., 2021; Ding et al., 2020), gamma-Fe(2)O(3) (Gisin et al., 2002; Hwang et al., 2022), and Fe(3)O(4) (Li et al., 2019; Liu et al., 2020). This article portrays the super huge scope amalgamation of monodisperse nanocrystals utilizing modest, non-poisonous metal salts as reactants. We had the option to create up to 40 g of monodisperse nanocrystals in a solitary response without the need of a size-arranging strategy.

To control molecule size, changing the trial setup is likewise conceivable. It has been effectively used to create nanocrystals of a few progress metal oxides. The ongoing manufactured approach is profoundly flexible. Zhou and Sheng et al. (2017) recommended in light of the consequences of computations made utilizing thickness useful hypothesis, a hypothetical technique for building inorganic nonlinear optical (NLO) gems for second symphonious age (SHG) is given. To research the steady designs of the inorganic gem with given organizations and look at the no centrosymmetric stable designs and the second-request nonlinear optical properties, a particular developmental calculation (GA) is made in this strategy by processing the related SHG coefficients.

The presentation of coordination designs of the structure units to develop individual designs during the GA strategy, which can clearly build the viability and achievement pace of getting the steady construction of the inorganic precious stones, is the principal distinction between the ongoing technique and traditional GA procedures. AgGaS2 and LiAsSe2 gems are introduced as two instances of ternary mixtures. The calculations and optical attributes of other metastable stages have been analyzed notwithstanding the experimentally revealed structures. Our outcomes definitively show the way that the proposed approach can give a commonsense approach to making and enhancing imaginative inorganic NLO precious stones.

Wang et al. (2022) recommended Quantum specks enjoy characteristic benefits because of their job as a key part of optoelectronic gadgets. Different quantum spots with movable emanation frequencies are generally used to fluctuate the range and shade of white light-discharging diodes (WLEDs). Here, we utilize a corrosive reagent designing procedure to increase the development of numerous carbon quantum spots (CQDs). The rising electron-pulling out bunches on the outer layer of CQDs produced using corrosive reagents help to make sense of the quantum size impact, it grows their molecule estimates and blushes their photoluminescence frequency. These CQDs produce splendid, astoundingly stable full-variety fluorescence that ranges from blue to red and, surprisingly, white light.

Full-variety emissive polymer movies and all types of WLEDs with high variety delivering records can be created by blending various kinds of CQDs in the legitimate proportions. The all-inclusive electron-giving/pulling out bunch designing strategy for the combination of tunable emissive CQDs will help the advancement of carbon-based radiant materials for the production of modern movies and gadgets. Li and Huang et al. (2019) recommended it is generally acknowledged that two clients can impart in a totally protected way utilizing quantum key dissemination.

Tragically, there is still no generally acknowledged illustration of how secure it is when source, gadget, and channel clamors are available. Here, it is shown that it is feasible to give unrestricted security to quantum key dissemination across erratic significant distances using a practical boisterous direct within the sight of shortcoming open minded quantum PCs. The proof is transformed from a loud quantum plan to a quiet quantum plan, and afterward from a silent traditional framework to a boisterous quantum plot. This adjustment of the proof permits traditional likelihood hypothesis to deal with it.

3. QUANTUM COMPUTING: AN OVERVIEW

The application of quantum physics concepts to the creation of potent processing systems is a fast developing field known as quantum computing. In contrast to conventional computer, which use bits to store and process data, quantum computers use quantum bits, or qubits, which can exist in several states at once. As a result, some computations can be carried out by quantum computers far more quickly than by traditional computers (Khang & Quantum, 2023).

3.1 Principles of Quantum Mechanics

Physicists use the field of quantum mechanics to explain how atomic and subatomic particles behave. It outlines a number of the fundamental ideas underlying quantum computing: Particles can exist in several states at once according to superposition in quantum physics. Qubits can simultaneously be in a superposition of the states 0 and 1 in the context of quantum computing (Khang & Kali, 2023).

No matter how far apart two particles are from one another, a phenomenon known as "entanglement" occurs when their states start to correlate. When qubits are entangled, modifying the state of one instantly changes the state of the other, allowing for extremely parallel computations.

A quantum system collapses into a predetermined state when it is measured. The system is in a superposition of states before measurement, but it resolves to a single state with a fixed probability after measurement. Quantum algorithms make use of this characteristic to get insightful findings.

3.2 Quantum Bits (Qubits)

The qubit is the basic building block of information in quantum computing. While qubits can exist in a superposition of these states, classical bits can only be either 0 or 1. A physical system with quantum qualities, such as an electron's spin, a photon's polarization, or an atom's energy levels, is used to accomplish this.

Mathematically, a vector in a two-dimensional complex vector space can be used to represent qubits. A qubit may be in the state $|0 + |1$, where $|0$ and $|1$ are the classical states 0 and 1, respectively, and are complex numbers known as probability amplitudes. The squared magnitudes of the amplitudes represent the probabilities of measuring the qubit in either state.

3.3 Quantum Algorithms

Quantum algorithms are created to address particular computer problems more effectively than classical algorithms by utilizing the special characteristics of quantum physics. Shor's algorithm (Shor, 1994), one of the most well-known quantum algorithms, effectively factors big numbers, a task that is computationally challenging for classical computers.

Grover's method for unstructured search issues, quantum simulation techniques for simulating quantum systems, and the quantum approximate optimisation algorithm (QAOA) for optimisation issues are further significant quantum algorithms (Khang & Vugar et al., 2023).

3.4 Quantum Gates and Operations

Gates are used in traditional computing to manipulate bits. Similar to this, qubits are controlled by quantum gates in quantum computing. These gates transform the quantum state of the qubits in a unitary manner.

The Pauli gates (X, Y, and Z), the Hadamard gate (H), and the controlled-NOT gate (CNOT) are examples of common quantum gates. These gates enable qubit state manipulation, superposition, and entanglement generation. Complex quantum computations can be carried out by mixing various gates in a quantum circuit. Qubit measurements are a part of quantum operations in addition to gates. A qubit's superposition is collapsed by measurements into a single state, producing conventional results that can be input into a calculation (Ushaa, Khang et al., 2023).

It's vital to remember that quantum computing is still in its infancy and that developing real-world, fault-tolerant quantum computers still presents a formidable task. However, in order to fully utilise the capabilities of quantum computing for various applications, researchers and businesses are working diligently to build quantum hardware, quantum algorithms, and error correction methods.

4. VULNERABILITIES OF CLASSICAL CRYPTOGRAPHY TO QUANTUM ATTACKS

Traditional cryptography is based on mathematical puzzles that are challenging for traditional computers to solve. However, with the development of quantum computers, some encryption techniques can now be effectively broken. The safety of traditional cryptographic systems is seriously jeopardised by this. Here are some significant quantum attack weaknesses in conventional cryptography:

4.1 Shor's Algorithm and Integer Factorization

Shor's algorithm is a quantum method that effectively factors huge composite numbers; it was developed by Shor Peter in 1994. Many popular cryptographic techniques, like the RSA algorithm, are built on integer factorization. Shor's approach can decrypt public-key cryptosystems like RSA and others by locating the modulus' prime factors in polynomial time on a quantum computer. The security of communications secured by RSA encryption is seriously jeopardised by this.

4.2 Grover's Algorithm and Cryptanalysis

Grover's algorithm, developed by Grover in 1996, is a quantum algorithm that can search a database without any sorting in roughly the square root of the time needed for a classical search. Strictly speaking, brute-force assaults are affected by this algorithm's effects on symmetric key encryption. Grover's approach can shorten an effective key length by a square root and speed up the search for the secret key of a symmetric encryption system. A 128-bit symmetric key, for instance, would offer a same level of security to a 64-bit key against a Grover's algorithm-based attack.

4.3 Quantum Attacks on Symmetric Cryptography

There are still weaknesses in symmetric key algorithms, despite the fact that they are typically thought to be more resistant to quantum assaults than public-key algorithms. For instance, the hidden subgroup problem, the foundation of various cryptographic techniques based on finite groups, such as the Advanced Encryption Standard (AES), can be effectively solved using the quantum algorithm known as the quantum Fourier transform. AES and other symmetric encryption methods based on related ideas could potentially be cracked if a quantum computer is capable of performing the quantum Fourier transform efficiently.

4.4 Limitations of Current Cryptographic Systems

The aforementioned flaws demonstrate how vulnerable contemporary cryptography systems are to quantum assaults. These systems were created with the presumption that traditional computers would be the main source of attack computing power. These presumptions are invalid now that powerful quantum computers are available. Therefore, it is essential to create new cryptographic algorithms, often known as post-quantum cryptography, that are immune to quantum attacks.

Designing algorithms that can fend off attacks from both classical and quantum computers is the goal of post-quantum cryptography. Lattice-based cryptography, code-based cryptography, multivariate polynomial cryptography, and other post-quantum cryptographic techniques are among the ones being investigated. These strategies are based on supposedly challenging mathematical puzzles, even for quantum computers. To secure the long-term security of sensitive data in the presence of quantum computers, it is essential to create and implement post-quantum cryptography solutions.

5. POST-QUANTUM CRYPTOGRAPHY

Quantum-resistant cryptography, commonly referred to as post-Quantum cryptography (PQC), describes cryptographic methods that are made to be secure against assaults from both classical and quantum computers. PQC intends to offer a path towards algorithms that can withstand attacks by quantum computers because they have the potential to disrupt many of the widely used cryptographic techniques (Prabu, Khang et al., 2023).

5.1 Requirements for Post-Quantum Cryptography

Algorithms for post-quantum cryptography must meet a number of criteria in order to be deemed secure in the age of quantum computing. These prerequisites consist of:

- **Resistance to Quantum Attacks**: Both classical and quantum computers should be protected against attacks by post-quantum algorithms. They shouldn't be susceptible to well-known quantum algorithms that effectively defeat traditional cryptography systems, such Shor's algorithm (Shor, 1994).
- **Computational Efficiency**: While security is crucial, post-quantum algorithms also need to be quick to compute. They should be practical for real-world implementations and able to run on conventional computers in a reasonable amount of time.
- **Key Sizes**: Post-Quantum algorithms must offer big enough keys to guarantee security. Even with the help of quantum computers, brute-force attacks should be able to be resisted by using big enough key sizes.
- **Interoperability and compatibility**: Post-Quantum algorithms ought to be made to work with current cryptography systems and protocols. During the transition time, they should be able to coexist alongside traditional cryptographic algorithms and be compatible with common cryptographic interfaces.

5.2 Lattice-based Cryptography

A promising method of post-quantum cryptography that depends on the difficulty of specific lattice problems is known as lattice-based cryptography. A discrete grid-like structure called a lattice is created by repeatedly repeating mathematical vectors. The mathematical operations carried out on these lattices serve as the foundation for lattice-based cryptography methods.

Lattice-based cryptography's security depends on how challenging particular puzzles, such the Shortest Vector Problem (SVP) or the Learning With Errors (LWE) problem, are to solve. Even for quantum computers, these problems are thought to be computationally challenging. Numerous cryptographic primitives, such as encryption, digital signatures, key exchange, and safe multiparty computation, are available using lattice-based cryptography.

Lattice-based cryptography has several benefits, including its adaptability and the availability of different encryption protocols like NTRUEncrypt, Kyber, and Lizard. These systems offer various security levels and performance trade-offs to accommodate various applications and needs.

5.3 Code-Based Cryptography

Based on error-correcting codes, code-based cryptography is another subfield of post-quantum cryptography. Widespread usage of error-correcting codes helps to protect data integrity and enable error recovery during data transmission and storage.

The difficulty of decoding specific structured codes is taken advantage of in code-based encryption to offer security. The McEliece cryptosystem, put forth by Robert McEliece in 1978, is the most popular code-based encryption scheme. It is based on how challenging it is to decode a particular kind of linear code called a Goppa code (Valerii, 2023).

Code-based cryptography relies on the fact that the decoding problem is thought to be challenging, even for quantum computers, to maintain its security. Code-based algorithms offer defence against quantum attacks, along with relatively small key sizes and effective conventional computer implementation.

The size of the ciphertext and public key, which can affect the effectiveness of transmission and storage, is a problem with code-based encryption. To make code-based schemes more useful for real-world applications, research is still being done on how to improve them and shrink the length of their keys and ciphertext.

Lattice-based and code-based cryptography are both being actively researched as potential post-quantum cryptography options. These methods provide robust security assurances against intrusions from both classical and quantum computers, opening up a promising path for protecting communications in the quantum computing age.

6. QUANTUM KEY DISTRIBUTION (QKD)

6.1 Principles of QKD

A cryptographic method called quantum key distribution (QKD) uses the concepts of quantum physics to permit safe communication between two parties. The fundamental idea underlying QKD is to use quantum physics' features to create a shared secret key between Alice and Bob that can be used to detect any eavesdropping efforts.

The standards of QKD depend on two central properties of quantum mechanics: the Heisenberg vulnerability guideline and the no-cloning hypothesis. The vulnerability standard expresses that it is difficult to quantify specific sets of properties of a quantum molecule, like its situation and energy, with erratic accuracy. The no-cloning hypothesis expresses that it is difficult to make an indistinguishable duplicate of an obscure quantum state.

These ideas allow QKD techniques to take use of the fact that any effort to assess a particle's quantum state while it is in motion will eventually upset it. As a result, if Eve tries to listen in on the conversation, her measurements will introduce inaccuracies that Alice and Bob will be able to spot, signalling a possible security breach.

6.2 Quantum Key Distribution Protocols

Several QKD protocols have been developed to achieve secure key distribution. Some of the commonly used protocols include:

- **BB84 Protocol**: One of the most frequently used QKD protocols, the BB84 protocol was developed by Charles H. Bennett and Gilles Brassard in 1984. The secret key is encoded using the characteristics of quantum states, such as the polarization of photons. Bob receives the quantum states that Alice has encoded using two orthonormal bases (such as rectilinear and diagonal) at random. In order to identify eavesdropping efforts, Alice and Bob compare a portion of their broadcast and received bits. Bob uses compatible bases to randomly measure the received states.
- **E91 Protocol**: The E91 protocol, put forth by Artur Ekert in 1991, takes advantage of the quantum entanglement phenomena. By taking precise measurements, Alice and Bob can establish a shared key and share entangled particles. The protocol's security is based on the breach of Bell's inequality, which, if noticed, signals the existence of eavesdropping.
- **SARG04 Protocol**: The SARG04 protocol, put out by Scarani, Acin, Ribordy, and Gisin in 2004, leverages the idea of decoy states to increase the security of QKD. Along with the key-carrying photons, additional decoy photons of varying intensities must be sent. Alice and Bob can determine the presence of an eavesdropper by examining the properties of the decoy photons.

6.3 Challenges and Limitations of QKD

Despite its promise for secure communication, QKD faces several challenges and limitations:

- **Distance and Loss**: The maximum distance that QKD systems can securely disseminate keys over. Transmission losses in optical fibres or atmospheric conditions are only two examples of the many variables that can readily disrupt or destroy quantum states. This limits the application space for QKD systems.
- **Key Rate and Speed**: When compared to traditional key exchange techniques, QKD produces secure keys at a somewhat sluggish rate. The speed at which quantum states can be created, transferred, and monitored sets a limit on the key generation rate. An important research problem is to increase key rate while keeping security.
- **Practical Implementation**: To implement QKD systems, sophisticated hardware and exact quantum state control are needed. The cost and technical difficulty of creating and maintaining QKD systems may make their widespread use impractical.
- **Side-Channel Attacks**: In place of directly violating quantum principles, side-channel attacks take use of flaws in the implementation or physical components to compromise QKD systems. Attacks on the side channels have the potential to make the key exchange less secure.

6.4 Quantum-Safe Key Exchange

Given the difficulties and restrictions of QKD, quantum-safe key exchange, sometimes referred to as post-quantum cryptography or quantum-resistant cryptography, is an alternative strategy for achieving secure communication. Key exchange methods that are quantum-safe are made to withstand attacks from both conventional and quantum computers.

The majority of quantum-safe key exchange methods rely on mathematical issues that are thought to be challenging for both classical and quantum computers to resolve. The McEliece cryptosystem (1978), lattice-based cryptography, code-based cryptography, and multivariate polynomial cryptography are a few examples of quantum-safe algorithms. These methods offer some protection from assaults by both classical and quantum computers, guaranteeing long-term confidentiality even in the face of quantum computers that can defeat conventional encryption techniques.

Future communication system security depends heavily on the creation and use of quantum-safe key exchange algorithms, particularly as the development of powerful, fault-tolerant quantum computers advances.

7. QUANTUM-RESISTANT ALGORITHMS

Given that quantum computing has the ability to perform some computations far more quickly than conventional computing, traditional encryption techniques' security is called into question. Researchers have been working on creating quantum-resistant algorithms that can withstand attacks from quantum computers to address this problem. Hash functions, symmetric key algorithms, public key algorithms, and the switch to quantum-resistant algorithms must all be taken into account in this situation (Teja, Shaik et al., 2023).

7.1 Hash Functions

In cryptography, hash functions are essential, especially for data integrity and digital signatures. Hash functions that are immune to quantum assaults are designed to offer the same security guarantees as traditional hash functions. The "Quantum-Resistant Secure Hash Algorithm" (QRSHA), which is meant to withstand attacks from both classical and quantum computers, is an illustration of a quantum-resistant hash function.

7.2 Symmetric Key Algorithms

Data encryption frequently employs symmetric key techniques like the Advanced Encryption Standard (AES). Even though symmetric key algorithms are typically thought to be immune to attacks from quantum computers, they could nevertheless be at risk from attacks utilising algorithms created especially for these machines. In order to maintain long-term security, academics are striving to create quantum-resistant symmetric key algorithms, such as the "Post-Quantum Advanced Encryption Standard" (PQ-AES).

7.3 Public Key Algorithms

Elliptic Curve Cryptography (ECC) and other public key methods are susceptible to assaults from quantum computers. Shor's algorithm (Shor, 1994), a quantum algorithm, is effective at solving discrete logarithm problems and factoring huge numbers, which are the bases of RSA and ECC, respectively. Researchers have been looking on post-quantum public key algorithms, such as lattice-based, code-based, and multivariate-based methods, to overcome this problem. The "New Hope" key exchange and the "Falcon" digital signature system are two examples of public key algorithms that can withstand quantum effects.

7.4 Transitioning to Quantum-Resistant Algorithms

Because changing protocols, software, and hardware systems includes switching from conventional cryptographic algorithms to quantum-resistant algorithms, this transition must be carefully planned and coordinated. In order to ensure compatibility with both classical and quantum-resistant cryptographic systems during the transition phase, organisations need to evaluate the potential effects on their current infrastructure. In order to facilitate communication and data exchange between classical and quantum-resistant systems, backward compatibility must also be maintained. Quantum-resistant algorithms are being evaluated and standardised by standardisation organisations like the National Institute of Standards and Technology (NIST) to help with this transition.

In conclusion, it is essential to create and implement quantum-resistant algorithms to maintain the long-term security of cryptographic systems in the face of advances in quantum computing. To develop cryptographic solutions that can withstand attacks from quantum computers, researchers are concentrating on hash functions, symmetric key algorithms, and public key algorithms. To achieve a seamless and secure migration process, the switch to quantum-resistant algorithms necessitates meticulous preparation and cooperation.

8. SECURE MULTIPARTY COMPUTATION (SMC) WITH QUANTUM COMPUTING

8.1 Overview of Secure Multiparty Computation

A cryptographic method called Secure Multiparty Computation (SMC) enables many parties to collaboratively compute a function over their private inputs without disclosing those inputs to one another. It permits data analysis and cooperation while protecting privacy. Using cryptographic methods, classical SMC makes sure that no party can learn anything about the other parties' secret inputs.

8.2 Quantum Secure Multiparty Computation

Quantum Secure Multiparty Computation (QSMC) blends the capabilities of quantum computing with the principles of SMC. It seeks to make computations secure even when nefarious quantum adversaries are present. In order to accomplish secure computation with improved privacy and efficiency, QSMC protocols take advantage of the special qualities of quantum systems, such as superposition and entanglement.

For SMC, quantum computing opens up new possibilities and problems. Certain cryptographic primitives that support classical SMC can be broken by quantum algorithms like Shor's algorithm (Shor, 1994). To assure secure computation in a post-quantum era, researchers are creating quantum-resistant SMC protocols. To safeguard the confidentiality of contributors' inputs, quantum SMC protocols sometimes use quantum-resistant cryptographic techniques like lattice-based or code-based cryptography.

8.3 Applications of Quantum SMC

There are numerous disciplines where secure collaboration and computation are necessary, including quantum SMC. Several potential uses include:

- **Secure data analysis**: Secure data analysis allows several people to examine confidential datasets together without disclosing private information. This has uses in the social sciences, finance, and healthcare industries (Khang & Medicine, 2023).
- **Machine learning that protects privacy**: Quantum SMC can allow for collaborative training of machine learning models utilizing personal datasets while protecting the privacy of specific training data. Quantum SMC can support secure auctions and bidding processes so that participants can make their bids while keeping their bid amounts a secret from other bidders.
- **Blockchain and cryptocurrency**: Quantum SMC can improve the security and anonymity of blockchain systems by facilitating secure transactions and consensus mechanisms amongst numerous parties.

8.4 Challenges and Future Directions

There are a number of difficulties and potential directions for quantum SMC:

- **Scalability**: Creating scalable and effective quantum SMC protocols is a difficult task. Errors are common in quantum systems, and quantum computations can have significant overhead. It is essential to overcome these obstacles in order to make quantum SMC viable.
- **Post-Quantum Security**: It is crucial to make sure that quantum SMC protocols can withstand attacks from quantum adversaries as quantum computing technology develops. For safe quantum SMC, research is concentrated on creating and examining quantum-resistant cryptographic methods.
- **Practical Implementations**: Research is still being done in the area of translating theoretical quantum SMC protocols into actual applications. Real-world implementation of quantum SMC protocols necessitates consideration of hardware and software limitations as well as compatibility with current systems.
- **Standardization**: To guarantee interoperability and broad adoption, standardizing quantum SMC protocols is crucial. The development of rules and procedures for quantum SMC is being pursued by standardization organizations and research communities.

9. CURRENT STATE OF RESEARCH AND INDUSTRY INITIATIVES

9.1 Research Efforts in Quantum Computing and Computer Security

Research is now being done in the domain of quantum computing and how it relates to computer security. To expand our understanding of quantum computing and create quantum-resistant cryptographic solutions, numerous academic institutions, universities, and businesses are devoting resources in this area. Several noteworthy research projects include:

- **NIST, the National Institute of Standards and Technology**, has taken the lead in standardizing post-quantum cryptography. In order to choose quantum-resistant cryptographic algorithms, they started a competition in 2017, and the standardization's completion is anticipated in the following years.
- **Academic Institutions**: Quantum computing and quantum-resistant encryption are the subject of extensive research in academic institutions all over the world. They contribute to the growth of theory, the creation of algorithms, and the design of protocols.
- **Industrial Research**: Businesses currently investigating the potential of quantum computing and its implications for computer security include IBM, Google, Microsoft, and Intel. They make investments in both the creation of quantum hardware and the study of quantum-resistant algorithms.

9.2 Quantum Computing and National Security Agencies

Due to its potential impact on communication security and cryptography systems, quantum computing has drawn a lot of attention from national security agencies. Research into quantum computing and its implications for national security is being actively pursued by several organizations:

- **American National Security Agency (NSA)**: The NSA has been researching quantum computers for a long time. They are keeping a close eye on the advancement of quantum computing technology and have released reports and suggestions on switching to quantum-resistant cryptography.
- **Canada's Communications Security Establishment (CSE)**: The CSE is heavily engaged in the study of quantum computing and how it may affect communication security. To provide cryptographic solutions that are resistant to quantum technology, they work with international partners.
- **GCHQ, the UK's government communications headquarters**: The GCHQ makes investments in quantum computing research and develops cryptography that is quantum-resistant. They participate in global partnerships and advance the creation of quantum-safe standards.

9.3 Industry Initiatives and Quantum-Safe Standards

The sector is actively engaged in the creation of quantum-safe standards and the application of quantum-resistant cryptography techniques. These activities and organizations include the following:

- **The working group for quantum-safe security (QSS)**: The goal of this working group, which is coordinated by the Cloud Security Alliance (CSA), is to identify quantum-resistant cryptographic technologies and create standards for their adoption across all sectors of society.
- **Project OQS**: The Open Quantum Safe An open-source initiative called the OQS Project intends to create and advance quantum-resistant cryptography algorithms and protocols. It gives programmers the frameworks and resources they need to incorporate quantum-safe cryptography into their projects.
- **Collaboration and Consortia**: To promote cooperation, research, and the creation of quantum-safe standards, a number of industrial collaborations and consortia, like the Quantum Economic Development Consortium (QED-C), are bringing together stakeholders from academia, industry, and government.

- **Standardization Bodies**: Defining and standardizing quantum-safe cryptographic algorithms and protocols is a top priority for standardization organizations like NIST, ISO, and the European Telecommunications Standards Institute (ETSI).

10. CHALLENGES AND FUTURE DIRECTIONS

10.1 Technical Challenges

There are a number of technical obstacles to the development and application of quantum computing and quantum-resistant cryptography, including:

- **Hardware Development**: Qubit scaling up while retaining low error rates is a big problem for quantum computers, which are still in their early phases of development. It is essential for the actual implementation of quantum computing to get over these technical obstacles.
- **Algorithm Development**: It is difficult to create effective and safe quantum-resistant algorithms. Researchers must create algorithms that are effective while defending against attacks from both classical and quantum computers.
- **Scalability**: In order to effectively handle complicated tasks, quantum computing systems must show scalability. A major technical problem is creating large-scale quantum systems and improving their functionality.

Designing and examining quantum-resistant cryptographic algorithms and protocols is a current field of research in post-quantum cryptography. It entails developing new mathematical foundations and cryptographic fundamentals that are resistant to attacks from quantum computers (Debosree, Khang et al., 2023).

10.2 Standardization and Interoperability

In order to use quantum-resistant cryptography and guarantee system compatibility, standardization is essential. Among the major difficulties in this field are:

- **Algorithm Consensus**: There needs to be agreement on protocols and algorithms that are quantum-resistant. Quantum-resistant algorithms are being evaluated and chosen by standardization organizations like NIST, but widespread deployment requires industry-wide agreement.
- **Interoperability**: During the transition phase, compatibility between classical and quantum-resistant systems is essential. A problem that needs to be solved is ensuring compatibility and smooth communication between various cryptography systems.
- **Long-Term Security**: Long-Term Security should be taken into consideration when designing quantum-resistant standards. It is a difficulty that needs careful thought to anticipate future developments in quantum computing and ensure that quantum-resistant algorithms remain secure for a long time.

10.3 Transition Strategies

Changing from classical to quantum-resistant encryption has a number of difficulties that must be overcome:

- **Period**: It is difficult to estimate the ideal period for switching to quantum-resistant algorithms. Companies must strike a compromise between the necessity of safeguarding against potential quantum risks and the viability of deploying new algorithms and modernizing current systems.
- **Infrastructure and Compatibility**: To update cryptographic systems, infrastructure, protocols, and software must be changed. A big problem is ensuring compatibility with existing systems and minimizing interruption throughout the changeover process.

Managing various cryptographic keys is necessary when switching to quantum-resistant algorithms. To maintain the confidentiality and integrity of data, organizations must devise plans for safely keeping and transferring keys.

10.4 Quantum Computing and the Future of Computer Security

By posing risks and supplying new tools for cryptography, quantum computing has the potential to completely transform computer security. Future computer security must take into account a number of factors, including:

- **Threats posed by quantum technology**: Traditional cryptography systems will become more open to intrusion as quantum computing technology develops. As a result, it is crucial to create and use quantum-resistant cryptography to safeguard critical information and communications.
- **Quantum-Safe Standards**: In the post-quantum age, it will be essential to create and implement quantum-safe standards to guarantee the safety of digital systems. To create widely embraced quantum-safe standards, cooperation between business, academia, and standardization organizations is crucial.
- **Quantum-Inspired Cryptography**: New encryption methods can be influenced by the ideas of quantum computing. Quantum-inspired technologies that increase security include quantum random number generators (QRNG) and quantum key distribution (QKD).
- **Continuous Research and Adaptation**: The fields of computer security and quantum computing are developing quickly. To remain ahead of new threats and provide cutting-edge cryptographic solutions, ongoing research, collaboration, and adaptation are essential

11. CONCLUSION

In conclusion, quantum computing is an innovative technology that has the potential to revolutionize computation. Quantum computers have the potential to solve complicated problems much quicker than conventional computers by utilizing the concepts of quantum mechanics, such as superposition and entanglement. Building quantum systems and algorithms has come a long way, despite still being in its infancy. Transformational developments in numerous scientific, technological, and industrial sectors will be made possible by continued study and development in this area.

Quantum attacks can be made against classical encryption, which has long been the foundation of secure communication. The security of well-known cryptographic techniques like RSA and ECC is in danger due to the enormous processing capacity of quantum computers. The security of conventional cryptography systems will deteriorate as quantum computers become more powerful. Recognizing this weakness and making an investment in post-quantum cryptography is essential for organizations and researchers who want to protect the confidentiality and integrity of sensitive data.

A potential defense against the dangers posed by quantum computers is post-quantum cryptography. It includes a variety of quantum-resistant cryptographic algorithms that guarantee long-term security even in the presence of potent quantum adversaries. The development of post-quantum cryptographic algorithms that offer the same level of security as traditional encryption is the subject of extensive research. In order to guarantee the security and privacy of digital communications in the quantum era, post-quantum cryptography standards must be adopted (Khang & Shah et al., 2023).

Using the concepts of quantum physics, quantum key distribution (QKD) offers a secure way to distribute cryptographic keys. Quantum key distribution (QKD) permits the creation of uncrackable cryptographic keys and provides unconditional security by taking advantage of the special qualities of quantum states. Despite difficulties with distance restrictions and practical implementations, QKD has a lot of potential for secure communication channels, especially in situations where information confidentiality is crucial. Networks for secure and quantum-resistant communication will be made possible by ongoing improvements and standardization efforts in QKD.

To reduce the security threats posed by quantum computers, it is essential to build algorithms that are immune to quantum effects. Long-term security for various applications is guaranteed by these algorithms, which are built to fend off attacks from both classical and quantum opponents. To find and create quantum-resistant algorithms for various cryptographic primitives, including encryption, digital signatures, and key exchange, extensive research and collaboration are now being conducted. Maintaining the security and integrity of data and communication in the post-quantum age will depend heavily on the acceptance and integration of these algorithms into current systems (Khang & Kali et al., 2023).

A substantial amount of potential exists for distributed computation that protects privacy using secure multiparty computation (SMC) and quantum computing. SMC enables many participants to collaboratively compute a function while maintaining the privacy of their private inputs by making use of the strength of quantum superposition and entanglement. This has important ramifications for applications like data sharing, medical research, and financial transactions that require joint examination of sensitive data. In order to explore the practical application of SMC with quantum computing, additional research and development are required. These efforts should focus on overcoming issues with scalability, robustness, and efficiency (Khang & Muthmainnah et al., 2023).

Research and business endeavors in the area of quantum computing are on the rise. The state of quantum computing hardware, software, and algorithms is being actively advanced by academic institutions, research organizations, and technological firms. Progress is being made in fields like qubit stability, error correction, quantum simulation, and algorithm development thanks to collaborations and cooperative efforts. Additionally, industry projects are concentrating on investigating the real-world uses of quantum computing in a variety of industries, including finance, healthcare, logistics, and optimization. The current momentum in the field sets the stage for transformative breakthroughs and establishes a strong foundation for the future of quantum computing (Khang, Vugar and Abuzarova et al., 2023).

In order for quantum computing to be widely used, a number of issues and uncertainties must be resolved. These difficulties include enhancing qubit stability and coherence, lowering error rates, creating scalable and fault-tolerant quantum systems, and devising effective algorithms for practical issues. Significant challenges include the development of quantum-safe standards and protocols as well as the integration of quantum technology with current infrastructures. However, these obstacles can be addressed with continued research, cooperation, and funding. With the ability to solve currently unsolvable problems, transform entire sectors, and usher in a new era of computation and scientific discovery, quantum computing has a bright future (Shyam & Khang et al., 2023).

REFERENCES

Abomhara M. & Kien, G. M. (2015). Vulnerabilities, dangers, intrusions, and assaults related to cyber security and the internet of things. *J. Cyber Secur. Mobil.*

Artur, K. (1991). Ekert, Quantum cryptography based on Bell's theorem. *Phys. Rev. Lett. 67*, 661. https://journals.aps.org/prl/abstract/10.1103/PhysRevLett.67.661

Arya, K., Arute, F., & Babbush, R. (2019). Employing a programmable superconducting processor to achieve quantum dominance. *Nature.*

Azouit, K., Azouit, A., El Housni, H., & El Hassouni, M. (2021). A Quantum Secure Direct Communication Protocol. In *2021 International Conference on Wireless Technologies, Embedded and Intelligent Systems (WITS)* (pp. 1-6). IEEE.

Boskovic, M., Becker, S., & Dhama, A. (2006). Trustworthy Software Systems: Definition and Discussion of Basic Terms. *ACM SIGSOFT Softw. Eng. Notes, 31*(1).

Charles, H. (1984). *Bennett and Gilles Brassard, International Association for Cryptologic Research, CryptoDB - An Update on Quantum Cryptography.* CRYPTO. doi:10.1007/3-540-39568-7_39

Diffie–Hellman. (1976). *Whitfield Diffie and Martin Hellman. Diffie–Hellman key exchange.* https://en.wikipedia.org/wiki/Diffie%E2%80%93Hellman_key_exchange

Ding, J., Gentry, C., & Malik, M. (2020). Learning With Errors: Survey and Future Directions. *Journal of Cryptology, 33*(4), 1163–1239.

Eswaran, U. (2023). Role of Quantum Computing in The Era of Artificial Intelligence. Applications and Principles of Quantum Computing. IGI Global Press. https://doi.org/ doi:10.4018/979-8-3693-1168-4-ch003

Ghosh, D. & Khang, A. (2023). Quantum Computing Internet of Things: Unlocking New Frontiers for Connectivity, Security, and Efficiency in Data Processing. Applications and Principles of Quantum Computing. IGI Global Press. https://doi.org/ doi:10.4018/979-8-3693-1168-4-ch016

Gisin, N., Ribordy, G., Tittel, W., & Zbinden, H. (2002). Quantum cryptography. *Reviews of Modern Physics, 74*(1), 145–195. doi:10.1103/RevModPhys.74.145

Hwang, J., Park, T., Park, H., & Kim, H. (2022). Quantum Secure Multi-Party Computation Protocol Based on the Three-Phase Quantum Key Agreement. *Symmetry, 14*(4), 1–16.

Khang, A. (2023). Introduction to Quantum Computing and Integration of Quantum Computing. Applications and Principles of Quantum Computing. (Eds.) IGI Global Press. https://doi.org/ doi:10.4018/979-8-3693-1168-4-ch002

Khang, A. (2023). AI and IoT-Based Technologies for Precision Medicine. IGI Global Press. doi:10.4018/979-8-3693-0876-9

Khang A. (2023). *Quantum, Applications and Principles of Quantum Computing*. IGI Global Press. https://doi.org/ doi:10.4018/979-8-3693-1168-4

Khang, A. (2023). Quantum Mechanics Primer - Fundamentals and Quantum Computing. Applications and Principles of Quantum Computing. IGI Global Press. https://doi.org/ doi:10.4018/979-8-3693-1168-4-ch001

Khang, A. (2023). Enabling the Future of Manufacturing: Integration of Robotics and IoT to Smart Factory Infrastructure in Industry 4.0. AI-Based Technologies and Applications in the Era of the Metaverse. IGI Global Press. doi:10.4018/978-1-6684-8851-5.ch002

Khang, A. (2023). AI-Aided Data Analytics Tools and Applications for the Healthcare Sector. AI and IoT-Based Technologies for Precision Medicine. IGI Global Press. doi:10.4018/979-8-3693-0876-9.ch018

Khang, A., & Muthmainnah, M. (2023). AI-Aided Teaching Model for the Education 5.0 Ecosystem. AI-Based Technologies and Applications in the Era of the Metaverse. IGI Global Press. doi:10.4018/978-1-6684-8851-5.ch004

Khang, A., Shah, V., & Rani, S. (2023). AI-Based Technologies and Applications in the Era of the Metaverse. IGI Global Press. doi:10.4018/978-1-6684-8851-5

Li, X., Huang, X., Zhang, M., Wang, L., & Zhang, Z. (2019). A Quantum Intrusion Detection Method Based on Quantum Entanglement. *Symmetry*, *11*(12), 1–14.

Liu, L., Yu, Y., Lai, L., Yu, R., & Zhang, X. (2020). Quantum Adversarial Machine Learning-Based Intrusion Detection. *Future Internet*, *12*(9), 1–19.

Mitra, J., Bhattacharya, P., & Poray, S. (2017). *Overview of quantum cryptography, security concerns, and upcoming difficulties.* Optronix, 4th International Conference on Opto-Electronics and Applied Optics, Kolkata, India.

Prabu, B., & Khang, A. (2023). Quantum Computing Smart City: Fostering Sustainability and Green Infrastructure with Smart City. Applications and Principles of Quantum Computing. IGI Global Press. https://doi.org/ doi:10.4018/979-8-3693-1168-4-ch009

Sergey, B., Englbrecht, M., König, R., & Peard, N. (2018). Correcting coherent errors with surface codes. *Nature*. https://www.nature.com/articles/s41534-018-0106-y#citeas

Shor, P. W. (1996). Polynomial-Time Algorithms for Prime Factorization and Discrete Logarithms on a Quantum Computer. *SIAM Journal on Computing*, *26*(5), 1484–1509. doi:10.1137/S0097539795293172

Shyam, R., Sihare, & Khang, A. (2023). Effects of Quantum Technology on Metaverse. AI-Based Technologies and Applications in the Era of the Metaverse. IGI Global Press. doi:10.4018/978-1-6684-8851-5.ch009

Teja, S. B. B., Saheb, S. H., & Khang, A. (2023). Quantum Deep Learning: Detection of Drowsiness in Drivers Using Tensorflow Framework. Applications and Principles of Quantum Computing. IGI Global Press. https://doi.org/ doi:10.4018/979-8-3693-1168-4-ch015

Valerio, S., Antonio, A., Grégoire, R., & Nicolas, G. (2004). Quantum Cryptography Protocols Robust against Photon Number Splitting Attacks for Weak Laser Pulse Implementations. *Phys. Rev. Lett. 92*, 057901. https://journals.aps.org/prl/abstract/10.1103/PhysRevLett.92.057901

Zhou, L., Sheng, Y. B., Ding, D. S., Shi, S. B., & Guo, G. C. (2017). Quantum memory and direct secure transmission. *Physical Review Letters*, *118*, 220501. doi:10.1103/PhysRevLett.118.220501

Chapter 21
The Convergence of Quantum Computing and Blockchain:
Opportunities, Threats, and Protective Strategies

Busra Ozdenizci Kose
https://orcid.org/0000-0002-8414-5252
Gebze Technical University, Turkey

ABSTRACT

The integration of quantum computing and blockchain represents a pivotal moment in technological advancement, offering both unique synergies and new security challenges. Today, blockchain developers and practitioners are working hard to make sense of the threats and opportunities that quantum computing will bring. This chapter addresses the quantum computing-blockchain puzzle by examining the potential innovations, opportunities, and threats of these two transformative technologies. Moreover, the importance of post-quantum cryptography and quantum-resistant blockchain architectures, as well as open-source development efforts and educational initiatives, are highlighted as measures to both overcome challenges and strengthen opportunities. As the impact and importance of quantum technology grows, this study aims to provide guidance to practitioners and developers, helping them overcome and mitigate the complexities and challenges ahead.

1. INTRODUCTION

Today, quantum computing and blockchain are two revolutionary technologies that have the potential to transform numerous industries and redefine how we approach both computational and financial challenges of the 21st century.

Quantum computing is paving the way for groundbreaking developments in multiple fields. By utilizing the principles of quantum mechanics and offering unique computing power, it aims to solve complex problems that were previously difficult to solve for classical computers (Gyongyosi & Imre,

DOI: 10.4018/979-8-3693-1168-4.ch021

2019; Savchuk & Fesenko, 2019; Gill et al., 2022). In this context, from accelerating drug discovery by simulating complex molecular structures to developing encryption methods in the face of increasing cybersecurity threats, potential applications of quantum computing have the power to reshape industries.

Blockchain, on the other hand, offers a significant paradigm shift based on a decentralized environment in the methods of recording, verifying and sharing information. It aims to record data in a distributed ledger as uniquely transparent, secure and immutable records, without the need for central intermediaries (Li et al., 2020; Guo et al., 2020; Gamage et al., 2020). With its robust cryptographic algorithms, blockchain technology already has potential in many areas beyond its well-known applications in cryptocurrencies such as Bitcoin and Ethereum. Unlike traditional central authority architectures, it has the ability to transform industries and increase user trust in digital transactions, from supply chain management to voting systems.

In the ever-evolving digital transformation landscape, quantum computing and blockchain stand out as two important technologies poised to redefine the future. Although there are extensive studies on how these two important technologies will affect digital transformation, how these two revolutionary technologies will affect each other is equally important. Their integration brings exciting opportunities as well as complex challenges and highlights the multifaceted consequences of their integration into the digital world.

For instance, with the ability to exploit algorithms like Shor's, quantum computers can pose a real threat to the cryptographic security of the blockchain-driven platforms, potentially leading to the decryption of private keys and the creation of vulnerabilities such as double-spending attacks (Khalid & Askar, 2021; Unogwu eet al., 2022; Allende et al., 2023). But the convergence of quantum computing and blockchain is not just about risks; it also offers opportunities and benefits (Cui et al., 2020; Younan et al., 2021; Thanalakshmi et al. (2023). For example, quantum computing can increase transaction speeds and efficiency of smart contracts. In addition, today quantum key distribution offers a promising way to increase blockchain security through the principles of quantum mechanics.

Today, blockchain developers and practitioners are working particularly hard to recognize these threats and opportunities. Efforts are ongoing to develop and integrate strong post-quantum cryptographic algorithms against quantum threats. While the integration of these solutions brings its own challenges, the possibility of a future where blockchain and quantum technologies can provide benefits together is quite high. A comprehensive review of 'Quantum Computing-Blockchain' puzzle situation, with its many benefits and challenges, can provide a useful roadmap for practitioners and developers.

This chapter addresses the deep intersection of two significant technologies -quantum computing and blockchain-, each transformative on its own, but when used together, they hold potential for both unique synergies and critical security vulnerabilities. The purpose of this chapter is to shed light on the ways in which quantum computing and blockchain are merging by evaluating potential innovations, opportunities, and upcoming threats. As the impact of quantum technology increases, this chapter will provide guidance and reference in overcoming the upcoming challenges.

In the light of all this information, the chapter first outlines a brief information about quantum computing and blockchain technologies including their principles and general features. Then, potential innovations and opportunities, threats and challenges that may arise in the integration of these technologies are presented systematically. Moreover, the protective measures and strategies are reviewed and discussed. As the research landscape concerning quantum computing and blockchain continues to evolve, this study will provide valuable insights for the development of Quantum Blockchain driven systems, benefiting various stakeholders, including practitioners, system and technology developers.

2. BACKGROUND

Before discussing the converging path of the two technologies, this section explains the general technical features and principles of the two revolutionary technologies separately.

2.1 Understanding Quantum Computing

Quantum computing provides a distinct approach to computational challenges, allowing for solutions that are more efficient than those achievable with traditional computing methods. The potential of quantum computing first became evident in 1994, when Peter Shor developed a quantum algorithm that could factor large numbers exponentially faster than the best-known algorithms (e.g., Rivest-Shamir-Adleman, RSA algorithm) running on classical computers (Gyongyosi & Imre, 2019). Shor's prime factorization algorithm was a significant development, as the difficulty of factoring large numbers forms the basis of many encryption schemes. This development created the "big bang" effect, paving the way for the development of quantum computing technology and quantum computers.

Quantum computing is a multidisciplinary field that combines elements of computer science, physics, and mathematics, focuses on the development of computer-based technologies within the principles of quantum theory, and uses quantum mechanics to solve complex problems faster than classical computers. The field of quantum computing includes hardware research and application development (Gyongyosi & Imre, 2019; Savchuk & Fesenko, 2019; Gill et al., 2022).

Quantum theory, a subset of physics, explains the nature and behavior of energy and matter at the quantum (i.e., atomic and subatomic) scale. The principles of quantum theory, when applied to computers, offer significant opportunities for processing information. By exploiting phenomena such as quantum superposition and quantum interference, quantum computers can solve some problems faster than traditional computers (Gyongyosi & Imre, 2019). They show promise in areas such as machine learning, optimization, and simulating physical systems at atomic and subatomic scales.

Quantum computers operate on the principle of quantum information, where information is denoted by quantum states. Figure 1 from the study of Gyongyosi & Imre (2019) provides well-defined view of quantum computing context. At the bottom layer, the highlighted DiVincenzo criteria, set forth in 2000 by theoretical physicist David P. DiVincenzo, outline seven essential requirements for building a functional and scalable quantum computer. These criteria are essential benchmarks in the field of quantum computing which are a scalable physical system with well-defined qubits, qubits state initialization capabilities, long relevant decoherence times, a set of universal quantum gates, qubit-specific measurement capability, and reliable qubit transmission for quantum communication (i.e., the ability to interconvert stationary and flying qubits and the ability to faithfully transmit flying qubits between specified locations).

Quantum algorithms provide the computational procedures that quantum computers execute to solve complex problems faster than classical counterparts. Many algorithms have been proposed in this context. The developed algorithms can provide significant speedups; exponential, polynomial or superpolynomial. The most famous of these is the Shor algorithm, which has significant implications for encryption and security.

Quantum computers utilize some essential principles from quantum mechanics, including superposition, entanglement, interference, and coherence (Gyongyosi & Imre, 2019; Savchuk & Fesenko, 2019; Gill et al., 2022). Grasping these fundamental concepts offers crucial insights into the overarching landscape of quantum computing.

Figure 1. Conceptual diagram of quantum computing technology
(Source: Gyongyosi & Imre, 2019)

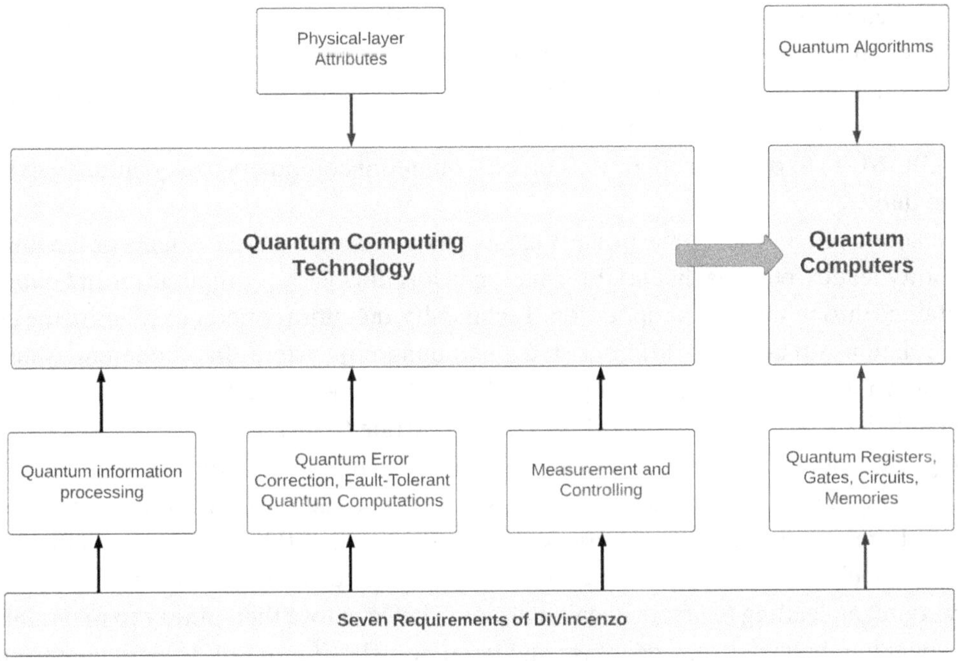

Quantum superposition refers to the remarkable capability of a quantum system to exist in several states simultaneously. Unlike classical computers, which use bits (either 0 or 1) as the smallest unit of data, quantum computers use quantum bits, or qubits, that can exist in a superposition of both 0 and 1 at the same time. This allows quantum computers to process a vast amount of information simultaneously. The power of superposition is harnessed in various quantum algorithms, enabling them to potentially outperform classical counterparts in certain tasks.

Quantum entanglement is a concept in which two or more quantum systems are inherently interconnected, such that the state of one is directly related to the state of the other. This connection persists regardless of the distance separating the entangled particles. Once an entangled particle is measured and its state determined, the state of its other partner is instantly known, even if they are light-years apart. This correlation allows quantum computers to process information in ways that classical systems cannot. Entanglement demonstrates the deeply interconnected nature of the quantum world and is crucial for advanced quantum computations and quantum information processing.

Quantum interference is another fundamental principle of quantum mechanics that allows quantum systems to combine their states in certain ways. In the context of quantum computing, interference is used strategically to manipulate the probabilities of qubit states. By carefully designing quantum algorithms, it is possible to constructively manipulate the desired outcomes (correct answers) and increase the probabilities. At the same time, it disruptively intervenes in unintended consequences (wrong answers), reducing their likelihood. This careful orchestration of constructive and destructive intervention is crucial in algorithms like Grover and Shor's, which allows quantum computers to eliminate possibilities more efficiently than classical machines.

The idea of reversible computation is crucial for quantum computation (Gyongyosi & Imre, 2019; Savchuk & Fesenko, 2019; Gill et al., 2022). In classical computing, most operations are irreversible. For example, when an AND gate outputs '0', it cannot be determined whether the inputs are '00', '01' or '10'. However, in quantum mechanics, the evolution of any closed system is characterized by unitary transformations that are inherently reversible. To realize principles of quantum mechanics, quantum computing uses reversible gates, known as quantum gates. Just as classical computers use logical gates (e.g., AND, OR, NOT) to perform operations on bits, quantum computers use quantum gates to perform operations on qubits.

For any quantum process, a reverse process exists that can cancel out the effects of the initial process. This reversibility feature ensures that no information is lost during the computation and quantum coherence is maintained throughout the computation. Technically, quantum coherence refers to the preservation of the phase relationship between different states in a quantum system. In a coherent system quantum states can exist in a superposition and their phase relationships are well defined. Common quantum gates include the Pauli-X, Y, Z gates, Hadamard gate, and controlled-NOT gate.

Accordingly, quantum decoherence is a significant challenge in quantum computing. It is the phenomenon when a quantum system loses coherence due to interactions with the external environment (Gyongyosi & Imre, 2019; Savchuk & Fesenko, 2019; Gill et al., 2022). Building a practical quantum computer is extremely challenging due to qubits' delicate nature. They can easily be disturbed by their external environment, leading to errors in calculations. They can lose their ability to maintain superpositions and its quantum behavior may decrease and begin to behave more like classical systems.

Accordingly, quantum computers have been created and tested by companies such as IBM, Google and other leading companies; but they are still in their early stages. Current quantum computers require extremely low temperatures to operate and are prone to errors because they are likely to interact with the external environment. Error-correcting codes and fault-tolerant designs can be identified as active research areas to solve these challenges. Undoubtedly, quantum computing technologies continue to hold great promise for upcoming computing and communications efforts.

2.2 Understanding Blockchain Technology

Blockchain technology as one of the striking and promising technological developments, has an important potential in the digital transformation of many industries. Blockchain technology comes to scene after the introduction of Bitcoin cryptocurrency by Satoshi Nakamoto in 2008 (Nakamoto, 2008). Since then, its applications have expanded far beyond cryptocurrencies, influencing sectors like finance, supply chain, logistics, healthcare, transportation, defense, public safety, and governance, offering solutions for transparency, traceability, and decentralized control.

As a distributed ledger technology (DLT), blockchain provides decentralized data storage network with a tamper-resistant shared ledger including chained blocks in distributed networks (Wang et al., 2019; Al-Jaroodi & Mohamed, 2019; Gamage et al., 2020); which can record and secure transactions using cryptography between two or more parties efficiently and in a verifiable and permanent way. It allows trustless users agree on an auditable, immutable and irreversible data without the need of a trusted third party as a referee. The key features enable blockchain technology receive extensive attention from various sectors and fields with the aim of secure data transfer and storage, identity management, authentication and access control, monitoring assets, time stamping and similar high security related services to enhance trust between entities (Fernández-Caramés & Fraga-Lamas, 2018).

Figure 2. Blockchain architecture
(Source: Fernández-Caramés & Fraga-Lamas, 2018; Zheng et al., 2018)

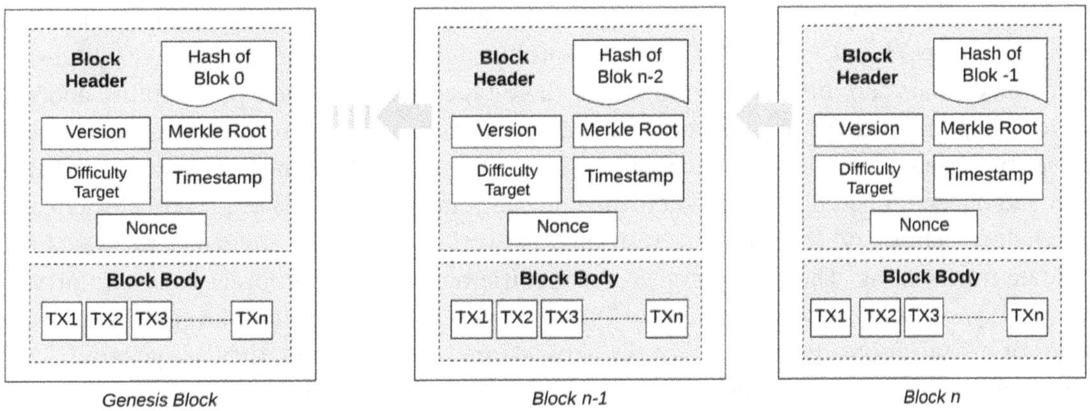

Blockchain technology integrates various core technologies and methods such as cryptographic hash functions, digital signature, consensus algorithms, and smart contracts. As illustrated in Figure 2, Blockchain is a chain of blocks linked together using cryptographic hash functions such as SHA-256, which is used in Bitcoin, ensuring data integrity and immutability throughout the chain. Each block contains hash of its previous block that allows users to agree on undeniable and immutable data (Zheng et al., 2018; Fernández-Caramés & Fraga-Lamas, 2018; Panarello et al., 2018).

The first block in a blockchain is known as the genesis block. Each block in the chain comprises a block header and a block body. The block header contains several fields including the hash value of the previous block, block version, timestamp, nonce, and more. The block body typically contains a list of transactions and may also have a counter indicating the number of transactions included in that block. Each block can be conceptualized as a container, storing details related to specific transactions, such as financial exchanges or other types of data transfers. As new transactions are created, the list keeps growing subsequently. These transactions are initiated by entities within the network, commonly referred to as nodes. Once a transaction is created by a node, the transaction is broadcasted by that node to all other nodes. Miner nodes then validate the transaction, and upon successful validation, the transaction is accepted by the other nodes and subsequently added to the blockchain.

To verify and record transactions, blockchain systems rely on a collective of participants known as nodes, which utilize consensus algorithms and mechanisms to reach agreement (Zheng et al., 2018; Gamage et al., 2020; Guo et al., 2022). They are vital for maintaining the integrity and security of decentralized systems by validating and agreeing on the content of new blocks, ensuring consistency across all nodes. Nakamoto's Proof of Work (PoW) was the pioneering consensus protocol, which is based on solving complex puzzles. Today there are other consensus mechanisms like Proof of Stake (PoS), Practical Byzantine Fault Tolerance (PBFT), Delegated Proof of Stake (DPoS), Proof of Bandwidth (PoB), Proof of Elapsed Time (PoET), and Proof of Authority (PoA) to achieve network consensus (Fernández-Caramés & Fraga-Lamas, 2018; Zheng et al., 2018).

Another important component of blockchain architecture is smart contract which is a self-executing contract with the terms of the agreement embedded in blockchain code (Mohanta et al., 2018; Zheng et al., 2018; Khan et al., 2021). Ethereum as a popular decentralized, open-source blockchain platform has introduced the concept of smart contracts. Once deployed, smart contracts automatically execute or enforce the required, certain actions when specific conditions are met, eliminating the need for intermediaries. They can be used in various applications, from simple fund transfers to complex decentralized applications (DApps) like lending platforms, decentralized exchanges, and governance systems.

Today, blockchains are often categorized into three types based on their accessibility and who has permission to participate in the network: public blockchains, private blockchains, and consortium (federated) blockchains (Kim, 2020; Singh et al., 2022). Public blockchains like Bitcoin and Ethereum, are open and decentralized platforms where anyone can join and participate. Private blockchains such as Hyperledger Fabric and R3Corda are restricted networks where only invited participants can join and validate transactions. They are designed for specific organizations in order to ensure privacy and control. Finally, consortium blockchains like Quorum are semi-private networks controlled by a group of entities or organizations. Participants of the network are predetermined; they are generally preferred in cases where multiple organizations need to share data but also need privacy. Each type offers unique advantages for different requirements of developers and practitioners.

Today, significant advances have been made on blockchain technology by giants such as IBM, Microsoft and other industry leaders, and the field is still maturing. Studies are being carried out to solve challenges such as scalability in contemporary blockchains, energy consumption in PoW systems and interoperability between different blockchain platforms. Solutions such as Layer 2 scaling, PoS consensus mechanisms and cross-chain bridges are among the leading research efforts to overcome these obstacles. Undoubtedly, blockchain technology will continue to offer tremendous potential to reshape various industries and enable greater transparency, trust and decentralized control in future digital ecosystems.

3. THE INTERSECTION OF BLOCKCHAIN AND QUANTUM COMPUTING

Today, two technological giants, blockchain and quantum computing, are on a path of unprecedented growth and discovery. Their convergence brings with it both opportunities and challenges that could redefine the digital landscape. In this section, these opportunities and challenges are first discussed comprehensively.

3.1 Strengths and Opportunities

The combination of quantum computing and blockchain technologies offers a transformative environment full of innovation. With potential applications and improvements, both technologies further leverage each other's strengths (Cui et al., 2020; Younan et al., 2021; Faruk et al., 2022; Dadhich & Tiwari, 2022; Shuaib et al., 2022; Unogwu et al., 2022; Allende et al., 2023; Thanalakshmi et al., 2023). Some of the key innovations and opportunities are explained hereunder:

(1) *Quantum-Powered Transaction Speeds:* As mentioned, unlike classical systems, quantum computers can process large volumes of data at unprecedented speeds. With this ability, quantum computing paves the way for the development of quantum-enhanced blockchain systems. More quickly than traditional computing, transactions can be verified and added to the chain on a blockchain network. For blockchains, which often struggle with scalability and transaction throughput limitations, this win represents a potential paradigm shift. Combining these two areas is not just about enabling

Figure 3. Understanding cross-chain interoperability
(Source: Cointelegraph,2023)

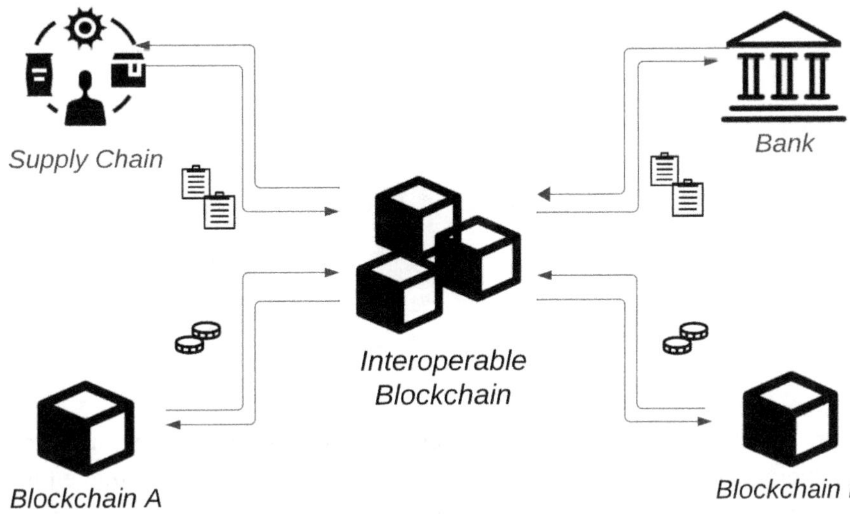

faster transaction verifications, it can also enable a deep redesign of the computation and verification frameworks that support decentralized ledgers.

(2) *Quantum-Secure Smart Contracts:* In the evolving landscape of blockchain technology, smart contracts are the pillar of automated trust and enforcement. The emergence of quantum computing could provide significant benefits for smart contracts. By intertwining quantum features with the basic architecture of smart contracts, quantum-secure smart contracts can be developed, paving the way for complex decentralized applications that operate with high security.

(3) *Quantum-Enhanced Consensus Mechanisms:* Consensus mechanisms are the heartbeats of blockchain networks, allowing that all participants agree on the validity of transactions. All existing consensus algorithms and solutions can be enhance with the advent of quantum computing. Quantum-enhanced consensus mechanisms can leverage the principle of superposition to reach consensus faster or use quantum entanglement to enable more synchronized and secure agreement processes between nodes.

(4) *Cross-Chain Interoperability with Quantum Entanglement:* Cross-chain interoperability can be defined as the ability of different blockchain platforms to communicate and share data seamlessly. This capability can allow a transaction or smart contract on one blockchain to trigger events on another blockchain network. The quantum entanglement principle can be used in the future for instantaneous cross-chain transactions and communications and can revolutionize how blockchains interoperate.

(5) *Blockchain-Supported Quantum Key Distribution:* One of the promising applications of quantum mechanics in secure communications is Quantum Key Distribution (QKD) (Scarani et al., 2009; Mailloux et al., 2015; Diamanti et al., 2016; Broadbent & Schaffner, 2016). QKD uses the principles of quantum mechanics to create a secure communication channel and to distribute cryptographic keys between two parties. This promising method can detect any eavesdropping attempts on the key exchange, offering an unprecedented level of security for communication. QKD can be used

Figure 4. Quantum key distribution
(Source: Mailloux et al., 2015)

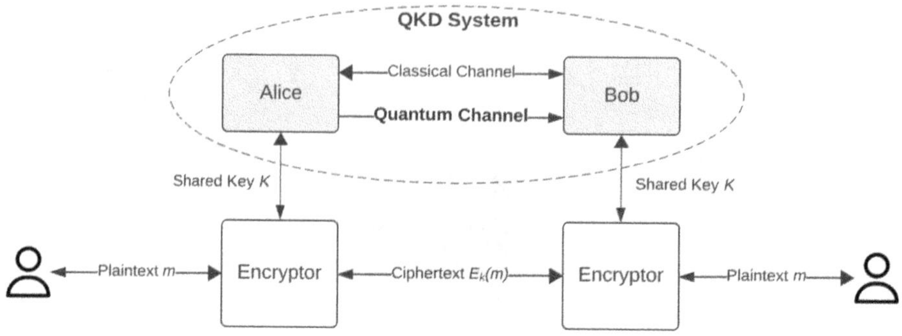

for ultra-secure communication channels, especially for sensitive data transmission, such as governmental or financial transactions. Integrating QKD with blockchains can make the transmission of data and value in blockchain networks even more secure than it is today. A blockchain could serve as a decentralized and tamper-proof ledger for these quantum keys. By storing quantum-generated cryptographic keys on a blockchain, the integrity and authenticity of the key distribution process can be ensured.

(6) *Blockchain-Supported Multi-party Quantum Computation:* Quantum networks enable multi-party quantum computations, where several parties can jointly perform a quantum computation on their combined data without revealing their individual data to the other parties (Crépeau et al., 2002; Wang & Alexander, 2020). Blockchain can be used to manage, verify and record these calculations in a decentralized manner, ensuring that transactions are carried out in a transparent and trustworthy manner.

(7) *Blockchain-Based Timestamping of Quantum Data:* Quantum systems are highly sensitive to external influences. Blockchain can act as decentralized timestamping tool to record quantum data measurements, ensuring that data has not been tampered with and providing a verifiable chronology of quantum data exchanges.

(8) *Blockchain-Based Identity and Access Management:* Quantum networks will require sophisticated identity and access management systems. Blockchain's decentralized and immutable nature can manage identities in quantum networks effectively, ensuring that only authorized entities can access and transmit quantum data.

Ultimately, the combination of blockchain and quantum networks offers a new frontier in digital innovation. By combining the powerful security measures of quantum mechanics with the decentralized integrity of the blockchain, it can shape a transformative paradigm for future communications and computing. This intersection not only promises improved security and efficiency, but also has the potential to redefine the foundations of our digital world.

Table 1. An assessment of existing cryptographic algorithms

Cryptographic Algorithm	Type	Purpose	Impact from Large-Scale Quantum Computer
AES	Symmetric key	Encryption	Larger key sizes needed
SHA-2, SHA-3	-	Hash functions	Larger output needed
RSA	Public key	Signatures, key establishment	No longer secure
ECDSA, ECDH (Elliptic Curve Cryptography)	Public key	Signatures, key establishment	No longer secure
DSA (Finite Field Cryptography)	Public key	Signatures, key establishment	No longer secure

3.2 Challenges and Threats

While the integration of quantum computing and blockchain offers immense potential, it also presents a set of challenges and threats that need to be addressed (Fedorov et al., 2018; Kiktenko et al., 2018; Shrivas et al., 2020; Ciu et al., 2020; Kearney & Perez-Delgado, 2021; Faruk et al., 2022; Shuaib et al., 2022; Unogwu et al., 2022; Thanalakshmi et al., 2023). Some of the challenges are explained hereunder:

(1) *Quantum Threat to Cryptography:* Most current blockchains rely on public-key cryptographic systems like RSA and ECC for security, particularly for digital signatures. Many existing encryption algorithms are potentially vulnerable to quantum attacks. Quantum computers, using algorithms like Shor's algorithm, can break these cryptographic systems in polynomial time. A powerful quantum computer can derive a user's private key from their public key, compromising the security of the user's funds or data on a blockchain. Table 1 from NIST IR 8105 "Report on Post-Quantum Cryptography," summarizes the impact of quantum computing on public-key and symmetric cryptography. In contrast with public-key encryption, the good news is that quantum computers do not significantly impact the security of symmetric cryptography. Existing symmetric algorithms – such as AES – can continue to be used with suitable key sizes.
This challenge encourages research into post-quantum cryptography (i.e., quantum-safe cryptography, quantum-resistant cryptography) which is explained in the following section. NIST defines the goal of quantum-safe cryptography as the development of "cryptographic systems that are secure against both quantum and classical computers, and can interoperate with existing communications protocols and networks. In 2017, NIST launched a competition to design, analyze and choose a set of quantum-safe algorithms for public-key cryptography. This is also known as the Post Quantum Cryptography (PQC) Standardization Challenge. The quantum-safe competition intends to develop cryptographic algorithms that will protect sensitive information even after the advent of quantum computers. The details regarding Quantum-Resistant Cryptography are presented in the next section.

(2) *Centralization by Quantum Computing:* If quantum computing becomes effective in blockchain operations, those with access to quantum technology may have an undue advantage. For instance, if quantum computers can solve PoW problems faster rates than classical computers, this could lead to mining centralization, which is antithetical to the decentralized concept of many blockchain networks.

Figure 5. 51% attack on blockchain

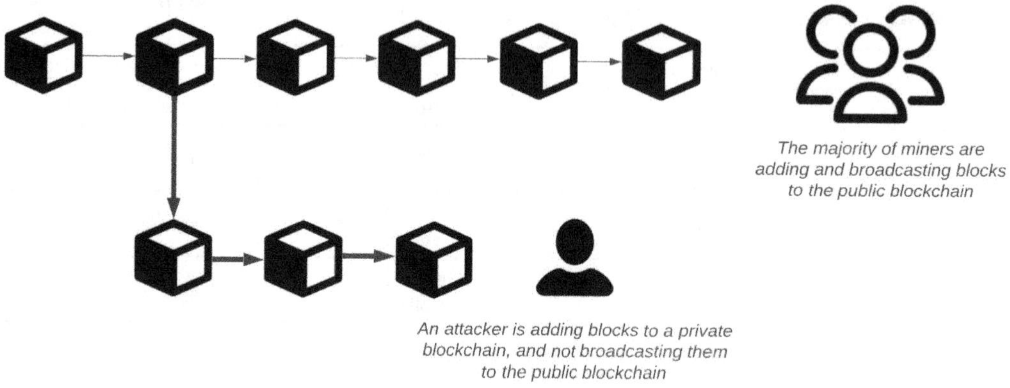

(3) *Quantum 51% Attacks:* A 51% attack happens when a single entity or coalition captures the majority (i.e., more than 50%) of the computing power of a blockchain network (Ye et al., 2018; Li et al., 2020; Leng et al., 2020). This attack targets PoW, the consensus mechanism in the blockchain network. It occurs when a miner or a community of miners controls more than 50% of the mining hash rate or computational power. The attack can compromise the security and privacy of the blockchain network in various ways. It can do double spending by consuming and then reversing the transaction, block transactions on the network and prevent users on the network from performing transactions, or illegally transfer digital coins on the network to target accounts, and lastly, manipulate the network for its benefit by changing consensus rules.
Quantum computers can significantly speed up the block discovery process in PoW systems due to their ability to process certain calculations exponentially faster than classical computers. This means that an entity with a sufficiently powerful quantum computer may control most of the block mining, paving the way for a 51% attack. A successful quantum-assisted majority attack can erode fundamental trust in the blockchain's immutability and decentralized nature, leading to a decline in adoption and a potential collapse in cryptocurrency value. Economic disruptions can occur due to double-spending scams, concentrated mining rewards, and skewed wealth distribution.

(4) *Smart Contract Vulnerabilities:* Quantum-secure smart contracts also present some challenges. The inherent complexity of quantum-secure smart contracts has the potential to unintentionally introduce new security vulnerabilities. As these contracts increase in complexity in incorporating quantum-resistant measures, they may become more difficult to audit and test comprehensively. This complexity can lead to the introduction of unforeseen gaps or errors. It may have various effects such as unauthorized access or changing the terms of the contract.

(5) *Quantum Hardware Limitations:* Incorporating quantum computing into blockchain networks presents some challenges arising from the current limitations of quantum hardware. Today's quantum computers are susceptible to errors due to the inherent variability of quantum states. Additionally, their dependence on extremely cold operating temperatures requires complex, high-maintenance installations. This may present scalability issues when it comes to creating large-scale, open blockchain networks. For blockchain to maintain the integrity and reliability of decentralized data, the

instabilities inherent in quantum computers must be addressed. Quantum hardware that will be compatible with the powerful demands of blockchain networks is required.

(6) *Quantum Data Storage Challenges:* Quantum data, governed by the principles of superposition and entanglement, requires special storage mechanisms to maintain its consistency and prevent data corruption. Integrating these sensitive quantum data mechanisms into the robust and permanent infrastructure of blockchains can pose significant challenges. Traditional blockchain architectures designed with classical data may not be equipped to handle the requirements of quantum information. If not adequately addressed, this can lead to potential data loss, inaccuracies in transaction records, and even systemic vulnerabilities. There is a compelling need for innovative solutions, frameworks and protocols that can seamlessly blend the capabilities of quantum data storage with the fundamental principles of blockchain technology.

(7) *Complexity of Transition to PQC*: The transition from classical cryptographic standards to quantum-resistant cryptographic algorithms is a significant shift. This transition includes algorithmic changes as well as complex system updates, compatibility checks and verification processes. With many blockchain-based systems relying on classical cryptographic methods, making them quantum secure requires examining both old and new code contents and ensuring backward compatibility. The use of the latest quantum security measures in combination with traditional encryption methods may cause potential incompatibility issues. This multidimensional revision brings significant requirements in terms of both intellectual capital and resource allocation. In this complex evolutionary stage in cryptography, the necessity of strategic planning, collaboration and methodological approach is essential. As new blockchains adopt quantum-resistant algorithms, they must also maintain a line of communication with their classical predecessors as non-quantum-resistant systems. This evolutionary process underlines the importance of interoperability, innovative quantum-driven designs and adaptable frameworks in the blockchain field.

4. PROTECTIVE MEASURES AND STRATEGIES

This section examines essential protective measures and strategies under three domains that can overcome the aforementioned challenges as well as enhance the opportunities and strengths of quantum-blockchain systems.

4.1 Post-Quantum Cryptography as Quantum-Resistant Cryptography

Today, as mentioned, many existing encryption algorithms, especially public key cryptosystems such as RSA and ECC, are potentially vulnerable to quantum attacks. This encourages research into post-quantum cryptography. The National Institute of Standards and Technology (NIST) has been leading an effort to standardize post-quantum cryptographic algorithms.

According to NIST, post-quantum cryptography (PQC), also known as quantum-resistant cryptography, is to develop cryptographic algorithms (usually public-key algorithms) that are secure against both quantum and classical computers, and can interoperate with existing communications protocols and networks. Classical public-key cryptosystems, such as RSA, Elliptic Curve Cryptography (ECC), and Diffie-Hellman, rely on the difficulty of certain mathematical problems (like factoring large numbers or computing discrete logarithms) for their security.

Quantum computers, particularly through algorithms like Shor's algorithm, can solve these problems exponentially faster than classical computers, thus potentially breaking these cryptosystems. The importance of post-quantum cryptography arises from the anticipation of functional, large-scale quantum computers becoming a reality in the future. Even though practical quantum computers that can break existing cryptography might still be years or decades away, transitioning to new security standards is a significant endeavor that requires considerable time.

Researchers and institutions worldwide are actively working on standardizing these post-quantum cryptographic methods (Kiktenko et al., 2018; Gyongyosi & Imre, 2019; Fernández-Caramés & Fraga-Lamas, 2020; Khalid & Askar, 2021; Thanalakshmi et al., 2023). Organizations like NIST have initiated processes to review, evaluate, and standardize PQC algorithms to prepare for a post-quantum era. Currently well-known post-quantum cryptography approaches are as follows (Figure 6):

- Lattice-based Cryptography: These algorithms are based on the hardness of certain problems in lattice-based structures. They are promising due to their efficiency, versatility, and potential resistance to quantum attacks. They have found applications in constructing encryption, digital signatures, and fully homomorphic encryption schemes.
- Code-based Cryptography: This method is based on the difficulty of decoding randomly generated linear codes. One of the oldest approaches to PQC, the McEliece encryption scheme, is based on this method.
- Multivariate Polynomial Cryptography: This method is based on the problem of finding solutions to systems such polynomials. This approach has been mainly used for constructing digital signatures.
- Hash-based Cryptography: This method uses hash functions to create secure digital signatures. Particularly for digital signatures, hash-based methods - like the Merkle signature scheme- have been studied for several decades and have the advantage due to their security guarantees.
- Isogeny-based Cryptography: This method is based on the computational difficulty of certain problems related to isogenies (morphisms) between elliptic curves. This is relatively new area focuses on the mathematical structure of elliptic curves and the functions (isogenies) that map between them.

4.2 Quantum-Resistant Blockchains

Quantum resistant blockchains, also known as quantum secure blockchains, refer to blockchain systems designed to withstand potential attacks from quantum computers (Kiktenko et al., 2018; Khalid & Askar, 2021; Zhang et al., 2021; Thanalakshmi et al., 2023). As mentioned, as quantum computers improve, they might pose a threat to traditional encryption schemes such as RSA and ECC commonly used in current blockchain systems. Therefore, the development of quantum-resistant blockchains as an important research area is vital to ensure the long-term security of distributed ledger, that even powerful quantum computers cannot easily break (Fernández-Caramés & Fraga-Lamas, 2020; Unogwu et al., 2022; Allende et al., 2023).

Quantum-resistant blockchains are primarily designed to protect against potential quantum computer attacks by combining post-quantum cryptographic algorithms such as lattice-based, hash-based, and code-based schemes. These blockchains can use Hybrid Systems that blend classical and post-quantum cryptographic methods for both immediate and future security. Another important feature is their agility, allowing them to adapt innovative cryptographic algorithms or modify existing ones, providing long-term security, reliability and competitive advantage.

Currently, some projects have started to be implemented in the field of quantum-resistant blockchains. One of the most well-known projects focused on creating a fully quantum-resistant blockchain is The Quantum Resistant Ledger (QRL) (https://www.theqrl.org). It is an open source blockchain platform designed explicitly to be secure against both classical and quantum computer threats. Developers and researchers can review, contribute to, and verify the platform's security. It employs the Extended Merkle Signature Scheme (XMSS), a NIST-approved post-quantum, hash-based digital signature system that is considered resistant to quantum computer attacks. QRL has transitioned from PoW to PoS consensus mechanism, making it more energy-efficient. It also provides a secure and user-friendly wallet for users to store and manage their QRL tokens.

Another important blockchain project is Mochimo (https://mochimo.org) that is designed to be resistant to quantum attacks and ensure the security of the blockchain even in the face of potential advances in quantum computing. The important aspect of this blockchain's network structure is that it uses Winternitz One-Time Signature Plus (WOTS+) Quantum Resistant Security approved by the EU funded PQCrypto research organization. This is a hash-based signature scheme that provides quantum resilience. WOTS+ plays a crucial role in Mochimo's quantum-proof design. As a single-use signature scheme, it ensures security by ensuring that each signature is valid for a single use. Built on cryptographic hash functions, WOTS+ provides resistance to quantum attacks, as no existing quantum algorithm can efficiently invert these hash functions. Another important feature includes the Trigg Algorithm, an energy-efficient proof-of-work consensus mechanism. Mochimo aims to be truly centralized, focusing on scalability, providing high speed and security, and preventing control from being concentrated in the hands of only a certain group of people.

The Quantum Secure Blockchains (QSB) (https://www.quantumblockchains.io) project within the European Union is also another considerable development that utilizes three critical elements of quantum cryptography which are QKD, quantum random number generators, and post-quantum cryptography. A special device is designed for post-quantum key distribution and a quantum entropy generator which functions as a QKD emulator, boasting a security level consistent with NIST-recommended post-quantum algorithms.

As the quantum computing field evolves, projects like QRL, Mochimo that prioritize quantum-proof designs will become increasingly crucial for ensuring the continued security and integrity of digital assets and transactions on blockchain platforms.

4.3 Open-Source Collaborative Efforts and Educational Initiatives

The evolution of quantum-resistant blockchain technologies and post-quantum cryptography approaches involves complex interaction between academia, industry, and government. Each of these organizations plays a critical role, and collaborations can significantly accelerate innovation, adoption, and standardization.

In terms of academia, universities and research institutions around the world should continue to explore new cryptographic methods, develop innovative PQC methods and algorithms, and create efficient quantum-resistant blockchain architectures. Research and development in the field of quantum-safe cryptographic algorithms should be promoted which will enhance more resilient quantum-resistant blockchains. Studies in this field should also include practical applications and the theoretical basis on which these applications are built.

Businesses, startups, and tech giants have the capacity to translate the fundamental work done in academia into practical applications, products, and services that meet market needs and drive technology adoption. Moreover, aware of the importance of continuous innovation, many industry players can invest more actively in the future of PQC and quantum-resistant blockchains by funding research projects, offering scholarships and even creating academic positions to promote collaboration. Moreover, open source development environment can accelerate innovation and problem solving; and also can facilitate efficient academia and industry collaborations and global cooperation.

Moreover, governments should also play a collaborative role in shaping the integration and development of blockchain and cryptographic technologies. They have the authority to create and enforce regulations and ensure that emerging quantum-resistant blockchains technologies uphold the principles of security, privacy, and justice for their citizens. Through international regulatory frameworks, governments can foster an environment of trust necessary for widespread adoption of these technologies. At the same time, the public sector and public services can also be an early adopter by integrating these solutions into various government projects.

Adopting unified strategies that streamline research, development, and deployment can optimize the quantum-blockchain evolution. In particular, standardization bodies such as the US-based National Institute of Standards and Technology provide an important nexus, bringing together academia and industry to create cryptographic approaches and promising protocols. Research consortiums by groups of universities, industry alliances, Public-Private Partnerships (PPP), cross-sector collaborations (i.e., collaboration of companies from different sectors like IT and finance) and joint ventures can play valuable role in evolution and widespread adoption of quantum-blockchain driven systems.

Furthermore, educational initiatives can also be implemented. Since both technologies are still newly developing, educational and training programs can be a valuable step in this field. It would be an appropriate strategy to educate developers, practitioners and all other stakeholders about the threats that may occur in quantum-blockchain integration and the innovative features it may bring, and to raise awareness about application areas that can overcome the difficulties. In this area, cybersecurity professionals play a crucial role as well and should be trained in quantum-safe practices.

By converging on these strategic foundations, the path to integrating quantum computing and blockchain can be realized successfully. Both technologies can leverage each other's strengths and mitigate inherent challenges. Thus, it could form the basis of a secure, decentralized digital future based on quantum computing.

5. CONCLUSION

Characterized by its decentralization, transparency and immutability, the blockchain ecosystem is on the verge of a transformative evolution with the emergence of quantum computing. Blockchain technology and quantum computing appear to have various powerful and compelling effects on each other. In this study, a wide range of powerful features and opportunities that can come to the fore when these two important technologies come together are discussed. It appears that blockchain technology can have positive impacts, from quantum-secure smart contracts and consensus algorithms to fast transaction transfers, cross-chain interoperability. Additionally, blockchain technology as a DLT can make significant contributions in terms of transparent, secure and immutable data in various quantum-oriented applications such as quantum key distribution, multi-party quantum computing, identity management and access control.

While the integration of quantum computing and blockchain offers tremendous potential, like all pioneering initiatives, it also presents a number of challenges and threats, from integration concerns to scalability issues. It can create significant security vulnerabilities, especially in the cryptography field of quantum computing. Cryptographic vulnerabilities of quantum technologies encourages research into post-quantum cryptography (i.e., quantum-safe cryptography, quantum-resistant cryptography) which is explained comprehensively as well. In addition, many important challenges such as quantum 51% attack, centralization issue on blockchain networks by quantum computing, smart contract vulnerabilities, and quantum hardware limitations need to be investigated and optimized. At this point, intellectual property issues and patent trolling should also be taken into consideration.

Considering all this, proactive research, development and collaboration will be vital as the quantum field continues to interact with the blockchain universe. It is necessary to ensure that the current blockchain ecosystem remains resilient and up-to-date in the quantum age. Decentralized Finance (DeFi), especially blockchain-based finance applications that are very popular today, is an important area that will be affected by quantum applications. As blockchain-based finance, De-Fi eliminates intermediaries such as banks and brokers and enables peer-to-peer transactions. The quantum-powered De-Fi space could bring challenges as well as unprecedented efficiencies (Orús et al., 2019; Egger et al., 2020; Lee et al., 2020).

Research and development in the fields of post-quantum cryptography and quantum-resistant blockchain are of great importance. The projects in question can be in many areas, from powerful quantum algorithms that can work in innovative blockchain networks, to quantum-supported smart contracts and consensus mechanisms, to architectures that will not centralize the distributed ledger structure in blockchain systems. It would be beneficial to accelerate collaborative work between academia, industry and governments and further encourage educational initiatives.

REFERENCES

Al-Jaroodi, J., & Mohamed, N. (2019). Blockchain in industries: A survey. *IEEE Access : Practical Innovations, Open Solutions*, *7*, 36500–36515. doi:10.1109/ACCESS.2019.2903554

Allende, M., León, D. L., Cerón, S., Pareja, A., Pacheco, E., Leal, A., Da Silva, M., Pardo, A., Jones, D., Worrall, D. J., Merriman, B., Gilmore, J., Kitchener, N., & Venegas-Andraca, S. E. (2023). Quantum-resistance in blockchain networks. *Scientific Reports*, *13*(1), 5664. doi:10.103841598-023-32701-6 PMID:37024656

Broadbent, A., & Schaffner, C. (2016). Quantum cryptography beyond quantum key distribution. *Designs, Codes and Cryptography*, *78*(1), 351–382. doi:10.100710623-015-0157-4 PMID:32226229

Cointelegraph (2023). *What is blockchain interoperability: A beginner's guide to cross-chain technology*. Coin Telegraph. https://cointelegraph.com/learn/what-is-blockchain-interoperability-a-beginners-guide-to-cross-chain-technology

Crépeau, C., Gottesman, D., & Smith, A. (2002, May). Secure multi-party quantum computation. In *Proceedings of the thiry-fourth annual ACM symposium on Theory of computing* (pp. 643-652). ACM. 10.1145/509907.510000

Cui, W., Dou, T., & Yan, S. (2020, July). Threats and opportunities: blockchain meets quantum computation. In *2020 39th Chinese control conference (CCC)* (pp. 5822-5824). IEEE. 10.23919/CCC50068.2020.9189608

Dadhich, M., & Tiwari, H. (2022). Quantum Blockchain for Smart Society: Applications, Challenges, and Opportunities. *Advancements in Quantum Blockchain With Real-Time Applications*, 178-198.

Diamanti, E., Lo, H. K., Qi, B., & Yuan, Z. (2016). Practical challenges in quantum key distribution. *npj Quantum Information, 2*(1), 1-12.

Egger, D. J., Gambella, C., Marecek, J., McFaddin, S., Mevissen, M., Raymond, R., Simonetto, A., Woerner, S., & Yndurain, E. (2020). Quantum computing for finance: State-of-the-art and future prospects. *IEEE Transactions on Quantum Engineering, 1*, 1–24. doi:10.1109/TQE.2020.3030314

Faruk, M. J. H., Tahora, S., Tasnim, M., Shahriar, H., & Sakib, N. (2022, May). A review of quantum cybersecurity: threats, risks and opportunities. In *2022 1st International Conference on AI in Cybersecurity (ICAIC)* (pp. 1-8). IEEE.

Fedorov, A. K., Kiktenko, E. O., & Lvovsky, A. I. (2018). Quantum computers put blockchain security at risk. *Nature, 563*(7732), 465–467. doi:10.1038/d41586-018-07449-z PMID:30451981

Fernández-Caramés, T. M., & Fraga-Lamas, P. (2018). A review on the use of blockchain for the internet of things. *IEEE Access: Practical Innovations, Open Solutions, 6*, 32979–33001. doi:10.1109/ACCESS.2018.2842685

Fernandez-Carames, T. M., & Fraga-Lamas, P. (2020). *Towards post-quantum blockchain: A review on blockchain cryptography resistant to quantum computing attacks*. IEEE.

Gill, S. S., Kumar, A., Singh, H., Singh, M., Kaur, K., Usman, M., & Buyya, R. (2022). Quantum computing: A taxonomy, systematic review and future directions. *Software, Practice & Experience, 52*(1), 66–114. doi:10.1002pe.3039

Guo, H., & Yu, X. (2022). A survey on blockchain technology and its security. *Blockchain: research and applications, 3*(2), 100067.

Gyongyosi, L., & Imre, S. (2019). A survey on quantum computing technology. *Computer Science Review, 31*, 51–71. doi:10.1016/j.cosrev.2018.11.002

Kearney, J. J., & Perez-Delgado, C. A. (2021). Vulnerability of blockchain technologies to quantum attacks. *Array (New York, N.Y.), 10*, 100065. doi:10.1016/j.array.2021.100065

Khalid, Z. M., & Askar, S. (2021). Resistant Blockchain cryptography to quantum computing attacks. *International Journal of Science and Business, 5*(3), 116–125.

Khan, S. N., Loukil, F., Ghedira-Guegan, C., Benkhelifa, E., & Bani-Hani, A. (2021). Blockchain smart contracts: Applications, challenges, and future trends. *Peer-to-Peer Networking and Applications, 14*(5), 2901–2925. doi:10.100712083-021-01127-0 PMID:33897937

Kiktenko, E. O., Pozhar, N. O., Anufriev, M. N., Trushechkin, A. S., Yunusov, R. R., Kurochkin, Y. V., Lvovsky, A. I., & Fedorov, A. K. (2018). Quantum-secured blockchain. *Quantum Science and Technology*, *3*(3), 035004. doi:10.1088/2058-9565/aabc6b

Kim, J. W. (2020). Blockchain technology and its applications: Case studies. *Journal of System and Management Sciences*, *10*(1), 83–93.

Lee, R. S., & Lee, R. S. (2020). Future trends in quantum finance. Quantum Finance: Intelligent Forecast and Trading Systems, 399-405.

Leng, J., Zhou, M., Zhao, J. L., Huang, Y., & Bian, Y. (2020). Blockchain security: A survey of techniques and research directions. *IEEE Transactions on Services Computing*, *15*(4), 2490–2510.

Li, X., Jiang, P., Chen, T., Luo, X., & Wen, Q. (2020). A survey on the security of blockchain systems. *Future Generation Computer Systems*, *107*, 841–853.

Mailloux, L. O., Grimaila, M. R., Hodson, D. D., Baumgartner, G., & McLaughlin, C. (2015). Performance evaluations of quantum key distribution system architectures. *IEEE Security and Privacy*, *13*(1), 30–40.

Mohanta, B. K., Panda, S. S., & Jena, D. (2018, July). An overview of smart contract and use cases in blockchain technology. In 2018 9th international conference on computing, communication and networking technologies (ICCCNT) (pp. 1-4). IEEE.

Nakamato, S. (2008). *Bitcoin: A Peer-to-Peer Electronic Cash System* [online].

Orús, R., Mugel, S., & Lizaso, E. (2019). Quantum computing for finance: Overview and prospects. *Reviews in Physics*, *4*, 100028.

Panarello, A., Tapas, N., Merlino, G., Longo, F., & Puliafito, A. (2018). Blockchain and iot integration: A systematic survey. *Sensors (Basel)*, *18*(8), 2575.

Savchuk, M. M., & Fesenko, A. V. (2019). Quantum computing: Survey and analysis. *Cybernetics and Systems Analysis*, *55*, 10–21.

Scarani, V., Bechmann-Pasquinucci, H., Cerf, N. J., Dušek, M., Lütkenhaus, N., & Peev, M. (2009). The security of practical quantum key distribution. *Reviews of Modern Physics*, *81*(3), 1301.

Shrivas, M. K., Dean, T. Y., & Brunda, S. S. (2020, January). The disruptive blockchain security threats and threat categorization. In 2020 First International Conference on Power, Control and Computing Technologies (ICPC2T) (pp. 327-338). IEEE.

Shuaib, M., Hassan, N. H., Usman, S., Alam, S., Sam, S. M., & Samy, G. A. N. (2022, July). Effect of quantum computing on blockchain-based electronic health record systems. In 2022 4th International Conference on Smart Sensors and Application (ICSSA) (pp. 179-184). IEEE.

Singh, S., Kumar, A., & Kathuria, M. (2022). Understanding the public, private and consortium consensus algorithms in blockchain technology. *International Journal of Blockchains and Cryptocurrencies*, *3*(3), 269–288.

Thanalakshmi, P., Rishikhesh, A., Marion Marceline, J., Joshi, G. P., & Cho, W. (2023). A Quantum-Resistant Blockchain System: A Comparative. *Analysis Mathematica*, *11*(18), 3947.

Unogwu, O. J., Doshi, R., Hiran, K. K., & Mijwil, M. M. (2022). Introduction to Quantum-Resistant Blockchain. In Advancements in Quantum Blockchain With Real-Time Applications (pp. 36-55). IGI Global.

Wang, L., & Alexander, C. A. (2020). Quantum science and quantum technology: Progress and challenges. *Am. J. Electr. Electron. Eng.*, *8*(2), 43–50.

Wang, L., Shen, X., Li, J., Shao, J., & Yang, Y. (2019). Cryptographic primitives in blockchains. *Journal of Network and Computer Applications*, *127*, 43–58.

Ye, C., Li, G., Cai, H., Gu, Y., & Fukuda, A. (2018, September). Analysis of security in blockchain: Case study in 51%-attack detecting. In 2018 5th International conference on dependable systems and their applications (DSA) (pp. 15-24). IEEE.

Younan, M., Elhoseny, M., Ali, A. A., & Houssein, E. H. (2021, December). Quantum Chain of Things (QCoT): A New Paradigm for Integrating Quantum Computing, Blockchain, and Internet of Things. In 2021 17th International Computer Engineering Conference (ICENCO) (pp. 101-106). IEEE

Zhang, P., Wang, L., Wang, W., Fu, K., & Wang, J. (2021). A blockchain system based on quantum-resistant digital signature. *Security and Communication Networks*, *2021*, 1–13.

Zheng, Z., Xie, S., Dai, H. N., Chen, X., & Wang, H. (2018). Blockchain challenges and opportunities: a survey

Compilation of References

Aakansha, C. Saxena, A., Ojha, D., & Sobti, K. A. (2023). Artificial Intelligence (AI) Centric Model in Metaverse Ecosystem," AI-Based Technologies and Applications in the Era of the Metaverse. (1 Ed.). IGI Global Press. doi:10.4018/978-1-6684-8851-5.ch001

Abd El-Aziz, R. M., Taloba, A. I., & Alghamdi, F. A. (2022). Quantum computing optimization technique for IoT platform using modified deep residual approach. *Alexandria Engineering Journal, 61*(12), 12497–12509. https://www.sciencedirect.com/science/article/pii/S1110016822004094. doi:10.1016/j.aej.2022.06.029

Abd El-Latif, A. A., Abd-El-Atty, B., Hossain, M. S., Rahman, M. A., Alamri, A., & Gupta, B. B. (2018, March). Efficient Quantum Information Hiding for Remote Medical Image Sharing. *IEEE Access : Practical Innovations, Open Solutions, 6*, 21075–21083. doi:10.1109/ACCESS.2018.2820603

Abd El-Latif, A. A., Abd-El-Atty, B., Mehmood, I., Muhammad, K., Venegas-Andraca, S. E., & Peng, J. (2021). Quantum-Inspired Blockchain-Based Cybersecurity: Securing Smart Edge Utilities in IoT-Based Smart Cities. Information Processing &amp. *Information Processing & Management, 58*(4), 102549. doi:10.1016/j.ipm.2021.102549

Abdi, H. (2007). *The Eigen-Decomposition: Eigenvalues and Eigenvectors*. UTD. https://www.utd.edu/

Abir, S. M. A. A., Anwar, A., Choi, J., & Kayes, A. S. M. (2021). IoT-Enabled Smart Energy Grid: Applications and Challenges. *IEEE Access : Practical Innovations, Open Solutions, 9*, 50961–50981. https://ieeexplore.ieee.org/abstract/document/9381850/. doi:10.1109/ACCESS.2021.3067331

Abomhara M. & Kien, G. M. (2015). Vulnerabilities, dangers, intrusions, and assaults related to cyber security and the internet of things. *J. Cyber Secur. Mobil.*

Abrams, D. S., & Lloyd, S. (1998). *A quantum algorithm providing exponential speed increase for finding eigenvalues and eigenvectors*. APS. https://journals.aps.org/prl/abstract/10.1103/PhysRevLett.83.5162

Acampora, G. (2019). Quantum machine intelligence: Launching the first journal in the area of quantum artificial intelligence. *Quantum Machine Intelligence, 1*(1-2), 1–3. https://link.springer.com/article/10.1007/s42484-019-00006-5. doi:10.100742484-019-00006-5

Acharya, R., Aleiner, I., Allen, R., Andersen, T. I., Ansmann, M., Arute, F., & Google Quantum, A. I. (2023). Suppressing quantum errors by scaling a surface code logical qubit. *Nature, 614*(7949), 676–681. doi:10.103841586-022-05434-1 PMID:36813892

Adebiyi, M. O., Fatinikun-Olaniyan, D., Osang, F., & Adebiyi, A. A. (2023). Quantum Theory Approach to Performance Enhancement in Machine Learning. *2023 International Conference on Science, Engineering and Business for Sustainable Development Goals (SEB-SDG)*, Omu-Aran, Nigeria. 10.1109/SEB-SDG57117.2023.10124582

Agarwal, A., & Mohanta, C. (2023). Development, Working and Standardization of Quantum Computing. *Journal of The Institution of Engineers (India): Series B, 104*(2), 523-531. doi:10.1007/s40031-023-00868-1

Agarwal, M. (2023). What Is Quantum gates and circuits. *Advancements in Quantum Computing. Exploring the Feasibility of Quantum Computing for Future Technology Applications.* LinkedIn. https://www.linkedin.com/pulse/what-quantum-gates-circuits-mayank-agarwal

Aggarwal, C. C. (2011). An introduction to social network data analytics. In C. C. Aggarwal (Ed.), *Social network data analytics* (pp. 1–15). Springer. https://link.springer.com/chapter/10.1007/978-1-4419-8462-3_1 doi:10.1007/978-1-4419-8462-3_1

Aggarwal, C. C. (2018). Big Data in Healthcare: Challenges, Opportunities, and Future Directions. *ACM Transactions on Management Information Systems, 9*(4), 1–7. https://www.sciencedirect.com/science/article/pii/S1474034616301938

Aggarwal, L., Sachdeva, S., & Goswami, P. (2023). Quantum healthcare computing using precision based granular approach. *Applied Soft Computing, 144*, 110458. doi:10.1016/j.asoc.2023.110458

Ahnefeld, F., Theurer, T., Egloff, D., Matera, J. M., & Plenio, M. B. (2022). *On the Role of Coherence in Shor's Algorithm.* arXiv preprint arXiv:2203.10632.

Aithal, P. (2023). Advances and new research opportunities in quantum computing technology by integrating it with other ICCT underlying technologies. *International Journal of Case Studies in Business* [IJCSBE]. *IT and Education, 7*(3), 314–358. doi:10.47992/IJCSBE.2581.6942.0304

Ajagekar, A., Humble, T., & You, F. (2020). Quantum computing based hybrid solution strategies for large-scale discrete-continuous optimization problems. *Computers & Chemical Engineering, 132*, 106630. doi:10.1016/j.compchemeng.2019.106630

Ajtai, M. (1996, July). Generating hard instances of lattice problems. In *Proceedings of the twenty-eighth annual ACM symposium on Theory of computing* (pp. 99-108). ACM. https://dl.acm.org/doi/pdf/10.1145/237814.237838

Akdemir, Z., Menekse, M., Hosseini, M., Nandi, A., & Furuya, K. (2021). Introducing quantum key distribution to high school students. *Science Teacher (Normal, Ill.), 88*, 44–51. https://www.researchgate.net/profile/Muhsin-Menekse-2/publication/349042469_Introducing_Quantum_Key_Distribution_to_High_School_Students/links/601c3b8392851c4ed54bc973/Introducing-Quantum-Key-Distribution-to-High-School-Students.pdf

Aktar, T. (2022). A Survey on IoT-Enabled Smart Grids: Technologies, Architectures, Applications, and Challenges. *sustainability, 15*(1), 717. https://www.mdpi.com/2071-1050/15/1/717

Alagic, G., Alperin-Sheriff, J., Apon, D., Cooper, D., Dang, Q., Kelsey, J., & Smith-Tone, D. (2020). Status report on the second round of the NIST post-quantum cryptography standardization process. *US Department of Commerce, NIST, 2.* https://rnbee.kiev.ua/goto/https://nvlpubs.nist.gov/nistpubs/ir/2020/NIST.IR.8309.pdf

Alam, M., Moroni, D., Pieri, G., Tampucci, M., Gomes, M., Fonseca, J., & Ferreira, J. (2018). Real-Time Smart Parking Systems Integration in Distributed ITS for Smart Cities. *Journal of Advanced Transportation,* 1–13. doi:10.1155/2018/1485652

AlexanderR.KanadeG.LiW.YangL.GanzhornM.PetrasG.CandelS. (2022). Qiskit: An open-source framework for quantum computing. Zenodo. https://joss.theoj.org/papers/10.21105/joss.05329.pdf

Al-Jaroodi, J., & Mohamed, N. (2019). Blockchain in industries: A survey. *IEEE Access : Practical Innovations, Open Solutions*, *7*, 36500–36515. doi:10.1109/ACCESS.2019.2903554

Allende, M., León, D. L., Cerón, S., Pareja, A., Pacheco, E., Leal, A., Da Silva, M., Pardo, A., Jones, D., Worrall, D. J., Merriman, B., Gilmore, J., Kitchener, N., & Venegas-Andraca, S. E. (2023). Quantum-resistance in blockchain networks. *Scientific Reports*, *13*(1), 5664. doi:10.103841598-023-32701-6 PMID:37024656

Al-Saggaf, A. A., Sheltami, T., Alkhzaimi, H., & Ahmed, G. (2022). Lightweight Two-Factor-Based User Authentication Protocol for IoT-Enabled Healthcare Ecosystem in Quantum Computing. *Arabian Journal for Science and Engineering*, 1–11. PMID:36164325

Alsuwian, T., Shahid Butt, A., & Amin, A. A. (2022). Smart Grid Cyber Security Enhancement: Challenges and Solutions—A Review. *Sustainability (Basel)*, *14*(21), 14226. https://www.mdpi.com/2071-1050/14/21/14226. doi:10.3390u142114226

Alvarez-Rodriguez, U., Sanz, M., Lamata, L., & Solano, E. (2018). Quantum artificial life in an IBM quantum computer. *Scientific Reports*, *8*(1), 14793. doi:10.103841598-018-33125-3 PMID:30287854

Ambika, N. (2020). Improved Methodology to Detect Advanced Persistent Threat Attacks. In Quantum Cryptography and the Future of Cyber Security. US: IGI Global. doi:10.4018/978-1-7998-2253-0.ch009

Ambika, N. (2022). Aiding IoT and Cloud to Control COVID-19: A Systematic Approach. In Husain M.S., Adnan M.H.B.M., Khan M.Z., Shukla S. and Khan F.U. (eds), Pervasive Healthcare, 349-365. Cham.: Springer. doi:10.1007/978-3-030-77746-3_21

Ambika, N. (2022). An Economical Machine Learning Approach for Anomaly Detection in IoT Environment. In S. A. Suresh, Y. Vimal, Harold Robinson, Dhinesh Kumar, R. Ramaswami and Udendhran (eds.), Bioinformatics and Medical Applications: Big Data Using Deep Learning Algorithms, 215-234. Hoboken, New Jersey: Wiley Publications. doi:10.1002/9781119792673.ch11

American Heritage. (2016). *The American Heritage® Dictionary of the English Language, Fifth Edition by the Editors of the American Heritage Dictionaries.* Houghton Mifflin Harcourt Publishing Company. https://www.ahdictionary.com/

Andreadou, N., Guardiola, M. O., & Fulli, G. (2016). Telecommunication Technologies for Smart Grid Projects with Focus on Smart Metering Applications. *Energies*, *9*(5), 375. https://www.mdpi.com/1996-1073/9/5/375. doi:10.3390/en9050375

Anh, P. T. N. (2024). AI Models for Disease Diagnosis and Prediction of Heart Disease with Artificial Neural Networks. Computer Vision and AI-integrated IoT Technologies in Medical Ecosystem. CRC Press. https://doi.org/ doi:10.1201/9781003429609-9

AQC. (2019). Advanced Quantum Computing: Interference and Entanglement. *Quantum Computing*. World Scientific (Europe). https://ieeexplore.ieee.org/abstract/document/8528677/

Ardehali, M., Chau, H. F., & Lo, H. K. (1998). Efficient quantum key distribution. *arXiv preprint quant-ph/9803007*. https://link.springer.com/article/10.1007/s00145-004-0142-y

Arner, D. W., Barberis, J., & Buckley, R. P. (2015). The evolution of Fintech: A new post-crisis paradigm. *SSRN, 47*, 1271–1285. https://www.zbw.eu/econis-archiv/handle/11159/270063. doi:10.2139srn.2676553

Arpita, N., Patnaik, A., & Satpathy, I. B. (2023). Incorporating Artificial Intelligence (AI) for Precision Medicine: A Narrative Analysis. AI and IoT-Based Technologies for Precision Medicine. IGI Global Press. doi:10.4018/979-8-3693-0876-9.ch002

Artur, K. (1991). Ekert, Quantum cryptography based on Bell's theorem. *Phys. Rev. Lett. 67*, 661. https://journals.aps.org/prl/abstract/10.1103/PhysRevLett.67.661

Arute, F. (2021). Quantum approximate optimization of non-planar graph problems on a planar superconducting processor. *Nature Physics*, *17*(3), 340–345. https://www.nature.com/articles/s41567-020-01105-x

Arute, F., Arya, K., Babbush, R., Bacon, D., Bardin, J. C., Barends, R., & Chen, Y. (2019). Quantum supremacy using a programmable superconducting processor. *Nature*, *574*(7779), 505–510. https://www.nature.com/articles/s41586%20019%201666%205. doi:10.103841586-019-1666-5 PMID:31645734

Arya, K., Arute, F., & Babbush, R. (2019). Employing a programmable superconducting processor to achieve quantum dominance. *Nature.*

Askr, H., Elgeldawi, E., Aboul Ella, H., Elshaier, Y. A. M. M., Gomaa, M. M., & Hassanien, A. E. (2023). Deep learning in drug discovery: An integrative review and future challenges. *Artificial Intelligence Review*, *56*(7), 5975–6037. doi:10.100710462-022-10306-1 PMID:36415536

AtomicStructure. (2023). *Atomic Structure - Discovery of Subatomic Particles*. JEEIIT, JEE, Study Material. https://byjus.com/jee/atomic-structure/

Avanzi, R., Bos, J., Ducas, L., Kiltz, E., Lepoint, T., Lyubashevsky, V., & Stehlé, D. (2017). Crystals-kyber. *NIST, Tech. Rep.* https://csrc.nist.gov/csrc/media/Presentations/2022/crystals-kyber-update/images-media/session-1-schwabe-crystals-kyber-pqc2022.pdf

Avanzi, R., Bos, J., Ducas, L., Kiltz, E., Lepoint, T., Lyubashevsky, V., & Stehlé, D. (2019). CRYSTALS-Kyber algorithm specifications and supporting documentation. *NIST PQC Round, 2*(4), 1-43. https://pq-crystals.org/kyber/data/kyber-specification-round3-20210131.pdf

Avramouli, M., Savvas, I. K., Vasilaki, A., & Garani, G. (2023). Unlocking the potential of quantum machine learning to advance drug discovery. *Electronics (Basel)*, *12*(11), 2402. doi:10.3390/electronics12112402

Ayanzadeh, R. (2018, February). Quantum artificial intelligence for natural language processing applications. In *Proceedings of the 49th ACM Technical Symposium on Computer Science Education* (pp. 273-273). ACM. https://dl.acm.org/doi/abs/10.1145/3159450.3162338

Ayoade, O., Rivas, P., & Orduz, J. (2022). Artificial intelligence computing at the quantum level. *Data*, *7*(3), 28. doi:10.3390/data7030028

Ayodele, T. O. (2010). Types of machine learning algorithms. *New advances in machine learning, 3*, 19-48.

Azad, U., Behera, B. K., Ahmed, E. A., Panigrahi, P. K., & Farouk, A. (2022). Solving Vehicle Routing Problem Using Quantum Approximate Optimization Algorithm. *IEEE Transactions on Intelligent Transportation Systems*. doi:10.1109/TITS.2022.3172241

Azouit, K., Azouit, A., El Housni, H., & El Hassouni, M. (2021). A Quantum Secure Direct Communication Protocol. In *2021 International Conference on Wireless Technologies, Embedded and Intelligent Systems (WITS)* (pp. 1-6). IEEE.

Badhwar, R. (2021). *The CISO's Next Frontier: AI, Post-Quantum Cryptography and Advanced Security Paradigms*. Springer. https://link.springer.com/chapter/10.1007/978-3-031-22032-6_7 doi:10.1007/978-3-030-75354-2

Bagozzi, R. P. (2007). The legacy of the technology Acceptance model and a proposal for a paradigm shift. *Journal of the Association for Information Systems*, *8*(4), 3. https://aisel.aisnet.org/jais/vol8/iss4/3/. doi:10.17705/1jais.00122

Baiardi, A., Christandl, M., & Reiher, M. (2023). Quantum computing for molecular biology. *ChemBioChem*, *24*(13), e202300120. doi:10.1002/cbic.202300120 PMID:37151197

Baldi, M., Bianchi, M., Chiaraluce, F., Rosenthal, J., & Schipani, D. (2016). Enhanced public key security for the McEliece cryptosystem. *Journal of Cryptology*, *29*(1), 1–27. https://link.springer.com/article/10.1007/s00145-014-9187-8. doi:10.100700145-014-9187-8

Balynsky, M., Chiang, H., Gutierrez, D., Kozhevnikov, A., Filimonov, Y., & Khitun, A. (2021). Quantum computing without quantum computers: Database search and data processing using classical wave superposition. *Journal of Applied Physics*, *130*(16), 164903. https://pubs.aip.org/aip/jap/article/130/16/164903/568206. doi:10.1063/5.0068316

Bano, I., Butt, U. D., & Mohsan, S. A. H. (2023). New challenges in drug discovery. In S. Das, S. Thomas, & P. P. Das (Eds.), *Novel Platforms for Drug Delivery Applications* (pp. 619–643). Woodhead Publishing. doi:10.1016/B978-0-323-91376-8.00021-5

Barbier, G., & Liu, H. (2011). Data mining in social media. In C. C. Aggarwal (Ed.), *Social network data analytics* (pp. 327–352). Springer. https://link.springer.com/chapter/10.1007/978-1-4419-8462-3_12 doi:10.1007/978-1-4419-8462-3_12

Barkoutsos, P. K., Nannicini, G., Robert, A., Tavernelli, I., & Woerner, S. (2020). Improving variational quantum optimization using CVaR. *Quantum*, *4*, 256. doi:10.22331/q-2020-04-20-256

Basak, U. P., Faruk, J. H., Nazim, M., Masum, M., Hossain, M. S., Uddin, G., Barzanjeh, S., Sheikh, I. A., & Rahman, A. (2022). *Evolution of Quantum Computing: A Systematic Survey on the Use of Quantum Computing Tools*. IEEE. https://ieeexplore.ieee.org/abstract/document/9842601/

Batra, K., Zorn, K. M., Foil, D. H., Minerali, E., Gawriljuk, V. O., Lane, T. R., & Ekins, S. (2021). Quantum Machine Learning Algorithms for Drug Discovery Applications. *Journal of Chemical Information and Modeling*, *61*(6), 2641–2647. doi:10.1021/acs.jcim.1c00166 PMID:34032436

Beach, G., Lomont, C., & Cohen, C. (2003). Quantum image processing (QuIP). Proc. - Appl. Imag. Pattern Recognit. IEEE. doi:10.1109/AIPR.2003.1284246

Beaver, D., Kumar, S., Li, H. C., Sobel, J., & Vajgel, P. (2010). Finding a needle in haystack: Facebook's photo storage. In *Proceedings of the nineth USENIX conference on operating systems design and implementation* (pp. 1–8). USENIX Association. https://www.usenix.org/event/osdi10/tech/full_papers/Beaver.pdf

Bechmann-Pasquinucci, H., & Gisin, N. (1999). Incoherent and coherent eavesdropping in the six-state protocol of quantum cryptography. *Physical Review A*, *59*(6), 4238–4248. https://journals.aps.org/pra/abstract/10.1103/PhysRevA.59.4238. doi:10.1103/PhysRevA.59.4238

Bechmann-Pasquinucci, H., & Peres, A. (2000). Quantum cryptography with 3-state systems. *Physical Review Letters, 85*(15), 3313–3316. https://journals.aps.org/prl/abstract/10.1103/PhysRevLett.85.3313. doi:10.1103/PhysRevLett.85.3313 PMID:11019329

Bechmann-Pasquinucci, H., & Tittel, W. (2000). Quantum cryptography using larger alphabets. *Physical Review A, 61*(6), 062308. https://journals.aps.org/pra/abstract/10.1103/PhysRevA.61.062308. doi:10.1103/PhysRevA.61.062308

BedÃn-Molina, J., Lopez, M., & Derpich, I. (2020). A home-based smart health model. *Advances in Mechanical Engineering, 12*(6), 168781402093528–. . doi:10.1177/1687814020935282

Bell, J. S. (1966). On the problem of hidden variables in quantum mechanics. *Reviews of Modern Physics, 38*(3), 447–452. https://link.springer.com/chapter/10.1007/978-3-0348-9259-9_21. doi:10.1103/RevModPhys.38.447

Bennett, C. H., & Brassard, G. (1984). IEEE Int. Conf. on Computers, Systems and Signal Processing. *IEEE.* https://iopscience.iop.org/article/10.1088/1367-2630/4/1/343/meta

Bennett, C. H. (1992). Quantum cryptography using any two nonorthogonal states. *Physical Review Letters, 68*(21), 3121–3124. https://journals.aps.org/prl/abstract/10.1103/PhysRevLett.68.3121. doi:10.1103/PhysRevLett.68.3121 PMID:10045619

Bennett, C. H., Brassard, G., & Mermin, N. D. (1992). Quantum cryptography without Bell's theorem. *Physical Review Letters, 68*(5), 557–559. https://journals.aps.org/prl/abstract/10.1103/PhysRevLett.68.557. doi:10.1103/PhysRevLett.68.557 PMID:10045931

Bernstein, D. J., Chuengsatiansup, C., Lange, T., & Van Vredendaal, C. (2016). NTRU Prime. *IACR Cryptol. ePrint Arch., 2016*, 461. http://hyperelliptic.org/tanja/vortraege/caen-ntruprime.pdf

Bernstein, D. J., Chuengsatiansup, C., Lange, T., & van Vredendaal, C. (2018). NTRU prime: reducing attack surface at low cost. In *Selected Areas in Cryptography–SAC 2017: 24th International Conference, Ottawa, ON, Canada*. Springer International Publishing. https://link.springer.com/chapter/10.1007/978-3-319-72565-9_12

Bernstein, D. J., Chuengsatiansup, C., Lange, T., & van Vredendaal, C. (2019). NTRU Prime: round 2. *Submission to the NIST PQC standardization process.* University of Adelaide. https://ntruprime.cr.yp.to

Bernstein, D. J., Lange, T., & Peters, C. (2008). Attacking and defending the McEliece cryptosystem. In *Post-Quantum Cryptography: Second International Workshop,* (pp. 31-46). Springer Berlin Heidelberg. https://link.springer.com/chapter/10.1007/978-3-540-88403-3_3

Bernstein, D. J. (2009). Introduction to post-quantum cryptography. In *Post-quantum cryptography* (pp. 1–14). Springer Berlin Heidelberg. https://link.springer.com/chapter/10.1007/978-3-540-88702-7_1 doi:10.1007/978-3-540-88702-7_1

Bethel, E., Amankwah, M. G., Balewski, J., Van Beeumen, R., Camps, D., Huang, D., & Perciano, T. (2023). Quantum computing and visualization: A disruptive technological change ahead. *arXiv preprint arXiv:2310.04937, 43*(6), 1-11. doi:/arXiv.2310.04937 doi:10.48550

Bharti, K., Cervera-Lierta, A., Kyaw, T. H., Haug, T., Alperin-Lea, S., Anand, A., & Guerreschi, G. G. (2022). Noisy intermediate-scale quantum algorithms. *Reviews of Modern Physics, 94*(1), 015004. https://journals.aps.org/rmp/abstract/10.1103/RevModPhys.94.015004. doi:10.1103/RevModPhys.94.015004

Bhat, H. A., Khanday, F. A., Kaushik, B. K., Bashir, F., & Shah, K. A. (2022). Quantum Computing: Fundamentals, Implementations and Applications. *IEEE Open Journal of Nanotechnology*, *3*, 61–77. doi:10.1109/OJNANO.2022.3178545

Bhatia, V., & Ramkumar, K. R. (2020). An Efficient Quantum Computing technique for cracking RSA using Shor's Algorithm. *2020 IEEE 5th International Conference on Computing Communication and Automation, ICCCA 2020.* Institute of Electrical and Electronics Engineers Inc. 10.1109/ICCCA49541.2020.9250806

Bhatia, A. S., Saggi, M. K., & Kais, S. (2023). Quantum machine learning predicting ADME-Tox properties in drug discovery. *Journal of Chemical Information and Modeling*, acs.jcim.3c01079. doi:10.1021/acs.jcim.3c01079 PMID:37603536

Bhattacharyya, R., Banerjee, M., Heiblum, M., Mahalu, D., & Umansky, V. (2019). Melting of interference in the fractional quantum Hall Effect: Appearance of neutral modes. *Physical Review Letters*, *122*(24), 246801. https://journals.aps.org/prl/abstract/10.1103/PhysRevLett.122.246801. doi:10.1103/PhysRevLett.122.246801 PMID:31322402

Bhavin, M., Tanwar, S., Sharma, N., Tyagi, S., & Kumar, N. (2021). Blockchain and quantum blind signature-based hybrid scheme for healthcare 5.0 applications. *Journal of Information Security and Applications*, *56*, 102673. doi:10.1016/j.jisa.2020.102673

Biamonte, J., Wittek, P., Pancotti, N., Rebentrost, P., Wiebe, N., & Lloyd, S. (2017). Quantum machine learning. *Nature*, *549*(7671), 195–202. https://www.nature.com/articles/nature23474. doi:10.1038/nature23474 PMID:28905917

Biham, E., Boyer, M., Boykin, P. O., Mor, T., & Roychowdhury, V. (2000, May). A proof of the security of quantum key distribution. In *Proceedings of the thirty-second annual ACM symposium on Theory of computing* (pp. 715-724). ACM. https://dl.acm.org/doi/pdf/10.1145/335305.335406

Bikmetov, R., Raja, M. Y. A., Kazi, K., Chowdhury, B., & Enslin, J. (2015). *Dynamic prediction capabilities of Smart Metering Infrastructure.* Proceedings of the 2015 North American Power Symposium (NAPS), Charlotte, NC, USA. https://ieeexplore.ieee.org/abstract/document/7335235/

Billewicz, K. (2016). Possibility of Internet of things technology implementation in smart power grids. *Energetyka (Warszawa)*, *5*, 264–270. https://www.infona.pl/resource/bwmeta1.element.baztech-8da0dc88-22bc-4963-bf28-d034e1a32d8d

Bisheh-Niasar, M., Azarderakhsh, R., & Mozaffari-Kermani, M. (2021). Instruction-set accelerated implementation of CRYSTALS-Kyber. *IEEE Transactions on Circuits and Systems. I, Regular Papers*, *68*(11), 4648–4659. https://ieeexplore.ieee.org/abstract/document/9525069/. doi:10.1109/TCSI.2021.3106639

Boger, Y. (2023, July 24). *Quantum computing and ai: A leap forward or a distant dream?* Datanami. https://www.datanami.com/2023/07/24/quantum-computing-and-ai-a-leap-forward-or-a-distant-dream/

Bonde, B., Patil, P., & Choubey, B. (2024). The Future of Drug Development with Quantum Computing. In A. Heifetz (Ed.), *High Performance Computing for Drug Discovery and Biomedicine* (pp. 153–179). Springer US., doi:10.1007/978-1-0716-3449-3_7

Borlase, S. (2017). *Smart Grids: Advanced Technologies and Solutions.* CRC Press. https://www.google.com/books?hl=en&lr=&id=VD4PEAAAQBAJ&oi=fnd&pg=PP1&dq=Smart+Grids:+Advanced+Technologies+and+Solutions.+CRC+Press&ots=uKXRz1EEFq&sig=lRRoKWYshe0alojZ0RlSvGwU4js

Bos, J. W., Ofner, M., Renes, J., Schneider, T., & van Vredendaal, C. (2021). The matrix reloaded: Multiplication strategies in FrodoKEM. In *Cryptology and Network Security: 20th International Conference, CANS 2021, Vienna, Austria, December 13-15, 2021, Proceedings 20* (pp. 72-91). Springer International Publishing. https://link.springer.com/chapter/10.1007/978-3-030-92548-2_5

Bos, J., Costello, C., Ducas, L., Mironov, I., Naehrig, M., Nikolaenko, V., & Stebila, D. (2016, October). Frodo: Take off the ring! Practical, quantum-secure key exchange from LWE. In *Proceedings of the 2016 ACM SIGSAC conference on computer and communications security* (pp. 1006-1018). ACM. https://dl.acm.org/doi/abs/10.1145/2976749.2978425

Bos, J., Ducas, L., Kiltz, E., Lepoint, T., Lyubashevsky, V., Schanck, J. M., & Stehlé, D. (2018, April). CRYSTALS-Kyber: a CCA-secure module-lattice-based KEM. In *2018 IEEE European Symposium on Security and Privacy (EuroS&P)* (pp. 353-367). IEEE. https://ieeexplore.ieee.org/abstract/document/8406610/

Boskovic, M., Becker, S., & Dhama, A. (2006). Trustworthy Software Systems: Definition and Discussion of Basic Terms. *ACM SIGSOFT Softw. Eng. Notes, 31*(1).

Botsinis, P., Alanis, D., Babar, Z., Nguyen, H. V., Chandra, D., Ng, S. X., & Hanzo, L. (2019, April). Quantum Search Algorithms for Wireless Communications. *IEEE Communications Surveys and Tutorials, 21*(2), 1209–1242. doi:10.1109/COMST.2018.2882385

Boucher, P. (2019). How artificial intelligence works. EPRS–European Parliamentary Research Service, In: europarl. Europa, 2.

Bourennane, M., Karlsson, A., & Björk, G. (2001). Quantum key distribution using multilevel encoding. *Physical Review A, 64*(1), 012306. https://journals.aps.org/pra/abstract/10.1103/PhysRevA.64.012306. doi:10.1103/PhysRevA.64.012306

Bova, F., Goldfarb, A., & Melko, R. (2021). Commercial applications of quantum computing. *EPJ Quantum Technol.* . doi:10.1140/epjqt/s40507-021-00091-1

Brijwani, G. N., Ajmire, P. E., & Thawani, P. V. (2023). Future of Quantum Computing in Cyber Security. In *Handbook of Research on Quantum Computing for Smart Environments* (pp. 267–298). IGI Global., doi:10.4018/978-1-6684-6697-1.ch016

Broadbent, A., & Schaffner, C. (2016). Quantum cryptography beyond quantum key distribution. *Designs, Codes and Cryptography, 78*(1), 351–382. doi:10.100710623-015-0157-4 PMID:32226229

BroughtonM.VerdonG.McCourtT.MartinezA. J.YooJ. H.IsakovS. V. (2003). TensorFlow Quantum: A Software Framework for Quantum Machine Learning . arXiv. https://arxiv.org/abs/2003.02989

Brummer, D. (2023). Quantum Computing and Artificial Intelligence. A Paradigm Shift for Investors. *OurCrowd.* https://www.ourcrowd.com/learn/quantum-computing-and-artificial-intelligence

Bruss, D., Erdélyi, G., Meyer, T., Riege, T., & Rothe, J. (2007). Quantum cryptography: A survey. *ACM Computing Surveys (CSUR), 39*(2), 6-es. https://dl.acm.org/doi/abs/10.1145/1242471.1242474

Bruß, D. (1998). Optimal eavesdropping in quantum cryptography with six states. *Physical Review Letters, 81*(14), 3018–3021. https://search.proquest.com/openview/63a2b9c913676ba69db4d45e5ba03e08/1?pq-origsite=gscholar&cbl=2026366&diss=y. doi:10.1103/PhysRevLett.81.3018

Bužek, V., & Hillery, M. (1996). Quantum copying: Beyond the no-cloning theorem. *Physical Review A*, *54*(3), 1844–1852. https://journals.aps.org/pra/abstract/10.1103/PhysRevA.54.1844. doi:10.1103/PhysRevA.54.1844 PMID:9913670

Campagne-Ibarcq, P., Eickbusch, A., Touzard, S., Zalys-Geller, E., Frattini, N. E., Sivak, V., & Schoelkopf, R. J. (2020). Quantum error correction of a qubit encoded in grid states of an oscillator. *Nature*, *584*(7821), 368–372. https://www.nature.com/articles/s41586-020-2603-3. doi:10.103841586-020-2603-3 PMID:32814889

Campbell, P., & Abhyankar, S. (1978). *Fractals, Form, Chance and Dimension*. Springer. https://link.springer.com/chapter/10.1007/978-3-642-70880-0_9 doi:10.1007/BF03023043

Canabarro, A., Mendonc, T. M., Nery, R., Moreno, G., Albino, A. S., de Jesus, G. F., & Chaves, R. (2022). *Quantum Finance: a tutorial on quantum computing applied to the financial market*. arXiv. org Papers, (2208.04382). https://ideas.repec.org/p/arx/papers/2208.04382.html

Cao, Y., Guerreschi, G. G., & Aspuru-Guzik, A. (2017). *Quantum neuron: an elementary building block for machine learning on quantum computers*. arXiv preprint arXiv:1711.11240

Cao, B., Fan, S., Zhao, J., Yang, P., Muhammad, K., & Tanveer, M. (2020). Quantum-enhanced multiobjective large-scale optimization via parallelism. *Swarm and Evolutionary Computation*, *57*, 100697. doi:10.1016/j.swevo.2020.100697

Cao, Y., Romero, J., Olson, J. P., Degroote, M., Johnson, P. D., Kieferová, M., & Aspuru-Guzik, A. (2019). Quantum chemistry in the age of quantum computing. *Chemical Reviews*, *119*(19), 10856–10915. https://pubs.acs.org/doi/abs/10.1021/acs.chemrev.8b00803. doi:10.1021/acs.chemrev.8b00803 PMID:31469277

Cárdenas-Martínez, H. M., Laverde-Forero, J. C., & Castang-Montiel, G. (2018). State of art in quantum computing (QC) with open source platforms. *Visión electrónica*, *1*(2), 321-332. http://revistas.udistrital.edu.co:8080/index.php/visele/article/view/18396

Carrascal, G., del Barrio, A. A., & Botella, G. (2021). First experiences of teaching quantum computing. *The Journal of Supercomputing*, *77*(3), 2770–2799. doi:10.100711227-020-03376-x

Cerezo, M., Verdon, G., Huang, H. Y., Cincio, L., & Coles, P. J. (2022). Challenges and opportunities in quantum machine learning. *Nature Computational Science*, *2*(9), 567–576. https://www.nature.com/articles/s43588-022-00311-3. doi:10.103843588-022-00311-3

Chaganti, S. K., Kesari, A. S., & Chowdhury, C. (2024). Execution of a quantum algorithm for calculating energies of some quantum mechanical systems. *Chemical Physics*, *576*, 112082. doi:10.1016/j.chemphys.2023.112082

Chakraborty, C., Bhattacharya, M., Dhama, K., & Lee, S.-S. (2023). Quantum computing on nucleic acid research: Approaching toward next-generation computing. *Molecular Therapy. Nucleic Acids*, *33*, 53–56. doi:10.1016/j.omtn.2023.06.007 PMID:37449046

Chalumuri, A., Kune, R., & Manoj, B. S. (2021). A hybrid classical-quantum approach for multi-class classification. *Quantum Information Processing*, *20*(3), 119. doi:10.100711128-021-03029-9

Charles, H. (1984). *Bennett and Gilles Brassard, International Association for Cryptologic Research, CryptoDB - An Update on Quantum Cryptography*. CRYPTO. doi:10.1007/3-540-39568-7_39

Chatterjee, P., & Chakraborty, R. (2020). A Brief Study on Quantum Computing. *International Journal of Innovative Research in Physics, 1*(4), 58–63. https://www.ingentaconnect.com/content/smart/ijiip/2020/00000001/00000004/art00007. doi:10.15864/ijiip.1407

Chawla, D., & Mehra, P. S. (2023). A Survey on Quantum Computing for Internet of Things Security. *Procedia Computer Science, 218*, 2191–2200. https://www.sciencedirect.com/science/article/pii/S1877050923001953. doi:10.1016/j.procs.2023.01.195

Chen, A., Peng, W., & Gui, T. (2013). *A variant of BB84 Protocol based on quantum phase*. Digital Library. https://digital-library.theiet.org/content/conferences/10.1049/cp.2013.2465

Chen, C. (2019). *A Social Network Based Approach for IoT Device Management and Service Composition*. IEEE. https://ieeexplore.ieee.org/abstract/document/7196496/

Chen, L., Jordan, S. P., Liu, Y. K., Moody, D., Peralta, R., Perlner, R., & Smith-Tone, D. (2016). *Report on post-quantum cryptography*. US Department of Commerce, National Institute of Standards and Technology. https://nvlpubs.nist.gov/nistpubs/ir/2016/nist.ir.8105.pdf

Cheng, H., Dinu, D., Großschädl, J., Rønne, P. B., & Ryan, P. Y. (2020). A lightweight implementation of NTRU Prime for the post-quantum internet of things. In *Information Security Theory and Practice: 13th IFIP WG 11.2 International Conference, WISTP 2019, Paris, France, December 11–12, 2019, Proceedings 13* (pp. 103-119). Springer International Publishing. https://link.springer.com/chapter/10.1007/978-3-030-41702-4_7

Cheng, C., Lu, R., Petzoldt, A., & Takagi, T. (2017, February). Securing the Internet of Things in a Quantum World. *IEEE Communications Magazine, 55*(2), 116–120. doi:10.1109/MCOM.2017.1600522CM

Chen, H., Chiang, R. H. L., & Storey, V. C. (2012). Business intelligence and analytics: From big data to big impact. *Management Information Systems Quarterly, 36*(4), 1165–1188. https://www.jstor.org/stable/41703503. doi:10.2307/41703503

Chen, S. Y.-C., & Yoo, S. (2021). Federated Quantum Machine Learning. *Entropy (Basel, Switzerland), 23*(4), 460. doi:10.3390/e23040460 PMID:33924721

Chen, X., Li, X. Z., & Wang, H. (2010). The Application of Smart Grid in IOT. *J. North China Power, 3*, 50–53. https://www.sciencedirect.com/science/article/pii/S1364032114003761

Choi, J., Oh, S., & Kim, J. (2020). The useful quantum computing techniques for artificial intelligence engineers. In *2020 International Conference on Information Networking (ICOIN),* (pp. 1-3). IEEE. 10.1109/ICOIN48656.2020.9016555

Cicconetti, C., Conti, M., & Passarella, A. (2022, June). Resource allocation in quantum networks for distributed quantum computing. In *2022 IEEE International Conference on Smart Computing (SMARTCOMP)* (pp. 124-132). IEEE. https://ieeexplore.ieee.org/abstract/document/9821091/

Ciliberto, C., Herbster, M., Ialongo, A. D., Pontil, M., Rocchetto, A., Severini, S., & Wossnig, L. (2018). Quantum machine learning: A classical perspective. *Proceedings - Royal Society. Mathematical, Physical and Engineering Sciences, 474*(2209), 20170551. https://royalsocietypublishing.org/doi/abs/10.1098/rspa.2017.0551. doi:10.1098/rspa.2017.0551 PMID:29434508

Clarke, R. B., Chefles, A., Barnett, S. M., & Riis, E. (2001). Experimental demonstration of optimal unambiguous state discrimination. *Physical Review A, 63*(4), 040305. https://journals.aps.org/pra/abstract/10.1103/PhysRevA.63.040305. doi:10.1103/PhysRevA.63.040305

CNOTGates. (2023) *Quantum Computing UK-CNOT gate tutorial with Code.* CNOTGates. https://quantumcomputinguk.org/tutorials/controlled-not-gate-program

CNOTGates. (2023). *Quantum Computing UK-CNOT gate tutorial with Code.* CNOTGates. https://quantumcomputinguk.org/tutorials/controlled-not-gate-program

Cointelegraph (2023). *What is blockchain interoperability: A beginner's guide to cross-chain technology.* Coin Telegraph. https://cointelegraph.com/learn/what-is-blockchain-interoperability-a-beginners-guide-to-cross-chain-technology

Corrado, C. A., & Hulten, C. R. (2010). How do you measure a "technological revolution"? *The American Economic Review, 100*(2), 99–104. https://pubs.aeaweb.org/doi/pdf/10.1257/aer.100.2.99. doi:10.1257/aer.100.2.99

Crépeau, C., Gottesman, D., & Smith, A. (2002, May). Secure multi-party quantum computation. In *Proceedings of the thiry-fourth annual ACM symposium on Theory of computing* (pp. 643-652). ACM. 10.1145/509907.510000

Cruz-Benito, J., Faro, I., Martín-Fernández, F., Therón, R., & García-Peñalvo, F. J. (2018). A Deep-Learning-Based Proposal to Aid Users in Quantum Computing Programming. In P. Zaphiris & A. Ioannou (Eds.), Lecture Notes in Computer Science. *Learning and Collaboration Technologies. Learning and Teaching. LCT 2018.* Springer. doi:10.1007/978-3-319-91152-6_32

Cui, W., Dou, T., & Yan, S. (2020, July). Threats and opportunities: blockchain meets quantum computation. In *2020 39th Chinese control conference (CCC)* (pp. 5822-5824). IEEE. 10.23919/CCC50068.2020.9189608

Cukier, K. (2010). Data, data everywhere: A special report on managing information. *The Economist.* https://www.economist.com/node/15557443

Da Silva, N., dos Santos, J., Gomes, R. L., Lins, R. A., & da Rocha, R. C. (2022). A Requirements Engineering Process for IoT Systems. *International Journal of Engineering Technologies and Management Research, 9.* https://www.academia.edu/download/63178921/Apr-2020-26-1587884210-A2520200502-113660-rma669.pdf

Dadhich, M., & Tiwari, H. (2022). Quantum Blockchain for Smart Society: Applications, Challenges, and Opportunities. *Advancements in Quantum Blockchain With Real-Time Applications*, 178-198.

Dang, V. B., Mohajerani, K., & Gaj, K. (2022). High-speed hardware architectures and FPGA benchmarking of crystals-kyber, ntru, and saber. *IEEE Transactions on Computers, 72*(2), 306–320. https://ieeexplore.ieee.org/abstract/document/9954215/. doi:10.1109/TC.2022.3222954

Dash, S., Priyadarshini, S., Mohanty, S. N., Priyadarshini, S., & Dulla, N. (2023). Quantum computing in netnomy: A networking paradigm in e-pharmaceutical setting. In Evolution and Applications of Quantum Computing (pp. 145-161). doi:10.1002/9781119905172.ch9

David Mermin, N. (2007). *Quantum Computer Science.* Cambridge Press. https://www.google.com/books?hl=en&lr=&id=q2S9APxFdUQC

De Barros, J. A., & Suppes, P. (2009). Quantum mechanics, interference, and the brain. *Journal of Mathematical Psychology, 53*(5), 306–313. https://www.sciencedirect.com/science/article/pii/S002224960900039X. doi:10.1016/j.jmp.2009.03.005

De Leon, N. P., Itoh, K. M., Kim, D., Mehta, K. K., Northup, T. E., Paik, H., Palmer, B. S., Samarth, N., Sangtawesin, S., & Steuerman, D. W. (2021). Materials challenges and opportunities for quantum computing hardware. *Science, 372*(6539), eabb2823. https://www.science.org/doi/abs/10.1126/science.abb2823. doi:10.1126cience.abb2823 PMID:33859004

De Maio, V., & Brandic, I. (2023). Accelerating Scientific Applications with the Quantum Edge: A Drug Design Use Case. In A. Bienz, M. Weiland, M. Baoulin, & C. Kruse (Eds.), *High Performance Computing, ISC High Performance 2023*. Springer Nature Singapore. doi:10.1007/978-3-031-40843-4_11

De Wolf, R. (2017). *The Potential Impact of Quantum Computers on Society*. arXiv. https://arxiv.org/abs/1712.05380

Deekshetha, H. R. & Tyagi, A. (2023). Chapter 16 - Automated and intelligent systems for next-generation-based smart applications. A. Tyagi, & A. Abraham (eds.), Data Science for Genomics. Academic Press. doi:10.1016/B978-0-323-98352-5.00019-7

Dejpasand, M. T., & Ghamsari, M. S. (2023). Research Trends in Quantum Computers by Focusing on Qubits as Their Building Blocks. *Quantum Reports, 5*(3), 597–608. doi:10.3390/quantum5030039

Denchev, V. S., Ding, N., Vishwanathan, S. V. N., & Neven, H. (2012). *Robust classification with adiabatic quantum optimization*. arXiv preprint arXiv:1205.1148.

Deng, F. G., Long, G. L., & Liu, X. S. (2003). Two-step quantum direct communication protocol using the Einstein-Podolsky-Rosen pair block. *Physical Review A, 68*(4), 042317. https://journals.aps.org/pra/abstract/10.1103/PhysRevA.68.042317. doi:10.1103/PhysRevA.68.042317

Deshmukh, A., Patil, D. S., Pawar, P. D., Kumari, S., & P., M. (2023). Recent Trends for Smart Environments With AI and IoT-Based Technologies: A Comprehensive Review. In A. Tyagi (Ed.), *Handbook of Research on Quantum Computing for Smart Environments* (pp. 435-452). IGI Global. doi:10.4018/978-1-6684-6697-1.ch023

Deshpande, S., Xu, C., Nawan, M., Nawaz, K., & Szefer, J. (2022). Fast and Efficient Hardware Implementation of HQC. *Cryptology ePrint Archive*. https://eprint.iacr.org/2022/1183

Deswal, Y., Asija, S., Tufail, A., Dubey, A., Deswal, L., Kumar, N., Saroya, S., Kirar, J. S., & Gupta, N. M. (2023). Instigating the in vitro antidiabetic activity of new tridentate Schiff base ligand appended M (II) complexes: From synthesis, structural characterization, quantum computational calculations to molecular docking, and molecular dynamics simulation studies. *Applied Organometallic Chemistry, 37*(4), e7050. doi:10.1002/aoc.7050

Deutsch, D. (1985). Quantum theory, the Church–Turing principle and the universal quantum computer. *Proceedings of the Royal Society of London. A. Mathematical and Physical Sciences, 400*(1818), 97-117. https://royalsocietypublishing.org/doi/abs/10.1098/rspa.1985.0070

Deutsch, D. (1985). *Quantum-Theory, The Church Turing Principle and the Universal Quantum Computer.* Royal Society Publishing. https://royalsocietypublishing.org/doi/abs/10.1098/rspa.1985.0070

Devitt, S. J. (2021). Building a fault-tolerant quantum computer: The hardware perspective. *npj Quantum Information, 7*(1), 1-12. https://www.nature.com/articles/s41534-021-00397-y

Diamanti, E., Lo, H. K., Qi, B., & Yuan, Z. (2016). Practical challenges in quantum key distribution. *npj Quantum Information, 2*(1), 1-12.

Diebold, F. X. (2012). *A personal perspective on the origin(s) and development of "big data": The phenomenon, the term, and the discipline* (Scholarly Paper No. ID 2202843). Social Science Research Network. http://papers.ssrn.com/sol3/papers.cfm?abstract id=2202843

Dieks, D. G. B. J. (1982). Communication by EPR devices. *Physics Letters. [Part A], 92*(6), 271–272. https://www.sciencedirect.com/science/article/pii/0375960182900846. doi:10.1016/0375-9601(82)90084-6

Diffie, W., & Hellman, M. E. (2022). New directions in cryptography. In *Democratizing Cryptography: The Work of Whitfield Diffie and Martin Hellman* (pp. 365-390). https://dl.acm.org/doi/pdf/10.1145/3549993.3550007

Diffie–Hellman. (1976). *Whitfield Diffie and Martin Hellman. Diffie–Hellman key exchange.* https://en.wikipedia.org/wiki/Diffie%E2%80%93Hellman_key_exchange

Ding, J., Gentry, C., & Malik, M. (2020). Learning With Errors: Survey and Future Directions. *Journal of Cryptology, 33*(4), 1163–1239.

DowlingJ. P.MilburnG. J. (2003). Quantum technology: the second quantum revolution. Philosophical Transactions of the Royal Society of London. Series A: Mathematical, Physical and Engineering Sciences, 361(1809), 1655-1674. https://arxiv.org/abs/quant-ph/0206091

Duan, H. (2022). The Principles, Algorithms and State-of-Art Applications of Quantum Computing. *Journal of Physics: Conference Series, 2386*(1), 012025. doi:10.1088/1742-6596/2386/1/012025

Ducas, L., Lyubashevsky, V., & Prest, T. (2014). Efficient identity-based encryption over NTRU lattices. In *Advances in Cryptology–ASIACRYPT 2014: 20th International Conference on the Theory and Application of Cryptology and Information Security, Kaoshiung, Taiwan, ROC, December 7-11, 2014, Proceedings, Part II 20* (pp. 22-41). Springer Berlin Heidelberg. https://link.springer.com/chapter/10.1007/978-3-662-45608-8_2

Dunjko, V., Taylor, J. M., & Briegel, H. J. (2016). Quantum-enhanced machine learning. *Physical Review Letters, 117*(13), 130501. https://journals.aps.org/prl/abstract/10.1103/PhysRevLett.117.130501. doi:10.1103/PhysRevLett.117.130501 PMID:27715099

Dyakonov, M. (2019). When will useful quantum computers be constructed? Not in the foreseeable future, this physicist argues. Here's why: The case against: Quantum computing. *IEEE Spectrum, 56*(3), 24–29. doi:10.1109/MSPEC.2019.8651931

Easttom, W. C., II. (2021). *Quantum computing fundamentals.* Addison-Wesley Professional. https://www.google.com/books?hl=en&lr=&id=otnPEAAAQBAJ&oi=fnd&pg=PT19&dq=Quantum+computing+fundamentals.+Addison-Wesley+Professional&ots=EUw6AzwVIS&sig=nPx4BVkAUuM2PSiqjhAHfvUuB4M

Egan, L., Deb, B., Noel, C., Risinger, A., Zhu, D., Biswas, D., & Monroe, C. (2021). Fault-tolerant operation of a quantum error-correction code. *Nature, 598*(7882), 281–286. https://arxiv.org/abs/2009.11482. doi:10.103841586-021-03928-y PMID:34608286

Egger, D. J., Gambella, C., Marecek, J., McFaddin, S., Mevissen, M., Raymond, R., & Yndurain, E. (2020). Quantum computing for Finance: State of the art and future prospects. *IEEE Trans. Quantum Eng*. IEEE. https://ieeexplore.ieee.org/abstract/document/9222275/

Egger, D. J., Gambella, C., Marecek, J., McFaddin, S., Mevissen, M., Raymond, R., Simonetto, A., Woerner, S., & Yndurain, E. (2020). Quantum computing for finance: State-of-the-art and future prospects. *IEEE Transactions on Quantum Engineering, 1*, 1–24. doi:10.1109/TQE.2020.3030314

EggerD. J.GutiérrezR. G.MestreJ. C.WoernerS. (2019). Credit Risk Analysis using Quantum Computers. arXiv. https://arxiv.org/abs/1907.03044

Egger, D. J., Gutiérrez, R. G., Mestre, J. C., & Woerner, S. (2020). Credit risk analysis using quantum computers. *IEEE Transactions on Computers, 70*(12), 2136–2145. https://ieeexplore.ieee.org/abstract/document/9259208/. doi:10.1109/TC.2020.3038063

Einstein. (1905). *Annus Mirabilis of Albert Einstein. The 1905 Papers*. LOC. https://guides.loc.gov/einstein-annus-mirabilis/1905-papers

Einstein, A., Podolsky, B., & Rosen, N. (1935). Can quantum-mechanical description of physical reality be considered complete? *Physical Review, 47*(10), 777–780. https://journals.aps.org/pr/abstract/10.1103/PhysRev.47.777. doi:10.1103/PhysRev.47.777

Ekert, A. K. (1991). Quantum cryptography based on Bell's theorem. *Physical Review Letters, 67*(6), 661–663. https://journals.aps.org/prl/abstract/10.1103/PhysRevLett.67.661. doi:10.1103/PhysRevLett.67.661 PMID:10044956

Ekert, A. K., & Huttner, B. (1994). Eavesdropping techniques in quantum cryptosystems. *Journal of Modern Optics, 41*, 2455–2466. https://journals.aps.org/prl/abstract/10.1103/PhysRevLett.84.4733

ElGamal, T. (1985). A public key cryptosystem and a signature scheme based on discrete logarithms. *IEEE Transactions on Information Theory, 31*(4), 469–472. https://elibrary.ru/item.asp?id=29749900. doi:10.1109/TIT.1985.1057074

Elías, F. (2023). A Practical Guide to Quantum Machine Learning and Quantum Optimization. Packt.

Elkhatib, R., Koziel, B., & Azarderakhsh, R. (2022, January). Faster Isogenies for Post-quantum Cryptography: SIKE. In *Cryptographers' Track at the RSA Conference* (pp. 49–72). Springer International Publishing. https://link.springer.com/chapter/10.1007/978-3-030-95312-6_3

EnergyExplained. (2023). *Forms of energy, What is energy?* U.S. Energy Information Administration. https://www.eia.gov/energyexplained/what-is-energy/forms-of-energy.php

Enzer, D. G., Hadley, P. G., Hughes, R. J., Peterson, C. G., & Kwiat, P. G. (2002). Entangled-photon six-state quantum cryptography. *New Journal of Physics, 4*(1), 45. https://iopscience.iop.org/article/10.1088/1367-2630/4/1/345/meta. doi:10.1088/1367-2630/4/1/345

Erhard, A., Wallman, J. J., Postler, L., Meth, M., Stricker, R., Martinez, E. A., & Blatt, R. (2019). Characterizing large-scale quantum computers via cycle benchmarking. *Nature Communications, 10*(1), 1–7. https://www.nature.com/articles/s41467-019-13068-7. doi:10.103841467-019-13068-7 PMID:31767840

Eskandarpour, R., Ghosh, K. J. B., Khodaei, A., Paaso, A., & Zhang, L. (2020). Quantum-enhanced grid of the future: A primer. *IEEE Access : Practical Innovations, Open Solutions, 8*, 188993–189002. https://ieeexplore.ieee.org/abstract/document/9226502/. doi:10.1109/ACCESS.2020.3031595

Eskandarpour, R., Gokhale, P., Khodaei, A., Chong, F. T., Passo, A., & Bahramirad, S. (2020, September). Quantum Computing for Enhancing Grid Security. *IEEE Transactions on Power Systems*, *35*(5), 4135–4137. doi:10.1109/TPWRS.2020.3004073

Essam, H., Abohashima, Z., Elhoseny, M., & Mohamed, W. (2022). Machine learning in the quantum realm: The state-of-the-art, challenges, and future vision. *Expert Systems with Applications, 194*. doi:10.1016/j.eswa.2022.116512

Eswaran, U. Khang A., & Eswaran, V. (2023). Role of Quantum Computing in The Era of Artificial Intelligence," Applications and Principles of Quantum Computing. (Eds.) IGI Global Press. https://doi.org/ doi:10.4018/979-8-3693-1168-4-ch003

Ewald, R. H. (2019). An Introduction to Quantum Computing and Its Application. In S. Feld & C. Linnhoff-Popien (Eds.), Lecture Notes in Computer Science: Vol. 11413. *Quantum Technology and Optimization Problems. QTOP 2019*. Springer. doi:10.1007/978-3-030-14082-3_1

Ezratty, O. (2023). Perspective on superconducting qubit quantum computing. *The European Physical Journal A*, *59*(5), 94. doi:10.1140/epja10050-023-01006-7

Fan, J., Han, F., & Liu, H. (2014). Challenges of big data analysis. *National Science Review*, *1*(2), 293–314. https://academic.oup.com/nsr/article-abstract/1/2/293/1397586. doi:10.1093/nsr/nwt032 PMID:25419469

Farhi, E., Goldstone, J., & Gutmann, S. (2014). A quantum approximate optimization algorithm. arXiv preprint arXiv:1411.4028.

Farhi, E., Goldstone, J., & Gutmann, S. (2014). *A Quantum Approximate Optimization Algorithm*. arXiv preprint arXiv:1411.4028. https://arxiv.org/abs/1411.4028

Faruk, M. J. H., Tahora, S., Tasnim, M., Shahriar, H., & Sakib, N. (2022). A review of quantum cybersecurity: threats, risks and opportunities. In *2022 1st International Conference on AI in Cybersecurity (ICAIC)* (pp. 1-8). IEEE. https://ieeexplore.ieee.org/abstract/document/9896970/

Faruk, M. J. H., Tahora, S., Tasnim, M., Shahriar, H., & Sakib, N. (2022). A review of quantum cybersecurity: threats, risks and opportunities. In *2022 1st International Conference on AI in Cybersecurity (ICAIC)*, (pp. 1-8). IEEE.

Faruk, M. J. H., Tahora, S., Tasnim, M., Shahriar, H., & Sakib, N. (2022, May). A review of quantum cyber-security: threats, risks and opportunities. In *2022 1st International Conference on AI in Cybersecurity (ICAIC)* (pp. 1-8). IEEE.

Fasnacht, D., & Straube, C. (2023). *Quantencomputing als Basistechnologie für den nächsten Konjunkturzyklus*. HMD Praxis der Wirtschaftsinformatik. doi:10.136540702-023-00969-x

Fastovets, D. V., Bogdanov, Yu. I., Bantysh, B. I., & Lukichev, V. F. (2018). Machine learning methods in quantum computing theory. *Proc. SPIE 11022, International Conference on Micro- and Nano-Electronics*. SPIE. 10.1117/12.2522427

Fedorov, A. K., Kiktenko, E. O., & Lvovsky, A. I. (2018). Quantum computers put blockchain security at risk. *Nature*, *563*(7732), 465–467. doi:10.1038/d41586-018-07449-z PMID:30451981

Fei, S.-M., Li, M., & Luo, S. (2023). Quantum Information and Computation. *Entropy (Basel, Switzerland)*, *25*(3), 463. doi:10.3390/e25030463 PMID:36981351

Feng, C., Chen, L., & Zhao, L.-J. (2023). Coherence and entanglement in Grover and Harrow–Hassidim–Lloyd algorithm. *Physica A, 626*, 129048. doi:10.1016/j.physa.2023.129048

Fernández Pérez, I., Prieta, F. l., Rodríguez-González, S., Corchado, J. M., & Prieto, J. (2023). Quantum AI: Achievements and Challenges in the Interplay of Quantum Computing and Artificial Intelligence. In: Julián, V., Carneiro, J., Alonso, R.S., Chamoso, P., Novais, P. (eds) *Ambient Intelligence—Software and Applications—13th International Symposium on Ambient Intelligence.* Springer, Cham. 10.1007/978-3-031-22356-3_15

Fernández-Caramés, T. M., & Fraga-Lamas, P. (2018). A review on the use of blockchain for the internet of things. *IEEE Access: Practical Innovations, Open Solutions, 6*, 32979–33001. doi:10.1109/ACCESS.2018.2842685

Fernandez-Carames, T. M., & Fraga-Lamas, P. (2020). Towards Post-Quantum Blockchain: A Review on Blockchain Cryptography Resistant to Quantum Computing Attacks. *IEEE Access : Practical Innovations, Open Solutions, 8*, 21091–21116. doi:10.1109/ACCESS.2020.2968985

Fernandez-Carames, T. M., & Fraga-Lamas, P. (2020). *Towards post-quantum blockchain: A review on blockchain cryptography resistant to quantum computing attacks.* IEEE.

Fernández-Martínez, M., Guirao, J. L. G., Sánchez-Granero, M. Á., & Segovia, J. E. T. (2019). *Fractal Dimension for Fractal Structures: With Applications to Finance* (Vol. 19). Springer. https://link.springer.com/article/10.1007/s00161-022-01116-5 doi:10.1007/978-3-030-16645-8

Flarend, A. & Hilborn, B. (2022). *Quantum Gates and Quantum Circuits.* Oxford Academic. . doi:10.1093/oso/9780192857972.003.0006

Flöther, F. (2023). *The state of quantum computing applications in health and medicine.* arXiv preprint arXiv:2301.09106

Frey, M. R. (2016). Quantum speed limits—Primer, perspectives, and potential future directions. *Quantum Information Processing, 15*(10), 3919–3950. https://link.springer.com/article/10.1007/s11128-016-1405-x. doi:10.100711128-016-1405-x

Gan, B. Y., Leykam, D., & Angelakis, D. G. (2022). Fock state-enhanced expressivity of quantum machine learning models. *EPJ Quantum Technology, 9*(1), 16. doi:10.1140/epjqt40507-022-00135-0

Gandomi, A., & Haider, M. (2015). Beyond the Hype: Big Data Concepts, Methods, and Analytics. *International Journal of Information Management, 35*(2), 137–144. https://www.sciencedirect.com/science/article/pii/S0268401214001066. doi:10.1016/j.ijinfomgt.2014.10.007

Garani, S. S., Nadkarni, P. J., & Raina, A. (2023). Theory behind quantum error correcting codes: An overview. *Journal of the Indian Institute of Science, 103*(2), 449–495. doi:10.100741745-023-00392-7

Gentry, C., & Szydlo, M. (2002). Cryptanalysis of the revised NTRU signature scheme. In *Advances in Cryptology—EUROCRYPT 2002: International Conference on the Theory and Applications of Cryptographic Techniques Amsterdam, The Netherlands, April 28–May 2, 2002 Proceedings 21* (pp. 299-320). Springer Berlin Heidelberg. https://link.springer.com/chapter/10.1007/3-540-46035-7_20

Gharehchopogh, F. S. (2023). Quantum-inspired metaheuristic algorithms: Comprehensive survey and classification. *Artificial Intelligence Review, 56*(6), 5479–5543. doi:10.100710462-022-10280-8

Ghasempour, A. (2019). Internet of Things in Smart Grid: Architecture, Applications, Services, Key Technologies, and Challenges. *Inventions (Basel, Switzerland)*, *4*(1), 22. https://www.mdpi.com/2411-5134/4/1/22. doi:10.3390/inventions4010022

Gheorghiu, V. (2021). Quantum Computing: An Overview of Recent Progress. *Annual Review of Condensed Matter Physics*, *12*, 247–271. https://www.annualreviews.org/doi/abs/10.1146/annurev-conmatphys-121919-050502

Ghosh, D. & Khang A. (2023). Quantum Computing Internet of Things: Unlocking New Frontiers for Connectivity, Security, and Efficiency in Data Processing. Applications and Principles of Quantum Computing. IGI Global Press. https://doi.org/ doi:10.4018/979-8-3693-1168-4-ch016

Ghosh, D. & Khang, A. (2023). Quantum Computing Internet of Things: Unlocking New Frontiers for Connectivity, Security, and Efficiency in Data Processing. Applications and Principles of Quantum Computing. IGI Global Press. https://doi.org/ doi:10.4018/979-8-3693-1168-4-ch017

Ghosh, S., Opala, A., Matuszewski, M., Paterek, T., & Liew, T. C. (2019). Quantum reservoir processing. *npj Quantum Information, 5*(1), 35. https://www.nature.com/articles/s41534-019-0149-8

Gilliam, A., Venci, C., Muralidharan, S., Dorum, V., May, E., Narasimhan, R., & Gonciulea, C. (2019). *Foundational patterns for efficient quantum computing*. arXiv preprint. https://arxiv.org/abs/1907.11513

Gill, S. S., Kumar, A., Singh, H., Singh, M., Kaur, K., Usman, M., & Buyya, R. (2022). Quantum computing: A taxonomy, systematic review and future directions. *Software, Practice & Experience*, *52*(1), 66–114. https://onlinelibrary.wiley.com/doi/abs/10.1002/spe.3039. doi:10.1002pe.3039

Gil, Y., & Zhou, M. (2019). Cloud AI: Making Artificial Intelligence Accessible to Everyone. *IEEE Intelligent Systems*, *34*(5), 2–5. https://ieeexplore.ieee.org/abstract/document/6547979/

Ginsberg, C. (2023). *How is Quantum Computing Used in Data Analytics?* Noble Desktop. https://www.nobledesktop.com/classes-near-me/blog/quantum-computing-in-data-analytics

Gisin, N., Ribordy, G., Tittel, W., & Zbinden, H. (2002). Quantum cryptography. *Reviews of Modern Physics*, *74*(1), 145–195. doi:10.1103/RevModPhys.74.145

Glendinning, I. (2005). European Centre For Parallel Computing At Vienna. The Bloch Sphere. http://andcorporation.com/bloch-sphere.pdf

Goldenberg, L., & Vaidman, L. (1995). Quantum cryptography based on orthogonal states. *Physical Review Letters*, *75*(7), 1239–1243. https://journals.aps.org/prl/abstract/10.1103/PhysRevLett.75.1239. doi:10.1103/PhysRevLett.75.1239 PMID:10060243

Golestan, S., Habibi, M. R., Mousazadeh Mousavi, S. Y., Guerrero, J. M., & Vasquez, J. C. (2023). Quantum computation in power systems: An overview of recent advances. *Energy Reports*, *9*, 584–596. doi:10.1016/j.egyr.2022.11.185

Gomathi, L., Mishra, A. K., & Tyagi, A. K. (2023). *Blockchain and Machine Learning Empowered Internet of Things Applications: Current Issues, Challenges and Future Research Opportunities*. 2023 4th International Conference on Smart Electronics and Communication (ICOSEC), Trichy, India. 10.1109/ICOSEC58147.2023.10276385

Gomathi, L., Mishra, A. K., & Tyagi, A. K. (2023). *Industry 5.0 for Healthcare 5.0: Opportunities, Challenges and Future Research Possibilities*. 2023 7th International Conference on Trends in Electronics and Informatics (ICOEI), Tirunelveli, India. 10.1109/ICOEI56765.2023.10125660

Gonçalves, C. P. (2016). *Quantum Neural Machine Learning Back propagation and Dynamics*. arXiv preprint arXiv:1609.06935.

Gope, P., & Sikdar, B. (2021). A Privacy-Aware Reconfigurable Authenticated Key Exchange Scheme for Secure Communication in Smart Grids. *IEEE Transactions on Smart Grid, 12*(6), 5335–5348. https://ieeexplore.ieee.org/abstract/document/9518386/. doi:10.1109/TSG.2021.3106105

Goyal, D. & Tyagi, A. (2020). *A Look at Top 35 Problems in the Computer Science Field for the Next Decade*. Taylor and Francis. . doi:10.1201/9781003052098-40

Goy, G., Loiseau, A., & Gaborit, P. (2022, September). A new key recovery side-channel attack on HQC with chosen ciphertext. In *International Conference on Post-Quantum Cryptography* (pp. 353–371). Springer International Publishing. https://link.springer.com/chapter/10.1007/978-3-031-17234-2_17 doi:10.1007/978-3-031-17234-2_17

Griffiths, D. & Schroeter, D. (2018). *Introduction to Quantum Mechanics*. Cambridge University Press. . doi:10.1017/9781316995433

Grinin, L., & Grinin, A. (2020). The Cybernetic Revolution and the Future of Technologies. In *The 21st Century Singularity and Global Futures* (pp. 377–396). Springer. https://link.springer.com/chapter/10.1007/978-3-030-33730-8_17 doi:10.1007/978-3-030-33730-8_17

Grover, L. K. (1996). A fast quantum mechanical algorithm for database search. *Proceedings of the twenty-eighth annual ACM symposium on Theory of computing*, (pp. 212-219). ACM. https://dl.acm.org/doi/pdf/10.1145/237814.237866

Gruska, J. (2005). *1999. Quantum computing. Vol.* McGraw-Hill.

Guerreschi, G. G., & Matsuura, A. Y. (2022). QX: A high-performance quantum computer simulation platform. *Quantum Science and Technology, 7*(4), 045003. https://ieeexplore.ieee.org/ abstract/document/7927034/. doi:10.1088/2058-9565/aacf0b

Gunduz, M. Z., & Das, R. (2020). Cyber-security on smart grid: Threats and potential solutions. *Computer Networks, 169*, 107094. https://www.sciencedirect.com/science/article/pii/S1389128619311235. doi:10.1016/j.comnet.2019.107094

Guo, H., & Yu, X. (2022). A survey on blockchain technology and its security. *Blockchain: research and applications, 3*(2), 100067.

Guo, Q., Hlauschek, C., Johansson, T., Lahr, N., Nilsson, A., & Schröder, R. L. (2022). Don't reject this: Key-recovery timing attacks due to rejection-sampling in HQC and BIKE. *IACR Transactions on Cryptographic Hardware and Embedded Systems*, 223-263. https://tches.iacr.org/index.php/TCHES/article/view/9700

Guo, Q., Johansson, T., & Nilsson, A. (2020, August). A key-recovery timing attack on post-quantum primitives using the Fujisaki-Okamoto transformation and its application on FrodoKEM. In *Annual International Cryptology Conference* (pp. 359–386). Springer International Publishing. https://link.springer.com/chapter/10.1007/978-3-030-56880-1_13 doi:10.1007/978-3-030-56880-1_13

Gupta, S., Mohanta, S., Chakraborty, M., & Ghosh, S. (2017). Quantum machine learning-using quantum computation in artificial intelligence and deep neural networks: Quantum computation and machine learning in artificial intelligence. In *2017 8th Annual Industrial Automation and Electromechanical Engineering Conference* (pp. 268-274). IEEE.

Gupta, N., Jati, A., Chauhan, A. K., & Chattopadhyay, A. (2020). Pqc acceleration using gpus: Frodokem, newhope, and kyber. *IEEE Transactions on Parallel and Distributed Systems*, *32*(3), 575–586. https://ieeexplore.ieee.org/abstract/document/9201530/. doi:10.1109/TPDS.2020.3025691

Gupta, S., Modgil, S., Bhatt, P. C., Chiappetta Jabbour, C. J., & Kamble, S. (2023). Quantum computing led innovation for achieving a more sustainable Covid-19 healthcare industry. *Technovation*, *120*, 102544. doi:10.1016/j.technovation.2022.102544

Gupta, V. K. (2002). Quantum to quantum computing. *IETE Technical Review*, *19*(5), 333–347. https://www.tandfonline.com/doi/abs/10.1080/02564602.2002.11417048. doi:10.1080/02564602.2002.11417048

Guth, A. (1979). *Waiting for the Big Bang: Innovators.* National Geographic. https://www.nationalgeographic.com/ science/article/140630-alan-guth-profile-inflation-cosmology-science

Gyongyosi, L., & Imre, S. (2019). A survey on quantum computing technology. *Computer Science Review*, *31*, 51–71. https://www.sciencedirect.com/science/article/pii/S1574013718301709. doi:10.1016/j.cosrev.2018.11.002

HadamardGate. (2023). *What is a Hadamard gate?* Hadmard Gate. https://pennylane.ai/qml/ glossary/what-is-a-hadamard-gate

HadamardGate. (n.d.). *What is a Hadamard gate?* Hadamard Gate. https://pennylane.ai/qml/glossary/what-is-a-hadamard-gate

HadamardGate. (n.d.). *What is a Hadamard gate?* Penny Lane. https://pennylane.ai/qml/glossary/what-is-a-hadamard-gate

Hajimahmud, V. A., Khang, A., Hahanov, V., Litvinova, E., Chumachenko, S., & Alyar, V. A. (2022). Autonomous Robots for Smart City: Closer to Augmented Humanity. AI-Centric Smart City Ecosystems: Technologies, Design and Implementation (1st Ed.), 7 (12), CRC Press. doi:10.1201/9781003252542-7

Hamze, F., King, A. D., Raymond, J., Roy, A. P., Israel, R., Andriyash, E., & Ranjbar, M. (2018). *U.S. Patent No. 9,881,256*. U.S. Patent and Trademark Office. https://heinonline.org/hol-cgi-bin/get_pdf.cgi?handle=hein.journals/jstl27&section=5

Han, K. H., & Kim, J. H. (2002, December). Quantum-inspired evolutionary algorithm for a class of combinatorial optimization. *IEEE Transactions on Evolutionary Computation*, *6*(6), 580–593. doi:10.1109/TEVC.2002.804320

Harwood, S., Gambella, C., Trenev, D., Simonetto, A., Bernal Neira, D., & Greenberg, D. (2021, January). Formulating and Solving Routing Problems on Quantum Computers. *IEEE Transactions on Quantum Engineering*, *2*, 1–17. doi:10.1109/TQE.2021.3049230

Hassan, T., & Ahmed, F. (2019). Transaction and Identity Authentication Security Model for E-Banking: Confluence of Quantum Cryptography and AI. In *Intelligent Technologies and Applications: First International Conference*. Springer Singapore. https://link.springer.com/chapter/10.1007/978-981-13-6052-7_29

Hassan, M., Rehmani, M. H., & Chen, J. (2019). Privacy preservation in blockchain-based IoT systems: Integration issues, prospects, challenges, and future research directions. *Future Generation Computer Systems*, *97*, 512–529. https://www.sciencedirect.com/science/article/pii/S0167739X18326542. doi:10.1016/j.future.2019.02.060

Havlíček, V., Córcoles, A. D., Temme, K., Harrow, A. W., Kandala, A., Chow, J. M., & Gambetta, J. M. (2019). Supervised learning with quantum-enhanced feature spaces. *Nature*, *567*(7747), 209–212. https://www.nature.com/articles/s41586-019-0980-2. doi:10.103841586-019-0980-2 PMID:30867609

Heffernan, S. (2005). *Modern banking*. John Wiley & Sons.

He, G., Zhu, J., & Zeng, G. (2006). Quantum secure communication using continuous variable Einstein-Podolsky-Rosen correlations. *Physical Review A, 73*(1), 012314. https://journals.aps.org/pra/abstract/10.1103/PhysRevA.73.012314. doi:10.1103/PhysRevA.73.012314

Hellman, M. (1976). New directions in cryptography. *IEEE Transactions on Information Theory, 22*(6), 644–654. https://dl.acm.org/doi/pdf/10.1145/3549993.3550007. doi:10.1109/TIT.1976.1055638

Hema, T., & Arowolo, M. O. (2023). Quantum Computing Significance on Multidimensional Data: R-Tree Search Based on Grover's Search Algorithm. In A. K. Tyagi (Ed.), *Handbook of Research on Quantum Computing for Smart Environments* (pp. 217–230). IGI Global. doi:10.4018/978-1-6684-6697-1.ch012

Hempel, C., Maier, C., Romero, J., McClean, J., Monz, T., Shen, H., & Aspuru-Guzik, A. (2018). Quantum chemistry calculations on a trapped-ion quantum simulator. *Physical Review X, 8*(3), 031022. https://journals.aps.org/prx/abstract/10.1103/PhysRevX.8.031022. doi:10.1103/PhysRevX.8.031022

Henderson, M., Gallina, J., & Brett, M. (2021). Methods for accelerating geospatial data processing using quantum computers. *Quantum Machine Intelligence, 3*(1), 4. https://link.springer.com/article/10.1007/s42484-020-00034-6. doi:10.100742484-020-00034-6

Hendrickx, N. W., Franke, D. P., Sammak, A., Kouwenhoven, M., Sabbagh, D., Yeoh, L., & Tagliaferri, M. L. V. (2021). A four-qubit germanium quantum processor. *Nature, 591*(7849), 580–585. https://www.nature.com/articles/s41586-021-03332-6. doi:10.103841586-021-03332-6 PMID:33762771

Herman, D., Shaydulin, R., Sun, Y., Chakrabarti, S., Hu, S., Minssen, P., & Pistoia, M. (2022). Portfolio optimization via quantum zeno dynamics on a quantum processor. arXiv preprint arXiv:2209.15024.

He, W., Zha, S., & Li, L. (2013). Social media competitive analysis and text mining: A case study in the pizza industry. *International Journal of Information Management, 33*(3), 464–472. https://www.sciencedirect.com/science/article/pii/S0268401213000030. doi:10.1016/j.ijinfomgt.2013.01.001

Hidary, J. (2020). *Quantum Computing: An Applied Approach*. Springer. https://link.springer.com/content/pdf/10.1007/978-3-030-83274-2.pdf

Hidary, J. D., & Hidary, J. D. (2021). A brief history of quantum computing. *Quantum Computing: An Applied Approach*, 15-21. Springer. https://link.springer.com/content/pdf/10.1007/978-3-030-83274-2.pdf

Hilmani, A., Maizate, A., & Hassouni, L. (2020). Automated Real-Time Intelligent Traffic Control System for Smart Cities Using Wireless Sensor Networks. *Wireless Communications and Mobile Computing*. . doi:10.1155/2020/8841893

Hoffstein, J., Howgrave-Graham, N., Pipher, J., Silverman, J. H., & Whyte, W. (2003, February). NTRUSIGN: Digital signatures using the NTRU lattice. In *Cryptographers' track at the RSA conference* (pp. 122-140). Springer Berlin Heidelberg. https://link.springer.com/chapter/10.1007/3-540-36563-X_9

Hoffstein, J., Pipher, J., & Silverman, J. H. (1998, June). NTRU: A ring-based public key cryptosystem. In *International algorithmic number theory symposium* (pp. 267-288). Berlin, Heidelberg: Springer Berlin Heidelberg. https://link.springer.com/chapter/10.1007/BFb0054868

Hoffstein, J., Pipher, J., & Silverman, J. H. (2001). NSS: An NTRU lattice-based signature scheme. In *Advances in Cryptology—EUROCRYPT 2001: International Conference on the Theory and Application of Cryptographic Techniques Innsbruck,* (pp. 211-228). Springer. https://www.google.com/books?hl=en&lr=&id=Ie-qCAAAQBAJ

Huang, H. Y., Broughton, M., Mohseni, M., Babbush, R., Boixo, S., Neven, H., & McClean, J. R. (2021). Power of data in quantum machine learning. *Nature Communications*, *12*(1), 2631. doi:10.103841467-021-22539-9 PMID:33976136

Hu, F., Wang, B.-N., Wang, N., & Wang, C. (2019). Quantum machine learning with D-wave quantum computer. *Quantum Engineering*, *1*(2), e12. doi:10.1002/que2.12

Hughes, C., Isaacson, J., Perry, A., Sun, R. F., & Turner, J. (2021). *Quantum Computing for the Quantum Curious.* Springer. https://library.oapen.org/handle/20.500.12657/48236

Humble, T. (2018). Consumer applications of quantum computing. *IEEE Consumer Electronics Magazine*, 1–5. https://ieeexplore.ieee.org/abstract/document/8482518/

Humble, T. S., McCaskey, A., Lyakh, D. I., Gowrishankar, M., Frisch, A., & Monz, T. (2021). Quantum computers for high-performance computing. *IEEE Micro*, *41*(5), 15–23. doi:10.1109/MM.2021.3099140

Huttner, B., Muller, A., Gautier, J. D., Zbinden, H., & Gisin, N. (1996). Unambiguous quantum measurement of nonorthogonal states. *Physical Review A*, *54*(5), 3783–3789. https://journals.aps.org/pra/abstract/10.1103/PhysRevA.54.3783. doi:10.1103/PhysRevA.54.3783 PMID:9913923

Hwang, J., Park, T., Park, H., & Kim, H. (2022). Quantum Secure Multi-Party Computation Protocol Based on the Three-Phase Quantum Key Agreement. *Symmetry*, *14*(4), 1–16.

IBM. (2023). *What is quantum computing?* IBM. https://www.ibm.com/topics/quantum-computing

Iosifescu-Enescu, I., Matthys, C., Gkonos, C., Iosifescu-Enescu, C. M., & Hurni, L. (2017). Cloud-Based Architectures for Auto-Scalable Web Geoportals towards the Cloudification of the GeoVITe Swiss Academic Geoportal. *ISPRS International Journal of Geo-Information*, *6*(7), 192. doi:10.3390/ijgi6070192

Ivanovic, I. D. (1987). How to differentiate between non-orthogonal states. *Physics Letters. [Part A]*, *123*(6), 257–259. https://www.sciencedirect.com/science/article/pii/0375960187902222. doi:10.1016/0375-9601(87)90222-2

Izsák, R., Riplinger, C., Blunt, N. S., de Souza, B., Holzmann, N., Crawford, O., Camps, J., Neese, F., & Schopf, P. (2023). Quantum computing in pharma: A multilayer embedding approach for near future applications. *Journal of Computational Chemistry*, *44*(3), 406–421. doi:10.1002/jcc.26958 PMID:35789492

Jadhav, A., Rasool, A., & Gyanchandani, M. (2023). Quantum Machine Learning: Scope for real-world problems. *Procedia Computer Science*, *218*, 2612–2625. doi:10.1016/j.procs.2023.01.235

Jaeger, G. (2007). *A Short Introduction to Quantum Information and Quantum Computation.* Google Books. https://www.google.com/books?hl=en&lr=&id=eQIP1_fc7gYC&oi=fnd&pg=PR1&dq=A+Short+Introduction+to+Quantum+Information+and+Quantum+Computation&ots=jSEn_N2UMq&sig=tOoLdo_Yzz1nw-9fYgP5Z3dXY9M8

Jattana, M. S., Jin, F., De Raedt, H., & Michielsen, K. (2023). Improved variational quantum eigensolver via quasidynamical evolution. *Physical Review Applied*, *19*(2), 024047. doi:10.1103/PhysRevApplied.19.024047

Jayanthi, P., Rai, B. K., & Muralikrishna, I. (2022). The Potential of Quantum Computing in Healthcare. In *Technology Road Mapping for Quantum Computing and Engineering* (pp. 81–101). IGI Global. doi:10.4018/978-1-7998-9183-3.ch006

Jayaram, M. A. (2018). Quantum Computing: Some Percepts and Realms of Applications. *International Journal of Computer Applications, 975.*

Jennewein, T., Simon, C., Weihs, G., Weinfurter, H., & Zeilinger, A. (2000). Quantum cryptography with entangled photons. *Physical Review Letters, 84*(20), 4729–4732. https://journals.aps.org/prl/abstract/10.1103/PhysRevLett.84.4729. doi:10.1103/PhysRevLett.84.4729 PMID:10990782

Jerbi, S., Fiderer, L. J., Poulsen Nautrup, H., Kübler, J. M., Briegel, H. J., & Dunjko, V. (2023). Quantum machine learning beyond kernel methods. *Nature Communications, 14*(1), 517. doi:10.103841467-023-36159-y PMID:36720861

Jia, M. (2019). Adopting Internet of Things for the development of smart buildings: A review of enabling technologies and applications. *IEEE Access : Practical Innovations, Open Solutions, 7*, 113601–113616. https://www.sciencedirect.com/science/article/pii/S0926580518307064

Johnson, D. (2017). *The impact of IoT security labelling on consumer product choice and willingness to pay.* PLOS. https://journals.plos.org/plosone/article?id=10.1371/journal.pone.0227800

Jones, E., & Knaack, P. (2019). Global financial regulation: Shortcomings and reform options. *Global Policy, 10*(2), 193–206. https://onlinelibrary.wiley.com/doi/abs/10.1111/1758-5899.12656. doi:10.1111/1758-5899.12656

Jones, J. A., & Mosca, M. (1999). Approximate quantum counting on an NMR ensemble quantum computer. *Physical Review Letters, 83*(5), 1050–1053. doi:10.1103/PhysRevLett.83.1050

Jones, T., Brown, A., Bush, I., & Benjamin, S. C. (2019, December). QuEST and High Performance Simulation of Quantum Computers. *Scientific Reports, 9*(1), 10736. doi:10.103841598-019-47174-9 PMID:31341200

Jordan, S. (2011). *Home.* Quantum Algorithm Zoo. https://quantumalgorithmzoo.org/

Jordan, S. P. (2017). Fast quantum computation at arbitrarily low energy. *Physical Review. A, 95*(3), 032305. https://journals.aps.org/pra/abstract/10.1103/PhysRevA.95.032305. doi:10.1103/PhysRevA.95.032305

Joshi, S., & Moazeni, S. (2023). Scaling up superconducting quantum computers with cryogenic rf-photonics. *Journal of Lightwave Technology*, 1-10. (In press) doi:10.1109/JLT.2023.3311806

Jurcevic, P., Javadi-Abhari, A., Bishop, L. S., Lauer, I., Borgorin, D., Brink, M., & Blais, A. (2021). Demonstration of quantum volume 64 on a superconducting quantum computing system. *Quantum Science and Technology, 7*(1), 015020. https://iopscience.iop.org/article/10.1088/2058-9565/abe519/meta. doi:10.1088/2058-9565/abe519

Just, B. (2022). Quantum Gates on One Qubit. In B. Just (Ed.), *Quantum Computing Compact: Spooky Action at a Distance and Teleportation Easy to Understand* (pp. 69–82). Springer Berlin Heidelberg. doi:10.1007/978-3-662-65008-0_9

Kadowaki, T., & Nishimori, H. (1998). Quantum annealing in the transverse Ising model. *Physical Review E: Statistical Physics, Plasmas, Fluids, and Related Interdisciplinary Topics, 58*(5), 5355–5363. https://journals.aps.org/pre/abstract/10.1103/PhysRevE.58.5355. doi:10.1103/PhysRevE.58.5355

Kak, S. (2000). Active agents, intelligence and quantum computing. *Information Sciences*, *128*(1-2), 1–17. https://www.sciencedirect.com/science/article/pii/S0020025500000438. doi:10.1016/S0020-0255(00)00043-8

Kanwar, P. (2023). *Quantum computing in artificial intelligence around the corner*. eInfo Chips. https://www.einfochips.com/blog/quantum-computing-in-artificial-intelligence-around-the-corner/

Kanwar, P. (2023, May 8). *Quantum Computing in artificial intelligence around the corner*. Product Engineering Services. https://www.einfochips.com/blog/quantum-computing-in-artificial-intelligence-around-the-corner

Karl, P., Fritzmann, T., & Sigl, G. (2022, April). Hardware Accelerated FrodoKEM on RISC-V. In *2022 25th International Symposium on Design and Diagnostics of Electronic Circuits and Systems (DDECS)* (pp. 154-159). IEEE. https://ieeexplore.ieee.org/abstract/document/9770148/

Kaur, D., Uslu, S., & Durresi, A. (2023, June). Quantum Algorithms for Trust-Based AI Applications. In *Conference on Complex, Intelligent, and Software Intensive Systems* (pp. 1-12). Cham: Springer Nature. https://link.springer.com/chapter/10.1007/978-3-031-35734-3_1

Kaye, A. P., Laflamme, R., & Mosca, M. (2006). *Introduction to Quantum Computing*. Oxford University Press., https://www.google.com/books?hl=en&lr=&id=18_-rgkdQGIC doi:10.1093/oso/9780198570004.001.0001

Kearney, J. J., & Perez-Delgado, C. A. (2021). Vulnerability of blockchain technologies to quantum attacks. *Array (New York, N.Y.)*, *10*, 100065. doi:10.1016/j.array.2021.100065

Keerthika, K., Kannan, M., & Khang, A. (2023). Medical Data Analytics: Roles, Challenges, and Analytical Tools. AI and IoT-Based Technologies for Precision Medicine. IGI Global Press. doi:10.4018/979-8-3693-0876-9.ch001

Kenneth, S. (2018). Philosophy of Science. *Physical Chemistry*. Elsevier. https://shop.elsevier.com/books/physical-chemistry/schmitz/978-0-12-800513-2

Khalid, Z. M., & Askar, S. (2021). Resistant Blockchain cryptography to quantum computing attacks. *International Journal of Science and Business*, *5*(3), 116–125.

Khanal, B., Orduz, J., Rivas, P., & Baker, E. (2023). Supercomputing leverages quantum machine learning and Grover's algorithm. *The Journal of Supercomputing*, *79*(6), 6918–6940. doi:10.100711227-022-04923-4

Khang A. (2024). The Quantum Evolution: Application of AI and Robotics in the Future of Quantum Technology. CRC Press. https://doi.org/ doi:10.1201/978-1-032-64207-9

Khang, A. & Alyar, A., Khalilov, M., Murad, B., & Litvinova, E. (2023). Introduction to Quantum Computing and Its Integration Applications. Applications and Principles of Quantum Computing. (Eds.) IGI Global Press. https://doi.org/ doi:10.4018/979-8-3693-1168-4-ch002

Khang, A. (2022). Cyber-Physical-Social System and İncident Management. AI-Centric Smart City Ecosystems: Technologies, Design and Implementation. CRC Press. doi:10.1201/9781003252542-2

Khang, A. (2023). AI and IoT-Based Technologies for Precision Medicine. IGI Global Press. doi:10.4018/979-8-3693-0876-9

Khang, A. (2023). AI-Aided Data Analytics Tools and Applications for the Healthcare Sector. AI and IoT-Based Technologies for Precision Medicine. IGI Global Press. doi:10.4018/979-8-3693-0876-9.ch018

Khang, A. (2023). Applications and Principles of Quantum Computing. IGI Global Press. https://doi.org/ doi:10.4018/979-8-3693-1168-4

Khang, A. (2023). Quantum Mechanics Primer - Fundamentals and Quantum Computing. Applications and Principles of Quantum Computing. IGI Global Press. https://doi.org/ doi:10.4018/979-8-3693-1168-4-ch001

Khang, A. (2023). Revolutionizing Agriculture: Exploring Advanced Technologies for Plant Protection in the Agriculture Sector. Handbook of Research on AI-Equipped IoT Applications in High-Tech Agriculture. IGI Global. doi:10.4018/978-1-6684-9231-4.ch001

Khang, A. (2024). Using Big Data to Solve Problems in the Field of Medicine. Computer Vision and AI-integrated IoT Technologies in Medical Ecosystem. CRC Press. https://doi.org/ doi:10.1201/9781003429609-21

Khang, A. (2024). Advanced IoT Technologies and Applications in the Industry 4.0 Digital Economy (1 Ed.) CRC Press. https://doi.org/ doi:10.1201/9781003434269

Khang, A. (2024). Application of Computer Vision in the Healthcare Ecosystem. Computer Vision and AI-integrated IoT Technologies in Medical Ecosystem. (1 Ed.) (2024). CRC Press. https://doi.org/ doi:10.1201/9781003429609-1

Khang, A. (2024). Medical and BioMedical Signal Processing and Prediction. Computer Vision and AI-integrated IoT Technologies in Medical Ecosystem. CRC Press. https://doi.org/ doi:10.1201/9781003429609-7

Khang, A. Rath, K., Satapathy, S., Kumar, A., Das, S., & Panda, M. (2023). Enabling the Future of Manufacturing: Integration of Robotics and IoT to Smart Factory Infrastructure in Industry 4.0. AI-Based Technologies and Applications in the Era of the Metaverse. IGI Global Press. doi:10.4018/978-1-6684-8851-5.ch002

Khang, A., & Muthmainnah, M. (2023). AI-Aided Teaching Model for the Education 5.0 Ecosystem" AI-Based Technologies and Applications in the Era of the Metaverse. (1 Ed.) IGI Global Press. doi:10.4018/978-1-6684-8851-5.ch004

Khang, A., Gupta, S. K., Rani, S., & Karras, D. A. (2023). Smart Cities: IoT Technologies, Big Data Solutions, Cloud Platforms, and Cybersecurity Techniques. CRC Press. doi:10.1201/9781003376064

Khang, A., Hajimahmud, V. A., Gupta, S. K., Babasaheb, J., & Morris, G. (2023). AI-Centric Modelling and Analytics: Concepts, Designs, Technologies, and Applications. CRC Press. https://doi.org/ doi:10.1201/9781003400110

Khang, A., Misra, A., Gupta, S. K., & Shah, V. (2023). AI-aided IoT Technologies and Applications in the Smart Business and Production. CRC Press. https://doi.org/ doi:10.1201/9781003392224

Khang, A., Rana, G., Tailor, R. K., & Hajimahmud, V. A. (2023). Data-Centric AI Solutions and Emerging Technologies in the Healthcare Ecosystem. CRC Press. doi:10.1201/9781003356189

Khang, A., Shah, V., & Rani, S. (2023). AI-Based Technologies and Applications in the Era of the Metaverse. (1 Ed.) IGI Global Press. doi:10.4018/978-1-6684-8851-5

Khang, A. (2023). *Advanced Technologies and AI-Equipped IoT Applications in High-Tech Agriculture* (1st ed.). IGI Global Press. doi:10.4018/978-1-6684-9231-4

Khang, A., & Abdullayev, V. A. (2024). *Computer Vision and AI-integrated IoT Technologies in Medical Ecosystem* (1st ed.). CRC Press. doi:10.1201/9781003429609

Khang, A., Chowdhury, S., & Sharma, S. (2022). *The Data-Driven Blockchain Ecosystem: Fundamentals, Applications, and Emerging Technologies* (1st ed.). CRC Press. doi:10.1201/9781003269281

Khang, A., Rani, S., & Sivaraman, A. K. (2022). *AI-Centric Smart City Ecosystems: Technologies, Design and Implementation* (1st ed.). CRC Press. doi:10.1201/9781003252542

Khanh, H. H. & Khang, A. (2021). The Role of Artificial Intelligence in Blockchain Applications. *Re-inventing Manufacturing and Business Processes through Artificial Intelligence*, 2(20-40). CRC Press. doi:10.1201/9781003145011-2

Khan, S. N., Loukil, F., Ghedira-Guegan, C., Benkhelifa, E., & Bani-Hani, A. (2021). Blockchain smart contracts: Applications, challenges, and future trends. *Peer-to-Peer Networking and Applications*, *14*(5), 2901–2925. doi:10.100712083-021-01127-0 PMID:33897937

Khan, T. M., & Robles-Kelly, A. (2020). Machine learning: Quantum vs classical. *IEEE Access : Practical Innovations, Open Solutions*, *8*, 219275–219294. doi:10.1109/ACCESS.2020.3041719

Kiktenko, E. O., Pozhar, N. O., Anufriev, M. N., Trushechkin, A. S., Yunusov, R. R., Kurochkin, Y. V., Lvovsky, A. I., & Fedorov, A. K. (2018). Quantum-secured blockchain. *Quantum Science and Technology*, *3*(3), 035004. doi:10.1088/2058-9565/aabc6b

Kikuchi, Y., Mc Keever, C., Coopmans, L., Lubasch, M., & Benedetti, M. (2023). Realization of quantum signal processing on a noisy quantum computer. *NPJ Quantum Information*, *9*(1), 93. doi:10.103841534-023-00762-0

Kim, J. (2015). *A study on the adoption of IoT smart home service: using Value-based Adoption Model*. Taylor & Francis. https://www.tandfonline.com/doi/abs/10.1080/14783363.2017.1310708

Kimani, K., Oduol, V., & Langat, K. (2019). Cyber security challenges for IoT-based smart grid networks. *International Journal of Critical Infrastructure Protection*, *25*(1), 36–49. https://www.sciencedirect.com/science/article/pii/S1874548217301622. doi:10.1016/j.ijcip.2019.01.001

Kim, J. W. (2020). Blockchain technology and its applications: Case studies. *Journal of System and Management Sciences*, *10*(1), 83–93.

Kim, Y., Eddins, A., Anand, S., Wei, K. X., van den Berg, E., Rosenblatt, S., Nayfeh, H., Wu, Y., Zaletel, M., Temme, K., & Kandala, A. (2023). Evidence for the utility of quantum computing before fault tolerance. *Nature*, *618*(7965), 500–505. doi:10.103841586-023-06096-3 PMID:37316724

Kjaergaard, M., Schwartz, M. E., Braumüller, J., Krantz, P., Wang, J. I. J., Gustavsson, S., & Oliver, W. D. (2020). Superconducting qubits: Current state of play. *Annual Review of Condensed Matter Physics*, *11*(1), 369–395. https://www.annualreviews.org/doi/abs/10.1146/annurev-conmatphys-031119-050605. doi:10.1146/annurev-conmatphys-031119-050605

Knewtson, H. S., & Rosenbaum, Z. A. (2020). *Toward understanding FinTech and its industry*. Emerald. https://www.emerald.com/insight/content/doi/10.1108/MF-01-2020-0024/full/bitcoin.org

Koashi, M. (2009). Simple security proof of quantum key distribution based on complementarity. *New Journal of Physics*, *11*(4), 045018. https://iopscience.iop.org/article/10.1088/1367-2630/11/4/045018/meta. doi:10.1088/1367-2630/11/4/045018

Koczka, F., Prantner, C., & Bíro, C. (2023). Educational issues raised by the availability of the quantum computer. *Central-European Journal of New Technologies in Research. Education and Practice*, *5*(1), 1–17. doi:10.36427/CEJNTREP.5.1.5051

Ko, K.-K., & Jung, E.-S. (2021). Development of Cybersecurity Technology and Algorithm Based on Quantum Computing. *Applied Sciences (Basel, Switzerland), 11*(19), 9085. doi:10.3390/app11199085

Krauss, T., & McCollum, J. (2021). Solving the Network Shortest Path Problem on a Quantum Annealer. *IEEE Transactions on Quantum Engineering, 1*, 1–12. doi:10.1109/TQE.2020.3021921

Krenn, M. (2002). *Quantum Error Correction: Surface code vs. color code*. StaxExchange. https://physics.stackexchange.com/questions/169176/quantum-error-correction-surface-code-vs-color-code

Kribs, D. W. (2005). A quantum computing primer for operator theorists. *Linear Algebra and Its Applications, 400*, 147–167. https://www.sciencedirect.com/science/article/pii/S0024379504004793. doi:10.1016/j.laa.2004.11.010

Kribs, D. W., Pasieka, A., Laforest, M., Ryan, C., & da Silva, M. P. (2009). Research problems on numerical ranges in quantum computing. *Linear and Multilinear Algebra, 57*(5), 491–502. https://www.tandfonline.com/doi/abs/10.1080/03081080802677441. doi:10.1080/03081080802677441

Kuang, R., Lou, D., He, A., & Conlon, A. (2021, April). Quantum safe lightweight cryptography with quantum permutation pad. In *2021 IEEE 6th international conference on computer and communication systems (ICCCS)* (pp. 790-795). IEEE. https://ieeexplore.ieee.org/abstract/document/9449247/

Kumar, M., & Pattnaik, P. (2020, September). Post quantum cryptography (PQC)-An overview. In *2020 IEEE High Performance Extreme Computing Conference (HPEC)* (pp. 1-9). IEEE. https://ieeexplore.ieee.org/abstract/document/9286147/

Kumari, A., Gupta, R., & Tanwar, S. (2021). Amalgamation of blockchain and IoT for smart cities underlying 6G communication: A comprehensive review. *Computer Communications, 172*, 102–118. Advance online publication. doi:10.1016/j.comcom.2021.03.005

Kumar, N. (2019). Practically Feasible Robust Quantum Money with Classical Verification. *Cryptography, 3*(4), 26. doi:10.3390/cryptography3040026

Kuppusamy, P. (2022). Quantum Computing and Quantum Machine Learning Classification–A Survey. In *2022 IEEE 4th International Conference on Cybernetics, Cognition and Machine Learning Applications (ICCCMLA)*, (pp. 200-204). IEEE.

Labs, B., Avenue, M., & Nj, M. H. (1994). *Quantum Mechanics helps in searching for a needle in a haystack.*, https://journals.aps.org/prl/abstract/10.1103/PhysRevLett.79.325

Lau, T. W., Chung, C. Y., Wong, K. P., Chung, T. S., & Ho, S. L. (2009). Quantum-inspired evolutionary algorithm approach for unit commitment. *IEEE Transactions on Power Systems, 24*(3), 1503–1512. doi:10.1109/TPWRS.2009.2021220

Lee, R. S., & Lee, R. S. (2020). Future trends in quantum finance. Quantum Finance: Intelligent Forecast and Trading Systems, 399-405.

Lee, E., Kim, Y. S., No, J. S., Song, M., & Shin, D. J. (2019). Modification of frodokem using gray and error-correcting codes. *IEEE Access : Practical Innovations, Open Solutions, 7*, 179564–179574. https://ieeexplore.ieee.org/abstract/document/8931793/. doi:10.1109/ACCESS.2019.2959042

Le, L., & Nguyen, T. N. (2022). DQRA: Deep Quantum Routing Agent for Entanglement Routing in Quantum Networks. *IEEE Transactions on Quantum Engineering, 3*, 1–12. doi:10.1109/TQE.2022.3148667

Lemaitre, G. (1949). *Father of the Big Bang*. AMNH. https://www.amnh.org/learn-teach/curriculum-collections/cosmic-horizons-book/georges-lemaitre-big-bang

Leng, J., Zhou, M., Zhao, J. L., Huang, Y., & Bian, Y. (2020). Blockchain security: A survey of techniques and research directions. *IEEE Transactions on Services Computing, 15*(4), 2490–2510.

Lewis, R. P., Igict, P., & Zhou, Z. (2009). Assessment of communication methods for smart electricity metering in the U.K. *Proceedings of the 2009 IEEE PES/IAS Conference on Sustainable Alternative Energy (SAE),* (pp. 1-4). IEEE. https://ieeexplore.ieee.org/abstract/document/5534884/

Li, B. (2021). *The evolution of the Internet of Things (IoT) over the past 20 years*. Science Direct. https://www.sciencedirect.com/science/article/pii/S0360835221000784

Li, F., Liang, X., & Jin, X. (2017). Integration of Internet of Things and Smart Grid in Industrial Systems. In 2017 IEEE International Conference on Internet of Things (iThings) and IEEE Green Computing and Communications (GreenCom) and IEEE Cyber, Physical and Social Computing (CPSCom) and IEEE Smart Data (SmartData). IEEE., https://ieeexplore.ieee.org/abstract/document/8276863/

Liang, X., Li, M., Xiao, Q., Chen, J., Yang, C., An, H., & He, L. (2023). Deep learning representations for quantum many-body systems on heterogeneous hardware. *Machine Learning: Science and Technology, 4*(1), 015035. https://iopscience.iop.org/article/10.1088/2632-2153/acc56a/meta. doi:10.1088/2632-2153/acc56a

Li, M. W., Wang, Y. T., Geng, J., & Hong, W. C. (2021). Chaos cloud quantum bat hybrid optimization algorithm. *Nonlinear Dynamics, 103*(1), 1167–1193. doi:10.100711071-020-06111-6

Lindblad, G. (1999). A general no-cloning theorem. *Letters in Mathematical Physics, 47*(2), 189–196. https://link.springer.com/article/10.1023/A:1007581027660. doi:10.1023/A:1007581027660

Lindh, R., & Fdez Galván, I. (2023). Molecular structure optimizations with Gaussian process regression. In P. O. Dral (Ed.), *Quantum Chemistry in the Age of Machine Learning* (pp. 391–428). Elsevier. doi:10.1016/B978-0-323-90049-2.00017-2

Lipman, P., & How, Q. C. W. T. C. (2021). *Forbes Technology Council*. Forbes Technology Council. https://www.forbes.com/sites/forbestechcouncil/2021/01/04/how-quantum-computing-will-transform-cybersecurity/?sh=9a638b57d3fb

Liu, L., Yu, Y., Lai, L., Yu, R., & Zhang, X. (2020). Quantum Adversarial Machine Learning-Based Intrusion Detection. *Future Internet, 12*(9), 1–19.

Liu, Q. (2023). Comparisons of conventional computing and quantum computing approaches. *Highlights in Science. Engineering and Technology, 38*, 502–507. doi:10.54097/hset.v38i.5875

Li, X., Huang, X., Zhang, M., Wang, L., & Zhang, Z. (2019). A Quantum Intrusion Detection Method Based on Quantum Entanglement. *Symmetry, 11*(12), 1–14.

Li, X., Jiang, P., Chen, T., Luo, X., & Wen, Q. (2020). A survey on the security of blockchain systems. *Future Generation Computer Systems, 107*, 841–853.

Li, Y., Tian, M., Liu, G., Peng, C., & Jiao, L. (2020). Quantum Optimization and Quantum Learning: A Survey. *IEEE Access : Practical Innovations, Open Solutions, 8*, 23568–23593. doi:10.1109/ACCESS.2020.2970105

Li, Z., Chen, C., & Li, M. (2021). A Comprehensive Review of Artificial Intelligence in Medicine. *Artificial Intelligence in Medicine, 113*, 102043. https://www.sciencedirect.com/science/article/pii/S002604951730015X

Lo, H. K., & Chau, H. F. (1999). Unconditional security of quantum key distribution over arbitrarily long distances. *science, 283*(5410), 2050-2056. https://www.science.org/doi/abs/10.1126/science.283.5410.2050

Lockwood, O., & Si, M. (2020). *Quantum algorithm for reinforcement learning in infinite MDPs with policy advice*. arXiv preprint arXiv:2008.01082.

Lucero, E., Barends, R., Chen, Y., Kelly, J., Mariantoni, M., Megrant, A., & Nigg, D. (2012). Computing prime factors with a Josephson phase qubit quantum processor. *Nature Physics, 8*(10), 719–723. https://www.nature.com/articles/nphys2385. doi:10.1038/nphys2385

Lu, Y. (2019). Artificial intelligence: A survey on evolution, models, applications and future trends. *Journal of Management Analytics, 6*(1), 1–29. doi:10.1080/23270012.2019.1570365

Lv, Z., & Qiao, L. (2020). Analysis of healthcare big data. *Future Generation Computer Systems, 109*, 103–110. doi:10.1016/j.future.2020.03.039

Lyulyov, O., Vakulenko, I., Pimonenko, T., Kwilinski, A., Dzwigol, H., & Dzwigol-Barosz, M. (2021). Comprehensive Assessment of Smart Grids: Is There a Universal Approach? *Energies, 14*(11), 3497. https://www.mdpi.com/1996-1073/14/12/3497. doi:10.3390/en14123497

Madbouly, M. M., Mokhtar, Y. F., & Darwish, S. M. (2021). Quantum Game Application to Recovery Problem in Mobile Database. *Symmetry, 13*(11), 1984. doi:10.3390ym13111984

Ma, H., Liu, J., Shang, H., Fan, Y., Li, Z., & Yang, J. (2023). Multiscale quantum algorithms for quantum chemistry. *Chemical Science (Cambridge), 14*(12), 3190–3205. doi:10.1039/D2SC06875C PMID:36970085

Mahesh, V., & Shijo, S. (2023). Accelerating drug discovery with quantum computing. In S. N. Mohanty, R. Alvalu, & S. Mohanty (Eds.), *Evolution and Applications of Quantum Computing* (pp. 175–181). Wiley. doi:10.1002/9781119905172.ch11

Mailloux, L. O., Grimaila, M. R., Hodson, D. D., Baumgartner, G., & McLaughlin, C. (2015). Performance evaluations of quantum key distribution system architectures. *IEEE Security and Privacy, 13*(1), 30–40.

Makletsov, S. V., Opokina, N. A., & Shafigullin, I. K. (2019). Application of fractal analysis method for studying stock market. *Int. Trans. J. Eng. Manag. Appl. Sci. Technol., 11.* https://tuengr.com/V11/11A01E.pdf

Malov, D. (2020, August). Quantum Algebraic Machine Learning. In *2020 IEEE 10th International Conference on Intelligent Systems (IS)* (pp. 426-430). IEEE. https://ieeexplore.ieee.org/abstract/document/9199982/

Mandelbrot, B. B. (1975). *Les Objets Fractals*. Forme, Hasard et Dimension; Champs Flammarion: Paris, France. https://www.jstor.org/stable/24949950

Mandelbrot, B. B. (1982). *The Fractal Geometry of Nature*. W.H. Freeman and Co. https://www.cambridge.org/core/journals/mathematical-gazette/article/fractal-geometry-of-nature-by-benoit-b-mandelbrot-pp-460-2275-1982-isbn-0716711869-wh-freeman/D0714C165F4986ECE608B61347EBB625

Mandelbrot, B. B. (2013). *Fractals and Chaos: The Mandelbrot Set and Beyond*. Springer Science & Business Media. https://link.springer.com/book/10.1007/978-1-4757-4017-2

Mandelbrot, B. B. (2013). *Fractals and Scaling in Finance: Discontinuity, Concentration, Risk. Selecta* (Vol. E). Springer Science & Business Media. https://www.google.com/books?hl=en&lr=&id=H6jqBwAAQBAJ

Mandelbrot, B. B., & Hudson, R. L. (2010). *The (mis) Behavior of Markets: A Fractal View of Risk, Ruin and Reward*. Profile Books. https://www.google.com/books?hl=en&lr=&id=zg91TAIs6bgC

Mandviwalla, A., Ohshiro, K., & Ji, B. (2019). Implementing Grover's Algorithm on the IBM Quantum Computers. *IEEE International Conference on Big Data, Big Data 2018*. Institute of Electrical and Electronics Engineers Inc. 10.1109/BigData.2018.8622457

Mangini, S., Tacchino, F., Gerace, D., Bajoni, D., & Macchiavello, C. (2021). Quantum computing models for artificial neural networks. *Europhysics Letters*, *134*(1), 10002. doi:10.1209/0295-5075/134/10002

Martín-Lopo, M. M. (2020). A Literature Review of IoT Energy Platforms Aimed at End Users. Proceedings of the 2020 IEEE International Conference on Industrial Internet. ICII. https://www.sciencedirect.com/science/article/pii/S138912861931271X

Mavroeidakos, T., & Chaldeakis, V. (2020). Threat Landscape of Next Generation IoT-Enabled Smart Grids. Advances in Information and Communication Technology, AIAI. Springer. https://link.springer.com/chapter/10.1007/978-3-030-49190-1_11

Mayers, D. (2001). Unconditional security in quantum cryptography. [JACM]. *Journal of the Association for Computing Machinery*, *48*(3), 351–406. https://dl.acm.org/doi/abs/10.1145/382780.382781. doi:10.1145/382780.382781

McEliece, R. J. (1978). A public-key cryptosystem based on algebraic. *Coding Thv*, *4244*, 114-116. https://ntrs.nasa.gov/api/citations/19780016269/downloads/19780016269.pdf#page=123

McGraw-Hill Ryerson Limited. (2013). Quantum Theory and Atomic Structure. McGraw-Hill. https://mysite.science.uottawa.ca/sgambarotta/sites/default/files/CHM%201311F/slide%20show/Ch-6%20atom/7_lecture.pdf

Mehrpouya, M., Dehghanghadikolaei, A., Fotovvati, B., Vosooghnia, A., Emamian, S. S., & Gisario, A. (2019). The Potential of Additive Manufacturing in the Smart Factory Industrial 4.0: A Review. *Applied Sciences (Basel, Switzerland)*, *9*(18), 3865. doi:10.3390/app9183865

Menezes, P. van Oorschot, C., & Vanstone, S. (2018). *Handbook of Applied Cryptography*. CRC Press. http://galois.azc.uam.mx/mate/propaganda/Menezes.pdf

Metawei, M. A., Eldeeb, H., Nassar, S. M., & Taher, M. (2022). Quantum Computing Meets Artificial Intelligence: Innovations and Challenges. In M. Virvou, G. A. Tsihrintzis, N. G. Bourbakis, & L. C. Jain (Eds.), *Handbook on Artificial Intelligence-Empowered Applied Software Engineering. Artificial Intelligence-Enhanced Software and Systems Engineering* (Vol. 2). Springer. doi:10.1007/978-3-031-08202-3_12

Miceli, R. & McGuigan, M. (2018). *Quantum Computation and Visualization of Hamiltonians using Discrete Quantum Mechanics and IBM QISKit*. NYSDS. . doi:10.1109/NYSDS.2018.8538959

Miller, A. (2019). The intrinsically linked future for human and Artificial Intelligence interaction. *Journal of Big Data*, *6*(1), 38. doi:10.118640537-019-0202-7

Milonni, P. W., & Hardies, M. L. (1982). Photons cannot always be replicated. *Phys. Lett.;(Netherlands)*, *92*(7). https://www.osti.gov/etdeweb/biblio/5926625

Miranda, E. R. (2021). Quantum computer: Hello, music! *Handbook of artificial intelligence for music: Foundations, advanced approaches, and developments for creativity*, (pp. 963-994). Springer. https://link.springer.com/chapter/10.1007/978-3-030-72116-9_34

Mishra, N. (2021). Quantum Machine Learning: A Review and Current Status. In N. Sharma, A. Chakrabarti, V. E. Balas, & J. Martinovic (Eds.), *Data Management, Analytics and Innovation. Advances in Intelligent Systems and Computing* (Vol. 1175). Springer. doi:10.1007/978-981-15-5619-7_8

Mishra, P., & Reshef, A. (2019). How do central bank governors' matter? Regulation and the financial sector. *Journal of Money, Credit and Banking, 51*(2-3), 369–402. https://onlinelibrary.wiley.com/doi/abs/10.1111/jmcb.12578. doi:10.1111/jmcb.12578

Misoczki, R., & Barreto, P. S. (2009, August). Compact McEliece keys from Goppa codes. In *International Workshop on Selected Areas in Cryptography* (pp. 376–392). Springer Berlin Heidelberg. https://link.springer.com/chapter/10.1007/978-3-642-05445-7_24 doi:10.1007/978-3-642-05445-7_24

Mitarai, K., Negoro, M., Kitagawa, M., & Fujii, K. (2018). Quantum circuit learning. *Physical Review. A, 98*(3), 032309. https://journals.aps.org/pra/abstract/10.1103/PhysRevA.98.032309. doi:10.1103/PhysRevA.98.032309

Mitra, J., Bhattacharya, P., & Poray, S. (2017). *Overview of quantum cryptography, security concerns, and upcoming difficulties.* Optronix, 4th International Conference on Opto-Electronics and Applied Optics, Kolkata, India.

Mocrii, D. (2018, April). IoT-based smart homes: A review of system architecture, software, communications, privacy and security. *IEEE Internet of Things Journal, 5*(2), 812–835. https://www.sciencedirect.com/science/article/pii/S2542660518300477

Mohamed, B. S., Kumar, M. S., & Srinivasagan, K. G. (2022). Quantum Computing and Quantum Cryptography. In Privacy and Security Challenges in Cloud Computing (pp. 171-189). CRC Press. doi:10.1201/9781003219880-9

Mohanta, B. K., Panda, S. S., & Jena, D. (2018, July). An overview of smart contract and use cases in blockchain technology. In 2018 9th international conference on computing, communication and networking technologies (ICCCNT) (pp. 1-4). IEEE.

Mohr, N., Masiowski, M., Zesko, M., & Soller, H. (2022). *Quantum Technology Monitor*. McKinsey. https://pubs.acs.org/doi/abs/10.1021/ct1007235

Mohseni, M., Read, P., Neven, H., Boixo, S., Denchev, V., Babbush, R., & Smelyanskiy, V. (2017). Commercialize quantum technologies in five years. *NATNews, 543*(7644), 171. https://www.nature.com/articles/543171a PMID:28277529

Moll, N., Barkoutsos, P., Bishop, L. S., Chow, J. M., Cross, A., Egger, D. J., Filipp, S., Fuhrer, A., Gambetta, J. M., Ganzhorn, M., Kandala, A., Mezzacapo, A., Müller, P., Riess, W., Salis, G., Smolin, J., Tavernelli, I., & Temme, K. (2018). Quantum optimization using variational algorithms on near-term quantum devices. *Quantum Science and Technology, 3*(3), 030503. doi:10.1088/2058-9565/aab822

Montanaro, A. (2016). Quantum algorithms: an overview. *npj Quantum Information, 2*(1), 1-8. https://www.nature.com/articles/npjqi201524

Montanaro, A. (2016). Quantum algorithms: An overview. *npj Quantum Information, 2*. doi:10.1038/npjqi.2015.23

Mori, H. (1980). Fractal dimensions of chaotic flows of autonomous dissipative systems. *Progress of Theoretical Physics, 63*(3), 1044–1047. https://academic.oup.com/ptp/article-abstract/63/3/1044/1920090. doi:10.1143/PTP.63.1044

Mosteanu, N. R., & Faccia, A. (2021). Fintech Frontiers in Quantum Computing, Fractals, and Blockchain Distributed Ledger: Paradigm Shifts and Open Innovation. *Journal of Open Innovation*, *7*(1), 19. doi:10.3390/joitmc7010019

Moussa, C., Calandra, H., & Dunjko, V. (2020). To quantum or not to quantum: Towards algorithm selection in near-term quantum optimization. *Quantum Science and Technology*, *5*(4), 044009. doi:10.1088/2058-9565/abb8e5

Movassagh, R. (2023). (in press). The hardness of random quantum circuits. *Nature Physics*. Advance online publication. doi:10.103841567-023-02131-2

Mueller, F., Byrd, G., & Dreher, P. (2019, February). Programming quantum computers: a primer with IBM Q and D-Wave exercises. In *PPoPP* (p. 451). https://arcb.csc.ncsu.edu/~mueller/qc/qc-tut/NEW-ASPLOS-4-14-2019.pdf

Mummadi, S., & Rudra, B. (2023). Fundamentals of Quantum Computation and Basic Quantum Gates. In A. K. Tyagi (Ed.), *Handbook of Research on Quantum Computing for Smart Environments* (pp. 1–24). IGI Global. doi:10.4018/978-1-6684-6697-1.ch001

MykhailovaM. (2022). Developing Programming Assignments for Teaching Quantum Computing and Quantum Programming. arXiv. https://arxiv.org/abs/2211.16347 doi:10.1109/QCE53715.2022.00092

Nair, M. & Tyagi, A. (2023). Blockchain technology for next-generation society: current trends and future opportunities for smart era. *Blockchain Technology for Secure Social Media Computing*. IET. . doi:10.1049/PBSE019E_ch11

Nair, M. M., & Tyagi, A. K. (2023). Chapter 11 - AI, IoT, blockchain, and cloud computing: The necessity of the future. R. Pandey, S. Goundar, & S. Fatima, (eds.) Distributed Computing to Blockchain. Academic Press. doi:10.1016/B978-0-323-96146-2.00001-2

Nakahara M., & Ohmi, T. (2008). *Quantum Computing*. CRC Press. https://www.google.com/books?hl=en&lr=&id=VdHJsTdoyAMC

Nakamato, S. (2008). *Bitcoin: A Peer-to-Peer Electronic Cash System* [online].

Naresh, V. S., Nasralla, M. M., Reddi, S., & García-Magariño, I. (2020). Quantum Diffie–Hellman Extended to Dynamic Quantum Group Key Agreement for e-Healthcare Multi-Agent Systems in Smart Cities. *Sensors (Basel)*, *20*(14), 3940. doi:10.339020143940 PMID:32679823

Nasir, M., Shao, Z., Zhang, Y., Zhang, J., Yang, Z., & Vasilakos, A. V. (2021). The Role of Edge Computing in Internet of Things. *IEEE Network*, *35*(1), 68–75. https://ieeexplore.ieee.org/abstract/document/8450541/

Nath, R. K., Thapliyal, H., & Humble, T. S. (2021). A Review of Machine Learning Classification Using Quantum Annealing for Real-World Applications. *SN Computer Science*, *2*(5), 365. doi:10.100742979-021-00751-0

Nemoto, K. (2023). Our future with quantum computers. *JSAP Review*, *230212*. doi:10.11470/jsaprev.230212

Neukart, F., Compostella, G., Seidel, C., Von Dollen, D., Yarkoni, S., & Parney, B. (2017, January). Traffic flow optimization using a quantum annealer. In *Frontiers in ICT* (Vol. 4, p. 29). Frontiers. https://www.frontiersin.org/articles/10.3389/fict.2017.00029/full

Neven, H., Denchev, V. S., Rose, G., & Macready, W. G. (2012, November). Qboost: Large scale classifier training withadiabatic quantum optimization. In *Asian Conference on Machine Learning* (pp. 333-348). PMLR.

Nguyen, D. T., & Gaj, K. (2021, June). Optimized software implementations of CRYSTALS-Kyber, NTRU, and Saber using NEON-based special instructions of ARMv8. In *Proceedings of the NIST 3rd PQC Standardization Conference (NIST PQC 2021)*. https://csrc.nist.gov/CSRC/media/Events/third-pqc-standardization-conference/documents/accepted-papers/nguyen-optimized-software-gmu-pqc2021.pdf

Nian, X., Sun, Y., Li, X., & Wang, Z. (2019). Integration of IoT in Smart Grid: A Survey. *IEEE Access : Practical Innovations, Open Solutions*, *7*, 70563–70577. https://ieeexplore.ieee.org/abstract/document/8625421/

Niazi, S. K. (2023). The coming of age of ai/ml in drug discovery, development, clinical testing, and manufacturing: The FDA perspectives. *Drug Design, Development and Therapy, 17*(null), 2691-2725. doi:10.2147/DDDT.S424991

Nielsen, M. A., & Chuang, I. (2002). *Quantum computation and quantum information*. MIG. http://mjg.global.s3.amazonaws.com/static/projects/refs/Quantum_Computation_and_Quantum_Information.pdf

Nielsen, M. A., & Chuang, I. L. (2010). *Quantum Computation and Quantum Information*. Cambridge University Press., doi:10.1017/CBO9780511976667

Nimbe, P., Weyori, B. A., & Adekoya, A. F. (2021). Models in quantum computing: A systematic review. *Quantum Information Processing*, *20*(2), 80. https://link.springer.com/article/10.1007/s11128-021-03021-3. doi:10.100711128-021-03021-3

Nivelkar, M., Bhirud, S., Singh, M., Ranjan, R., & Kumar, B. (2023). Quantum Computing to Study Cloud Turbulence Properties. *IEEE Access : Practical Innovations, Open Solutions*, *11*, 70679–70690. doi:10.1109/ACCESS.2023.3289924

Njorbuenwu, M., Swar, B., & Zavarsky, P. (2019). A survey on the impacts of quantum computers on information security. In *2019 2nd International conference on data intelligence and security (ICDIS)*, (pp. 212-218). IEEE. 10.1109/ICDIS.2019.00039

O'Driscoll, L., Nichols, R., & Knott, P. A. (2019). A hybrid machine learning algorithm for designing quantum experiments. *Quantum Machine Intelligence*, *1*(1-2), 5–15. doi:10.100742484-019-00003-8

Omarini, A. (2020). FinTech: A New Hedge for a Financial Re-intermediation. Strategy and Risk Perspectives. *Frontiers in Artificial Intelligence*, *3*, 63. https://www.frontiersin.org/articles/10.3389/frai.2020.00063/full. doi:10.3389/frai.2020.00063 PMID:33733180

Orús, R., Mugel, S., & Lizaso, E. (2019). Quantum computing for finance: Overview and prospects. *Reviews in Physics*, *4*, 100028. https://www.sciencedirect.com/science/ article/pii/S2405428318300571. doi:10.1016/j.revip.2019.100028

Ou, Q., Zhen, Y., Li, Y., Zhang, Y., & Zeng, L. (2012). *Application of Internet of Things in Smart Grid Power Transmission*. Third FTRA International Conference on Mobile, Ubiquitous, and Intelligent Computing, Vancouver, BC, Canada. https://ieeexplore.ieee.org/abstract/document/6305831/

Oumarou, O., Paler, A., & Basmadjian, R. (2020). *QUANTIFY: A framework for resource analysis and design verification of quantum circuits*. arXiv. https://arxiv.org/abs/2007.10893 doi:10.1109/ISVLSI49217.2020.00032

Pal, S., Bhattacharya, M., Dash, S., Lee, S.-S., & Chakraborty, C. (2023). Future potential of quantum computing and simulations in biological science. *Molecular Biotechnology*. doi:10.100712033-023-00863-3 PMID:37717248

Pal, S., Bhattacharya, M., Lee, S.-S., & Chakraborty, C. (2023). (in press). Quantum computing in the next-generation computational biology landscape: From protein folding to molecular dynamics. *Molecular Biotechnology*. doi:10.100712033-023-00765-4 PMID:37244882

Pan, J., Jain, R., Paul, S., Vu, T., Saifullah, A., & Sha, M. (2015). An Internet of Things Framework for Smart Energy in Buildings: Designs. *IEEE Internet of things Journal*. https://ieeexplore.ieee.org/abstract/document/7061425/

Panarello, A., Tapas, N., Merlino, G., Longo, F., & Puliafito, A. (2018). Blockchain and iot integration: A systematic survey. *Sensors (Basel)*, *18*(8), 2575.

Pan, D., Li, K., Ruan, D., Ng, S. X., & Hanzo, L. (2020). Single-photon-memory two-step quantum secure direct communication relying on Einstein-Podolsky-Rosen pairs. *IEEE Access : Practical Innovations, Open Solutions*, *8*, 121146–121161. https://ieeexplore.ieee.org/abstract/document/9129730/. doi:10.1109/ACCESS.2020.3006136

Pande, M., & Mulay, P. (2020). Bibliometric Survey of Quantum Machine Learning. *Science & Technology Libraries*, *39*(4), 369–382. doi:10.1080/0194262X.2020.1776193

Pandey, J. P., Ojha, D. B., & Sharma, A. (2005). Enhance fuzzy commitment scheme: an approach for post quantum cryptosystem. *Journal of Applied and Theoretical Information Technology*, 16-19. https://www.academia.edu/download/7297632/3vol9no1.pdf

Pandey, R., Maurya, P., Singh, G. D., & Faiyaz, M. S. (2023). *Simulating Quantum Principles: Qiskit Versus Cirq. In Quantum Computing: A Shift from Bits to Qubits* (pp. 333-348). Springer Nature Singapore. https://link.springer.com/chapter/10.1007/978-981-19-9530-9_18

Parhami, B. (2020). Reliability Inversion: A Cautionary Tale. *Computer*, *53*(6), 28–33. doi:10.1109/MC.2019.2958907

Parvin, K., Hannan, M. A., Mun, L. H., Lipu, M. S. H., Abdolrasol, M. G. M., Ker, P. J., Muttaqi, K. M., & Dong, Z. Y. (2022). The future energy internet for utility energy service and demand-side management in smart grid: Current practices, challenges and future directions. *Sustainable Energy Technologies and Assessments*, *53*, 102648. https://www.sciencedirect.com/science/article/pii/S221313882200697X. doi:10.1016/j.seta.2022.102648

PauliGates. (2023). *Knowledge base*. Quantum Inspire. https://www.quantum-inspire.com/kbase/pauli-x/

Pechen, A., Eliot, L., Dutton, Z., Carrasco, J. A., Hunt, T., Kalay, A. E., & Dunkel, J. (2022). *Ethical considerations for quantum machine learning research and development.* arXiv preprint arXiv:2209.05204.

Peikert, C. (2016). A decade of lattice cryptography. *Foundations and trends® in theoretical computer science, 10*(4), 283-424. https://www.nowpublishers.com/article/Details/TCS-074

Peng, B. Y., Marotzke, A., Tsai, M. H., Yang, B. Y., & Chen, H. L. (2023). Streamlined NTRU prime on FPGA. *Journal of Cryptographic Engineering*, *13*(2), 167–186. https://link.springer.com/article/10.1007/s13389-022-00303-z. doi:10.100713389-022-00303-z

Perdomo-Ortiz, A., Fluegemann, J., Narasimhan, S., Biswas, R., & Smelyanskiy, V. (2015). *A quantum-inspired classical algorithm for recommendation systems*. arXiv preprint arXiv:1503.01334.

Perdomo-Ortiz, A., Benedetti, M., Realpe-Gómez, J., & Biswas, R. (2018). Opportunities and challenges for quantum-assisted machine learning in near-term quantum computers. *Quantum Science and Technology*, *3*(3), 1–13. doi:10.1088/2058-9565/aab859

Perdomo-Ortiz, A., Feldman, A., Ozaeta, A., Isakov, S. V., Zhu, Z., O'Gorman, B., Katzgraber, H. G., Diedrich, A., Neven, H., de Kleer, J., Lackey, B., & Biswas, R. (2019). Readiness of quantum optimization machines for industrial applications. *Physical Review Applied, 12*(1), 014004. doi:10.1103/PhysRevApplied.12.014004

Peruzzo, A., McClean, J., Shadbolt, P., Yung, M. H., Zhou, X. Q., Love, P. J., & O'brien, J. L. (2014). A variational eigenvalue solver on a photonic quantum processor. *Nature Communications, 5*(1), 1–7. https://www.nature.com/articles/ncomms5213. doi:10.1038/ncomms5213 PMID:25055053

Peters, E., Caldeira, J., & Ho, A. (2021). Machine learning of high dimensional data on a noisy quantum processor. *npj Quantum Inf., 7*, 161. . doi:10.1038/s41534-021-00498-9

Petschnigg, C., Brandstötter, M., Pichler, H., Hofbaur, M., & Dieber, B. (2019, May). Quantum computation in robotic science and applications. In *2019 International Conference on Robotics and Automation* (ICRA) (pp. 803-810). IEEE. https://ieeexplore.ieee.org/abstract/document/8793768/

Philippon, T. The fintech opportunity (No. w22476). Natl. Bur. Econ. Res. 2016, 1–25. https://www.nber.org/papers/w22476

Physics. (2023). *Gravitational Potential Energy, Work, Energy, and Energy Resources*. Lumen Learning. https://courses.lumenlearning.com/suny-physics/chapter/7-3-gravitational-potential-energy/

Plenio, M. B., & Virmani, S. S. (2014). An introduction to entanglement theory. In *Quantum Information and Coherence* (pp. 173–209). Springer. https://link.springer.com/chapter/10.1007/978-3-319-04063-9_8 doi:10.1007/978-3-319-04063-9_8

Poggel, B., Quetschlich, N., Burgholzer, L., Wille, R., & Lorenz, J. M. (2023, July). Recommending solution paths for solving optimization problems with quantum computing. In *2023 IEEE International Conference on Quantum Software (QSW)* (pp. 60-67). IEEE. 10.1109/QSW59989.2023.00017

Polanyi Festschrift Committee (Ed.). (2015). *The Logic of Personal Knowledge: Essays Presented to M. Polanyi on His Seventieth Birthday*. Routledge. https://www.google.com/books?hl=en&lr=&id=BzHLCQAAQBAJ

Pooranam, N. (2023). Quantum computing: future of artificial intelligence and its applications. *Quantum Computing and Artificial Intelligence: Training Machine and Deep Learning Algorithms on Quantum Computers, 163*.

Potential and Kinetic Energy (Conservation of Energy). 2002. http://www.csun.edu/~psk17793/S9CP/S9%20potential_and_kinetic_energy.htm

Prabu, B., & Khang, A. (2023). Quantum Computing Smart City: Fostering Sustainability and Green Infrastructure with Smart City. Applications and Principles of Quantum Computing. (Eds.) IGI Global Press. https://doi.org/ doi:10.4018/979-8-3693-1168-4-ch009

Prabu, B., & Khang, A. (2023). Quantum Computing Smart City: Fostering Sustainability and Green Infrastructure with Smart City. Applications and Principles of Quantum Computing. IGI Global Press. https://doi.org/ doi:10.4018/979-8-3693-1168-4-ch010

Pramod, A. (2022). *Emerging Innovations in the Near Future Using Deep Learning Techniques, Book: Advanced Analytics and Deep Learning Models*. Wiley Scrivener. doi:10.1002/9781119792437.ch10

Prasad, N. K., Wille, R., Broekema, P. C., Rahman, N., Sung, J., Bishop, L. S., & Amin, M. H. (2021). *The future of quantum information processing*. arXiv preprint arXiv:2103.06269.

Prasad, K. K., & Aithal, P. S. (2015). Massive Growth of Banking Technology with the Aid of 5G Technologies. [IJMIE]. *Int. J. Manag. IT Eng., 5*, 616–627. https://www.academia.edu/download/46356634/58.pdf

Preskill, J. (2018). Quantum Computing in the NISQ era and beyond. *Quantum, 2*, 79. https://quantum-journal.org/papers/q-2018-08-06-79/

Preskill, J. (2018). Quantum Computing in the NISQ era and beyond. *Quantum, 2*, 79. https://quantum-journal.org/papers/q-2018-08-06-79/?fbclid=IwAR0FRb9N2fas7ETWu2M40OS6prXB5QvFME_WRELpm2CAU-cccVIEzA_UmLn4

Preskill, J. (2018). Quantum Computing and the Entanglement Frontier. *Notices of the American Mathematical Society, 65*(4), 378–388. https://www.ams.org/journals/notices/201804/rnoti-p378.pdf

Quiroga, D., Date, P., & Pooser, R. (2021). Discriminating Quantum States with Quantum Machine Learning. *2021 IEEE International Conference on Quantum Computing and Engineering (QCE)*, Broomfield, CO, USA. 10.1109/QCE52317.2021.00088

Rakesh, P. (2022). *Data Analytics and Quantum Computing*. LinkedIn. https://www.linkedin.com/pulse/data-analytics-quantum-computing-rakesh-p

Ramalingam, H., & Venkatesan, V. P. (2019). Conceptual analysis of Internet of Things use cases in Banking domain. In *IEEE Region 10 Conference (TENCON)* (pp. 2034–2039). IEEE. doi:10.1109/TENCON.2019.8929473

Ramezani, S. B., Sommers, A., Manchukonda, H. K., Rahimi, S., & Amirlatifi, A. (2020, July). Machine learning algorithms in quantum computing: A survey. In *2020 International joint conference on neural networks (IJCNN)* (pp. 1-8). IEEE. https://ieeexplore.ieee.org/abstract/document/9207714/

Ramezani, S. B., Sommers, A., Manchukonda, H. K., Rahimi, S., & Amirlatifi, A. (2020). Machine Learning Algorithms in Quantum Computing: A Survey. *2020 International Joint Conference on Neural Networks (IJCNN)*, Glasgow, UK. 10.1109/IJCNN48605.2020.9207714

Rana, G., Khang, A., Sharma, R., Goel, A. K., & Dubey, A. K. (2021). Reinventing Manufacturing and Business Processes through Artificial Intelligence. CRC Press. doi:10.1201/9781003145011

Rani, S., Kumar Pareek, P., Kaur, J., Chauhan, M., & Bhambri, P. (2023). Quantum machine learning in healthcare: Developments and challenges. *2023 IEEE International Conference on Integrated Circuits and Communication Systems (ICICACS)*. IEEE. 10.1109/ICICACS57338.2023.10100075

Raschka, S., Patterson, J., & Nolet, C. (2020). Machine Learning in Python: Main Developments and Technology Trends in Data Science, Machine Learning, and Artificial Intelligence. *Information (Basel), 11*(4), 193. doi:10.3390/info11040193

Ratan, R., Shukla, M. K., & Oruç, A. Y. (2007). Quantum Switching Networks with Classical Routing. IEEE. https://ieeexplore.ieee.org/abstract/document/4298416/

Rawat, B., Mehra, N., Bist, A. S., Yusup, M., & Yulia, P. A. S. (2022). Quantum computing and ai: Impacts & possibilities. *ADI Journal on Recent Innovation, 3*(2), 202–207.

Raychev, N. (2016). Interactive Quantum Development Environment (IQDE). *J. Quantum Inform. Sci., 6*(2), 105–120. https://www.scirp.org/html/4-1300196_67808.htm. doi:10.4236/jqis.2016.62010

Remanan, S. (2021, April 9). *Beginner's Guide to Quantum Machine Learning.* Paperspace Blog. https://blog.paperspace.com/beginners-guide-to-quantum-machine-learning/

Repka, M., & Zajac, P. (2014). Overview of the McEliece cryptosystem and its security. *Tatra Mountains Mathematical Publications, 60*(1), 57-83. https://sciendo.com/article/10.2478/tmmp-2014-0025

Rietsche, R., Dremel, C., Bosch, S., Steinacker, L., Meckel, M., & Leimeister, J. M. (2022). Quantum computing. *Electronic Markets, 32*(4), 2525-36. https://link.springer.com/article/10.1007/s12525-022-00570-y

Rietsche, R., Dremel, C., Bosch, S., Steinacker, L., Meckel, M., & Leimeister, J. M. (2022). Quantum computing. *Electronic Markets, 32*(4), 2525–2536. https://link.springer.com/article/10.1007/s12525-022-00570-y. doi:10.100712525-022-00570-y

Rivest, R. L., Shamir, A., & Adleman, L. (1978). A method for obtaining digital signatures and public-key cryptosystems. *Communications of the ACM, 21*(2), 120–126. https://dl.acm.org/doi/abs/10.1145/357980.358017. doi:10.1145/359340.359342

Rodríguez-González. S. (2023). Quantum AI: Achievements and Challenges in the Interplay of Quantum Computing and Artificial Intelligence. In *Ambient Intelligence—Software and Applications—13th International Symposium on Ambient Intelligence, 603*. Springer Nature.

Rose, G., Gildert, S., Macready, W. G., & Walliman, D. C. (2019). *U.S. Patent No. 10,318,881*. U.S. Patent and Trademark Office. https://heinonline.org/hol-cgi-bin/get_pdf.cgi?handle=hein.journals/jstl27&section=5

Rothe, J. (2005). *Complexity theory and cryptology: an introduction to cryptocomplexity* (Vol. 91). Springer. https://link.springer.com/book/10.1007/3-540-28520-2

Rusca, D., Boaron, A., Curty, M., Martin, A., & Zbinden, H. (2018). Security proof for a simplified Bennett-Brassard 1984 quantum-key-distribution protocol. *Physical Review. A, 98*(5), 052336. https://journals.aps.org/pra/abstract/10.1103/PhysRevA.98.052336. doi:10.1103/PhysRevA.98.052336

Sagingalieva, A., Kordzanganeh, M., Kenbayev, N., Kosichkina, D., Tomashuk, T., & Melnikov, A. (2023). Hybrid quantum neural network for drug response prediction. *Cancers (Basel), 15*(10), 2705. doi:10.3390/cancers15102705 PMID:37345042

Sahni, V. (2007), *Quantum computing*. Tata Mc Graw Hill Publishing Company Limited. https://www.mitindia.edu/images/mit/Departments/EI/syllabus/BE_EIE-R2019-_A-I.pdf#page=24

Sahota, N. (2023, June 7). *Quantum computing and its impact on AI.* Neil Sahota. https://www.neilsahota.com/quantum-computing-and-its-impact-on-ai-unlocking-limitless-potential/

Saini, S., Khosla, P., Kaur, M., & Singh, G. (2020). Quantum Driven Machine Learning. *International Journal of Theoretical Physics, 59*(12), 4013–4024. doi:10.100710773-020-04656-1

Sarkar, S., & Bertels, K. (2018). *The Computer Architecture of Quantum Computing*. IEEE. https://ieeexplore.ieee.org/abstract/document/9076341/

Sarkar, A., Al-Ars, Z., & Bertels, K. (2021). Estimating Algorithmic Information Using Quantum Computing for Genomics Applications. *Applied Sciences (Basel, Switzerland), 11*(6), 2696. doi:10.3390/app11062696

Sasi, P., Gulshan, S., Tyagi, A. K., Kakulapati, V., Shyam Mohan, J. S., & Singh, R. k. (2023). Quantum Computing and the Qubit: The Future of Artificial Intelligence. In A. K. Tyagi (Ed.), *Handbook of Research on Quantum Computing for Smart Environments* (pp. 231–244). IGI Global. doi:10.4018/978-1-6684-6697-1.ch013

Savchuk, M. M., & Fesenko, A. V. (2019). Quantum computing: Survey and analysis. *Cybernetics and Systems Analysis*, *55*, 10–21.

Scarani, V., Bechmann-Pasquinucci, H., Cerf, N. J., Dušek, M., Lütkenhaus, N., & Peev, M. (2009). The security of practical quantum key distribution. *Reviews of Modern Physics*, *81*(3), 1301.

Scherer, W. (2019). *Mathematics of Quantum Computing: An Introduction*. Springer International Publishing. doi:10.1007/978-3-030-12358-1

Schöffel, M., Feldmann, J., & Wehn, N. (2022, October). Code-based Cryptography in IoT: A HW/SW Co-Design of HQC. In *2022 IEEE 8th World Forum on Internet of Things (WF-IoT)* (pp. 1-7). IEEE. https://ieeexplore.ieee.org/abstract/document/10152031/

Schuld, M. & Petruccione, F. (2018). *Supervised Learning with Quantum Computers*. Springer. https://link.springer.com/content/pdf/10.1007/978-3-319-96424-9.pdf

Schuld, M., & Petruccione, F. (2018). Supervised learning with quantum computers. *Quantum Science and Technology*. doi:10.1007/978-3-319-96424-9

Schuld, M., & Petruccione, F. (2021). *Machine Learning with Quantum Computers* (2nd ed.). Springer. https://link.springer.com/content/pdf/10.1007/978-3-030-83098-4.pdf doi:10.1007/978-3-030-83098-4

Schuld, M., Sinayskiy, I., & Petruccione, F. (2015). An introduction to quantum machine learning. *Contemporary Physics*, *56*(2), 172–185. doi:10.1080/00107514.2014.964942

Selvam, A. M. (2009). Fractal fluctuations and statistical normal distribution. *Fractals*, *17*(3), 333–349. https://www.worldscientific.com/doi/abs/10.1142/S0218348X09004272. doi:10.1142/S0218348X09004272

Seo, H., Sanal, P., Jalali, A., & Azarderakhsh, R. (2020). Optimized implementation of SIKE round 2 on 64-bit ARM Cortex-A processors. *IEEE Transactions on Circuits and Systems. I, Regular Papers*, *67*(8), 2659–2671. https://ieeexplore.ieee.org/abstract/document/9036884/. doi:10.1109/TCSI.2020.2979410

Sergey, B., Englbrecht, M., König, R., & Peard, N. (2018). Correcting coherent errors with surface codes. *Nature*. https://www.nature.com/articles/s41534-018-0106-y#citeas

Shaari, J. S., Lucamarini, M., & Wahiddin, M. R. B. (2006). Deterministic six states protocol for quantum communication. *Physics Letters. [Part A]*, *358*(2), 85–90. https://www.sciencedirect.com/science/article/pii/S0375960106007006. doi:10.1016/j.physleta.2006.05.007

Shah, V., & Khang, A. (2023). Metaverse-Enabling IoT Technology for a Futuristic Healthcare System. AI-Based Technologies and Applications in the Era of the Metaverse. IGI Global Press. doi:10.4018/978-1-6684-8851-5.ch008

Shahwar, T., Zafar, J., Almogren, A., Zafar, H., Rehman, A. U., Shafiq, M., & Hamam, H. (2022). Automated Detection of Alzheimer's via Hybrid Classical Quantum Neural Networks. *Electronics (Basel)*, *11*(5), 721. doi:10.3390/electronics11050721

Shaib, A., Naim, M. H., Fouda, M. E., Kanj, R., & Kurdahi, F. (2023). Efficient noise mitigation technique for quantum computing. *Scientific Reports*, *13*(1), 3912. doi:10.103841598-023-30510-5 PMID:36890156

Shaikh, T. A., & Ali, R. (2016). Quantum Computing in Big Data Analytics: A Survey. *2016 IEEE International Conference on Computer and Information Technology (CIT)*. IEEE. 10.1109/CIT.2016.79

Shamshad, S., Riaz, F., Riaz, R., Rizvi, S. S., & Abdulla, S. (2023). *Quantum Computing Supremacy in the Internet of Things*. Encyclopedia. https://encyclopedia.pub/entry/34723

Shannon, K., Towe, E., & Tonguz, O. K. (2020). *On the use of quantum entanglement in secure communications: a survey.* arXiv preprint arXiv:2003.07907.

Shan, T., Tay, F. R., & Gu, L. (2021). Application of artificial intelligence in dentistry. *Journal of Dental Research, 100*(3), 232–244. doi:10.1177/0022034520969115 PMID:33118431

Sharma, A. (2021). *A Textbook on Modern Quantum Mechanics*. Taylor & Francis. . doi:10.1201/9781003154457

Sharma, S., Ramkumar, K. R., Kaur, A., Hasija, T., Mittal, S., & Singh, B. (2023). Postquantum Cryptography: A Solution to the Challenges of Classical Encryption Algorithms. In R. Agrawal, C. Kishore Singh, A. Goyal, & D. K. Singh (Eds.), *Modern Electronics Devices and Communication Systems*. Springer Nature Singapore., doi:10.1007/978-981-19-6383-4_3

Sheth, A., Anantharam, P., Henson, C., & Sahoo, S. (2020). Next Generation Computing Paradigms: Cloud, Edge, and Fog/Edge Computing for Internet of Things and Artificial Intelligence. *IEEE Internet of Things Journal, 7*(12), 10977–10980. https://link.springer.com/chapter/10.1007/978-3-030-74402-1_13

Shinde, A. R., & Bendale, S. P. (2023). Evolution of Quantum Machine Learning and an Attempt of Its Application for SDN Intrusion Detection. In R. Pandey, N. Srivastava, N. K. Singh, & K. Tyagi (Eds.), *Quantum Computing: A Shift from Bits to Qubits* (pp. 437–456). Springer Nature Singapore. doi:10.1007/978-981-19-9530-9_22

Shor, P. W. (1995). Scheme for reducing decoherence in quantum computer memory. *Physical Review, 52*(4), R2493–R2496. https://journals.aps.org/pra/abstract/10.1103/PhysRevA.52.R2493. doi:10.1103/PhysRevA.52.R2493 PMID:9912632

Shor, P. W. (1996). Polynomial-Time Algorithms for Prime Factorization and Discrete Logarithms on a Quantum Computer. *SIAM Journal on Computing, 26*(5), 1484–1509. doi:10.1137/S0097539795293172

Shor, P. W. (1997). Polynomial-time algorithms for prime factorization and discrete logarithms on a quantum computer. *SIAM Review, 41*(2), 303–332. https://search.proquest.com/openview/9335e070823d763bdbdf578ed5271127/1?pq-origsite=gscholar&cbl=47353. doi:10.1137/S0036144598347011

Shor, P. W., & Preskill, J. (2000). Simple proof of security of the BB84 quantum key distribution protocol. *Physical Review Letters, 85*(2), 441–444. https://journals.aps.org/prl/abstract/10.1103/PhysRevLett.85.441. doi:10.1103/PhysRevLett.85.441 PMID:10991303

Shrivas, M. K., Dean, T. Y., & Brunda, S. S. (2020, January). The disruptive blockchain security threats and threat categorization. In 2020 First International Conference on Power, Control and Computing Technologies (ICPC2T) (pp. 327-338). IEEE.

Shuaib, M., Hassan, N. H., Usman, S., Alam, S., Sam, S. M., & Samy, G. A. N. (2022, July). Effect of quantum computing on blockchain-based electronic health record systems. In 2022 4th International Conference on Smart Sensors and Application (ICSSA) (pp. 179-184). IEEE.

Shyam, R. (2023). Effects of Quantum Technology on Metaverse," AI-Based Technologies and Applications in the Era of the Metaverse. IGI Global Press. doi:10.4018/978-1-6684-8851-5.ch009

Siddiqui, S., Islam, M. J., & Shehab, O. (2014). *Five Quantum Algorithms Using Quipper*. arXiv. https://arxiv.org/abs/1406.4481

Singh, J., & Bhangu, K. S. (2023). Contemporary quantum computing use cases: Taxonomy, review and challenges. *Archives of Computational Methods in Engineering*, *30*(1), 615–638. doi:10.100711831-022-09809-5

Singh, K., Bradley, C. E., Anand, S., Ramesh, V., White, R., & Bernien, H. (2023). Mid-circuit correction of correlated phase errors using an array of spectator qubits. *Science*, *380*(6651), 1265–1269. doi:10.1126cience.ade5337 PMID:37228222

Singh, S., Kumar, A., & Kathuria, M. (2022). Understanding the public, private and consortium consensus algorithms in blockchain technology. *International Journal of Blockchains and Cryptocurrencies*, *3*(3), 269–288.

Slussarenko, S., & Pryde, G. J. (2019). Photonic quantum information processing: A concise review. *Applied Physics Reviews*, *6*(4), 41303. https://pubs.aip.org/aip/apr/article/6/4/041303/997349. doi:10.1063/1.5115814

Smith, A. (2020). *User values and the development of a cybersecurity public policy for the IoT.* Science Direct. https://www.sciencedirect.com/science/article/pii/S0268401219313301

Softley, T. P. (2023). Cold and ultracold molecules in the twenties. *Proceedings - Royal Society. Mathematical, Physical and Engineering Sciences*, *479*(2274), 20220806. doi:10.1098/rspa.2022.0806

Soljanin, E. (2020). Quantum Information Processing: An Essential Primer. *IEEE Journal on Selected Areas in Information Theory*, *1*(2), 351–366. https://ieeexplore.ieee.org/abstract/document/9177097/. doi:10.1109/JSAIT.2020.3019340

Solmeyer N., Dixon, R. & Balu, R. (2017). *Quantum routing games*. IOP Science. . doi:10.1088/1751-8121/aae31f

Song, Y. E., Liu, Y., Fang, S., & Zhang, S. Research on Applications of the Internet of Things in the Smart Grid. In Proceedings of the 7th International Conference on Intelligent Human-Machine Systems and Cybernetics, 26–27 August 2015, Hangzhou, China, 2015. https://ieeexplore.ieee.org/abstract/document/7334945/

Soni, V. S. A., G., & Tyagi, A. K. (2023). A Review on Recent Trends in Quantum Computation Technology. In A. Tyagi (Ed.), Handbook of Research on Quantum Computing for Smart Environments (pp. 48-64). IGI Global. doi:10.4018/978-1-6684-6697-1.ch003

Sood, V., & Chauhan, R. P. (2023). (in press). Archives of quantum computing: Research progress and challenges. *Archives of Computational Methods in Engineering*. doi:10.100711831-023-09973-2

Spiller, T. P. (1998). Basic elements of quantum information technology. In *Introduction to Quantum Computation and Information* (pp. 1-28). Wold Scientific. https://www.worldscientific.com/doi/abs/10.1142/9789812385253_0001

Stallings, W. (2006). *Cryptography and network security, 4/E.* Pearson Education India. https://iopscience.iop.org/article/10.1088/1757-899X/981/2/022028/meta

Steane, A. 1996, Steane code, https://errorcorrectionzoo.org/c/steane

Steinmüller, P., Schulz, T., Graf, F., & Herr, D. (2022). *eXplainable AI for Quantum Machine Learning.* arXiv preprint arXiv:2211.01441.

Stilck França, D., & Garcia-Patron, R. (2021). Limitations of optimization algorithms on noisy quantum devices. *Nature Physics*, *17*(11), 1221–1227. doi:10.103841567-021-01356-3

Stinson, D. R., & Paterson, M. (2018). *Cryptography: theory and practice*. CRC press. https://api.taylorfrancis.com/content/books/mono/download?identifierName=doi&identifierValue=10.1201/9781420057133&type=googlepdf

Strand, M. (2021, May). A Status Update on Quantum Safe Cryptography. In *2021 International Conference on Military Communication and Information Systems (ICMCIS)* (pp. 1-7). IEEE. https://ieeexplore.ieee.org/abstract/document/9486413/

Su, H. Y. (2020). Simple analysis of security of the BB84 quantum key distribution protocol. *Quantum Information Processing, 19*(6), 169. https://link.springer.com/article/10.1007/s11128-020-02663-z. doi:10.100711128-020-02663-z

Sultana, F. (2020). Paradigm shift and diversity in finance. *Journal of Financial Accounting Research, 2*, 94–113. https://www.mdpi.com/2199-8531/7/1/19

Sutter, P. (2019). *Physicists are still trying to reconcile two different worlds: the quantum and the macro. Do We Live in a Quantum World?* Space. https://www.space.com/do-we-live-in-quantum-world.html

Svítek, M. (2022). Emergent Intelligence in Generalized Pure Quantum Systems. *Computation (Basel, Switzerland), 10*(6), 88. doi:10.3390/computation10060088

Tacchino, F., Mangini, S., Barkoutsos, P. K., Macchiavello, C., Gerace, D., Tavernelli, I., & Bajoni, D. (2021). Variational learning for quantum artificial neural networks. *IEEE Transactions on Quantum Engineering, 2*, 1–10. https://ieeexplore.ieee.org/abstract/document/9364892/. doi:10.1109/TQE.2021.3062494

Taha, A. M. (2018). An IoT Architecture for Assessing Road Safety in Smart Cities. *Wireless Communications and Mobile Computing.* . doi:10.1155/2018/8214989

Tailor, R. K., Pareek, R., & Khang A. (2022). Robot Process Automation in Blockchain. CRC Press. doi:10.1201/9781003269281-8

Tao, J. (2019). IoT-Based Intelligent Perception and Access of Manufacturing Resource Toward Cloud Manufacturing. In Proceedings of the 2019 15th International Conference on Computational Intelligence and Security. CIS. https://dl.acm.org/doi/abs/10.1145/3334480.3375032

Taylor, K. (2023). IoT and Quantum Computing: A Futuristic Convergence. *Hitech Nectar.* https://www.hitechnectar.com/blogs/iot-and-quantum-computing-a-futuristic-convergence/

Taylor, R. D. (2020). Quantum artificial intelligence: A "Precautionary" US approach? *Telecommunications Policy, 44*(6), 101909. doi:10.1016/j.telpol.2020.101909

Teja, S. B. B., & Saheb, S. H. & Khang A. (2023). Quantum Deep Learning: Detection of Drowsiness in Drivers Using Tensorflow Framework. Applications and Principles of Quantum Computing. IGI Global Press. https://doi.org/ doi:10.4018/979-8-3693-1168-4-ch015

Teja, V., Banerjee, P., Sharma, N. N., & Mittal, R. K. (2007). Quantum cryptography: State-of-art,challenges and future perspectives. *2007 7th IEEE Int. Conf. Nanotechnol. - IEEE-NANO*. IEEE. 10.1109/NANO.2007.4601420

Terhal, B. M. (2015). Quantum error correction for quantum memories. *Reviews of Modern Physics, 87*(2), 307–346. https://journals.aps.org/rmp/abstract/10.1103/RevModPhys.87.307. doi:10.1103/RevModPhys.87.307

Tesch, C. M., & De Vivie-Riedle, R. (2004, December). Vibrational molecular quantum computing: Basis set independence and theoretical realization of the Deutsch-Jozsa algorithm. *The Journal of Chemical Physics, 121*(24), 12158–12168. doi:10.1063/1.1818131 PMID:15606234

Thanalakshmi, P., Rishikhesh, A., Marion Marceline, J., Joshi, G. P., & Cho, W. (2023). A Quantum-Resistant Blockchain System: A Comparative. *Analysis Mathematica*, *11*(18), 3947.

Thapliyal, H., & Ranganathan, N. (2009). Design of efficient reversible binary subtractors based on a new reversible gate. *Proc. 2009 IEEE Comput. Soc. Annu. Symp.* IEEE. 10.1109/ISVLSI.2009.49

Tharrmashastha, S. A. (2020). *Quantum Algorithms for Cryptographically Significant Boolean Functions - An IBMQ Experience.* Springer. https://link.springer.com/content/pdf/10.1007/978-981-16-3061-3.pdf

Thomas, R. K. (2023). Quantum Neural Networks: Bridging Quantum Computing and Machine Learning. *Asian Journal of Research in Computer Science*, *16*(3), 1–10. https://www.tandfonline.com/doi/pdf/10.1080/00018735700101156. doi:10.9734/ajrcos/2023/v16i3340

Tonekaboni, S., Joshi, S., McCradden, M. D., & Goldenberg, A. (2019). What clinicians want: contextualizing explainable machine learning for clinical end use. In *Machine Learning for Healthcare Conference* (pp. 359-380). PMLR. https://proceedings.mlr.press/v106/tonekaboni19a.html

Torgler, B. (2020). Big data, artificial intelligence, and quantum computing in sports. *21st Century Sports: How Technologies Will Change Sports in the Digital Age.*

Tosh, D., Galindo, O., Kreinovich, V., & Kosheleva, O. (2020). *How Quantum Cryptography and Quantum Computing Can Make Cyber-Physical Systems More Secure.* Scholar Works. https://scholarworks.utep.edu/cs_techrep/1397/

Treleaven, P. (2015). Financial regulation of FinTech. *J. Financ. Perspect.*, *3*, 1–17. https://www.emerald.com/insight/content/doi/10.1108/978-1-78973-545-120191018/

Tsurumaru, T., & Tamaki, K. (2008). Security proof for quantum-key-distribution systems with threshold detectors. *Physical Review A*, *78*(3), 032302. https://journals.aps.org/pra/abstract/10.1103/PhysRevA.78.032302. doi:10.1103/PhysRevA.78.032302

Tyagi, A. & Khang, A. (2023). Quantum Computing: Evolution, Advancements, Opportunities and Challenges for Future Generation. Applications and Principles of Quantum Computing. IGI Global Press. https://doi.org/doi:10.4018/979-8-3693-1168-4-ch019

Tyagi, A. (2022). *Handbook of Research on Technical, Privacy, and Security Challenges in a Modern World.* IGI Global. doi:10.4018/978-1-6684-5250-9

Tyagi, A. (2023). Chapter 2 - Decentralized everything: Practical use of blockchain technology in future applications. R. Pandey, S. Goundar, & S. Fatima (eds.), Distributed Computing to Blockchain. Academic Press. doi:10.1016/B978-0-323-96146-2.00010-3

Tyagi, A. K. (Ed.). (2023). *Handbook of Research on Quantum Computing for Smart Environments.* IGI Global. doi:10.4018/978-1-6684-6697-1

Tyagi, A. K., Dananjayan, S., Agarwal, D., & Thariq Ahmed, H. F. (2023). Blockchain—Internet of Things Applications: Opportunities and Challenges for Industry 4.0 and Society 5.0. *Sensors (Basel)*, *23*(2), 947. doi:10.339023020947 PMID:36679743

Tyagi, A., Kukreja, S., Nair, M. M., & Tyagi, A. K. (2022). Machine Learning: Past, Present and Future. *NeuroQuantology : An Interdisciplinary Journal of Neuroscience and Quantum Physics*, *20*(8). doi:10.14704/nq.2022.20.8.NQ44468

Tychola, K. A., Kalampokas, T., & Papakostas, G. A. (2023). Quantum Machine Learning—An Overview. *Electronics (Basel), 12*(11), 2379. https://www.researchgate.net/profile/Kyriaki-Tychola/publication/371001049_Quantum_Machine_Learning-An_Overview/links/646e1f426a0082273fa86d0f/Quantum-Machine-Learning-An-Overview.pdf. doi:10.3390/electronics12112379

Uehara, G. S., Spanias, A., & Clark, W. (2021). *Quantum Information Processing Algorithms with Emphasis on Machine Learning.* 2021 12th International Conference on Information, Intelligence, Systems & Applications (IISA), Chania Crete, Greece. 10.1109/IISA52424.2021.9555570

Ugwuishiwu, C. H., Orji, U. E., Ugwu, C. I., & Asogwa, C. N. (2020). An overview of quantum cryptography and shor's algorithm. *Int. J. Adv. Trends Comput. Sci. Eng, 9*(5). https://www.academia.edu/download/84737888/ijatcse82952020.pdf

Ullah, Z., Al-Turjman, F., Mostarda, L., & Gagliardi, R. (2020). Applications of artificial intelligence and machine learning in smart cities. *Computer Communications, 154*, 313–323. https://www.sciencedirect.com/science/article/pii/S0140366419320821. doi:10.1016/j.comcom.2020.02.069

Unogwu, O. J., Doshi, R., Hiran, K. K., & Mijwil, M. M. (2022). Introduction to Quantum-Resistant Blockchain. In Advancements in Quantum Blockchain With Real-Time Applications (pp. 36-55). IGI Global.

Unruh, D. (2018) *Quantum Cryptography*. UT. https://courses.cs.ut.ee/all/MTAT.07.024/2017_fall/uploads/

Ur Rasool, R., Ahmad, H. F., Rafique, W., Qayyum, A., Qadir, J., & Anwar, Z. (2023). Quantum computing for healthcare: A review. *Future Internet, 15*(3), 94. https://www.mdpi.com/1999-5903/15/3/94. doi:10.3390/fi15030094

Uslar, M., Rohjans, S., Neureiter, C., Andrén, F. P., Velasquez, J., Steinbrink, C., Efthymiou, V., Migliavacca, G., Horsmanheimo, S., & Brunner, H. (2019). Applying the Smart Grid Architecture Model for Designing and Validating System-of-Systems in the Power and Energy Domain: A European Perspective. *Energies, 12*(2), 258. https://www.mdpi.com/1996-1073/12/2/258. doi:10.3390/en12020258

Utzerath, C., & Fernández, G. (2017). Shaping science for increasing interdependence and specialization. *Trends in Neurosciences, 40*(3), 121–124. https://www.cell.com/trends/neurosciences/fulltext/S0166-2236(16)30192-8. doi:10.1016/j.tins.2016.12.005 PMID:28126248

Valerio, S., Antonio, A., Grégoire, R., & Nicolas, G. (2004). Quantum Cryptography Protocols Robust against Photon Number Splitting Attacks for Weak Laser Pulse Implementations. *Phys. Rev. Lett. 92*, 057901. https://journals.aps.org/prl/abstract/10.1103/PhysRevLett.92.057901

Van der Schaar, M. (2019). Machine Learning and AI for Healthcare. *IEEE Journal of Selected Topics in Signal Processing, 13*(2), 239–241. https://link.springer.com/content/pdf/10.1007/978-1-4842-6537-6.pdf

Vasiljeva, T., & Lukanova, K. (2016). Commercial banks and FINTECH companies in the digital transformation: Challenges for the future. *J. Bus. Manag.* https://journals.riseba.eu/index.php/jbm/article/view/91

Villalba-Diez, J., Gutierrez, M., Grijalvo Martín, M., Sterkenburgh, T., Losada, J. C., & Benito, R. M. (2021). Quantum JIDOKA. Integration of Quantum Simulation on a CNC Machine for In–Process Control Visualization. *Sensors (Basel), 21*(15), 5031. doi:10.339021155031 PMID:34372267

Wang, C., Li, X., Liu, Y., & Wang, H. (2014). *The research on development direction and points in IoT in China power grid.* In International Conference on Information Science, Electronics and Electrical Engineering, Sapporo, Japan. https://ieeexplore.ieee.org/abstract/document/6948106/

Wang, J. (2014). Manifold: A parallel simulation framework for multicore systems. *ISPASS 2014 - IEEE Int. Symp. Perform. Anal. Syst. Softw.* IEEE. 10.1109/ISPASS.2014.6844466

Wang, L., & Alexander, C. A. (2020). Quantum science and quantum technology: Progress and challenges. *Am. J. Electr. Electron. Eng.*, *8*(2), 43–50.

Wang, L., Shen, X., Li, J., Shao, J., & Yang, Y. (2019). Cryptographic primitives in blockchains. *Journal of Network and Computer Applications*, *127*, 43–58.

Wang, P. H., Chen, J. H., Yang, Y. Y., Lee, C., & Tseng, Y. J. (2023). Recent advances in quantum computing for drug discovery and development. *IEEE Nanotechnology Magazine*, *17*(2), 26–30. doi:10.1109/MNANO.2023.3249499

Wang, S., Pei, Z., Wang, C., & Wu, J. (2022). Shaping the Future of the Application of Quantum Computing in Intelligent Transportation System. *Intell. Converg. Networks*, *2*(4), 259–276. doi:10.23919/ICN.2021.0019

Wang, Y. F., Lin, W. M., Zhang, T., & Ma, Y. Y. (2012). Research on application and security protection of Internet of Things in Smart Grid. In *Proceedings of the IET International Conference on Information Science and Control Engineering (ICISCE 2012)*, Shenzhen, China. 10.1049/cp.2012.2311

Wang, Y., Yu, X., Zhao, Y., Nag, A., & Zhang, J. (2022, December). Pre-established entanglement distribution algorithm in quantum networks. *Journal of Optical Communications and Networking*, *14*(12), 1020–1033. doi:10.1364/JOCN.465432

Wan, L., Zheng, F., Fan, G., Wei, R., Gao, L., Wang, Y., & Dong, J. (2022, September). A novel high-performance implementation of crystals-kyber with ai accelerator. In *European Symposium on Research in Computer Security* (pp. 514–534). Springer Nature Switzerland. https://link.springer.com/chapter/10.1007/978-3-031-17143-7_25 doi:10.1007/978-3-031-17143-7_25

Wecker, D. (2020). Progress and Prospects for Quantum Supremacy in Computing. *Frontiers in Physics*, *8*, 660. https://www.frontiersin.org/articles/10.3389/fphy.2020.00660/full

Wei, L., Liu, H., Xu, J., Shi, L., Shan, Z., Zhao, B., & Gao, Y. (2023). Quantum machine learning in medical image analysis: A survey. *Neurocomputing*, *525*. doi:10.1016/j.neucom.2023.01.049

Weigert, S. (2009). No-cloning theorem. In *Compendium of quantum physics* (pp. 404–405). Springer Berlin Heidelberg. https://link.springer.com/article/10.1007/s00500-022-07043-6 doi:10.1007/978-3-540-70626-7_124

Wiebe, N., Kapoor, A., & Svore, K. M. (2014). *Quantum deep learning*. arXiv preprint arXiv:1412.3489.

Will, K. (2022). *An Infinity of Worlds: Cosmic Inflation and the Beginning of the Universe*. GoodReads. https://www.goodreads.com/en/book/show/58275961

Williams, C. P., & Clearwater, S. H. (1998). *Explorations in quantum computing*. Springer.

Wittek, P. (2014). *Quantum Machine Learning – What Quantum Computing Means to Data Mining*. Elsevier. https://www.google.com/books?hl=en&lr=&id=92hzAwAAQBAJ

Wodkiewicz, K., Knight, P. L., Buckle, S. J., & Barnett, S. M. (1987). Squeezing and superposition states. *Physical Review A*, *35*(6), 2567–2577. https://journals.aps.org/pra/abstract/10.1103/PhysRevA.35.2567. doi:10.1103/PhysRevA.35.2567 PMID:9898446

Wong, H. Y. (2024a). Quantum Gate Introduction: NOT and CNOT Gates. In H. Y. Wong (Ed.), *Introduction to Quantum Computing: From a Layperson to a Programmer in 30 Steps* (pp. 133–141). Springer International Publishing. doi:10.1007/978-3-031-36985-8_15

Wong, H. Y. (2024b). Shor's Algorithm. In H. Y. Wong (Ed.), *Introduction to Quantum Computing: From a Layperson to a Programmer in 30 Steps* (pp. 289–298). Springer International Publishing., doi:10.1007/978-3-031-36985-8_29

Wootters, W. K., & Zurek, W. H. (1982). A single quantum cannot be cloned. *Nature*, *299*(5886), 802–803. https://www.nature.com/articles/299802a0. doi:10.1038/299802a0

Wright, K., Beck, K. M., Debnath, S., Amini, J. M., Nam, Y., Grzesiak, N., & Figgatt, C. (2019). Benchmarking an 11-qubit quantum computer. *Nature Communications*, *10*(1), 1–6. https://www.nature.com/articles/s41467-019-13534-2. doi:10.103841467-019-13534-2 PMID:31784527

Xia, C., Li, H., & Hu, J. (2022). Semi-quantum digital signature protocol based on Einstein–Podolsky–Rosen steering. *Journal of Physics. A, Mathematical and Theoretical*, *55*(32), 325302. https://iopscience.iop.org/article/10.1088/1751-8121/ac7f6d/meta. doi:10.1088/1751-8121/ac7f6d

Xiang, Y., Su, X., Mišta, L. Jr, Adesso, G., & He, Q. (2019). Multipartite Einstein-Podolsky-Rosen steering sharing with separable states. *Physical Review. A*, *99*(1), 010104. https://journals.aps.org/pra/abstract/10.1103/PhysRevA.99.010104. doi:10.1103/PhysRevA.99.010104

Xing, Y., & Li, S. (2021). A compact hardware implementation of CCA-secure key exchange mechanism CRYSTALS-KYBER on FPGA. *IACR Transactions on Cryptographic Hardware and Embedded Systems*, 328-356. Ruhr Universitat. https://tches.iacr.org/index.php/TCHES/article/view/8797

Xu, G., Mao, J., Sakk, E., & Wang, S. P. (2023, March). An Overview of Quantum-Safe Approaches: Quantum Key Distribution and Post-Quantum Cryptography. In *2023 57th Annual Conference on Information Sciences and Systems* (CISS) (pp. 1-6). IEEE. https://ieeexplore.ieee.org/abstract/document/10089619/

Xu, S. (2014). *The internet of things: a survey*. Science Direct. https://www.sciencedirect.com/science/article/pii/S1389128610001568

Yang, H. (2017). *The Internet of Things for Smart Manufacturing: A Review*. Taylor and Francis. https://www.tandfonline.com/doi/abs/10.1080/24725854.2018.1555383

Yang, Z., Zolanvari, M., & Jain, R. (2023). A Survey of Important Issues in Quantum Computing and Communications. *IEEE Communications Surveys and Tutorials*, *25*(2), 1059–1094. doi:10.1109/COMST.2023.3254481

Yanofsky, N. S., & Mannucci, M. A. (2008). *Quantum computing for computer scientists*. Cambridge University Press. https://books.google.com.vn/books?hl=en&lr=&id=U1chAwAAQBAJ&redir_esc=y doi:10.1017/CBO9780511813887

Yao, Y., Li, Y., Han, R., & Wang, X. (2018). Internet of Things for Smart Grid: Technologies, Standards and Challenges. *IEEE Communications Surveys and Tutorials*, *20*(1), 674–707. https://ieeexplore.ieee.org/abstract/document/9139976/

Ye, C., Li, G., Cai, H., Gu, Y., & Fukuda, A. (2018, September). Analysis of security in blockchain: Case study in 51%-attack detecting. In 2018 5th International conference on dependable systems and their applications (DSA) (pp. 15-24). IEEE.

Ying, M. (2010, February). Quantum computation, quantum theory and AI. *Artificial Intelligence*, *174*(2), 162–176. doi:10.1016/j.artint.2009.11.009

Yingngam, B. (2023). New drug discovery. In M. S. Aslam & M. M. Ahmad (Eds.), *Multidisciplinary Applications of Natural Science for Drug Discovery and Integrative Medicine* (pp. 134–184). IGI Global. doi:10.4018/978-1-6684-9463-9.ch005

Younan, M., Elhoseny, M., Ali, A. A., & Houssein, E. H. (2021, December). Quantum Chain of Things (QCoT): A New Paradigm for Integrating Quantum Computing, Blockchain, and Internet of Things. In 2021 17th International Computer Engineering Conference (ICENCO) (pp. 101-106). IEEE

Zeadally, S. (2020). Toward efficient smartification of the Internet of Things (IoT) services. In *Proceedings of the 2020 18th International Conference on Advances in Mobile Computing & Multimedia (MoMM),* (pp. 392-397). Science Direct. https://www.sciencedirect.com/science/article/pii/S0167739X17317326

Zeguendry, A., Jarir, Z., & Quafafou, M. (2023). Quantum Machine Learning: A Review and Case Studies. *Entropy (Basel, Switzerland)*, *25*(2), 287. doi:10.3390/e25020287 PMID:36832654

Zeilinger, A. (1999). Experiment and the foundations of quantum physics. *APS*. https://journals.aps.org/rmp/abstract/10.1103/RevModPhys.71.S288

Zeng, B., Chen, X., Zhou, D. L., & Wen, X. G. (2019). *Quantum Information Meets Quantum Matter*. Springer.

Zhang, L. (2018). *IoT-Based Intelligent Perception and Access of Manufacturing Resource Toward Cloud Manufacturing*. IEEE. https://ieeexplore.ieee.org/abstract/document/6742605/

Zhang, X., Wang, L., Helwig, J., Luo, Y., Fu, C., & Xie, Y. (2023). *Artificial intelligence for science in quantum, atomistic, and continuum systems.* arXiv. https://arxiv.org/abs/2307.08423

Zhang, C., & Zhang, X. (2019). An Overview of Quantum Computing: Technologies, Applications, and Challenges. *Future Generation Computer Systems*, *97*, 271–281.

Zhang, L., & Zhang, L. (2022). Artificial intelligence for remote sensing data analysis: A review of challenges and opportunities. *IEEE Geoscience and Remote Sensing Magazine*, *10*(2), 270–294. doi:10.1109/MGRS.2022.3145854

Zhang, P., Wang, L., Wang, W., Fu, K., & Wang, J. (2021). A blockchain system based on quantum-resistant digital signature. *Security and Communication Networks*, *2021*, 1–13.

Zhang, Y., & Ni, Q. (2020). Recent advances in quantum machine learning. Quantum. *Quantum Engineering*, 2(1), e34. doi:10.1002/que2.34

Zhao, X., Li, M., Xiao, Q., Chen, J., Wang, F., Shen, L., & Liang, X. (2022, November). AI for quantum mechanics: high performance quantum many-body simulations via deep learning. In *SC22: International Conference for High Performance Computing, Networking, Storage and Analysis* (pp. 1-15). IEEE. https://ieeexplore.ieee.org/abstract/document/10046063/

Zhao, L. J., Zhang, F. Z., & He, H. X. (2013). *The Application of Internet of Things in Smart Grid," in Advanced Materials Research.* Trans Tech Publications, Ltd. https://www.mdpi.com/2071-1050/15/1/717

Zheng, Z., Xie, S., Dai, H. N., Chen, X., & Wang, H. (2018). Blockchain challenges and opportunities: a survey

Zheng, C., & Long, G. (2014). Quantum secure direct dialogue using Einstein-Podolsky-Rosen pairs. *Science China. Physics, Mechanics & Astronomy*, *57*(7), 1238–1243. https://link.springer.com/article/10.1007/s11433-014-5461-x. doi:10.100711433-014-5461-x

Zhong, H. S., Wang, H., Deng, Y. H., Lu, M., Peng, L. C., Luo, Y. H., & Pan, J. W. (2020). Quantum computational advantage using photons. *Science*, *370*(6523), 1460–1463. doi:10.1126cience.abe8770 PMID:33273064

Zhou, W. (2020). *Internet of Things (IoT) in high-risk Environment, Health and Safety (EHS) industries: A comprehensive review*. Science Direct. https://www.sciencedirect.com/science/article/pii/S0167923618300344

Zhou, L., Sheng, Y. B., Ding, D. S., Shi, S. B., & Guo, G. C. (2017). Quantum memory and direct secure transmission. *Physical Review Letters*, *118*, 220501. doi:10.1103/PhysRevLett.118.220501

Zhou, L., Wang, S. T., Choi, S., Pichler, H., & Lukin, M. D. (2020). Quantum approximate optimization algorithm: Performance, mechanism, and implementation on near-term devices. *Physical Review X*, *10*(2), 021067. doi:10.1103/PhysRevX.10.021067

Zhou, Z.-H. (2022). Open-environment machine learning. *National Science Review*, *9*(8), nwac123. doi:10.1093/nsr/nwac123 PMID:35992239

Zhu, D., Linke, N. M., Benedetti, M., Landsman, K. A., Nguyen, M., Alderete, C. H., & Monroe, C. (2019). Training of quantum circuits on a hybrid quantum computer. *Science Advances*, *5*(10), eaaw9918. https://www.science.org/doi/abs/10.1126/sciadv.aaw9918. doi:10.1126ciadv.aaw9918 PMID:31667342

Zou, D., Pan, N., Chen, T., Sun, H., & Zhang, X. (2023). Experimental simulation of topological quantum computing with classical circuits. *Advanced Intelligent Systems*, 2300354. doi:10.1002/aisy.202300354

About the Contributors

Alex Khang, Professor in Information Technology, D.Sc., D.Litt., Universities and Institutions of Science and Technology in Vietnam, India and United States, AI and Data Scientist, Department of AI and Data Science, Global Research Institute of Technology and Engineering, Fort Raleigh, North Carolina, 27612, United States. He has 28+ years of working experience, teaching experience, and research experience in Information Technology (Software Development, Database Technology, AI Engineering, Data Engineering, Data Science, Data Analytics, IoT-based Technologies, and Cloud Computing) at the Universities of Technology and Science in USA, EU, India, and Vietnam. He has published 46 Documents Indexed Scopus, 52 authored books (Vietnamese), 02 authored Software development books (English), 45 book chapters in AI and Data science, 10 edited books in AI, IoT and Data science, and 15 edited books (calling for book chapters) in the field of AI Ecosystem (AI, ML, DL, Robotics, Data Science, Big Data, and IoT), Smart City Ecosystem, Healthcare Ecosystem, Fintech technology, and Blockchain technology.

***

Kiran Bellam holds a Ph.D. in Computer Science from Auburn University and has served as a faculty member at Prairie View A&M University, where she currently holds the position of Associate Dean for the College of Engineering. Dr. Bellam's research has significantly contributed to fields like cloud computing, energy-efficient storage disks, and Engineering education. She secured substantial research grants to support her research. Her work has been widely published and recognized, making her a notable figure in the field of computer science.

Sonam Chauhan is an independent author with a passion for exploring the frontiers of technology and science. With a background in fields as diverse as quantum cryptography, biometrics, and machine learning, Dr. Chauhan has dedicated her career to pushing the boundaries of knowledge.

Miloni Ganatra, Indus University, Ahmadabad, Gujarat, India. Prof. Miloni Ganatra is currently pursuing Ph.D. at Gujarat Technological University under the Faculty of Engineering and Technology in Electronics and Communication Engineering & working as an Assistant Professor in the Electronics & Communication Engineering Department at Indus university. Ahmadabad, Gujarat & . She has 13 years of experience in academics & 1 year of experience at Industry . Her area of interest is VLSI Design & Biomedical Signal Processing. She also holds the membership of IETE, ISTE etc.

Debosree Ghosh, Shree Ramkrishna Institute of Science and Technology, Kolkata, West Bengal, India.

Sandhia G.K. is working as an Assistant Professor in the Department of Computing Technologies, SRM Institute of Science and Technology, Kattankulathur Campus, Chennai, India. She has over 12+ years of experience in Teaching and research. She has authored many research articles which was published in various journals and Conferences.

Varun Kesavan, research scholar in Finance, Accounts and Economics division in VIT Business School, VIT University, Vellore Campus. A voracious and prolific writer and intensive researcher by all means. Till date he has published forty-one books, also he has published more than three hundred and around thousand articles in leading newspaper, journals, and in websites on a various managerial domain.

Rakesh Kumar Sheoran is a faculty of Physics and Teacher Education at State Institute of Advanced Studies in Teacher Education, Gurugram and durrently persuing Ph.D. in Physics from Banasthali Vidyapith, Rajasthan under supervision of Prof. Seema Verma in the areaof quantum error correction.

Gayathri M. is an Assistant Professor in Department of Computing Technologies, School of Computing, S.R.M Institute of Science and Technology, Kattankulathur campus, Chennai, India. She earned Ph.D. in Computer Science & Engineering from S.R.M Institute of Science and Technology, Chennai. She has over twelve years of experience in Teaching and research. Her areas of interest include Image processing, Quantum Computing, Machine Learning, Biometrics and Internet of things. She has published many research papers in international conferences and reputed journals. Has obtained the certificate of Excellence in Samsung Prism research project worklet titled" Multi camera frame Stitching". Has completed two research project worklets for Samsung prism. One of the instructors in the course "Introduction to Quantum Circuits "launched by Linux foundation on EDX platform. Have coordinated and launched a live certification course for industry leaders " Quantum Computing for Managers" during march 2023. Have received two research grants from MEITY QCAL worth around 68000 dollar credits.

Ambika N. is a MCA, MPhil, Ph.D. in computer science. She completed her Ph.D. from Bharathiar university in the year 2015. She has 16 years of teaching experience and presently working for St.Francis College, Bangalore. She has guided BCA, MCA and M.Tech students in their projects. Her expertise includes wireless sensor network, Internet of things, cybersecurity. She gives guest lectures in her expertise. She is a reviewer of books, conferences (national/international), encyclopaedia and journals. She is advisory committee member of some conferences. She has many publications in National & international conferences, international books, national and international journals and encyclopaedias. She has some patent publications (National) in computer science division.

N. Krishnaraj is working as a Associate Professor, School of Computing, SRM Institute of Science and Technology, Kattankulathur, Tamilnadu, India. He is having 16 years of experience in teaching and research, his research areas are Biometrics, Wireless sensor networks, Internet of Things, Medical image processing. He has completed one funded research project supported by DST, India. He is Cisco certified Routing and switching professional. He has published more than 65 articles in reputed international journals and 15 articles presented in international conferences. He is being serving as a Editorial board member of MAT Journal, IRED and Allied Academic Sciences. He has delivered several special lectures in workshops and seminars. He is a professional society member for ISTE, IEI and IAENG.

Busra Ozdenizci Kose is an Associate Professor of Management Information Systems at Gebze Technical University. She teaches courses in a wide range of topics such as Management Information Systems, Computer Science, Business Analysis, Software Design and Development, Business Process Management, Mobile and Wireless Technologies, Near Field Communication (NFC), Internet of Things (IoT) and Blockchain. She co-authored Near Field Communication: From Theory to Practice, (2012) and Professional NFC Application Development for Android, (2013). She gives importance of academia & industry relationship, and takes roles such as project manager, researcher, and consultant for national and international companies in this manner.

M. Pushpalatha is working as Professor,Head of Department, Department of Computing Technologies, SRM Institute of Science and Technology,Kattankulathur. Have around more than 20 years of teaching Experience. Have been Principal Investigator for DST,DRDO projects and have published good research articles under refereed high quality journals.

Roheen Qamar, Quaid-e-Awam University of Engineering, Sciences & Technology, Haqani town Nawabshah, Shaheed Benazirabad, 67450, Pakistan. Her research areas include: Artificial Intelligence, DDOS Block chain, Internet of Things Soft Computing, Convolutional Neural Network, Shallow Neural Network, Case Cade Neural Network, Fit Net Neural Network, Recursive Neural Network and Image Processing so on. She works as a reviewer in many international journals.

Thilagavathy R. is working in the Department of Computing Technologies in SRM Institute of Science and Technology, Kattankulathur campus, chennai, India. She has 17+ years of teaching and research experience. Apartment from academic teaching, She played different roles in her academic career. She is an active member of ACM and ISTE. She has authored many research articles in various peer reviewed journals and conferences. She has authored 3 textbooks and 2 book chapters.

Savita Kumari Sheoran is full professor of Computer Science & Engineering and currently serving as Chairperson of Department, Dean of Faculty of Engineering & Technology at Indira Gandhi University Meerpur, Rewari. Apart from teaching computer science at UG and PG level she is actively involved in supervising doctoral students. She published more than 70 research papers and 05 books and has been granted one Indian patent.

Nidhi Sindhwani is currently working as an Assistant Professor in Amity Institute of Information Technology, Amity University, Noida, India. She has done her Ph.D. (ECE) from Punjabi University, Punjab, India. She has teaching experience of more than 17 years. She is a Life Member of Indian Society for Technical Education (ISTE) and also a Member of IEEE. She has Google Scholar h-index of 17 and i10-index of 30. She has 48 papers in Scopus/SCIE/SCI in Indexed Journals/ Conferences / book chapters and 4 Patents. She is the Editor of 4 reputed Edited Books. She has presented various research papers in National & International Conferences. She has got the Best Paper Presentation Award in an International Conference. She has chaired a Session in 15 International Conferences. She has been a reviewer in many of the reputed indexed Journals. Her research / interest areas include Wireless Communication, Image Processing, Optimization, Machine Learning, IoT etc.

Ramveer Singh is currently working as Professor in CSE Deptt. at Galgotias College of Engineering & Technology, Greater Noida. His qualification is Ph.D (CSE), M. Tech (CSE), B.E (IT). He has 20 years excellence experience in Academics and Administrative. He has served to reputed institutes of NCR like Greater Noida Institute of Technology, Greater Noida, Raj Kumar Goel Institute of Technology, Ghaziabad and Radha Govind Engineering College, Meerut. Dr. Singh has published various patents and more than 45 research papers in reputed International and National journals / Conferences. He worked as Ph.D thesis evaluator and Supervisor in various universities, Research paper reviewer of reputed journals. He is also active senior member of various professional societies (CSI, IAENG, CSTA etc).

Sakthi Srinivasan K., Professor in Finance, Accounts and Economics division in VIT Business School, VIT University, Vellore Campus. He has 28 years of teaching experience and published research papers in the area of Economics, Institutional Investments, Machine Learning, Predictive Analytics.

Priyanga Subbiah received the B.Tech degree in Information Technology from New Prince Shri Bhavani College of Engineering and Technology, Chennai, India, in 2015 and the M.E. degree in Computer Science and Engineering from Thiagarajar College of Engineering, Madurai, in 2020. He is currently pursuing the Ph.D. degree in Computer Science and Engineering at SRM Institute of Science and Technology, Chengalpattu, Tamil Nadu, India. From 2022, she was a Research Scholar with the Computer Science and Engineering. His research interest includes the processing of images and detection of disease in the plant using Artificial Intelligence.

T. Nithya B.E, M.TECH., Ph.D. is currently working as an Assistant Professor in the Department of Computer Science and Engineering in Rajalakshmi Institute of Technology,since March 2023.She Obtained B.E(CSE) from KINGS COLLEGE OF ENGINEERING and M.TECH(CSE) from Ponnaiyah Ramajayam Institute of Science and Technology and P.hD in Alagappa University.She is University Topper in M.Tech.She has been in teaching profession for the past 9 years.Her areas of interest include Data Science,Database Management System,Cyber Security,Intrusion Detection, and Machine Learning. She has published 3 papers in various international conferences.She has attended many workshops.She is Data science Centre Head in Rajalakshmi Institute of Technology.She is active member in organizing various curricular and cocurricular activities of various events and Programs.

Saraswathi T. is currently working as an Assistant professor in easwari engineering college, chennai. With Teaching experience of 11 years she has published 2 books with international book publishers and has presented papers in international conferences and published papers in scopus indexed journals and also filed patents. she is currently pursuing her Ph.D in Anna University, Chennai.

Seema Verma is a full Professor in Department of Physical Sciences at Banasthali Vidyapith, Rajasthan. She has published about 140 research papers in international and national journals and conferences and received an Indian Patent of communications system.

Bancha Yingngam, an Associate Professor at Ubon Ratchathani University, Thailand, specializes in Pharmaceutical Sciences and Technology. He earned his pharmacy degree in 2005 and a Ph.D. in 2011 from the same institution. During a postdoctoral fellowship at Karl-Franzens-Universität Graz, Austria, from 2013 to 2021, he expanded his research to include optimization and machine learning algorithms, earning a Scopus H-index of 10. He has published over 20 scholarly works and currently focuses on artificial intelligence applications in green chemistry and pharmacy. Dr. Yingngam has received the "Outstanding Research Award in Health Science" twice and has mentored 11 graduate students in Austria. He has also reviewed over 100 Scopus-listed articles, underscoring his commitment to advancing his field.

Baqar Ali Zardari, Quaid-e-Awam University of Engineering, Sciences & Technology, Haqani town Nawabshah, Shaheed Benazirabad, Sindh 67450, Pakistan, Received (B.S) degree in Computer Science, (M.S) in Information Technology and (PhD) From QUEST Nawabshah, His Area of Interest Includes HCI, Distributed Databases, User Experience, Artificial Intelligence, DDOS and Shallow Neural Network.

Index

S

T

www.ingramcontent.com/pod-product-compliance
Lightning Source LLC
LaVergne TN
LVHW062105300926
849161LV00010B/2376
* 9 7 9 8 3 6 9 3 1 1 6 8 4 *